COMPARATIVE GROWTH OF MAMMALIAN, INSECT AND PLANT CELLS

COMPARATIVE GROWTH OF MAMMALIAN, INSECT AND PLANT CELLS

Edited by:

Carlos Tello Lacal

Delve Publishing

www.delvepublishing.com

Comparative Growth of Mammalian, Insect and Plant Cells

Carlos Tello Lacal

Delve Publishing
2010 Winston Park Drive,
2nd Floor
Oakville, ON L6H 5R7
Canada
www.delvepublishing.com
Tel: 001-289-291-7705
001-905-616-2116
Fax: 001-289-291-7601
Email: orders@arclereducation.com

ISBN: 978-1-77361-050-4 (Hardcover)

DECLARATION

Some content or chapters in this book are open access copyright free published research work, which is published under Creative Commons License and are indicated with the citation. We are thankful to the publishers and authors of the content and chapters as without them this book wouldn't have been possible.

DECLARATION

Some content or chapters in this book are open access copyright free published research work, which is published under Creative Commons License and are indicated with the citation. We are thankful to the publishers and authors of the content and chapters as without them this book wouldn't have been possible.

ABOUT THE EDITOR

Carlos Tello Lacal

Carlos Tello (Valencia, Spain, 1981) studied Agricultural Engineering with specialization in Biotechnology at the Universidad Politécnica de Valencia, where he graduated with a BSc thesis on the characterization of an Arabidopsis mutant with increased tolerance to seed ageing and salt stress. He moved afterwards to Sevilla to complete his MSc and PhD at the Instituto de Recursos Naturales y Agrobiología, where he studied the mechanisms employed by plants in their adaptation to abiotic stress and more specifically the role of the Arabidopsis Na+/H+ antiporter SOS1 in the regulation of Na+ and K+ homeostasis. Since 2014, he is based in Zürich and develops several freelance science-related jobs. He has recently retrained himself in the field of cell culture in bioreactors by attending an advanced training course at the Zürcher Hochschule für Angewandte Wissenschaften.

TABLE OF CONTENT

LIST OF CONTRIBUTORS

Carlos Tello
Instituto de Recursos Naturales y Agrobiología de Sevilla

Rishi Man Chugh
Cell Biology Laboratory, National Institute of Pathology (ICMR), New Delhi,

Madhusudan Chaturvedi
Cell Biology Laboratory, National Institute of Pathology (ICMR), New Delhi,

Lakshmana Kumar Yerneni
Cell Biology Laboratory, National Institute of Pathology (ICMR), New Delhi,

Shahla Shahsavandi
Razi Vaccine & Serum Research Institute, P.O. Box 31975-148, Karaj 31976 19751, Iran

Mohammad Majid Ebrahimi
Razi Vaccine & Serum Research Institute, P.O. Box 31975-148, Karaj 31976 19751, Iran

Shahin Masoudi
Razi Vaccine & Serum Research Institute, P.O. Box 31975-148, Karaj 31976 19751, Iran

Hasan Izadi
Razi Vaccine & Serum Research Institute, P.O. Box 31975-148, Karaj 31976 19751, Iran

Galina Gulis
Institute of Biological Sciences, Department of Cell Biology, University of Brasilia, Campus Universitário Darcy Ribeiro

Kelly Cristina Rodrigues Simi
Institute of Biological Sciences, Department of Cell Biology, University of Brasilia, Campus Universitário Darcy Ribeiro

Renata Rodrigues de Toledo
Institute of Biological Sciences, Department of Cell Biology, University of Brasilia, Campus Universitário Darcy Ribeiro

Andrea Queiroz Maranhao
Institute of Biological Sciences, Department of Cell Biology, University of Brasilia, Campus Universitário Darcy Ribeiro

Marcelo Macedo Brigido
Institute of Biological Sciences, Department of Cell Biology, University of Brasilia, Campus Universitário Darcy Ribeiro

Chi-Yen Lin
Department of Chemistry, and Center for Neuroscience Research, University at Albany, SUNY, Albany, New York, 12222, United States of America

Zhen Huang
Department of Chemistry, and Center for Neuroscience Research, University at Albany, SUNY, Albany, New York, 12222, United States of America

Wei Wen
Department of Chemistry, and Center for Neuroscience Research, University at Albany, SUNY, Albany, New York, 12222, United States of America

Andrew Wu
Department of Chemistry, and Center for Neuroscience Research, University at Albany, SUNY, Albany, New York, 12222, United States of America

Congzhou Wang
Department of Chemistry, and Center for Neuroscience Research, University at Albany, SUNY, Albany, New York, 12222, United States of America

Li Niu
Department of Chemistry, and Center for Neuroscience Research, University at Albany, SUNY, Albany, New York, 12222, United States of America

Maria Wolun-Cholewa
Department of Cell Biology, Poznan University of Medical Science, Poznan, Poland,

Krzysztof Langer
Laboratory for Materials Physicochemistry and Nanotechnology, Adam Mickiewicz University, Srem, Poland,

Krzysztof Szymanowski
Department of Mother's and Child's Health, Poznan University of Medical Science, Poznan, Poland,

Aleksandra Glodek
Department of Cell Biology, Poznan University of Medical Science, Poznan, Poland,

Anna Jankowska
Department of Cell Biology, Poznan University of Medical Science, Poznan, Poland,

Wojciech Warchol
Department of Biophysics, Poznan University of Medical Science, Poznan, Poland

Jerzy Langer
Laboratory for Materials Physicochemistry and Nanotechnology, Adam Mickiewicz University, Srem, Poland,

Erica J. Fratz
Department of Molecular Medicine, Morsani College of Medicine, Tampa, Florida, United States of America,

Gregory A. Hunter
Department of Molecular Medicine, Morsani College of Medicine, Tampa, Florida, United States of America,

Gloria C. Ferreira
Department of Molecular Medicine, Morsani College of Medicine, Tampa, Florida, United States of America,
Department of Chemistry, University of South Florida, Tampa, Florida, United States of America

Doris Cassio
Inserm UMR-S757, Orsay 91405, France
Université Paris-Sud, Orsay 91405, France

David G. García Münzer
[1]Biological Systems Engineering Laboratory, Centre for Process Systems Engineering, Department of Chemical Engineering, Imperial College London, London, United Kingdom,

Margaritis Kostoglou
Department of Chemistry, Aristotle University of Thessaloniki, Thessaloniki, Greece,

Michael C. Georgiadis
Department of Chemical Engineering, Aristotle University of Thessaloniki, Thessaloniki, Greece

Efstratios N. Pistikopoulos
Biological Systems Engineering Laboratory, Centre for Process Systems Engineering, Department of Chemical Engineering, Imperial College London, London, United Kingdom,

Athanasios Mantalaris
Biological Systems Engineering Laboratory, Centre for Process Systems Engineering, Department of Chemical Engineering, Imperial College London, London, United Kingdom,

Yvonne E. Thomassen
Institute for Translational Vaccinology, Process Development, PO BOX 450, Bilthoven 3720 AL, The Netherlands

Olaf Rubingh
Institute for Translational Vaccinology, Process Development, PO BOX 450, Bilthoven 3720 AL, The Netherlands

René H. Wijffels
Wageningen University, Bioprocess Engineering, PO BOX 8129, Wageningen 6700 EV, The Netherlands

Leo A. van der Pol
Institute for Translational Vaccinology, Process Development, PO BOX 450, Bilthoven 3720 AL, The Netherlands

Wilfried A.M. Bakker
Institute for Translational Vaccinology, Process Development, PO BOX 450, Bilthoven 3720 AL, The Netherlands

Yoshifumi Hashimoto
Boyce Thompson Institute at Cornell University, Tower Road, Ithaca NY 14853 USA

Sheng Zhang and Gary W Blissard
Boyce Thompson Institute at Cornell University, Tower Road, Ithaca NY 14853 USA

Kai Yu
Biodynamic Optical Imaging Center, School of Life Science, Peking University, Beijing 100871, China
State Key Laboratory of Protein and Plant Gene Research, Peking University, Beijing 100871, China

Yang Yu
Biodynamic Optical Imaging Center, School of Life Science, Peking University, Beijing 100871, China
State Key Laboratory of Protein and Plant Gene Research, Peking University, Beijing 100871, China

Xiaoyan Tang
Biodynamic Optical Imaging Center, School of Life Science, Peking University, Beijing 100871, China
State Key Laboratory of Protein and Plant Gene Research, Peking University, Beijing 100871, China

Huimin Chen
Biodynamic Optical Imaging Center, School of Life Science, Peking University, Beijing 100871, China
State Key Laboratory of Protein and Plant Gene Research, Peking University, Beijing 100871, China

Junyu Xiao
State Key Laboratory of Protein and Plant Gene Research, Peking University, Beijing 100871, China
Peking-Tsinghua Center for Life Sciences, Peking University, Beijing 100871, China

Xiao-Dong Su
Biodynamic Optical Imaging Center, School of Life Science, Peking University, Beijing 100871, China
State Key Laboratory of Protein and Plant Gene Research, Peking University, Beijing 100871, China

Anton Strunov
Institute of Molecular and Cellular Biology, Siberian Branch of RAS, Novosibirsk, 630090, Russia
Institute of Cytology and Genetics, Siberian Branch of RAS, Novosibirsk, 630090, Russia

Lidiya V. Boldyreva
Institute of Molecular and Cellular Biology, Siberian Branch of RAS, Novosibirsk, 630090, Russia

Gera A. Pavlova
Institute of Molecular and Cellular Biology, Siberian Branch of RAS, Novosibirsk, 630090, Russia
Kazan Federal University, Kazan, 420008, Russia

Alexey V. Pindyurin
Institute of Molecular and Cellular Biology, Siberian Branch of RAS, Novosibirsk, 630090, Russia
Institute of Cytology and Genetics, Siberian Branch of RAS, Novosibirsk, 630090, Russia
Novosibirsk State University, Novosibirsk, 630090, Russia

Maurizio Gatti
Institute of Molecular and Cellular Biology, Siberian Branch of RAS, Novosibirsk, 630090, Russia
IBPM CNR and Department of Biology and Biotechnology, Sapienza University of Rome, Rome, 00185, Italy

Elena Kiseleva
Institute of Molecular and Cellular Biology, Siberian Branch of RAS, Novosibirsk, 630090, Russia
Institute of Cytology and Genetics, Siberian Branch of RAS, Novosibirsk, 630090, Russia

Maeva Giraudo
Institut National de la Recherche Agronomique, UMR 1301 Interactions Biotiques et Sante´ Ve´ge´tale, Centre National de la Recherche Scientifique, UMR 6243, Univer-

site´ de Nice Sophia Antipolis, Sophia-Antipolis, France,
UMR 6023 CNRS-Universite´ Blaise Pascal, Ba^t. Biologie A – Campus des Ce´zeaux, Aubie`re, France,

Je´ ro^ me Califano
Institut National de la Recherche Agronomique, UMR 1301 Interactions Biotiques et Sante´ Ve´ge´tale, Centre National de la Recherche Scientifique, UMR 6243, Universite´ de Nice Sophia Antipolis, Sophia-Antipolis, France,
De´partement des affaires re´glementaires, Grasse, France,

Fre´de´ rique Hilliou
Institut National de la Recherche Agronomique, UMR 1301 Interactions Biotiques et Sante´ Ve´ge´tale, Centre National de la Recherche Scientifique, UMR 6243, Universite´ de Nice Sophia Antipolis, Sophia-Antipolis, France,

Trang Tran
Institut National de la Recherche Agronomique, UMR 1301 Interactions Biotiques et Sante´ Ve´ge´tale, Centre National de la Recherche Scientifique, UMR 6243, Universite´ de Nice Sophia Antipolis, Sophia-Antipolis, France,
Lanaud Gestion-Po^le de Lanaud, Boisseuil, France,

Nathalie Taquet
Institut National de la Recherche Agronomique, UMR 1301 Interactions Biotiques et Sante´ Ve´ge´tale, Centre National de la Recherche Scientifique, UMR 6243, Universite´ de Nice Sophia Antipolis, Sophia-Antipolis, France,
Bioimagerie, Villeneuve Loubet, France

Rene´ Feyereisen
Institut National de la Recherche Agronomique, UMR 1301 Interactions Biotiques et Sante´ Ve´ge´tale, Centre National de la Recherche Scientifique, UMR 6243, Universite´ de Nice Sophia Antipolis, Sophia-Antipolis, France,

Gae lle Le Goff
Institut National de la Recherche Agronomique, UMR 1301 Interactions Biotiques et Sante´ Ve´ge´tale, Centre National de la Recherche Scientifique, UMR 6243, Universite´ de Nice Sophia Antipolis, Sophia-Antipolis, France,

Stefan W Metz
Laboratory of Virology, Wageningen University, Droevendaalsesteeg 1, 6708 PB Wageningen, the Netherlands.

Corinne Geertsema
Laboratory of Virology, Wageningen University, Droevendaalsesteeg 1, 6708 PB Wageningen, the Netherlands.

Byron E Martina
Erasmus MC, Department of Virology, 3000 CA Rotterdam, the Netherlands.

Paulina Andrade
Laboratory of Virology, Wageningen University, Droevendaalsesteeg 1, 6708 PB Wageningen, the Netherlands.

Jacco G Heldens
Nobilon International BV, Boxmeer, the Netherlands.

Monique M van Oers
Laboratory of Virology, Wageningen University, Droevendaalsesteeg 1, 6708 PB Wageningen, the Netherlands.

Rob W Goldbach
Laboratory of Virology, Wageningen University, Droevendaalsesteeg 1, 6708 PB Wageningen, the Netherlands.

Just M Vlak
Laboratory of Virology, Wageningen University, Droevendaalsesteeg 1, 6708 PB Wageningen, the Netherlands.

Gorben P Pijlman
Laboratory of Virology, Wageningen University, Droevendaalsesteeg 1, 6708 PB Wageningen, the Netherlands.

Milos Tanurdzic
Cold Spring Harbor Laboratory, Cold Spring Harbor, New York, United States of America,

Matthew W. Vaughn
Cold Spring Harbor Laboratory, Cold Spring Harbor, New York, United States of America,

Hongmei Jiang
Department of Statistics, Northwestern University, Evanston, Illinois, United States of America,

Tae-Jin Lee
Department of Horticultural Science, North Carolina State University, Raleigh, North Carolina, United States of America,

R. Keith Slotkin
[1]Cold Spring Harbor Laboratory, Cold Spring Harbor, New York, United States of America,

Bryon Sosinski
Department of Horticultural Science, North Carolina State University, Raleigh, North Carolina, United States of America,

William F. Thompson
Departments of Plant Biology, Genetics, and Crop Science, North Carolina State University, Raleigh, North Carolina, United States of America,

R. W. Doerge
Departments of Statistics and Agronomy, Purdue University, West Lafayette, Indiana, United States of America

Robert A. Martienssen
Cold Spring Harbor Laboratory, Cold Spring Harbor, New York, United States of America,

Eva Nocarova
Department of Plant Physiology, Charles University in Prague, Faculty of Science

Lukas Fischer
Department of Plant Physiology, Charles University in Prague, Faculty of Science

Shilpa Ramani
Department of Biochemistry, Indian Institute of Science, Bangalore – 560012, India

Chelliah Jayabaskaran
Department of Biochemistry, Indian Institute of Science, Bangalore – 560012, India

Nikolay Vasilev
Department Plant Biotechnology, Fraunhofer Institute for Molecular Biology and Applied Ecology IME, Aachen, Germany,

Christian Schmitz
Department Plant Biotechnology, Fraunhofer Institute for Molecular Biology and Applied Ecology IME, Aachen, Germany,

Ulrike Gro¨ mping
Department II–Mathematics, Physics and Chemistry, Beuth University of Applied Sciences, Berlin, Germany,

Rainer Fischer
Department Plant Biotechnology, Fraunhofer Institute for Molecular Biology and Applied Ecology IME, Aachen, Germany,
Institute for Molecular Biotechnology, RWTH Aachen University, Aachen, Germany,

Stefan Schillberg
Department Plant Biotechnology, Fraunhofer Institute for Molecular Biology and Applied Ecology IME, Aachen, Germany,

Institute for Phytopathology and Applied Zoology, Justus-Liebig University Giessen, Giessen, Germany

Li-Fen Huang
Graduate School of Biotechnology and Bioengineering, Yuan Ze University, 135 Yuan-Tung Road, Taoyuan, Taiwan, ROC,

Chia-Chun Tan
Graduate School of Biotechnology and Bioengineering, Yuan Ze University, 135 Yuan-Tung Road, Taoyuan, Taiwan, ROC,

Ju-Fang Yeh
Department of Life Science, National Central University, Taoyuan, Taiwan, ROC,

Hsin-Yi Liu
Department of Life Science, National Central University, Taoyuan, Taiwan, ROC,

Yu-Kuo Liu
Graduate Institute of Biochemical and Biomedical Engineering, Chang Gung University, Taoyuan, Taiwan, ROC,

Shin-Lon Ho
Department of Agronomy, National Chi-Yi University, Chiayi, Taiwan, ROC

Chung-An Lu
Department of Life Science, National Central University, Taoyuan, Taiwan, ROC,

Sangram K Lenka
Department of Biology, University of Massachusetts, Amherst, MA 01003, USA.

Nadia Boutaoui
Department of Biology, University of Massachusetts, Amherst, MA 01003, USA.

Bibin Paulose
Department of Biochemistry and Molecular Biology, University of Massachusetts, Amherst, MA 01003, USA.

Kham Vongpaseuth
Department of Chemical Engineering, University of Massachusetts, Amherst, MA 01003, USA.
Plant Biology Graduate Program, University of Massachusetts, Amherst, MA 01003, USA.

Jennifer Normanly
Department of Biochemistry and Molecular Biology, University of Massachusetts, Amherst, MA 01003, USA.
Plant Biology Graduate Program, University of Massachusetts, Amherst, MA 01003, USA.

Susan C Roberts
Department of Chemical Engineering, University of Massachusetts, Amherst, MA 01003, USA.
Plant Biology Graduate Program, University of Massachusetts, Amherst, MA 01003, USA.

Elsbeth L Walker
Department of Biology, University of Massachusetts, Amherst, MA 01003, USA.
Plant Biology Graduate Program, University of Massachusetts, Amherst, MA 01003, USA.

PREFACE

Cell culture is extensively employed in the biotechnological and pharmaceutical industries for the production of antiviral vaccines, monoclonal antibodies, recombinant proteins, secondary metabolites and *in vitro* cultivated cells. This technique is successfully applied to the growth of cell lines isolated from different species of mammals, insects and plants. In order to optimize cell growth and product yield, it is essential to study the metabolism of each cell line to allow for the adjustment of the growth conditions and culture medium composition accordingly. Through the compilation of open access articles, the present book provides numerous examples of the *in vitro* cultivation of different mammalian, insect and plant cell lines, as well as their biotechnological applications.

In Chapter number 1, the editor discusses the composition of mammalian, insect and plant cell culture media based on the metabolic requirements of these organisms. The first block of nine chapters presents cell culture experiments with different mammalian cell lines. The authors of the study shown in Chapter 2 assayed three different 3T3 fibroblast subculture schemes to investigate their effect on the proliferative feeder contamination of target cells. In Chapter 3, the obtaining of low pathogenic influenza virus replication in BHK-21 cells is achieved through the expression of a chicken embryo factor X. The optimized production of human immunoglobulin G in CHO cells under doxycycline induction is investigated in Chapter 4. In Chapter 5, the effect of temperature on recombinant protein production is studied in HEK-293 cells. The authors of the study presented in Chapter 6 cultured HeLa cells in 3D through the electrospinning of a nanostructured polymer grid. In Chapter 7, the erythroid-specific ALAS isozyme is expressed in K562 cells to study the accumulation of the heme precursor PPIX, as well as the cell death rate caused by this protein. In Chapter 8, the effect of long-term culture of MDCK cells on the number of chromosomes is investigated. A mathematical model for the GS-NS0 cell cycle progression is described in Chapter 9. Finally, different Vero cell cultivation methods are assayed to optimize poliovirus D-antigen yields in the study presented in Chapter 10. The second block of five chapters deals with insect cell culture. The authors of the study shown in Chapter 11 generated primary cell cultures and individual cell lines from eggs of the moth *Ascalapha odorata* and measured the production of recombinant alkaline phosphatase and β-galactosidase in this system. A transcriptome analysis of High-Five cells aimed at optimizing the secretion of recombinant proteins by using the baculovirus expression system is presented in Chapter 12. In Chapter 13, a method for the ultrastructural analysis of mitosis in S2 cells is described. The effect of the hormone agonists methoxyfenozide and methoprene on *Sf9* proliferation is examined in Chapter

14. Finally, the study presented in Chapter 15 shows the production of Chikunguya virus E1 and E2 glycoproteins in *Sf*21 cells. The last block of six chapters explores the *in vitro* culture and biotechnological applications of plant cells. In Chapter 16, the epigenetic instability of immortalized Arabidopsis cells is investigated. The cloning of BY-2 cells is employed to reduce heterogeneous expression of transgenes in Chapter 17. In Chapter 18, *Catharanthus roseus* cells are treated with UV-B to increase the production of catharanthine and vindoline. In Chapter 19, a large-scale statistical experiment is performed to identify the cultivation factors that most severely affect geraniol production in tobacco NN cells. In Chapter 20, several signaling peptides are tested in order to optimize recombinant protein secretion in rice cells. Finally, the molecular genetics of the anticancer agent paclitaxel (Taxol®) are investigated in *Taxus cuspidata* cells through the identification of genes with altered expression in response to the elicitor methyl jasmonate. The present book provides college students, teachers, researchers, workers of the pharmaceutical and biotechnological industries and other readers interested in cell biology and biotechnology with a detailed overview of the biotechnological applications of mammalian, insect and plant cells and the factors influencing cell growth and recombinant protein yield.

Editor

CHAPTER

1

GROWTH MEDIA FOR MAMMALIAN, INSECT AND PLANT CELL CULTURE

Carlos Tello

Instituto de Recursos Naturales y Agrobiología de Sevilla

INTRODUCTION

Cell culture is the growth of eukaryotic cells under controlled conditions for the biotechnological production of proteins and tissues. Aided by the fascinating developments achieved during the last decades, mammalian cell culture technology has become a major field showing an interdisciplinary interplay between medicine, biology and engineering (Kelly *et al.*, 1993; Howaldt *et al.*, 2005). Products such as antiviral vaccines, monoclonal antibodies, interferons, therapeutic proteins, cells as product for cell therapy, functional tissues for transplantation or bioartificial organs are all obtained through mammalian cell culture by the biopharmaceutical industry (Otzurk and Hu, 2006; Eibl *et al.*, 2009; Pörtner, 2009). Insect cells in conjunction with the baculovirus expression vector system (BEVS) are commonly applied in the routine production of recombinant proteins as well. Insect cells have similar requirements to their mammalian counterparts, but are grown at lower temperature (28°C), and pH (6.2-6.9), require higher rates of specific oxygen uptake and tolerate higher levels of free amino acids and glucose without shifting to overflow metabolism, which allows for a faster cell growth and the design of richer media. Although human-like posttranscriptional modifications such as glycosylation have been described for specific cell lines, terminal sialylation remains insufficient (Schlaeger, 1996a; Ikonomou *et al.*, 2003; Tomiya *et al.*, 2003; Legardinier *et al.*, 2005; Weber and Fussenegger, 2005, 2009). Plant cell culture is employed over traditionally grown wild type or transgenic plants for the sterile production of metabolites under defined, controlled conditions. Although recombinant therapeutic proteins such as human serum albumin, the AAT antibody (for lung diseases), EPO (for anemia), GM-CSF (for stimulating the bone marrow), HBsAg (for hepatitis B prevention), VEGF (for tumor diagnostics) or the PMP vaccine against *Newcastle Disease Virus* have been successfully produced in plant cells (Sijmons *et al.*, 1990; Ma *et al.*, 2003; Gorr and Wagner, 2005; Schillberg *et al.*, 2005; Evans, 2006), their main application is the obtaining of secondary metabolites such as rosmarinic acid (food additive), shinokin (antimicrobial pigment), ginsenoids (food additives and cosmetics), taxol, camptothecin, podophyllotoxin, vincristine, vinblastine and colchicine (potential anticancer agents) (Parnham and Kesselring, 1985; Hibino and Ushiyama, 1999; Kreis *et al.*, 2001; Vanisree *et al.*, 2004; Wink *et al.*, 2005). When compared to their animal counterparts, plant cells require lower temperatures (23-29°C) and pH values (5.0-6.0), but are more demanding in terms of CO_2, light and growth regulators (Fowler, 1988; Endress, 1994; Kieran *et al.*, 1997; Eibl and Eibl, 2009).

In order to meet the specific requirements for each cell line type, the accurate formulation of a growth medium supplying all the necessary nutrients is of vital importance. The present review focuses on this aspect of cell culture and describes the effects of the main components on the growth and metabolism of mammalian, insect and plant cells.

Culture media for mammalian cells

A growth medium for mammalian cells should provide all the necessary nutrients required for growth and product formation, have a certain buffer capacity to stabilize the pH at 7.0-7.3 and approximate osmolality of 300-350 mOsm/l to protect the cell membrane (Eagle, 1959; Butler and Jenkins, 1989; Oh *et al.*, 1995; Ryu and Lee, 1997; Zanghi *et al.*, 1999;

Zengotita *et al.*, 2002; Zhu *et al.*, 2005; Pörtner, 2009). The first minimal essential medium was formulated by Eagle (1955) and contained amino acids, glucose and vitamins. A medium based on this one, Dulbecco's Modified Eagle's Medium (DMEM), is still used to maintain primary cell cultures and lines (van der Valk *et al.*, 2010). Fetal bovine serum (FBS) is commonly employed to maintain cells alive for longer time periods, given its abundant content in growth factors, proteins, vitamins, trace elements and hormones. However, major flaws of FBS include batch-to-batch variations in its composition, likelihood of contamination by viruses and the collection method, which causes unnecessary suffering to unborn calf (Anon, 1993; Wessman and Levings, 1999; van der Valk *et al.*, 2004; Schiff, 2005). Therefore, guidelines for good cell culture practice, as well as the ECVAM Scientific Advisory Committee (ESAC), strongly recommend the use of serum-free media with a chemically-defined composition (Hartung *et al.*, 2002; Coecke *et al.*, 2005; ESAC, 2008). Since the pioneering work of Hayashi and Sato (1976), in which serum was replaced by a selected mixture of hormones, more than 100 serum-free cell-specific media have been developed (Barnes and Sato, 1980a, 1980b; Taub, 1990; Bjare, 1992; Gstraunthaler, 2003; Grillberger *et al.*, 2009; Zähringer, 2009). New formulations are recommended to start with a 50:50 (v/v) combination of DMEM and Ham's F-12 nutrient mixture (Ham, 1965; Barnes and Sato, 1980a, 1980b; Jayme *et al.*, 1997). Additionally, they should contain the ITS (insulin, transferrin and selenium) supplement, since insulin has long been known to be an essential hormone in cell culture, transferrin transports iron into cells and selenium is a trace element acting in oxidative stress-protective proteins (Gey and Thalmier, 1924; Bjare, 1992; Helmy *et al.*, 2000). DMEM F12 is frequently supplemented with serum or replacement ingredients such as meat and plant hydrolysates to maintain cell viability and density, but efforts are being made to formulate chemically-defined media through the study of the metabolic requirements of the cell clones used (Keay, 1975; Chen and Harcum, 2006; Kim and Lee, 2009; Huang *et al.*, 2010; Kim do *et al.*, 2013; Duarte *et al.*, 2014; Rouiller *et al.*, 2014).

The main substrates for mammalian cell growth are glucose and glutamine (Duval *et al.*, 1991; Gódia an Cairó, 2006). Glucose, contained at concentrations of 10-25 mM in the culture medium, provides energy for cells through its metabolism to pyruvate through the glycolysis pathway, which produces 2 molecules of ATP, 2 of NADH and 1.1-1.7 of the growth inhibitor lactate per molecule of glucose. The pyruvate is further oxidatively decarbonized to acetyl-CoA and transformed through the TCA-cycle and the respiratory chain into water and CO_2, resulting in the production of 36 molecules of ATP per molecule of glucose. Additionally, the pentose-phosphate pathway uses glucose byproducts to build nucleotides (Zielke *et al.*, 1978; Reitzer *et al.*, 1979; Butler, 1987; Batt and Kompala, 1989; Gódia and Cairó, 2006; Pörtner, 2009). Glutamine is added at approximate concentrations of 1-5 mM to the culture medium and constitutes an essential precursor for the biosynthesis of purines, pyrimidines, amino sugars and asparagines. This amino acid is incorporated into the cell by uptake transporters and metabolized in the TCA cycle or converted into glutamate and aspartate. Its complete oxidation to CO_2 and incomplete oxidations to aspartate and lactate generate 21, 12 and 6 molecules of ATP per molecule of glutamine, respectively, as well as 0.7 molecules of the toxic compound ammonia (Zielke *et al.*, 1978; Reitzer *et al.*, 1979; Schneider *et al.*, 1996; Butler, 2004; Gódia and Cairó, 2006; Pörtner, 2009). The substitution of glutamine by the glutamine-containing dipeptides alanyl-glutamine and glycyl-glutamine, commercialized under the trademark GLUTMAX™, allows for not only the autoclaving of media containing this amino acid, but also the controlled release of glutamine depending on the activity of peptidases, which results in lower glutamine consumption and ammonia production (Christie and Butler, 1994).

The 13 essential amino acids required for mammalian cell culture, mostly for protein synthesis, are arginine (Arg), cysteine (Cys), glutamine (Gln), histidine (His), isoleucine (Ile), leucine (Leu), lysine (Lys), methionine (Met), phenylalanine (Phe), threonine (Thr), tryptophan (Trp), tyrosine (Tyr) and valine (Val). The DMEM medium provides them in concentrations ranging from 0.1 to 4 mM. Additionally, Ham's F-12 provides the seven non-essential amino acids alanine (Ala), asparagine (Asn), aspartic acid (Asp), glutamic acid (Glu), glycine (Gly), proline (Pro) and serine (Ser). While some (Arg, Asp, Cys, His, Ile, Leu, Lys, Met, Phe, Pro, Ser, Thr, Trp, Tyr and Val) are consumed, others (Ala, Gly, Glu) are partially produced by using essential amino acids as precursors (Pörtner, 2009; van der Valk *et al.*, 2010; Green *et al.*, 2016). It is of particular importance to combine metabolic profiling and flux analysis with stoichiometric analysis of metabolic pathways to determine the optimal concentrations of each amino acid in order to maintain the equilibrium of metabolic pathways (Chen and Harcum, 2006; Rouiller *et al.*, 2014). As an illustrative example, both Ser and Gly are involved in the metabolism of nucleic acid precursors through the tetrahydrofolate (THF) cycle. Supplementation with Gly in Ser-depleted conditions leads to Ser production, which draws metabolites away from the THF cycle and results in the inhibition of cell proliferation (Locasale *et al.*, 2013; Amelio *et al.*, 2014; Duarte *et al.*, 2014; Labuschagne *et al.*, 2014). The functionality, regulation of activity and correct level of expression of amino acid transporters is essential to maintain mammalian cell growth and metabolism, since these proteins largely govern the intracellular availability of amino acids (Lasorsa *et al.*, 2003; Saier *et al.*, 2014; Salazar *et al.*, 2016). Additionally, chemical properties such as solubility of the mixture under the intended storage conditions, interaction with metal ions present in the cell culture medium to form complexes, mixed crystal formation between different amino acids in the mixture or susceptibility to changes in environmental conditions should be considered when formulating a culture medium (Stadtman and Levine, 2003; Jover *et al.*, 2008; Tseng *et al.*, 2009; Remko *et al.*, 2010; van der Valk *et al.*, 2010; Lee *et al.*, 2013; Landauer, 2014; Salazar *et al.*, 2016). In chemically-non defined media, amino acids and small peptides can be delivered through protein hydrolysates, whose beneficial or detrimental effects on cell growth remain controversial (Schlaeger, 1996a; Burteau *et al.*, 2003; Keay, 2004).

Vitamins are required for the optimal macronutrient metabolism, cell viability and genome stability (Arigony *et al.*, 2013). Vitamin A, also referred to as retinol, retinoic acid, retinal, a- and b-carotene, lycopene, lutein, zeaxanthin, b-cryptoxanthin or astaxanthin, has been found to show protective effects against oxidative stress at low concentration while increased DNA damage at higher levels, consistent with the capacity of b-carotene to act as a prooxidant at high concentrations and high oxygen tension (Palozza *et al.*, 2003; Roehrs *et al.*, 2010; Azqueta and Collins, 2012). Vitamin B7, also known as biotin, acts as a cofactor for the enzymes pyruvate carboxylase, propionyl-CoA carboxylase, crotonyl-CoA carboxylase and two isozymes of acetyl-CoA carboxylase, which catalyze important steps of the cell metabolism such as the biosynthesis of fatty acids, gluconeogenesis and amino acid metabolism (Jitrapakdee and Wallace, 2003; Mayende *et al.*, 2012). While its excessive supplementation increases DNA strand breaks in cell cultures, their high intake *in vivo* combined with low consumption of multiple other nutrients has been associated with enhanced genomic stability. Additionally, it controls several aspects of cell biology and development by regulating gene expression (Rodríguez-Meléndez and Zempleni, 2003; Rodríguez-Meléndez *et al.*, 2004; Fenech *et al.*, 2005). Vitamin B9, also known as folate or folic acid, is a key carbon donor during nucleotide biosynthesis that prevents the occurrence of chromosomal breaks and DNA damage, in addition to maintaining the DNA methylation pattern, thereby promoting genome stability in cell culture (Blount and Ames, 1995; Lindahl and Wood, 1999; Li *et al.*, 2003; Stover, 2004; Salbaum and Kappen, 2012). Vitamin B12 or cyanocobalamin is required for the synthesis of methionine and the methyl donor S-adenosyl methionine, which is required to maintain the DNA methylation patterns that determine gene expression and DNA conformation (Zingg and Jones, 1997; Battaglia-Hsu *et al.*, 2009). Vitamin C, also referred to as ascorbate or ascorbic acid, serves as cofactor to numerous enzymes such as prolyl hydroxylases, which play an essential role in collagen biosynthesis and inhibition of the hypoxia-inducible factor (HIF) 1. This transcription factor regulates the expression of many genes involved in tumor growth, energy metabolism, neutrophil function and apoptosis. Additionally, vitamin C exhibits antioxidant properties (Traber and Stevens, 2011). Vitamin E, comprising compounds of the tocopherol and tocotrienol families, is a lipid-soluble chain-breaking antioxidant that scavenges hydroperoxyl radicals and prevents the propagation of free radicals in membranes and plasma lipoproteins (Traber and Atkinson, 2007; Traber and Stevens, 2011). Vitamin E has been demonstrated to selectively protect SDH (var^+) cells from oxidative damages and apoptosis, rebalance the redox cofactor NAD^+ and prevent the formation of the oxidation product 8-oxo-7,8-dihydro-2'-de-oxyguanosine from the DNA nucleoside deoxyguanosine, thus reinforcing its role in maintaining DNA integrity and stability (Bergstrom *et al.*, 2012; Ni and Eng, 2012). *In vitro* cell culture studies showed that a-tocopherol, both alone and combined with ascorbic acid, exhibits protective effects against DNA oxidative damage, apoptosis and autophagy (Stewart *et al.*, 1996; Sweetman *et al.*, 1997).

Cell culture media contain numerous inorganic ions, being Na^+, K^+, Cl^-, Ca^{2+}, Mg^{2+}, SO_4^- and PO_4^{2-} the most abundant. The major sources for these ions are the inorganic salts used in the formulation of media, such as NaCl, $NaHCO_3$, KCl, $CaCl_2$, $MgCl_2$, $MgSO_4$, NaH_2PO_4 and KH_2PO_4, which dissociate almost completely in aqueous solutions (Whitten, 1956; Burton, 1975; Biggers, 1998). Inorganic ions serve a large variety of functions in mammalian cells and homeostasis must be maintained to ensure that K^+ concentrations are high while those of Na^+ and Cl^+ are kept low. The ubiquitously expressed Na^+,K^+-ATPase mediates cellular ion homeostasis and establishes an ion gradient that maintains cell membrane potential and serves as energy source for metabolite, amino acid and ion transporters coupled to this gradient (Kaplan, 2002; Hediger *et al.*, 2004). The cytoplasmic concentration of osmolytes can be adjusted by the cell to control its intracellular osmotic pressure and regulate its volume (Hallows and Knauf, 1994; Hoffmann *et al.*, 2009). Upon cell volume decrease, the Na^+/H^+ antiporters from the NHE family are rapidly activated to stimulate Na^+ uptake. The cell pH rises as a result of the extrusion of H^+, which activates the HCO_3^-/Cl^- antiporters of the AE family that mediate Cl^- uptake. The increased intracellular osmotic pressure arising from the uptake of both ions causes the cell to swell to restore its initial volume. The efflux of HCO_3^- ions neutralizes that of H^+, thereby stabilizing the pH (Jiang *et al.*, 1997; Alexander and Grinstein, 2006). In cells with a high extracellular osmolality (e.g. kidney cells), the level of inorganic ions required to balance the high external osmotic pressure would be detrimental. Instead, they accumulate high concentrations of organic osmolytes such as amino acids, amino acid derivatives and sugars, which are compatible with the biochemical functions of the cell even at high concentrations. Because organic osmolytes are slowly synthesized, these cells also rely on inorganic ion transport to quickly regulate their volume (Yancey *et al.*, 1982; Yancey, 1994; Lang *et al.*, 1998). Additionally, mammalian cells need the presence of several trace elements, most of which are sufficiently supplied with Ham's F12 solution. Among them, Cu, Fe, Se and Zn are the most important (Ham, 1965). Copper serves as cofactor for enzymes involved in several processes and its homeostasis in the cell is tightly regulated. It plays a dual role in oxidative stress, since it can induce reactive oxygen species (ROS) formation when involved in Fenton-like or Haber-Weiss reactions, while many antioxidant enzymes use this metal as cofactor (Linder, 2012). Iron is a crucial element involved in essential processes such as electron transport, cellular respiration, proliferation and differentiation or the regulation of gene expression. It can undergo univalent redox reactions between its oxidized (Fe^{3+}) and reduced (Fe^{2+}) forms, which together with its capacity to participate in Fenton and Haber-Weiss reactions contributes to the formation of ROS. Alternatively, iron is involved in DNA repair mechanisms and enzymes such as the helicases ERCC2 and BACH1 contain Fe/S clusters (Aspinwall *et al.*, 1997; Boldt, 1999; Rudolf *et al.*, 2006; Prá *et al.*, 2011). Selenium plays an essential role in membrane integrity, energy metabolism and protection against DNA damage due to its role in redox regulation and antioxidant function as constituent of selenocysteine-containing selenoproteins (Kryukov *et al.*, 2002; Saito and Takahashi, 2002; Bräuer and Savaskan, 2004; Papp *et al.*, 2007; Muecke *et al.*, 2010). Finally, zinc is a component of over 300 proteins including DNA-binding proteins with zinc fingers, Cu/Zn superoxide dismutases, the estrogen receptor and the synaptic transmission protein (Ames, 1999).

Although some cell types can be maintained in basal media, most need supplements such as hormones, growth factors, protease inhibitors or shear force protectants to survive, proliferate and differentiate. Media have been traditionally supplied with approximately 5-10% serum such as fetal bovine serum (FBS) or horse serum (HS) in order to provide these elements, but they are progressively being replaced by other compounds to overcome the disadvantages previously described (Bettger and McKeehan, 1986; Griffiths, 1987; Buttler and Jenkins, 1989). Hormones such as insulin, glucocorticoids, triiodothyronine and steroids are often employed in serum-free formulations (Hayashi and Sato, 1976; Barnes and Sato, 1980a, 1980b; Lindl and Gstraunthaler, 2008; van der Valk *et al.*, 2010). Growth factors are generally cell type-specific and need to be added to increase cell proliferation and stimulate certain functions. For instance, fibroblast growth factor-2 is employed in chondrocyte culture (Mandl *et al.*, 2004). Some cells can produce them during cultivation, thereby promoting their own growth and development (Gospodarowicz and Moran, 1976). During cell turnover, lysosomal peptidases and trypsin can be released into the growth medium. To prevent protein damage, the protease inhibitors a_1-antitrypsin and a_2-macroglobulin can be added (Gstraunthaler, 2003). Cells growing in serum-free media are unprotected from the shear stress arising from the growth in bioreactors and perfusion cultures. The detergent Pluronic F-68 has a protective effect similar to that of serum, although its addition is not strictly necessary (Zhang *et al.*, 1992, 1995; Al-Rubeai *et al.*, 1992). In order to facilitate cell adhesion and carry low molecular weight molecules, proteins such as albumin can be provided. This protein acts as a carrier of essential fatty acids and lipids, which are also commercially available as water-soluble supplements (Taub, 1990; Keenan *et al.*, 2006). Finally, antibiotics such as penicillin and streptomycin are often added on a laboratory scale to prevent contamination, although their use on production scale is not recommended given the adverse effects that antibiotics may have on cell growth and the possibility to develop antibiotic-resistant microorganisms (Kuhlmann, 1996; Pörtner, 2009).

Culture media for insect cells

The growth of insect cells (IC), together with their transformation with genetically engineered baculovirus expression vector systems (BEVS), is very well established and can be employed in the robust, scalable and stable production of viruses and recombinant proteins to manufacture biopesticides, vaccines, therapeutics and vectors for gene therapy (Ikonomou *et al.*, 2003; Palomares *et al.*, 2006; Agathos, 2010). Although the culture medium is composed of the same basal elements (carbohydrates, amino acids, salts), insect cell metabolism shows some notable differences when compared to that of mammalian cells and the concentration of these compounds must be adjusted accordingly (Neerman and Wagner, 1996; Drugmand *et al.*, 2012).

As in the case of mammalian cells, glucose is the preferred energy and carbon source. Insect cells can grow on media containing only glucose as carbohydrate, but some lines can consume other sugars such as fructose or maltose upon glucose depletion (Landureau, 1969; Reuveny *et al.*, 1992; Bédard *et al.*, 1993; Drews *et al.*, 1995; Ikonomou *et al.*, 2003). Sucrose, which is added to adjust osmolarity, is only consumed by Sf-9 cells during baculovirus infection (Wang *et al.*, 1993a; Sugiura and Amann, 1996). Although trehalose is the most abundant carbohydrate in hemolymph, it is not consumed if other carbon sources are present (Vaughn, 1973). 80% of the pyruvate formed from the glucose metabolism is channeled into the TCA cycle in Sf-9 cells. Most of it contributes to cell growth through its diversion to acetyl-CoA, while a small part leads to alanine and lipid formation. The growth of Sf-9 and High-Five cells has been shown to stop under glucose deprivation, while excessive supply of this carbohydrate results in the detrimental accumulation of lactate and alanine (Bédard *et al.*, 1993; Öhman *et al.*, 1995; Mendoça *et al.*, 1999; Drews *et al.*, 2000; Bernal *et al.*, 2009).

The main amino acid used as energy source is Gln. At glucose-limiting conditions, this amino acid is transformed into Glu, which is further converted to a-ketoglutarate by glutamate dehydrogenase and fed into the TCA cycle (Wang et al., 1993; Öhman et al., 1996; Drews et al., 2000). High-Five cells consume Asn extensively, while Sf-9 only use this amino acid under Gln deprivation (Danner *et al.*, 1995; Yang *et al.*, 1996; Rhiel *et al.*, 1997). The rest of amino acids are consumed to a lesser extent and the lines commonly employed differ in their requirements. While Sf-9 can use Arg, Asn, Asp, Gln, Glu, Met and Ser, High-Five employs Arg, Asp, Cys, Glu, His, Leu, Lys, Met and Thr for energy production (Drews *et al.*, 1995; Drugmand, 2007). Additionally, the group of amino acids comprising Arg, Asn, Asp, Gln, Glu, Gly and Ser is required to sustain high growth rate in both Sf-9 and High-Five, as well as during the infection phase to produce high protein yields. The absence of Ile, Lys, His, Leu or Thr stops cell growth in Sf-9, while Cys and Gly are required in cells lacking a functional methylene tetrahydrofolase and Met and Tyr are recommendable to delay cell death (Tremblay *et al.*, 1992; Ferrance *et al.*, 1993; Wang *et al.*, 1993; Radford *et al.*, 1997; Rhiel *et al.*, 1997; Mendoça *et al.*, 1999).

Insect cells have limited capacity to synthesize, desaturate and elongate fatty acids and cannot synthesize cholesterol, but they metabolize lipids to produce energy (Luukkonen *et al.*, 1973; Mitsuhashi, 1989; Öhman *et al.*, 1995). Because lipid deprivation produces deficient baculoviruses and cell degeneration, insect cells require lipid supplementation (Goodwin, 1991).

Insect cells accumulate low levels of lactate when compared to their mammalian counterparts. While Sf-9 do not produce this inhibitor if sufficient oxygen is supplied, High-Five usually produce 10-20 mM lactate and up to 45 in non-adapted media or under stress conditions (Sugiura and Amann, 1996; Yang *et al.*, 1996; Rhiel *et al.*, 1997; Ikonomou *et al.*, 2003; Drugmand *et al.*, 2005; Bernal *et al.*, 2009). As in the case of mammalian cells, the toxicity of lactate is due to medium acidification and increased osmolality. Nevertheless, concentrations up to 10 mM do not affect the growth of Sf-9 cells and necrosis is not induced below 12.5 mM in this line and High-Five (Palomares and Ramírez, 1996; Cruz *et al.*, 2000; Drugmand, 2007). Ammonia and Ala are produced as byproducts from the catabolism of amino acids, being the former suspected to exert its toxic effects through the acidification of the cytoplasm and mitochondria and the competition with K^+ for the transport by Na^+,K^+-ATPases, as is the case for mammalian cells (Martinelle *et al.*, 1996; Schneider *et al.*, 1996). Although ammonia production is higher in insect cells, they are less sensitive to this inhibitor than mammalian cells. Concentrations of 10 mM do not affect the growth of Sf-9 cells, which can use it to produce Glu and a-ketoglutarate. Ala is produced under glucose excess and concentrations up to 100 mM can be withstood by the cells (Bédard *et al.*, 1993; Öhman *et al.*, 1995, 1996). High-Five cells produce 20 mM ammonia derived from the Gln and Asn metabolism. At 30 mM, it has been shown to affect cell growth only slightly, but protein production severely (Sugiura and Amann, 1996; Yang *et al.*, 1996; Chico and Jäger, 2000).

The composition of the first insect cell culture media, such as Grace's, TNM-FH and TC-100 media, was based on that of hemolymph. They were supplemented with 5-10% FBS in order to provide essential components such as growth factors, detoxifying and carrier proteins, protease inhibitors and shear protectants (Grace, 1962; Hink, 1970; Gardiner and Stockdale, 1975; Drugmand *et al.*, 2012). The first serum-free medium, called CDM, was formulated by Wilkie and collaborators (1980) and contained Yeastolate, an ultrafiltered yeast extract providing B complex vitamins and bioactive peptides (Mitsuhashi, 1989; Mendoça *et al.*, 2007). IPL-41, also supplemented with Yeastolate plus a lipid mixture emulsified in Pluronic F-68, was developed one year after (Weiss *et al.*, 1981). In the following years, modifications of both media with Yeastolate, egg yolk, lactalbumin or Ex-Cyte were developed to sustain the growth of Sf-9, Sf-21, Tn-386 and SL-2 lines (Röder, 1982; Maiorella *et al.*, 1988; Hink, 1991; Schlaeger, 1996a). SF-1, a medium sustaining the growth of High-Five cells, contains the enzymatic digest of animal tissue Primatone, which delays apoptosis, thereby prolonging the stationary phase (Schlaeger *et al.*, 1993; Schlaeger, 1996b). Öhman and collaborators created the KDM-10 medium, specifically adapted to Sf-9, in 1995. In the last two decades, numerous commercial media supporting the growth of Sf and Tn cells, such as Sf-900 II, Sf-900 III and Express-Five™ SFM from Invitrogen Life Technologies, Insect-XPRESS from Lonza Ex-Cell 400, 405 and 420 from JRH Biosciences or HyQ SFX-Insect and HYQ CCM3 from Hyclone, have appeared on the market and are routinely used (Agathos, 1996, 2007; Drugmand *et al.*, 2012). In the IC-BEVS, which has different, specific metabolic needs for the growth and infection/production phases, the use of commercial media of unknown composition is inviable. Instead, low-cost, in-house serum- and protein-free media can be formulated through empirical, statistical or genetic algorithms (Agathos, 2007). For instance, the YPR and YSD media, IPL-41 derivatives supplemented with lipids, Primatone and Yeastolate or dextran, soy protein hydrolysate and Yeastolate, respectively, were obtained through these procedures (Ikonomou *et al.*, 2001; Agathos, 2010). Culture media for insect cells are typically buffered with phosphate and therefore do not require a controlled CO_2 environment. The media are formulated to an osmolality of 350 mOsm/l, slightly higher than that required for mammalian cells. Because insect cells are more sensitive to shear stress than their mammalian counterparts, supplementation with Pluronic F-68 is essential (Palomares *et al.*, 2000; van Oers and Lynn, 2010).

Culture media for plant cells

The plant cell shows several characteristics that differentiate it from its animal counterparts. Its large size (up to 100 mm), results in lower growth rates (doubling times of several days) and growth processes represented by a sigmoidal curve with lag, exponential, delay, stationary growth and lethal phases (Takebe *et al.*, 1971; Zenk *et al.*, 1975; Endress, 1994). The presence of a relatively inflexible cell wall confers increased robustness to plant cells, which show only minor changes in morphology, reductions in metabolite yield and biomass production as a result of shear stress (Kieran *et al.*, 1997; Eibl and Eibl, 2009). Finally, plant cells possess a central vacuole that increases size during the growth phase and is the main site of product accumulation. While secondary metabolites accumulate intracellularly, recombinant proteins and molecules over 20 kDa are generally secreted into the culture medium through the cell wall.

A large cell mass during the correct cell cycle stage is essential for the optimized yield of growth-associated products (Misawa, 1994; Kreis *et al*., 2001; Fischer *et al*., 2004).

About 80% of all plant cell cultivations are performed on Murashige and Skoog (MS) basal medium. This formula was initially developed for tobacco tissue culture and provides the main inorganic macronutrients required by plant cells through the salts NH_4NO_3, KNO_3, $CaCl_2{\cdot}2H_2O$, KH_2PO_4, $MgSO_4{\cdot}7H_2O$, $MnSO_4{\cdot}4H_2O$, $FeSO_4{\cdot}7H_2O$, H_3BO_3, KI, $CuSO_4{\cdot}5H_2O$, $Na_2MoO_4{\cdot}2H_2O$ and $CoCl_2{\cdot}6H_2O$ (Murashige and Skoog, 1962; Murashige, 1973; Evans *et al*., 1983). It differs from other media such as Gamborg et al.'s B5 medium (1968) in its high concentration of nitrate, potassium and ammonia. Cell growth and product formation are strongly influenced by the nutrient concentration of the medium, as well as the presence of plant growth regulators, and it is therefore common to modify its composition for an optimized production (Endress, 1994; Kreis *et al*., 2001; Satio and Mukazami, 2002; Eibl and Eibl, 2009). As illustrative examples, the highest anthocyanin yield in grape (*Vitis vinifera* L.) cell suspension was obtained when the nitrate content was reduced from 25 to 6.25 mM and sucrose was increased from 88 to 132 mM, while that of podophyllin was achieved in *Podophyllum hexandrum* cells at 60 g/l glucose, 1.25 mM phosphate and 60 mM of nitrogen with a combination of ammonia and nitrate in the 1:2 ratio (Bao Do and Cormier, 1991; Chattopadhyay *et al*., 2003).

Sucrose is the most widely used carbon source and constitutes 2-5% of the plant cell culture medium. The extracellular invertase hydrolyzes this disaccharide to the monosaccharides glucose and fructose. Although an increase in the initial concentration of sucrose can improve product yield as described by Petersen and collaborators (1992), opposite results were obtained by Oksman-Caldentey *et al*. (1994) and Chattopadhyay *et al*. (2003). In the latter cases, high sugar concentrations were speculated to have caused cell plasmolysis, thereby reducing the synthesis of secondary metabolites. Glucose and fructose can replace sucrose, while other sugars are poor carbon sources. Although it is not essential, m-inositol is commonly added owing to its role in promoting callus growth. Supplementation with amino acids is not required, but 0.05-0.1% casein hydrolysate or 2-10 mM Glu can be added when inorganic nutrients are insufficient. The only vitamin that may be required is thiamine (0.1-1 mg/l), while inositol, nicotinic acid and pyridoxine can enhance cell growth (Gamborg *et al.*, 1968, 1976; Murashige, 1973).

Growth regulators affect cell growth and development. The ratio of auxins, such as indole-3-acetic acid (IAA), 2,4-dichlorophenoxyacetic acid (2,4-D), napthaleneacetic acid (NAA), indole-3-butyric acid (IBA) or 2,4,5-trichlorophenoxyacetic acid (2,4,5-T), to cytokinins, such as kinetin, benzylaminopurine (BAP), zeatin, purine and adenine, is critical, since low values of this parameter typically stimulate cell growth, while high ratios promote cell division (Eibl and Eibl, 2009). It is also important to select the appropriate phytohormone. 2,4-D (1-10 mM) promotes rapid callus induction, but may cause abnormalities and mutations during long-term cultivation and inhibit the regeneration process. Addition of a cytokinin (kinetin, zeatin or benzyladenine at 0.1-1 mM) can also benefit callus production. In turn, a combination of NAA and benzyladenine, zeatin or isopentenyl adenine is preferred to induce morphogenesis, which is repressed by 2,4-D. IAA can be used instead of NAA, but it should be filter-sterilized and may be degraded by cellular enzymes. Gibberellins may be beneficial to calli and organ cultures of certain species but are generally non-essential (Murashige, 1973; Gamborg *et al.*, 1976; Hess, 1992). Finally, supplements such as polyvinylpyrrolidone, gelatin, bovine serum albumin and sodium chloride must be added to stabilize secreted recombinant proteins (Shadwick and Doran, 2004).

REFERENCES

1. Agathos, S.N. (1996) Insect cell bioreactors. Cytotechnology. 20, 173-189
2. Agathos, S.N. (2007) Development of serum-free media for lepidopteran insect cell lines. In: Murhammer DW, editor. Baculovirus Expression Protocols. Totowa, NJ: Humana Press, pp. 155-185
3. Agathos, S.N. (2010) Insect cell culture. In: Baltz RH, Davies JE, Demain AL, editors. Manual of Industrial Microbiology and Biotechnology. Washington, DC: ASM Press, pp. 212-222
4. Al-Rubeai, M., Emery, A.N, Chalder, S. (1992) The effect of Pluronic F-68 on hybridoma cells in continuous cell culture. Appl Microbiol Biotechnol. 37, 44-45

5. Alexander, R.T., Grinstein, S. (2006) Na^+/H^+ exchangers and the regulation of volume. Acta Physiol (Oxf). 187, 159-167

6. Amelio, I., Cutruzzola, F., Antonov, A., Agostini, M., Melino, G. (2014) Serine and glycine metabolism in cancer. Trends Biochem Sci. 39, 191-198

7. Ames, B.N. (1999) Micronutrient deficiencies. A major cause of DNA damage. Annals of the New York Academy of Sciences. 889, 87-106

8. Anon (1993) Points to consider in the characterization of cell lines used to produce biologicals. Available from: <http://www.fda.gov/downloads/BiologicsBlood Vaccines/GuidanceComplianceRegulatoryInformation/ OtherRecommendations forManufacturers/UCM062745.pdf> (accessed 16.02.10)

9. Arigony, A.L., de Oliveira, I.M., Machado, M., Bordin, D.L., Bergter, L., Prá, D., Henriques, J.A. (2013) The influence of micronutrients in cell culture: a reflection of viability and genome stability. Biomed Res Int. 2013, 597282

10. Aspinwall, R., Rothwell, D.G., Roldán-Arjona, T., Anselmino, C., Ward, C.J., Cheadle, J.P., Sampson, J.R., Lindahl, T., Harris, P.C., Hickson, I.D. (1997) Cloning and characterization of a functional homolog of Escherichia coli endonuclease III. Proc Natl Acad Sci USA. 94, 109-114

11. Azqueta, A., Collins, A.R. (2012) Carotenoids and DNA damage. Mutation Research. 733, 4-13

12. Bao Do, C., Cormier, F. (1991) Effects of low nitrate and high sugar concentrations on anthocyanin content and composition of grape (***Vitis vinifera*** L.) cell suspension. Plant Cell Rep. 9, 500-504

13. Barnes, D., Sato, G. (1980a) Methods for growth of cultured cells in serum-free medium. Analytical Biochemistry. 102, 255-270

14. Barnes, D., Sato, G. (1980b) Serum-free cell culture: a unifying approach. Cell. 22, 649–655

15. Batt, C.B., Kompala, S.K. (1989) A structured modeling framework for the dynamics of hybridoma growth in continuous suspension cultures. Biotechnol Bioeng. 34, 515-531

16. Battaglia-Hsu, S.F., Akchiche, N., Noel, N., Alberto, J.M., Jeannesson, E., Orozco-Barrios, C.E., Martinez-Fong, D., Daval, J.L., Guéant, J.L. (2009) Vitamin B12 deficiency reduces proliferation and promotes differentiation of neuroblastoma cells and up-regulates PP2A, proNGF, and TACE. Proc Natl Acad Sci USA. 106, 21930-21935
17. Bédard, C., Tom, R., Kamen, A. (1993) Growth, nutrient consumption, and end-product accumulation in Sf-9 and BTI-EAA insect cell cultures: insights into growth limitation and metabolism. Biotechnol Prog. 9, 615-624
18. Bergstrom, T., Ersson, C., Bergman, J., Moller, L. (2012) Vitamins at physiological levels cause oxidation to the DNA nucleoside deoxyguanosine and to DNA—alone or in synergism with metals. Mutagenesis. 27, 511-517
19. Bernal, V., Carinhas, N., Yokomizo, Y., Carrondo, M.J.T., Alves, P.M. (2009) Cell density effect in the baculovirus-insect cells system: a quantitative analysis of energetic metabolism. Biotechnol Bioeng. 104, 162-180
20. Bettger, W.J., McKeehan, W.L. (1986) Mechanisms of cellular nutrition. Physiological Reviews. 66, 1-35
21. Biggers, J.D. (1998) Reflections on the culture of the preimplantation embryo. Int J Dev Biol. 42, 879-884
22. Bjare, U. (1992) Serum-free cell culture. Pharmacology and Therapeutics. 53, 355-374
23. Blount, B.C., Ames, B.N. (1995) DNA damage in folate deficiency. Baillieres Clin Haematol. 8, 461-478
24. Boldt, D.H. (1999) New perspectives on iron: an introduction. AmericanJournal of theMedical Sciences. 318, 207-212
25. Bräuer, A.U., Savaskan, N.E. (2004) Molecular actions of selenium in the brain: neuroprotective mechanisms of an essential trace element. Reviews in the Neurosciences. 15, 19-32
26. Burteau, C.C., Verhoeye, F.R., Mols, J.F., Ballez, J.S., Agathos, S.N., Schneider, Y.J. (2003) Fortification of a protein-free cell culture medium with plant peptones improves cultivation and productivity of an interferon-gamma-producing CHO cell line. In Vitro Cellular and Developmental Biology Animal. 39, 291-296

27. Burton, R.F. (1975) Appendix A: osmolality and osmotic coef fi cients. In: Ringer solutions and physiological salines. Wright Scientechnica, Bristol

28. Buttler, M. (1987) Growth limitations in microcarrier cultures. Adv Biochem Eng. 34, 57-84

29. Buttler, M. (2004) Animal cell culture technology – the basics (2nd edition). Oxford University Press, New York

30. Butler, M., Jenkins, H. (1989) Nutritional aspects of the growth of animal cells in culture. J Biotechnol. 12, 97-110

31. Chattopadhyay, S., Mehra, R.S., Srivastava, A.K., Bhojwani, S.S., Bisaria, V.S. (2003) Effect of major nutrients on podophyllotoxin production in ***Podophyllum hexandrum*** suspension cultures. Appl Microbiol Biotechnol. 60, 541-546

32. Chen, P., Harcum, S.W. (2006) Effects of elevated ammonium on glycosylation gene expression in CHO cells. Metab Eng. 8, 123-132

33. Chico, E., Jäger, V. (2000) Perfusion culture of baculovirus-infected BTI-Tn-5B1-4 insect cells: a method to restore cell-specific b-trace glycoprotein productivity at high cell density. Biotechnol Bioeng. 70, 574-586

34. Christie, A., Butler, M. (1994) Growth and metabolism of a murine hybridoma in cultures containing glutamine-based dipeptides. GIBCO Focus. 16, 9-13

35. Cruz, H.J., Freitas, C.M., Alves, P.M., Moreira, J.L. (2000) Carrondo MJT. Effects of ammonia and lactate on growth, metabolism, and productivity of BHK cells. Enzyme Microb Technol. 27, 43-52

36. Danner, D.J., Godwin, G.P., Kassay, K., Gorfien, S.F. (1995) Growth of BTI-TN-5B1-4 cells in a new serum-free medium. Baculovirus and Insect Cell Gene Expression Conference. March 26–30. Pinehurst, N.C.

37. Drews, M., Doverskog, M., Öhman, L., Chapman, B.E., Jacobsson, U., Kuchel, P.W., Häggström, L. (2000) Pathways of glutamine metabolism in ***Spodoptera frugiperda*** (Sf9) insect cells: evidence for the presence of the nitrogen assimilation system, and metabolic switch by 1H/15N NMR. J Biotechnol. 78, 23-37

38. Drews, M., Paalme, T., Vilu, R. (1995) The growth and nutrient

utilization of the insect cell line ***Spodoptera frugiperda*** Sf9 in batch and continuous culture. J Biotechnol. 40, 187-198

39. Drugmand, J.-C. (2007) Study of the metabolism and physiology of High-Five insect cells for the development of processes for the production of recombinant protein. [Ph.D.Thesis]. Louvain-la-Neuve, Belgium: Université Catholique de Louvain.
40. Drugmand, J.-C., Schneider, Y.-J., Agathos, S.N. (2005) Environmental effects of lactate on High-Five™ insect cell metabolism. In: Godia F, Fussenegger M, editors. Animal Cell Technology Meets Genomics. The Netherlands: Springer, pp. 91-94
41. Drugmand, J.C., Schneider, Y.J., Agathos, S.N. (2012) Insect cells as factories for biomanufacturing. Biotechnol Adv. 30, 1140-1157
42. Duarte, T.M., Carinhas, N., Barreiro, L.C., Carrondo, M.J., Alves, P.M., Teixeira, A.P. (2014) Metabolic responses of CHO cells to limitation of key amino acids. Biotechnol Bioeng. 111, 2095-2106
43. Duval, D., Demangel, C., Munier-Jolain, K., Miosecc, S., Geahel, I. (1991) Factors controlling cell proliferation and antibody production in mouse hybridoma cells: I. Influence of amino acid supply. Biotechnol Bioeng. 38, 561-570
44. Eagle, H. (1959) Amino acid metabolism in mammalian cell cultures. Science. 120, 432-437
45. Eibl, R., Eibl, D. (2009) Plant cell-based bioprocessing. . In: Eibl, R., Eibl, D., Pörtner, R., Catapano, G., Czermak, P. (eds). Cell and Tissue Reaction Engineering. Springer, Dordrecht, Heidelberg, London, New York, pp 315-356
46. Eibl, D., Eibl, R., Pörtner, R. (2009) Mammalian cell culture technology: an emerging field. In: Eibl, R., Eibl, D., Pörtner, R., Catapano, G., Czermak, P. (eds). Cell and Tissue Reaction Engineering. Springer, Dordrecht, Heidelberg, London, New York, pp 3-12
47. Endress, R. (1994) Plant Cell Biotechnology. Springer, Berlin, Heidelberg, New York
48. Evans, J. (2006) Plant-derived drug. http://www.rsc.org/chemistryworld/ News/ 2006/ February/ 07020602.asp. Cited 10 Apr 2007

49. Evans, D.A., Coleman, J., Kearns, A. (2003) Plant Cell Culture. BIOS Scientific. Abingdon

50. Fenech, M., Baghurst, P., Luderer, W., Turner, J., Record, S., Ceppi, M., Bonassi, S. (2005) Low intake of calcium, folate, nicotinic acid, vitamin E, retinol, β-carotene and high intake of pantothenic acid, biotin and riboflavin are significantly associated with increased genome instability— results from a dietary intake and micronucleus index survey in South Australia. Carcinogenesis. 26, 991-999

51. Ferrance, J.P., Goel, A., Ataai, M.M. (1993) Utilization of glucose and amino acids in insect cells cultures: quantifying the metabolic flows within the primary pathways and medium development. Biotechnol Bioeng. 42, 697-707

52. Fischer, R., Stoger, E., Schillberg, S., Christou, P., Twyman, R.M. (2004) Plant-based production of biopharmaceuticals. Curr Opin Plant Biol. 7, 152-158

53. Fowler, M.W. (1998) Problems in commercial exploitation of plant cell cultures. In: Bock, B., Marsh, J. (eds). Application of plant cell and tissue culture. Ciba Foundation Symposium 137. Wiley-Chichester, pp 239-253

54. Gamborg, O.L., Miller, R.A., Ojimak, K. (1968) Nutrient requirements of suspension cultures of soybean root cells. Exp Cell Res. 50, 151-158

55. Gamborg, O.L., Murashige, T., Thorpe, A., Vasil, I.K. (1976) Plant tissue culture media. In Vitro. 12, 473-478

56. Gardiner, G.R., Stockdale, H. (1975) Two tissue culture media for production of lepidopteran cells and nuclear polyhedrosis viruses. J Invert Path. 25, 363-370

57. Gey, G.O., Thalhimer, W. (1924) Observations on the effects of insulin introduced into the medium of tissue cultures. JAMA. 82(20), 1609

58. Gódia, F., Cairó, J.J. (2006) Cell metabolism. In: Otzurk, S.S. and Hu, W.-S. (eds). Cell culture technology for pharmaceutical and cell-based therapies. Taylor & Francies, New York.

59. Goodwin, R.H. (1991) Replacement of vertebrate serum with lipids and other factors in the culture of invertebrate cells, tissues, parasites and pathogens. In Vitro Cell Dev Biol-Anim. 27, 470-478

60. Gorr, G., Wagner, S. (2005) Humanized glycosylation: production of biopharmaceuticals in a moss bioreactor. In: Knäblein, J. (ed). Modern Biopharmecauticals, vol 3. Wiley VCH, Weinheim, Germany, pp 919-929

61. Gospodarowicz, D., Moran, J.S. (1976) Growth factors in mammalian cell culture. Annual Review of Biochemistry. 45, 531-558

62. Grace, T.D.C. (1962) Establishment of four strains of cells from insect tissues grown ***in vitro***. Nature. 195, 788-789

63. Green, C.R., Wallace, M., Divakaruni, A.S., Phillips, S.A., Murphy, A.N., Ciaraldi, T.P., Metallo, C.M. (2016) Branched-chain amino acid catabolism fuels adipocyte differentiation and lipogenesis. Nat Chem Biol. 12, 15-21

64. Grillberger, L., Kreil, T.R., Nasr, S., Reiter, M. (2009) Emerging trends in plasma-free manufacturing of recombinant protein therapeutics expressed in mammalian cells. Biotechnology Journal. 4, 186-201

65. Griffiths, J.G. (1987) Serum and growth factors in cell culture media: an introduction review. Dev Biol Stand. 66, 155-160

66. Gstraunthaler, G. (2003) Alternatives to the use of fetal bovine serum: serum-free cell culture. ALTEX. 20, 275-281

67. Hallows, K.R., Knauf, P.A. (1994) Principles of cell volume regulation. In: Strange K (ed) Cellular and molecular physiology of cell volume regulation. CRC, Boca Raton, FL

68. Hartung, T., Balls, M., Bardouille, C., Blanck, O., Coecke, S., Gstraunthaler, G., Lewis, D. (2002) Good cell culture practice. ECVAM good cell culture practice task force report 1. Alternatives to Laboratory Animals. 30, 407-414

69. Hayashi, I., Sato, G.H. (1976) Replacement of serum by hormones permits growth of cells in a defined medium. Nature. 259, 132-134

70. Hediger, M.A., Romero, M.F., Peng, J.B., Rolfs, A., Takanaga, H., Bruford, E.A. (2004) The ABCs of solute carriers: physiological, pathological and therapeutic implications of human membrane transport proteins: introduction. Pflugers Arch. 447, 465-468

71. Hess, D. (1992) Biotechnologie der Pflanzen. Eugen Ulmer, Stuttgart

72. Hink, W.F. (1970) Established insect cell line from the cabbage looper, ***Trichoplusia ni***. Nature 229, 466-467

73. Hink, W.F. (1991) A serum-free medium for the culture of insect cells and production of recombinant proteins. In Vitro Cell Dev Biol. 27, 397-401

74. Hoffmann, E.K., Lambert, I.H., Pedersen, S.F. (2009) Physiology of cell volume regulation in vertebrates. Physiol Rev. 89, 193-277

75. Huang, Y.M., Hu, W., Rustandi, E., Chang, K., Yusuf-Makagiansar, H., Ryll, T. (2010) Maximizing productivity of CHO cell-based fed-batch culture using chemically defined media conditions and typical manufacturing equipment. Biotechnol Prog. 26, 1400-1410

76. ESAC (2008) ESAC statement on the use of FCS and other animal-derived supplements. Available from:http://ecvam.jrc.it/publication/ESAC28_statement_FCS_20080508.pdf> (accessed 16.02.10).

77. Helmy, M.H., Ismail, S.S., Fayed, H., El-Bassiouni, E.A. (2000) Effect of selenium supplementation on the activities of glutathione metabolizing enzymes in human hepatoma Hep G2 cell line. Toxicology. 144, 57-61

78. Hibino, K., Ushiyama, K. (1999) Commercial production of ginseng by plant tissue culture technology. In: Fu, T.J., Curtis, W.R. (eds). Plant cell and tissue culture for the production of food ingredients. Kluwer, New York, pp 215-224

79. Howaldt, M., Walz, F., Kempken, R. (2005) Kultur von Tierzellen. In: Chmiel, H. (ed). Bioprozesstechnik, Spektrum Akademischer Verlag

80. Ikonomou, L., Bastin, G., Schneider, Y.-J., Agathos, S.N. (2001) Design of an efficient medium for insect cell culture and recombinant protein production. In Vitro Cell Dev Biol-Anim. 37, 549-559

81. Ikonomou, L., Schneider, Y.J., Agathos, S.N. (2003) Insect cell culture for industrial production of recombinant proteins. Appl Microbiol Biotechnol. 62, 1-20

82. Jiang, L., Chernova, M.N., Alper, S.L. (1997) Secondary regulatory volume increase conferred on ***Xenopus oocytes*** by expression of AE2 anion exchanger. Am J Physiol. 272, C191–C202

83. Jitrapakdee, S., Wallace, J.C. (2003) The biotin enzyme family:

conserved structural motifs and domain rearrangements. Current Protein and Peptide Science. 4, 217-229

84. Jover, J., Bosque, R., Sales, J. (2008) A comparison of the binding affinity of the common amino acids with different metal cations. Dalton Trans. 45, 6441-6453
85. Kaplan, J.H. (2002) Biochemistry of Na^+,K^+-ATPase. Annu Rev Biochem. 71, 511-535
86. Keay, L. (1975) Autoclavable low cost serum-free cell culture media. The growth of L cells and BHK cells on peptones. Biotechnol Bioeng. 17, 745-764
87. Keenan, J., Pearson, D., Clynes, M. (2006) The role of recombinant proteins in the development of serum-free media. Cytotechnology. 50, 49-56
88. Kelly, B.D., Chiou, T.W., Rosenberg, M., Wang, D.I.C. (1993) Industrial animal cell culture. In: Stephanopolous, G. (ed). Biotechnology, Vol. 3 (Bioprocessing), VCH Verlagsgesellschaft, Weinheim.
89. Kieran, P.M., MacLoughlin, P.F., Malone, D.M. (1997) Plant cell suspension cultures: some engineering considerations. J Biotechnol. 59, 39-52
90. Kim, S.H., Lee, G.M. (2009) Development of serum-free medium supplemented with hydrolysates for the production of therapeutic antibodies in CHO cell cultures using design of experiments. Appl Microbiol Biotechnol. 83, 639-648
91. Kim do, Y., Chaudhry, M.A., Kennard, M.L., Jardon, M.A., Braasch, K., Dionne, B., Butler, M., Piret, J.M. (2013) Fed-batch CHO cell t-PA production and feed glutamine replacement to reduce ammonia production. Biotechnol Prog. 29, 165-175
92. Kreis, W., Baron, D., Stoll, G. (2001) Biotechnologie der Arzneistoffe. Deutscher Apotheke Verlag, Stuttgart
93. Kryukov, G.V., Kumar, R.A., Koc, A., Sun, Z., Gladyshev, V.N. (2004) Selenoprotein R is a zinc-containing stereo-specific methionine sulfoxide reductase. Proc Natl Acad Sci USA. 99, 4245-4250
94. Kuhlmann, I. (1996) The prophylactic use of antibiotics in cell

culture. Cytotechnology. 19, 95-105

95. Labuschagne, C.F., van den Broek, N.J., Mackay, G.M., Vousden, K.H., Maddocks, O.D. (2014) Serine, but not glycine, supportsone-carbon metabolism and proliferation of cancer cells. Cell Rep. 7, 1248-1258
96. Landauer, K. (2014) Designing media for animal cell culture: CHO cells, the industrial standard. Methods Mol Biol (Clifton, NJ) 1104, 89-103
97. Landureau, J.C. (1969) Etude des exigences d'une lignée de cellules d'insectes (souche EPa) II. Vitamines hydrosolubles. Exp Cell Res. 54, 399-402
98. Lang, F., Busch, G.L., Volkl, H. (1998) The diversity of volume regulatory mechanisms. Cell Physiol Biochem. 8, 1-45
99. Lasorsa, F.M., Pinton, P., Palmieri, L., Fiermonte, G., Rizzuto, R., Palmieri, F. (2003) Recombinant expression of the Ca^{2+}-sensitive aspartate/glutamate carrier increases mitochondrial ATP production in agonist-stimulated Chinese hamster ovary cells. J Biol Chem. 278, 38686-38692
100. Lee, C.-Y., Chen, J.-T., Chang, W.-T., Shiah, I.M. (2013) Effect of pH on the solubilities of divalent and trivalent amino acids in water at 298.15 K. Fluid Phase Equilib. 343, 30-35
101. Legardinier, S., Duonor-Cerutti, M., Devauchelle, G., Combarnous, Y, Cahoreau, C. (2005) Biological activities of recombinant equine luteinizing hormone/chorionic gonadotropin (eLH/CG) expressed in Sf9 and Mimic insect cell lines. J Mol Endocrinol. 34, 47-60
102. Li, G.M., Presnell, S.R., Gu, L. (2003) Folate deficiency, mismatch repair-dependent apoptosis, and human disease. J Nutr Biochem. 14, 568-575
103. Lindahl, T., Wood, R.D. (1999) Quality control by DNA repair. Science. 286, 1897-1905
104. Linder, M.C. (2012) The relationship of copper to DNA damage and damage prevention in humans," Mutation Research. 733, 83-91
105. Lindl, T., Gstraunthaler, G. (2008) Zell- und Gewebekultur. Von den Grundlagen zur Laborbank. Spektrum Akademischer Verlag, Heidelberg

106. Locasale, J.W. (2013) Serine, glycine and one-carbon units: cancer metabolism in full circle. Nat Rev Cancer. 13, 572-583

107. Luukkonen, A., Brummer-Korvenkotio, M., Renkonen, O. (1973) Lipids of cultured mosquito cells (***Aedes albopictus***). Comparison with cultured mammalian fibroblasts (BHK 21cells). Biochim Biophys Acta. 326, 256-261

108. Ma, J.K.C., Drake, P.M.W., Christou, P. (2003) The production of recombinant pharmaceutical proteins in plants. Nat Rev Genet. 4, 794-805

109. Maiorella, B., Inlow, D., Shauger, A., Harano, D. (1988) Large-scale insect cell-culture for recombinant protein production. Bio/ Technol. 6, 1406-1410

110. Mandl, E.W., Jahr, H., Koevoet, J.L., van Leeuwen, J.P., Weinans, H., Verhaar, J.A., van Osch, G.J. (2004) Fibroblast growth factor-2 in serum-free medium is a potent mitogen and reduces dedifferentiation of human ear chondrocytes in monolayer culture. Matrix Biology. 23, 231-241

111. Martinelle, K., Westlund, A., Häggström, L. (1996) Ammonium ion transport — a cause of cell death. Cytotechnology. 22, 251-254

112. Mayende, L., Swift, R.D., Bailey, L.M., Soares da Costa, T.P., Wallace, J.C., Booker, G.W., Polyak, S.W. (2012) A novel molecular mechanism to explain biotin-unresponsive holocarboxylase synthetase deficiency. J Mol Med (Berl). 90, 81-88

113. Mendonça, R.Z., de Oliveira, E.C., Pereira, C.A., Lebrun, I. (207) Effect of bioactive peptides isolated from yeastolate, lactalbumin and NZCase in the insect cell growth. Bioprocess Biosyst Eng. 30, 157-164

114. Mendonça, R.Z., Palomares, L.A., Ramírez, O.T. (1999) An insight into insect cell metabolism through selective nutrient manipulation. J Biotechnol. 72, 61-75

115. Misawa, M. (1994) Plant tissue culture: An alternative for production of useful metabolites. FAO Agricultural Services Bulletin N 108. Food and Agriculture Organiation of the United Nations, Rome, p 95

116. Mitsuhashi, J. (1989) Nutritional requirements of insect cells in vitro. In: Mitsuhashi J, editor. Invertebrate Cell System Applications.

Boca Raton, Florida: CRC Press, pp. 3-20

117. Muecke, R., Schomburg, L., Buentzel, J., Kisters, K., Micke, O. (2010) Selenium or no selenium-that is the question in tumor patients: a new controversy. Integrative Cancer Therapies. 9, 136-141

118. Murashige, T. (1973) Nutrition of plant cells and organs ***in vitro***. In Vitro. 9, 81-85

119. Murashighe, T., Skoog, F. (1962) A revised medium for rapid growth and bioassays with tobacco tissue cultures. Physiol Plant. 15, 473-479

120. Neermann, J., Wagner, R. (1996) Comparative analysis of glucose and glutamine metabolism in transformed mammalian cell lines, insect and primary liver cells. J Cell Physiol. 166, 152-169

121. Ni, Y., Eng, C. (2012) Vitamin E protects against lipid peroxidation and rescues tumorigenic phenotypes in cowden/cowden-like patient-derived lymphoblast cells with germline SDHx variants. Clinical Cancer Research. 18, 4954-4961

122. van Oers, M.M., Lynn, D.E. (2010) Insect cell culture. In: Encyclopedia of Life Sciences. Chichester: Wiley

123. Oh, S.K.W., Chua, F.K.F., Choo, A.B.H. (1995) Intracellular responses of productive hybridomas subjected to high osmotic pressure. Biotechnol Bioeng. 46, 525-535

124. Öhman, L., Alarcon, M., Ramqvist, A.-K., Häggström, L. (1996) Glutamine is not an essential amino acid for Sf-9 insect cells. Biotechnol Lett. 18, 765-770

125. Öhman, L., Ljunggren, J., Häggström, L. (1995) Induction of a metabolic switch in insect cells by substrate-limited fed batch cultures. Appl Microbiol Biotechnol. 43, 1006-1013

126. Oksman-Caldentey. K.M., Sevón, K.M., Vanhala, L., Hiltunen, R. (1994) Effect of nitrogen and sucrose on the primary and secondary metabolism of transformed root cultures of ***Hyoscyamus muticus***. Plant Cell Tissue Org Cult. 45, 57-69

127. Otzurk, S.S., Hu, W.S. (eds) (2006) Cell Culture Technology for Pharmaceutical and Cell-Based Therapies. Taylor & Francis, New York.

128. Palomares, L.A., Estrada, S., Ramírez, O.T. (2006) Principles and applications of the insect cellbaculovirus expression vector system. In: Ozturk SS, Hu W-S, editors. Cell Culture Technology for Pharmaceutical and Cell-Based Therapies. New York, U.S.A: Taylor & Francis, pp 627–692

129. Palomares, L.A., González, M., Ramírez, O.T. (2000) Evidence of Pluronic F-68 direct interaction with insect cells: impact on shear protection, recombinant protein and baculovirus production. Enzyme Microb Technol. 26, 324-331

130. Palomares, L.A., Ramírez, O.T. (1996) The effect of dissolved oxygen tension and the utility of oxygen uptake rate in insect cell culture. Cytotechnology. 22, 225-237

131. Palozza, P., Serini, S., Di Nicuolo, F., Piccioni, E., Calviello, G. (2003) Prooxidant effects of β-carotene in cultured cells. Molecular Aspects of Medicine. 24, 353-362

132. Papp, L.V., Lu, J., Holmgren, A., Khanna, K.K. (2007) From selenium to selenoproteins: synthesis, identity, and their role in human health. Antioxidants and Redox Signaling. 9, 775-806

133. Parnham, M.J., Kesselring, K. (1985) Rosmarinic acid. Drugs Future. 10, 756-757

134. Petersen, M., Dombrowski, K., Gertlowski, C., Haeusler, E., Kawatzki, B., Meinhard, J., Alfermann, A.W. (1992) The use of plant cell cultures to study natural product biosynthesis. In: Oono, K., Hirabayashi, T., Kikuchi, S., Handa, H., Kajiwara, K. (eds) Plant tissue culture and gene manipulatin for breeding and formation of phytochemicals. NAIR, Tsukuba, pp 293-296

135. Pörtner, R. (2009) Characteristics of mammalian cells and requirements for cultivation. . In: Eibl, R., Eibl, D., Pörtner, R., Catapano, G., Czermak, P. (eds). Cell and Tissue Reaction Engineering. Springer, Dordrecht, Heidelberg, London, New York, pp 13-54

136. Prá, D., Rech-Franke, S.I., Pegas-Henriques, J.A., Fenech, M. (2011) A possible link between iron deficiency and gastrointestinal carcinogenesis. Nutr Cancer. 61, 415-426

137. Radford, K.M., Reid, S., Greenfield, P.F. (1997) Substrate limitation

in the baculovirus expression vector system. Biotechnol Bioeng. 56, 32-44

138. Reitzer, L.J., Wice, B.M., Kenell, D. (1979) Evidence that glutamine, not sugar, is the major energy source for cultured HeLa cells. J Biol Chem. 254, 2669-2676

139. Remko, M., Fitz, D., Rode, B.M. (2010) Effect of metal ions (Li^+, Na^+, K^+, Mg^{2+}, Ca^{2+}, Ni^{2+}, Cu^{2+} and Zn^{2+}) and water coordination on the structure and properties of l-histidine and zwitterionic l-histidine. Amino Acids. 39, 1309-1319

140. Reuveny, S., Kemp, C.W., Eppstein, L., Shiloach, J. (1992) Carbohydrate metabolism in insect cell cultures during cell growth and recombinant protein production. Ann NY Acad Sci. 665, 230-237

141. Rhiel, M., Mitchell-Logean, C.M., Murhammer, D.W. (1997) Comparison of ***Trichoplusia ni*** BTI-Tn-5B1-4 (High Five™) and ***Spodoptera frugiperda*** Sf-9 insect cell line metabolism in suspension cultures. Biotechnol Bioeng. 55, 909-920

142. Röder, A. (1982) Development of a serum-free medium for cultivation of insect cells. Naturwissenschaften. 69, 92-93

143. Rodríguez-Meléndez, R., Griffin, J.B., Zempleni, J. (2004) Biotin supplementation increases expression of the cytochrome P 450 1B1 gene in Jurkat cells, increasing the occurrence of single-stranded DNA breaks. J Nutr. 134, 2222-2228

144. Rodríguez-Meléndez, R., Zempleni, J. (2003) Regulation of gene expression by biotin. Journal of Nutritional Biochemistry. 14, 680-690

145. Roehrs, R., Freitas, D.R., Masuda, A., Henriques, J.A., Guecheva, T.N., Ramos, A.L., Saffi, J. (2010) Effect of vitamin A treatment on superoxide dismutase-deficient yeast strains. Arch Microbiol. 192, 221-228

146. Rouiller, Y., Perilleux, A., Vesin, M.N., Stettler, M., Jordan, M., Broly, H. (2014) Modulation of mAb quality attributes using microliter scale fed-batch cultures. Biotechnol Prog. 30, 571-583

147. Rudolf, J., Makrantoni,V., Ingledew, W.J., Stark, M.J.R., White, M.F. (2006) The DNA repair helicases XPD and FancJ have essential iron-sulfur domains. Molecular Cell. 23, 801-808

148. Ryu, J.S., Lee, G.M. (1997) Effect of hypoosmotic stress on hybridoma cell growth and antibody production. Biotechnol Bioeng. 55, 565-570

149. Saier, M.H. Jr, Reddy, V.S., Tamang, D.G., Vastermark, A. (2014) The transporter classification database. Nucl Acids Res. 42((Databaseissue)), D251–D258

150. Saito, K., Mikuzami, H. (2002) Plant cell cultures as producers of secondary metabolites. In: Oksman-Caldentey, K.M., Barz, W.H. (eds) Plant Biotechnology and Transgenic Plants. Marcel Dekker, New York, pp 77-109

151. Saito, Y., Takahashi, K. (2002) Characterization of selenoprotein P as a seleniumsupply protein. European Journal of Biochemistry. 269, 5746-5751

152. Salazar, A., Keusgen, M., von Hagen, J. (2016) Amino acids in the cultivation of mammalian cells. Amino Acids. 48, 1161-1171

153. Salbaum, J.M., Kappen, C. (2012) Genetic and epigenomic footprints of folate. Prog Mol Biol Transl Sci. 108, 129-158

154. Schiff, L.J. (2005) Review: production, characterization, and testing of banked mammalian cell substrates used to produce biological products. In Vitro Cellular and Developmental Biology Animal. 41, 65–70

155. Schillberg, S., Twyman, R.M., Fischer, R. (2005) Biopharmaceutical production in cultured plant cells. In: Knäblein, J. (ed). Modern biopharmaceuticals, vol 3. Wiley VCH, Weinheim, Germany, pp 949-965

156. Schlaeger, E.J. (1996a) Medium design for insect cell culture. Cytotechnology. 20, 57-70

157. Schlaeger, E.-J. (1996b) The protein hydrolysate, Primatone RL, is a cost-effective multiple growth promotor of mammalian cell culture in serum-containing and in serum-free media and displays anti-apoptosis properties. J Immunol Meth. 194, 191-199

158. Schneider, M., Marison, I.W., vonStockar, U. (1996) The importance of ammonia in mammalian cell culture. Journal of Biotechnology. 46, 161-185

159. Shadwick, F.S., Doran, P.M. (2004) Foreign protein expression

using plant cell suspension and hairy root cell and organ cultures. In: Fischer, R., Schillberg, S. (eds) Molecular Farming. Wiley-VCH, Weinheim, pp 13-36

160. Sijmons, P.C., Dekker, B.M.M., Schrammeijer, B., Verwoerd, T.C., van den Elzen, P.J.M. Hoekema, A. (1990) Production of correctly processed human serum albumin in transgenic plants. Bio/Technology. 8, 217-221

161. Stadtman, E.R., Levine, R.L. (2003) Free radical-mediated oxidation of free amino acids and amino acid residues in proteins. Amino Acids. 25, 207–218

162. Stewart, M.S., Cameron, G.S., Pence, B.C. (1996) Antioxidant nutrients protect against UVB-induced oxidative damage to DNA of mouse keratinocytes in culture. Journal of Investigative Dermatology. 106, 1086-1089

163. Stover, P.J. (2004) Physiology of folate and vitamin B12 in health and disease. Nutrition Reviews. 62, S3-S13

164. Sugiura, T., Amann, E. (1996) Properties of two insect cell lines useful for the baculovirus expression system in serum-free culture. Biotechnol Bioeng. 51, 494-499

165. Sweetman, S.F., Strain, J.J., McKelvey-Martin, V.J. (1997) Effect of antioxidant vitamin supplementation on DNA damage and repair in human lymphoblastoid cells. Nutrition and Cancer. 27, 122-130

166. Takebe, J., Labib, G., Melchers, G. (1971) Regeneration of whole plants from isolated mesophyll protoplasts of tobacco. Naturwiss. 5, 318-320

167. Taub, M. (1990) The use of defined media in cell and tissue culture. Toxicol in Vitro. 4, 213-225

168. Tomiya, N., Betenbaugh, M.J., Lee, Y.C. (2003) Humanization of lepidopteran insect-cell-produced glycoproteins. Acc Chem Res. 36, 613-620

169. Traber, M.G., Atkinson, J. (2007) Vitamin E, antioxidant and nothing more. Free Radical Biology and Medicine. 43, 4-15

170. Traber, M.G., Stevens, J.F. (2011) Vitamins C and E: beneficial effects from a mechanistic perspective. Free Radical Biology and Medicine. 51, 1000-1013

171. Tremblay, G., Mejia, N., MacKenzie, A. (1992) The NADP-dependent methylenetetrahydrofolate synthetase is not expressed in ***Spodoptera frugiperda*** cells. J Biol Chem. 267, 8281-8285

172. Tseng, H.-C., Lee, C.-Y., Weng, W.-L., Shiah, I.M. (2009) Solubilities of amino acids in water at various pH values under 298.15 K. FluidPhase Equilib. 285, 90-95

173. van der Valk, J., Brunner, D., De Smet, K., Fex-Svenningsen, A., Honegger, P., Knudsen, L.E., Lindl, T., Noraberg, J., Price, A., Scarino, M.L., Gstraunthaler, G. (2010) Optimization of chemically defined cell culture media-replacing fetal bovine serum in mammalian in vitro methods. Toxicol in Vitro. 24, 1053-1063

174. van der Valk, J., Mellor, D., Brands, R., Fischer, R., Gruber, F., Gstraunthaler, G., Hellebrekers, L., Hyllner, J., Jonker, F.H., Prieto, P., Thalen, M., Baumans, V. (2004) The humane collection of fetal bovine serum and possibilities for serum-free cell and tissue culture. Toxicol in Vitro. 18, 1-12

175. Vanisree, M., Lee, C.Y., Lo, S.F., Nalawade, S.M., Lin, C.Y., Tsay, H.S. (2004) Studies on the production of some important secondary metabolits from medicinal plants by plant tissue cultures. Bot Bull Acad Sin. 45, 1-22

176. Vaughn, J.L. (1973) Insect cell nutrition: emphasis on sterols and fatty acids. In Vitro. 1973, 122-128

177. Wang, M.-Y., Kwong, S., Bentley, W.E. (1993) Effects of oxygen/glucose/glutamine feeding on insect cell baculovirus protein expression: a study on epoxide hydrolase production. Biotechnol Prog. 9, 355-361

178. Weber, W., Fussenegger, M. (2005) Baculovirus-based production of biopharmaceuticals using insect cell culture processes. In: Knäblein, J. (ed). Modern Biopharmaceuticals. Wiley-VCH, Weinheim, Germany, pp 1045-1062

179. Weber, W., Fussenegger, M. (2009) Insect cell-based recombinant protein production. . In: Eibl, R., Eibl, D., Pörtner, R., Catapano, G., Czermak, P. (eds). Cell and Tissue Reaction Engineering. Springer, Dordrecht, Heidelberg, London, New York, pp 263-314

180. Weiss, S.A., Smith, G.C., Kalter, S.S., Vaughn, J.L. (1981) Improved method for the production of insect cell cultures in large volume. In

Vitro. 17, 495-502

181. Wessman, S.J., Levings, R.L. (1999) Benefits and risks due to animal serum used in cell culture production. Developments in Biological Standardization. 99, 3-8
182. Whitten, W.K. (1956) Culture of tubal mouse ova. Nature. 177, 96
183. Wilkie, G.E.I., Stockdale, H., Pirt, S.V. (1980) Chemically-defined media for production of insect cells and viruses. In Vitro Cell Dev Biol. 46, 29-37
184. Wink, M., Alfermann, A.W., Franke, R., Wetterauer, B., Distl, M., Windhoevel, J., Jrohn, O., Fuss, E., Garden, H., Mohagheghzadeh, A., Wildi, E., Ripplinger, P. (2005) Sustainable production of phytochemicals by plant in vitro cultures: anticancer agents. Plant Genet Res. 3, 90-100
185. Yancey, P.H. (1994) Compatible and counteracting solutes. In: Strange K (ed) Cellular and molecular physiology of cell volume regulation. CRC, Boca Raton, FL
186. Yancey, P.H., Clark, M.E., Hand, S.C., Bowlus, R.D., Somero, G.N. (1982) Living with water stress: evolution of osmolyte systems. Science. 217, 1214-1222
187. Yang, J.D., Gecik, P., Collins, A., Czarnecki, S., Hsu, H.H., Ladsun, A., Sundaram, R., Muthukumar, G., Silberklang, M. (1996) Rational scale-up of a baculovirus-insect cell batch process based on medium nutritional depth. Biotechnol Bioeng. 52, 696-706
188. Zähringer, H. (2009) Leckerlis für die Zellen. Laborjournal. 4, 74-81
189. Zanghi, J., Schmelzer, A., Mendoza, R., Knop, R., Miller, W. (1999) Bicarbonate concentration and osmolality are key determinatns in the inhibition of CHO cell polysialylation under elevated pCO_2 or pH. Biotechnol Bioeng. 65, 182-191
190. Zengotita de, V.M., Schmelzer, A., Mendoza, R., Knop, R., Miller, W. (1999) Characterization of hybridoma cell responses to elevated pCO_2 and osmolality: Intracellular pH, cell size, apoptosis, and metabolism. Biotechnol Bioeng. 77, 369-380

191. Zenk, M.H., El-Shagi, H., Schule, U. (1975) Anthraquinone production by cell suspension cultures of ***Morinda citrifolia***. Planta Med Suppl. 79-101

192. Zhang, Z., Al-Rubeai, M., Thomas, C.R. (1992) Effect of pluronic F-68 on the mechanical properties of mammalian cells. Enzyme and Microbial Technology. 14, 980-983

193. Zhang, Z., Chisti, Y., Moo-Young, M. (1995) Effects of the hydrodynamic environment and shear protectants on survival of erythrocytes in suspension. J Biotechnol Bioeng. 77, 369-380

194. Zhu, M.M., Goyal, A., Rank, D.L., Gupta, S.K., Vanden-Boom, T., Lee, S.S. (2005) Effects of elevated pCO2 and osmolality on gowth of CHO cells and production of antibody-fusion protein B1: a case study Biotechnol Prog. 21, 70-77

195. Zielke, H.R., Oznard, P.T., Tildon, J.T., Sevdalain, D.A., Cornblath, M. (1978) Reciprocal regulation of glucose and glutamine utilization by cultured human diploid fibroblasts. J Cell Phys. 95, 41-48

196. Zingg, J.M., Jones, P.A. (1997) Genetic and epigenetic aspects of DNA methylation on genome expression, evolution, mutation and carcinogenesis. Carcinogenesis. 18, 869-882

CHAPTER

2

OCCURRENCE AND CONTROL OF SPORADIC PROLIFERATION IN GROWTH ARRESTED SWISS 3T3 FEEDER CELLS

Rishi Man Chugh, Madhusudan Chaturvedi, Lakshmana Kumar Yerneni

Cell Biology Laboratory, National Institute of Pathology (ICMR), New Delhi, India

ABSTRACT

Growth arrested Swiss mouse embryonic 3T3 cells are used as feeders to support the growth of epidermal keratinocytes and several other target cells. The 3T3 cells have been extensively subcultured owing to their popularity and wide distribution in the world and, as a consequence selective inclusion of variants is a strong possibility in them. Inadvertently selected variants expressing innate resistance to mitomycin C may continue to proliferate even after treatment with such growth arresting agents. The failure of growth arrest can lead to a serious risk of proliferative feeder contamination in target cell cultures. In this study, we passaged Swiss 3T3 cells (CCL-92, ATCC) by different seeding densities and incubation periods. We tested the resultant cultures for differences in anchorage-independent growth, resumption of proliferation after mitomycin C treatment and occurrence of proliferative feeder contaminants in an epidermal keratinocyte co-culture system. The study revealed subculture dependent differential responses. The cultures of a particular subculture procedure displayed unique cell size distribution and disintegrated completely in 6 weeks following mitomycin C treatment, but their repeated subculture resulted in feeder regrowth as late as 11 weeks after the growth arrest. In contrast, mitomycin C failed to inhibit cell proliferation in cultures of the other subculture schemes and also in a clone that was established from a transformation focus of super-confluent culture. The resultant proliferative feeder cells contaminated the keratinocyte cultures. The anchorage-independent growth appeared in late passages as compared with the expression of mitomycin C resistance in earlier passages. The feeder regrowth was prevented by identifying a safe subculture protocol that discouraged the inclusion of resistant variants. We advocate routine anchorage-independent growth assay and absolute confirmation of feeder disintegration to qualify feeder batches and caution on the use of fetal bovine serum.

INTRODUCTION

Large quantities of cultured epithelial autografts (CEA) for clinical use in the treatment of extensively burned patients are speedily grown from the adult epidermal keratinocytes over the growth arrested Swiss mouse embryonic 3T3 dermal fibroblasts [1]. These cells are superior in supporting the growth of other target cells as well [2, 3]. The original inactivation method involved γ-irradiation, although a more convenient option has been the treatment with mitomycin C (MC) [3]. The growth arrested 3T3 fibroblasts reportedly survived in CEA and elicited immunogenicity in recipient resulting in complete graft breakdown [4]. Reasonably the viable feeders can result either from the mitotically inactive yet surviving feeders or the proliferating ones. Although, there is evidence of proliferation in other growth arrested mouse embryonic feeders, but there are no specific studies to link the persistence of the viable 3T3 feeders with the failure of growth arrest [5].

The 3T3 cells have the potential to undergo spontaneous transformation depending on subculture, confluence state, and type and concentration of serum [6, 7]. Repeated and inconsistent passaging of cell cultures leads to the accumulation of specific transformed variants and display of altered characteristics [8]. Selective accumulation of such variants, particularly in late passage cultures of 3T3 is a strong possibility as they have been extensively subcultured due to their popularity and wide distribution through several channels in the world [8]. But indications of transformation such as loss of contact inhibition and presentation of phenotypic differences may not readily be apparent, when the transformed variants are less frequent. However, few variants with innate resistance to growth arrest may continue to even after exposure to MC. Such proliferative feeders then become visible and contaminate the target cell cultures. We hypothesize that the presence of such variants in 3T3 cell cultures is a potential cause for failure of growth arrest.

We, therefore, propose to investigate if the proliferative feeder contamination of target cells is dependent on the adopted subculture protocol for 3T3 cells and identify preventive strategies. The identified solutions can help in eliminating apprehensions on feeder dependant culture system [9], which is the most efficient and economical method to culture stem cells compared with feeder-free systems [10]. We observed that it was essential to validate each lot of the growth arrested 3T3 cells through confirmation of the complete disintegration of feeders before qualifying them as safe feeders.

MATERIALS & METHODS

3T3 fibroblast culture

The frozen SWISS 3T3 cells (CCL-92, ATCC) supplied at 115th passage (designated as zero passage) were quickly thawed and grown in 3T3-CBS medium consisting of DMEM with 1.5 grams of sodium bicarbonate per liter and 10% calf bovine serum (CBS) in a humidified 5% CO_2 atmosphere at 37°C. The cells were serially subcultured until 6 passages to establish cryopreserved master and working banks (S1 Fig.). The frozen cells were incubated for 4 days to compensate for the initial slow growth while the subsequent cultures were passaged at uniform intervals of 3 days. The cultures were negative for Mycoplasma [11].

Subculture schemes

The working bank cultures were subjected to specific subculture schemes which were determined after several rounds of preliminary experiments. Initially, 7th passage (P7) cultures were setup in T75 flasks from working bank by seeding 3000 cells per cm^2 and were subcultured by three schemes

(S2 Fig.). Two of the schemes denoted as 3K3D and 3K4D, represented incubation of 3000 plated cells per cm^2 for 3 and 4 days, respectively. The third scheme of 4K3D involved 3 days incubation of 4000 cells plated per cm^2. The resultant cells were tested for anchorage-independent growth and stability of growth arrest by MC treatment. Further, the 3K3D cells that exhibited irreversible growth arrest and no anchorage-independent growth were repeatedly subcultured by 3K3D procedure to determine the number of passages that these characters persisted. Simultaneously, the cells were also passaged serially by the 4K3D scheme for comparison to match with the same subculture interval. The subcultures were continued in T25 flasks until the cells exhibited an obvious 3-dimensional growth in suspension.

Generation of spontaneously transformed clone

A transformed clone of 3T3 cells was derived from a spontaneously induced transformation focus in a confluent culture (S2 Fig.) that was established by incubating 3000 cells per cm^2 in 3T3-FBS medium containing 10% Fetal Bovine Serum [6]. The incidence of foci was increased by sub-culturing a large focus which was scraped with a cell scrapper. Subsequent culture was trypsinized and thousand cells were incubated in methyl cellulose for two weeks. The discrete spheres formed by the transformed variants were subjected to the single sphere cloning in 24-well plates. One of the clone cultures established from a single sphere was expanded by two passages before experimentation.

Growth characteristics and cell size measurements

The P6 cell output was subcultured as per the 3K3D and 4K3D schemes in T25 flasks and counted on each day until confluence. Lag period, doubling time and saturation density were estimated by the standard growth curve method. Similar estimations were made for the cells of P5 and the clone after plating them at a density of 3000 cells per cm^2. The saturation density of 3K3D and 4K3D cells was additionally estimated in T75 flasks.

The 3rd day cells were analyzed for cell size using Cellometer Vision (Nexcelom Bioscience LLC, Lawrence, MA, USA) which was previously calibrated with beads of known sizes. In brief, the trypsinized cells were pipetted into the disposable counting chambers, bright-field images were automatically captured and the pixel area of the captured cells was converted to corresponding cell diameter.

Growth arrest of 3T3

The cultures intended for growth arrest were treated for two hours with 4 μg per ml MC (Catalogue No. M4287, Sigma-Aldrich) which was prepared in 3T3-CBS medium by diluting a stock solution of 200 μg per ml of HEPES Buffered Earl's Salt (HBES). Subsequently, cultures were washed 3-times with 3T3-CBS medium with 15-minutes incubation at each step, detached using 0.25% trypsin + 0.03% EDTA solution and counted by trypan blue exclusion. The cells were replated alone at a density of 14000 cells per cm^2 or co-cultured with epidermal keratinocytes in triplicate T25 flasks and incubated in keratinocyte growth medium (KGM) which was renewed every 3 days. The cultures beyond 8th passage were exposed to MC at every passage, washed thrice and continued incubation in KGM. The responsiveness of serial cultures to MC was tested until the cells showed anchorage-independent growth for two consecutive passages.

Assessment of feeder cell fate

The MC treated cells were microscopically assessed on a weekly basis, assigned as completely disintegrated or proliferating and the duration elapsed was recorded. Criteria considered for complete disintegration were the absence of healthy cells and presence of a very low density of cells that exhibited vacuolated cytoplasm and/or loss of membrane integrity. The visual assessment was performed by three observers. Complete disintegration was verified by transferring the remaining attached cells by trypsinization to a fresh flask and observing for cell attachment. Conversely, cultures designated as proliferating if one or more proliferative foci were observed were continued incubation until confluence.

Anchorage-independent growth assay

10^3 to 10^5 3T3 cells at every passage were suspended in methylcellulose (Methocel, Sigma-Aldrich) which was prepared at a final concentration of 0.8% in 3T3-CBS medium, poured over a base of 0.6% agar in 35 mm dish and incubated under standard culture conditions in triplicate for 2 weeks with periodic examination under Nikon inverted phase contrast microscope.

Keratinocyte—feeder co-culture

Primary Human epidermal keratinocytes (Cat No.PH10205A, www.genlantis.com) were co-cultured with feeder cells by the basic Rheinwald-Green (1975) technique. The keratinocyte growth medium (KGM) consisted of DMEM with 1/3 volume Ham's F12, 10% FBS, adenine (1.8 x 10^{-4}M), transferrin (5 μg/ml), triiodo-L-thyronine (2 x 10^{-9}M), dexamethasone (1μg/ml), hydrocortisone (0.4 μg/ml), insulin (5 μg/ml), Cholera Toxin (1 x 10^{-10}M), L-serine (20 μg/ml) and L-glutamine (100 μg/ml). EGF (10ηg/ml) was added to the medium 2 days after culture initiation. Flasks containing 14,000 cells per cm^2 feeder cells were inoculated with 2000 cells per cm^2 keratinocytes and incubated at standard culture conditions for 7–10 days. The cultures were then treated with 0.03% EDTA to remove feeder cells followed by detachment of keratinocytes using 0.08% trypsin + 0.01% EDTA + 0.025% glucose [1] and the viable cells were counted by trypan blue exclusion.

Assessment of feeder contamination

The non-proliferative feeder cell contamination was assessed by plating 1000 detached keratinocytes per 60-mm dish and incubated overnight in KGM. To assess proliferative feeder contaminants, 50,000 keratinocytes were plated in T25 flasks and incubated in KGM for 3 weeks. The cultures were fixed in a chilled solution of three parts of methanol and one part of glacial acetic acid stained with 0.125 μg Hoechst 33258 (Sigma, H-6024) per ml of Hank's balanced salt solution in the dark for 10 minutes, washed with distilled water and observed in an inverted fluorescence microscope (Nikon Diaphot 300) fitted with excitation filter of 330–380 nm and emission filter of 420 nm. The 3T3 cells were distinguished from Keratinocytes on the basis of nuclear size, morphology and fluorescence pattern [12].

Statistics

The experiments of growth arrest, co-culture, anchorage-independent growth assays and feeder cell contamination assays were performed in triplicate. The cell yields among different groups were compared by Student's t test while the linearity of cell yields in serial subcultures was tested by regression. Skew, Kurtosis, Confidence level for mean of cell size distribution were calculated in Microsoft Excel (2007).

RESULTS

Cell banking

The split ratio that worked out during the establishment of master bank ranged from 1:2 to 1:6.4 while it was 1:8 during expansion for working bank and an average seeding density of 3028±113 cells per cm^2 was maintained throughout

Subculture schemes and response to MC

The P6 cultures yielded 1,771,851 ± 36,695 cells that corresponded to 40% confluence. The subsequent subcultures as per 3K3D, 3K4D and 4K3D schemes yielded 1,759,950 ± 64,089, 2,276,700 ± 63,580 and 2,103,675 ± 27,152 cells, respectively (Fig. 1) which corresponded to confluence levels of 42%, 52% and 48%, respectively. As expected for the higher seeding or longer incubation, the P7 cell yields of 4K3D and 3K4D schemes, were significantly higher ($P<0.05$) than the 3K3D scheme.

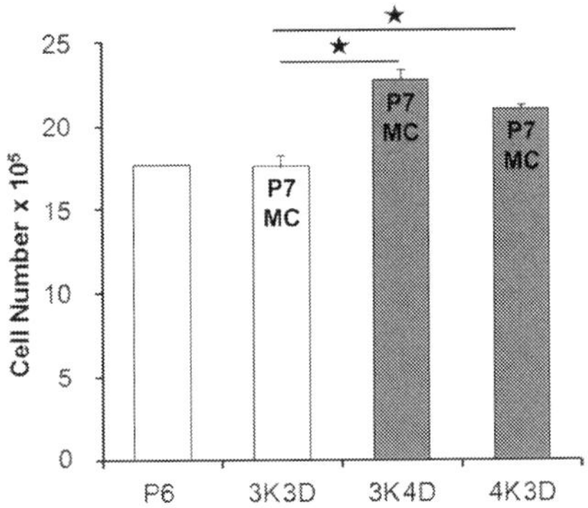

Figure 1: P6 and P7 cell yields and responsiveness to Mitomycin C.

Column chart representing the influence of subculture schemes on Swiss 3T3 cell outputs in T75 flasks at 7th passage (P7) and responsiveness to Mitomycin C (MC). The uniformly set up 6th passage (P6) cultures from a working bank were subcultured to P7 by schemes of 3K3D, 3K4D and4K3D (* $P<0.05$). Shaded columns represent exhibition of resistance to MC and each column represents a mean value from 3 flasks with the standard deviation.

There was a resumption of cell division in MC treated 3K4D and 4K3D flasks as evident by the formation of proliferative foci, whereas 3K3D cultures exhibited irreversible growth arrest (Table 1). The formation of proliferation foci in 4K3D cells was rapid which was noticed as early as 1 week after re-plating. However, it was noticed in 3K4D only after 4 weeks by which time the disintegration of susceptible cells was extensive (Fig. 2A). Initially, the foci in reviving cultures were less apparent and showed diffuse zones of bipolar cells among a mix of large attached cells with or without vacuolated cytoplasm and floating cellular debris. Most of such zones, but not all, progressively turned into distinct proliferative foci of polygonal cells with high nucleus to cytoplasm ratio (Fig. 2B).

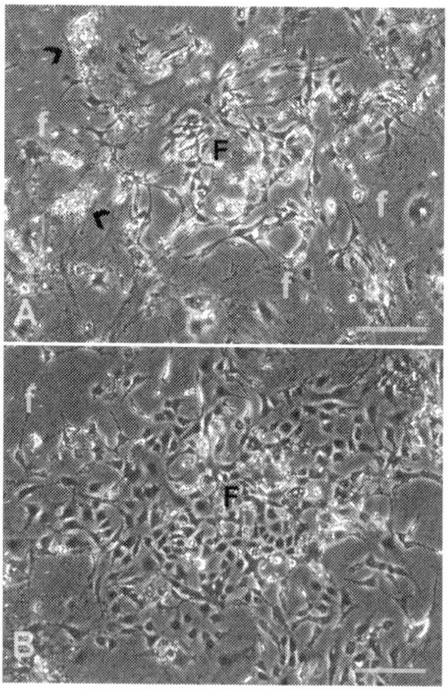

Figure 2: Feeder regrowth in 3K4D culture after mitomycin C treatment.

The newly formed bipolar cells (F) appeared amidst the attached MC treated large cells (f) and the floating cellular debris (arrowhead, A) after four weeks of post-treatment period. Five weeks later, the same field showed a distinct large collection of multipolar cells with high nucleus to cytoplasm ratio (B). Scale bar: 100 μm.

Table 1: Influence of various subculture schemes at P7 of Swiss 3T3 cells on anchorage- independent growth and responsiveness to Mitomycin C (MC) treatment.

Subculture Scheme	Growth in methylcellulose	Post-MC incubation period (Weeks)	Cells attached after subculture	Post-MC status of culture
3K3D	Nil	6	No	D
3K4D	Nil	4	-	P
4K3D	Nil	1	-	P

D = Disintegrated cultures,
P = Proliferative cultures.

doi:10.1371/journal.pone.0122056.t001

The frequency of proliferative foci at any given time point appeared to vary among replicates. But, it was not possible to accurately estimate the number of foci because they were more frequent at the periphery of culture flasks making the observation difficult. Not all proliferative zones progressed into distinct foci, but a few remained stable or even disintegrated. Several new foci appeared intermittently while a number of old ones grew larger and coalesced. All cultures showing progressive proliferation with the new cellular phenotypes became confluent in additional 8–12 weeks. On the other hand, the MC-exposed fibroblasts from 3K3D showed progressive irreversible disintegration and presented extensive vacuolization and completely disintegrated in 6 weeks (Fig. 3 top left; Table 1). So far, we have validated the 3K3D subculture scheme on 22 out of 40 working bank stocks for irreversible growth-arrest.

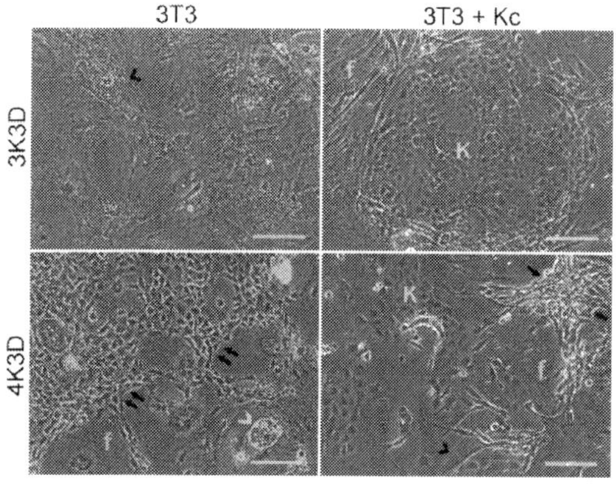

Figure 3: Influence of subculture schemes on responsiveness to Mitomycin C.

The 3K3D feeders plated either alone (3T3) or co-cultured with human epidermal keratinocytes (3T3 + Kc) showed degeneration of feeders (arrowhead) and a large well circumscribed keratinocyte (K) colony surrounded by inactivated feeders. The 4K3D feeders plated alone exhibited newly formed compact proliferative feeder cells with high nucleus to cytoplasm ratio (arrows) over a background of enlarged (f) and degenerating cells (white arrow head). The co-culture with human epidermal keratinocytes showed the proliferative foci (arrows). The 3T3 alone and the co-cultures were 2 and 1 weeks old, respectively. Scale bar: 100 μm.

Keratinocyte—feeder co-culture

One week old co-cultures initiated with the 3K3D feeder cells revealed discrete keratinocyte colonies surrounded by the non-proliferating feeder cells which presented signs of disintegration (Fig. 3). But the 4K3D feeders that showed failed growth arrest as early as one week after they were plated alone, similarly revived and produced several proliferative foci in co-cultures. The newly proliferated 3T3 fibroblasts were easily identified by their characteristic narrow cell bodies compared with the large disintegrating cells that presented cytoplasmic vacuolization or nuclear fragmentation (Fig. 3 bottom left).

Feeder cell contamination

The Hoechst-stained feeder contaminants displayed large nuclei with brightly stained coarse chromatin as compared with the small and dull nuclei of keratinocytes (Fig. 4). The frequency of non-proliferative feeder contamination with 3K3D and 4K3D feeders was uniform at 28 ± 4 per 1000 cells counted. However, when the proliferative capability was assessed, it was found that the 3K3D contaminants progressively disintegrated, whereas the keratinocytes showed a limited proliferation (Fig. 4 Left panel). In contrast, the 4K3D feeder contaminants proliferated to generate a large population of small cells (Fig. 4 Right panel). The invading new feeder cells, presenting several dividing cells, tightly enveloped the keratinocyte colonies and the few large non-proliferative feeder cells.

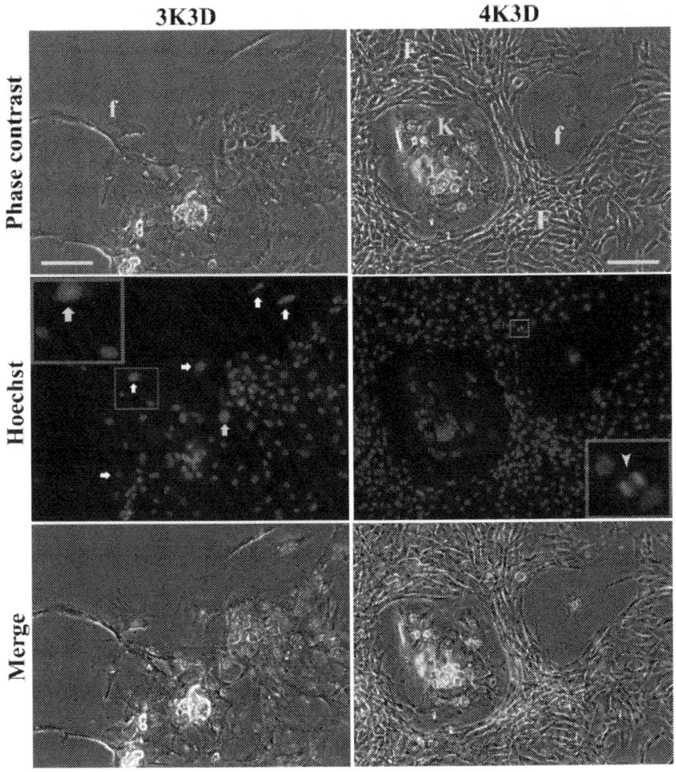

Figure 4: Proliferative feeder cell contamination.

Keratinocyte cultures plated alone without the feeders showing cytological features of attached keratinocytes (K) and the contaminating non-proliferative (f) and proliferative feeders (F). The 3K3D feeder cells that came along with the keratinocytes as contaminants (Left panel) presented no dividing cells and were broad with large nuclei in phase contrast and showed coarse chromatin aggregates (arrows) that stained bright in Hoechst (inset of Hoechst image). The keratinocytes comprised of a mix of small polygonal or broad terminally differentiated cells which presented the dull small nuclei. The 4K3D contaminating feeder cells (Right panel) consisted of well circumscribed keratinocyte colonies (K) enveloped by numerous newly formed narrow-bodied feeder cells (F) with several cell divisions (inset of Hoechst image) in addition to a few broad non-proliferating feeder cells (f) showing vesicular nuclei. Scale bar: 100 μm; Left inset: 200 μm; Right inset: 450 μm.

Anchorage-independent growth

No cultures until P7 showed anchorage-independent growth in any of the dishes of methyl cellulose (Table 1). Similarly, the serially passaged cultures from P8 to P14 of 3K3D scheme and P8 to P10 of 4K3D scheme failed to grow in suspension (Table 2). However, the 3K3D cells showed linear and 3-dimensional growth at P15 and P16, respectively (Fig. 5) while the 4K3D cells produced similar growth at earlier passages of P11 and P12, respectively. The cells of these cultures exhibited angular outlines with a shallow contrast, as opposed to the smooth outlines of single cells in non-proliferating cultures. The growth of cells in dishes plated with 50,000 cells was observed within 4–6 days of incubation as compared with two weeks in at least one of the dishes plated with 1000 cells.

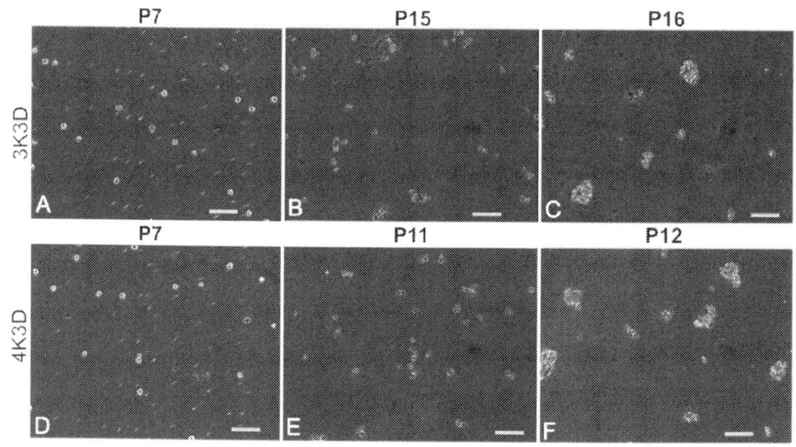

Figure 5: Anchorage-independent growth assay.

Swiss 3T3 fibroblasts from different passages during serial subculture by the 3K3D and 4K3D schemes were plated in methylcellulose. Discrete cells from 7th Passage (P7) cultures of 3K3D (A) and 4K3D (D) subculture schemes showed smooth outline. Cells with angular outlines having a shallow contrast were observed at P15 of 3K3D (B) and P11 of 4K3D (E) with occasional linear aggregates. Conspicuous 3-dimensional spheres appeared in the subsequent passages of P16 (C) and P12 (F). Scale bar: 100 μm.

Table 2: Influence of serial subculture schemes at various passages of Swiss 3T3 cells on anchorage-independent growth and responsiveness to Mitomycin C (MC) treatment.

Subculture Scheme	Passage	Growth in methylcellulose	Post-MC incubation period (Weeks)	Cell attachment after subculture	Post-MC status of culture
3K3D	8	Nil	8	No	D
	9–11	Nil	10	No	D
	12	Nil	11	-	P
	13	Nil	19	Yes*	D
	14	Nil	4	-	P
	15	Linear	3	-	P
	16	3-D	2	-	P
4K3D	8–10	Nil	1	-	P
	11	Linear	1	-	P
	12	3-D	1	-	P

D = Disintegrated cultures;
P = Proliferative cultures;
* Few loosely attached vacuolated cells completely disintegrated in 7 days.

Serial subculture and response to MC

The cell yields of serial cultures passaged by either 3K3D ($R^2 = 0.125$) or 4K3D ($R^2 = 0.012$) scheme did not exhibit any significant trend (Fig. 6). The average 4K3D cell output of $6.62 \pm 0.69 \times 10^5$ was significantly ($P<0.001$) higher than $4.84 \pm 0.44 \times 10^5$ of 3K3D (Fig. 6 inset) and these values corresponded to 72% and 51% confluence, respectively.

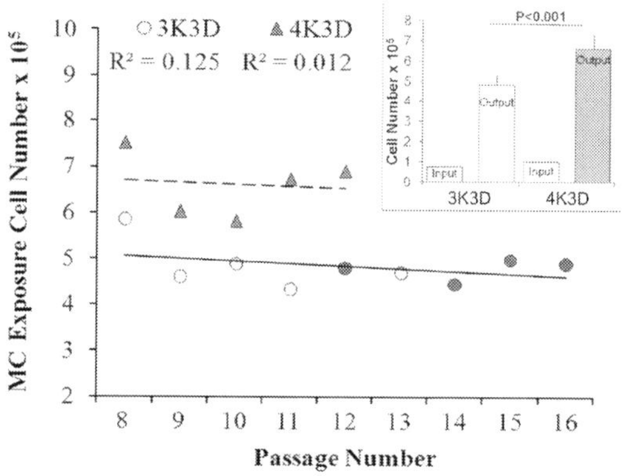

Figure 6: Cell yields during serial subculture.

Scattered plot showing the influence of serial subculture of Swiss 3T3 fibroblasts by schemes of 3K3D and 4K3D on cell outputs and responsiveness to Mitomycin C treatment at each passage. The 3K3D cultures were serially subcultured either by the same scheme or the 4K3D scheme. Shaded markers denote passages at which MC treated cells showed proliferative foci. The columns in the inset represent the number of cells seeded (Input) and yield (Output). Each column represents average cell number of all the passages from each scheme. The shaded column represents consistent resistance to Mitomycin C at all passages.

The serial cultures of the 3K3D scheme from P8 to P11 exposed to MC exhibited total cellular disintegration that took 8–10 weeks (Table 2). But the subsequent subcultures exhibited the conspicuous proliferation foci with the exception of P13, which disintegrated slowly over more than 19 weeks of post-MC incubation. The time period for the proliferating cells to appear during post-MC incubation gradually decreased with increase in passage number. It was 11, 4, 3 and 2 weeks for P12, P14, P15 and P16 cultures, respectively. On the other hand, the 4K3D cultures persistently resumed post-MC cell proliferation at all the tested passages within a week.

Transformed clone and response to MC

A clone was established from a transformation focus in a super-confluent culture. Initially, few foci of heterogeneous cells appeared in a 3K3D culture after two weeks of incubation in FBS containing medium (Fig. 7 A-C). The cells from a large focus isolated by scraping and trypsinization were able to induce more of such foci in a subsequent culture. Each focus was a discrete collection of small cells characterized by high nucleus to cytoplasm ratio (Fig. 7 D-F). The single cell suspension from this culture incubated in methyl cellulose grew into three-dimensional spheres (Fig. 7 G). A single cloned sphere produced a large homogeneous colony with irregular margins after ten days of incubation (Fig. 7 H-K). The colony was further expanded by two subcultures in culture flasks. The resultant cell population predominantly consisted of narrow bipolar and tripolar cells (Fig. 7 L).

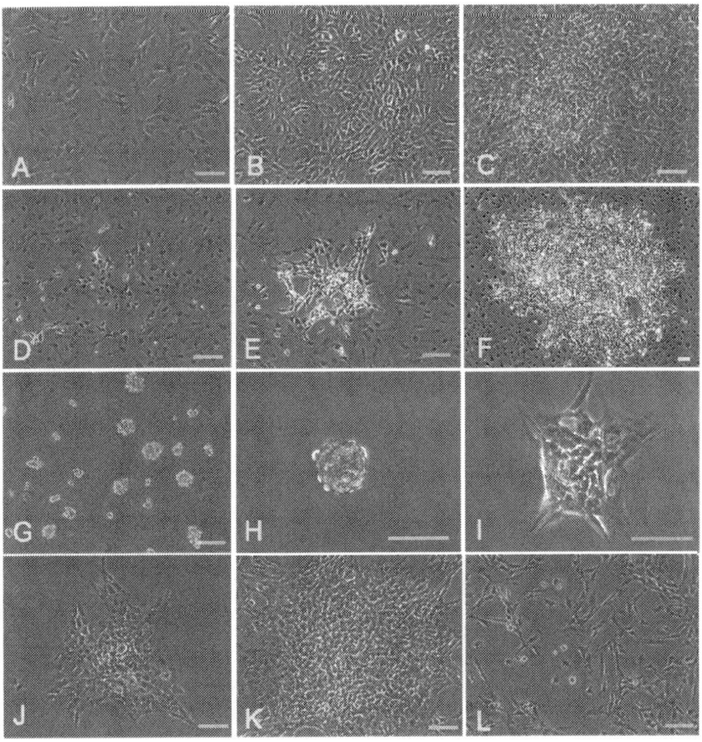

Figure 7: Establishment of spontaneously transformed clone.

The culture of 7th passage Swiss 3T3 fibroblasts initiated with 3000 cells per cm^{2}after 3 days showed cells with uniform morphology (A) which reached confluence in five days (B) and formed few transformation foci in ten days (C). After two weeks, the cells from such foci were scraped, trypsinized briefly and the single cells were re-plated in a fresh flask. Several discrete clusters of small cells with compact cell bodies and high nucleus to cytoplasm ratio appeared in two days (D) which grew steadily forming conspicuous colonies in four days (E) and enlarged further in ten days (F) over a background of normal looking broad cells. Two weeks later, the trypsinized cells from this culture formed discrete spheres in methyl cellulose (G) which were subjected to the single sphere cloning in 24-well plates (H). The spheres readily anchored overnight (I), spread out to form small colonies in three days (J) and established as large dense colonies in ten days (K). One such colony further expanded in flasks presented narrow bipolar and tripolar cells (L). Scale bar: 100 µm.

The MC-exposed clone cells after two weeks showed several pockets of newly formed small cells in the midst of enlarged and degenerating cells (Fig. 8 A). Similarly, the epidermal keratinocytes co-cultured with the MC-treated clone cells showed the presence of the newly proliferating and the disintegrating cells. Both the cell types were easily distinguishable from the discrete keratinocyte colonies within a week under the phase contrast (Fig. 8 B).

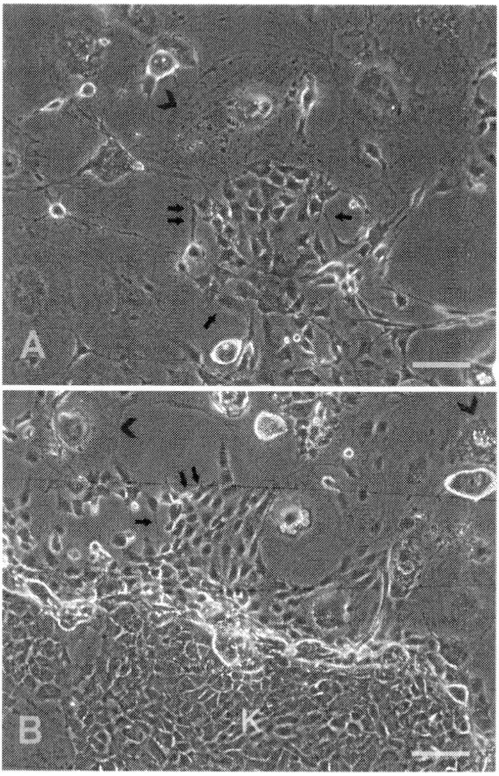

Figure 8. Failure of growth arrest in spontaneously transformed clone.

The 2nd passage culture of the clone exposed to Mitomycin C and replated alone (A) exhibited the proliferative focus consisting of the newly formed compact cells (small arrows) surrounded by the broad cells (arrowhead) after 14 days of incubation. The co-culture of the growth-arrested clone cells with the human epidermal keratinocytes (B) presented the well-spread out degenerating cells (arrowhead) and the newly formed cells (small arrows) at the periphery of a large colony of keratinocytes (K) after 7 days. Scale bar: 100 μm.

Growth characteristics

The growth curves of 3K3D and 4K3D were more or less similar (Fig. 9). The lag phases of 3K3D and 4K3D cells were 32.4 and 30.0 hours, respectively; the doubling times were 38.2 and 38.5 hours, respectively; the saturation densities in T25 flasks were 37,800 and 36,700 per cm^2, respectively while they were 1.48 times higher at 56,000 and 1.6 times higher at 58,700 per cm^2 in T75 flasks, respectively (Table 3). This was in contrast to the P5 cells which exhibited 26.5 hours of lag phase; 17.7 hours of doubling time and T25 saturation density of 26,800 cells per cm^2, while it was 1.81 times higher at 48,500 in T75. On the other hand, the spontaneously transformed clone cells exhibited the longest lag phase of 67.7 hours with a prolonged pre-exponential phase, a doubling time of 26.7 hours and attained a much higher density of 159,200 cells per cm^2 after 10 days in culture exhibiting transformation foci.

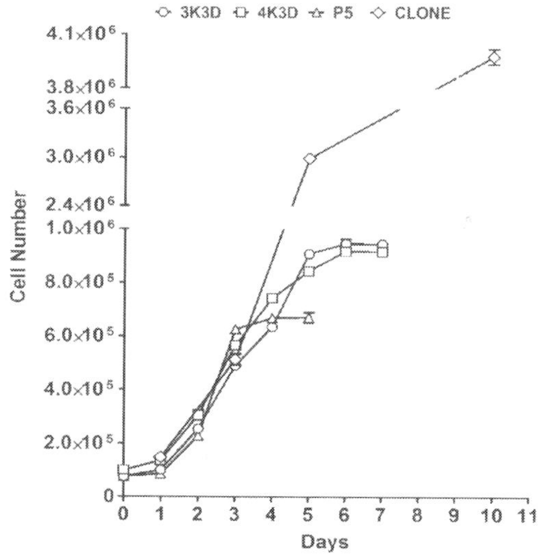

Figure 9: Growth curves.

Growth curves of Swiss 3T3 cells of P5, spontaneously transformed clone and the P7 cells produced by subculture schemes of 3K3D and 4K3D.

Table 3. Influence of subculture on growth characteristics and cell size in Swiss 3T3 cells.

Characteristics		3K3D	4K3D	P5	Clone
Lag phase (Hr)		32.4	30.00	26.5	67.7
Doubling time		36.2	38.5	17.7	26.7
Saturation density—cells/cm^2 (T25)		37,800	36,700	26,800	>1,59,200*
Saturation density—cells/cm^2 (T75)		56,000	58,700	48,500	
Cell size (µm)	Sample size	1,924	3,053	2,600	2,619
	Minimum	5.04	5.04	5.45	5.45
	Maximum	33.89	45.17	51.83	51.91
	Mean	11.92	15.44	20.48	21.21
	Median	11.27	15.13	17.59	17.95
	Mode	7.28	14.92	15.40	15.40
	SD	4.09	4.19	7.27	7.73
	Coefficient of variation	0.34	0.27	0.36	0.36
	Skew	1.48	1.86	2.47	2.73
	Kurtosis	0.72	2.31	5.8	7.67
	Confidence Level for mean (95%)	0.009	0.01	0.009	0.009

Cell size measurements

The cell size statistics of 3 days old cultures of 3K3D, 4K3D, clone and P5 cells are given in Table 3. The 3K3D and 4K3D cells showed maximum cell sizes of 33.89 and 45.17, and the mean cells sizes were 11.92 and 15.44, respectively while maintaining a uniform minimum cell size of 5.04 µm. On the other hand, the maximum cell dimensions of P5 and clone cells were higher at 51.83 and 51.91 µm with mean cell sizes of 20.48 µm and 21.21 µm, respectively, while the smallest cell in both the groups measured 5.45 µm. The data from all the groups indicated highly skewed nature of cell size distribution, out of which 3K3D showed the least skewness of 1.48 compared to 1.86, 2.47 and 2.73 for 4K3D, P5 and the clone, respectively. Similarly, the MC sensitive 3K3D cells showed a tendency to a higher uniformity of cell sizes with a smaller kurtosis of 0.72 against 2.31, 5.8 and 7.67 for 4K3D, P5, and the clone, respectively. The deviation in cell size distribution of 3K3D cells compared to other tested groups is apparent from the histogram (Fig. 10).

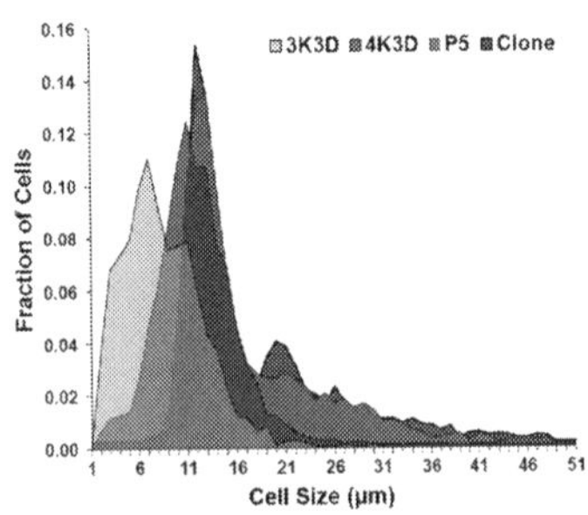

Figure 10. Graphical representation of cell size distribution.

Area plot showing Swiss 3T3 cells of P5, spontaneously transformed clone and the P7 cells produced by subculture schemes of 3K3D and 4K3D.

DISCUSSION

Murine Swiss embryonic 3T3 cells have been popularly used as feeder cells for large-scale expansion of epidermal keratinocytes for clinical exploitation in the treatment of burns [1, 13, 14]. The basic technique involved co-culture of epithelial cells with the 3T3 feeder cells growth arrested by μ-irradiation or treatment with mitomycin C. The epithelial cells are isolated after selectively removing the persistent feeder cells with EDTA treatment. However, the procedure always allowed a small proportion of feeder cells to remain as contaminants [4, 12]. The problem might be aggravated further if the feeders resumed mitotic activity because foci of proliferation are known to occur in growth arrested murine embryonic fibroblasts [5].

We occasionally observed sporadic formation of such proliferation foci occurring more preferentially in certain feeder batches (Data not shown). We noticed that the different subculture seeding densities employed to generate these feeder batches remained well within the narrow range of supplier's recommendations. We presumed that the feeder regrowth was perhaps the reflection of subculture dependent distribution of MC resisting cells. The randomness may be due to initial low frequency of such cells which upon serial passaging might become regular. Therefore, we investigated the relationship of the growth arrest failure with the variables of seeding density, the culture duration, and the serial subculture. This study revealed that the MC resistance was specific to subculture procedure. It underlined the need to establish a safe protocol independently for each supplied culture, which possibly contained low level of such variants. The 3T3 cells, originally established by Todaro and Green [15], have been extensively circulated in the world through various repositories for several decades. Each lot of cells has since been repeatedly subcultured leading to the accumulation of variants [8]. It is likely that the cultures resisting growth arrest facilitated a subculture dependent accumulation of specific genetic variants with innate resistance to MC.

Considering that the sensitivity to MC is specific to each feeder cell type [16], the presence of MC-resistant variants may necessitate the use of higher concentrations of MC to ensure irreversible growth arrest. Correspondingly, use of 4 μg/ml was tested to perform optimally in earlier studies [17, 18], while higher concentrations of MC were reported in certain later studies [19, 20, 21, 22, 23]. But the use of higher concentrations can leave harmful residues in target cells [24]. We were able to use a lower concentration yet contained the proliferation of feeder cells by identifying and validating a favorable subculture scheme that perhaps avoided the inclusion of MC-resisting variants.

The other notable point is the induction of clone by prolonged confluence state and exposure to the growth factor rich fetal bovine serum. FBS has been known to stimulate anchorage-independent growth and transformation foci in 3T3 cells [6, 25]. But the use of FBS in Swiss 3T3 cell culture has been widely reported [8, 18, 26, 27] despite the recommendation of the suppliers to use calf bovine serum in culture medium [28]. The existence of several cell lines with the same acronym 3T3 requiring FBS has further complicated the issue.

An important guidance for 3T3 cells is to subculture at 80 percent confluence or less to discourage variant selection [28]. We found that the cultures of all the experimental groups reached a far less confluence state than 80%, yet produced passage scheme dependent feeder regrowth. Another key directive is to limit the passages of 3T3 cells for use in bioengineering of cultured epidermis [29], because the cultured cells respond differently to DNA damaging agents in a passage number dependent manner [30]. Consistently, we found that the serial passaging even by the safer scheme of 3K3D led to MC resistance from P12 onwards, although it was intermittently absent at P13.

There is an utmost need to establish fool-proof growth arrest by observing absolute feeder disintegration because feeder regrowth occurred in P12 cells of 3K3D scheme after 11 weeks, but it took 19 weeks to the P13 cells. Therefore, the standard practice of visually assessing the regrowth of feeders under the microscope for a fixed period of 10–14 days for mitotic activity [5], may not be sufficient to rule out the feeder regrowth. Additionally, it is essential to validate visual observation by transferring the remaining cells for attachment and survival.

Under the normal growth conditions, the 3T3 cells do not show anchorage-independent growth in methylcellulose/soft agar [25]. Therefore, the presentation of three-dimensional growth is indicative of selection of transformants. But this expression was apparent only during late passages as compared with the display of MC resistance at relatively earlier passages. It suggests that the variants responsible for both these presentations are probably not the same. But, on the whole, the methylcellulose assay is of considerable value for different lots of Swiss 3T3 cells as an advance warning for a deviant culture.

The MC sensitive 3K3D cells were the smallest and showed a tendency to decrease in excess asymmetry (skew) in size distribution as compared with the 4K3D cells although both the groups presented comparable growth curves. Moreover, the presentation of the lowest skew and Kurtosis by 3K3D cells suggests phenotypic divergence. Additionally, while considering the doubling time and the saturation density of the P5 cells as near normal [15, 28], the 3K3D cells presented the highly deviated growth characteristics and the cell size distribution parameters. It suggests that the cells of 3K3D, like those of 4K3D and the clone, may not represent a truly normal population. But the presentation of absolute irreversible growth arrest qualifies them as the ideal feeders. It is possible that the 3K3D subculture might have facilitated the accumulation of variants other than those that resisted MC thereby altering the growth characteristics. Normally, the repositories perform batch-testing of cell lines for basic growth characteristics and accordingly issue guidance on subculture procedures [28], but do not test for MC resistance. Therefore, several combinations of the recommended seeding densities and incubation periods need to be included during subculture and tested for MC resistance. Once a validated procedure is established it is likely to work safe for the rest of the cell batches generated by a two-tiered banking system [29].

Lastly, the other candidate feeder cell types which are currently in use may also need to be evaluated similarly for responsiveness to growth arresting methods including μ-irradiation because growth arrest failure is also known to occur in such cells [5].

CONCLUSION

We recommend the adoption of tests to rule out (1) MC-resistance (2) proliferative feeder cell contamination of target cells and (3) growth in methylcellulose before qualifying the feeder batches as safe. The subculture of the already over passaged Swiss 3T3 cells should be limited and use of FBS in place of bovine calf serum is an avoidable risk factor. Additionally, the MC resisting clone and the late passage cultures are useful to study molecular pathways involved in chemoresistance and post-chemotherapy initiation of cell cycling in cancer cells.

AUTHOR CONTRIBUTIONS

Conceived and designed the experiments: LKY. Performed the experiments: RMC MC. Analyzed the data: LKY RMC MC. Wrote the paper: LKY RMC MC.

REFERENCES

1. Rheinwald JG, Green H. Serial cultivation of strains of human epidermal keratinocytes: the formation of keratinizing colonies from single cells. Cell. 1975; 6: 331–344. pmid:1052771
2. Lechner JF, Haugen A, Autrup H, McClendon IA, Trump BF, Harris CC. Clonal Growth of Epithelial Cells from Normal Adult Human Bronchus Cancer Res. 1981; 41: 2294–2304. pmid:7016311
3. Navasaria HA, Sexton C, Bouvard V, Leigh IM. Growth of keratinocytes with a 3T3 feeder layer: basic techniques In: Leigh IM and Watt FM editors. The Keratinocytes Hand Book. Cambridge: Cambridge University Press; 1994. p. 5–12.
4. Hultman CS, Brinson GM, Siltharm S, DeSerres S, Cairns BA, Peterson HD et al. Allogeneic fibroblasts used to grow cultured epidermal autografts persist *in vivo* and sensitize the graft recipient for accelerated second-set rejection. The Journal of Trauma. 1996; 41:51–58. pmid:8676424
5. Connor DA. Mouse embryo fibroblast (MEF) feeder cell preparation. Current Protocols in Molecular Biology. 2000; 51: 23.2.1–23.2.7.
6. Rubin H, Xu K. Evidence for the progressive and adaptive nature of spontaneous transformation in the NIH 3T3 cell line. Proc Nat Acad

Sci USA. 1989; 86: 1860–1864. pmid:2928308

7. Bowen-pope DF, Vogel A, Ross R. Production of platelet-derived growth factor-like molecules and reduced expression of platelet-derived growth factor receptors accompany transformation by a wide spectrum of agents. PNAS (USA). 1994; 81: 2396–2400.
8. Takeuchi T, Wang L, Mori S, Nakagawa K, Hiroshi Y, Kanda T. Characterization of mouse 3T3-Swiss albino cells available in Japan: Necessity of quality control when used as feeders. Jpn J Infect Dis. 2008; 61: 9–12. pmid:18219127
9. Mallon BS, Park KY, Chen KG, Haliton RS, McKay RDG. Toward xeno-free culture of human embryonic stem cells. Int J Biochem Cell Biol. 2006; 38:1063–1075. pmid:16469522
10. Lamb R, Ambler CA. Keratinocytes Propagated in Serum-Free, Feeder-Free Culture Conditions Fail to Form Stratified Epidermis in a Reconstituted Skin Model. PLoS ONE. 2013; 8(1): e52494. pmid:23326335
11. Kumar A, Ali A, Yerneni LK. Tandem use of immunofluorescent and DNA staining assays to validate nested PCR detection of mycoplasma. In Vitro Dev Biol Anim. 2008; 44: 189–192. pmid:18528736
12. Alitalo K, Kuismanen E, Myllyla R, Kiistala U, Asko-Selgavaara S, Vaheri A. Extracellular matrix proteins of human epidermal keratinocytes and feeder 3T3 cells. J Cell Biol. 1982; 94: 497–505. pmid:6182145
13. O›Connor NE, Mulliken JB, Banks-Schlegel S, Kehinde 0, Green H. Grafting of burns with cultured epithelium prepared from autologous epidermal cells. Lancet. 1981; 1(8211):75–78. pmid:6109123
14. Atiyeh BS, Costagliola M, Hayek SN, Dibo SA. Effect of silver on burn wound infection control and healing: Review of the literature. Burns. 2007; 33: 139–148. pmid:17137719
15. Todaro GJ, Green H. Quantitative studies of the growth of mouse embryo cells in culture and their development in to established lines. J. Cell Biology. 1963; 17: 299–313. pmid:13985244
16. Ponchio L, Duma L, Oliviero B, Gibelli N, Pedrazzoli P, Cuna GRD. Mitomycin C as an alternative to irradiation to inhibit the feeder layer growth in long-term culture assays Cytotherapy. 2000;

2: 281–286. pmid:12042037

17. Watt FM. Selective migration of terminally differentiating cells from the basal layer of cultured human epidermis. J Cell Biology. 1984; 98: 16–21. pmid:6546761
18. Hoffmann BLA, Rheinwald JG. Polycyclic aromatic hydrocarbon mutagenesis of human epidermal keratinocytes in culture Proc. Natl. Acad. Sci. USA. 1984; 81: 7802–7806. pmid:6440145
19. Berry RD, Powell SC, Paraskeva C. In vitro culture of human foetal colonic epithelial cells and their transformation with origin minus SV40 DNA. Br. J. Cancer. 1988; 57: 287 289. pmid:2833300
20. JV , Pepper MS, Nakamura T, Orci L, Montesano R. Hepatocyte growth factor stimulates extensive development of branching duct-like structures by cloned mammary gland epithelial cells. Journal of Cell Science. 1995; 108: 413–430. pmid:7768990
21. Epstein SP, Wolosin JM, Asbell PA. P63 expression levels in side population and low light scattering ocular surface epithelial cells. Trans Am Ophthalmol Soc. 2005; 103:187–199. pmid:17057802
22. Qin J, Tang DG. Prostate cancer stem cells. In: Guan XY editor. Cancer stem cells. Transworld Research Network. 2010; p. 37–56.
23. Lu R, Bian F, Lin J, Su Z, Qu Y, Pflugfelder SC et al. Identification of Human Fibroblast Cell Lines as a Feeder Layer for Human Corneal Epithelial Regeneration. PLoS ONE. 2012; 7: e38825. pmid:22723892
24. Zhou D, Lin G, Zeng SC, Xiong B, Xie PY, Cheng DH et al. Trace levels of mitomycin C disrupt genomic integrity and lead to DNA damage response defect in long-term cultured human embryonic stem cells. Arch Toxicol. 2014;
25. Peehl DM, Stanbridge EJ. Anchorage-independent growth of normal human fibroblasts. Proc. Natl. Acad. Sci. USA. 1981; 78: 3053–3057. pmid:6942414
26. Pallen CJ, Tong PH. Elevation of membrane tyrosine phosphatase activity in density-dependent growth-arrested fibroblasts. Proc Natl Acad Sci USA. 1991; 88: 6996–7000. pmid:1651489
27. Bray LJ, Heazlewood CF, Atkinson K, Hutmacher DW, Harkin DG. Evaluation of methods for cultivating limbal mesenchymal stromal

cells. Cytotherapy. 2012; 14: 936–947. pmid:22587591

28. American Type Culture Collection (ATCC) USA, CCL-92, Culture method. Available: www.atcc.org/products/all/ccl-92.aspx#culturemethod. Accessed 15 September 2014.
29. Japan Tissue Engineering Co., Ltd. review report (August 6, 2007) Pharmaceuticals and Medical Devices Agency. Available: http://www.pmda.go.jp/english/service/pdf/medical_devices/jace_oct2007_e.pdf. Accessed 15 September 2014.
30. Chang-Liu CM, Woloschak GE. Effect of passage number on cellular response to DNA-damaging agents: cell survival and gene expression. Cancer Letters. 1997; 113: 77–86. pmid:9065805

CHAPTER

3

EXPRESSION OF FACTOR X IN BHK-21 CELLS PROMOTES LOW PATHOGENIC INFLUENZA VIRUSES REPLICATION

Shahla Shahsavandi, Mohammad Majid Ebrahimi, Shahin Masoudi, and Hasan Izadi

Razi Vaccine & Serum Research Institute, P.O. Box 31975-148, Karaj 31976 19751, Iran

ABSTRACT

A cDNA clone for factor 10 (FX) isolated from chicken embryo inserted into the mammalian cell expression vector pCDNA3.1 was transfected into the baby hamster kidney (BHK-21) cell line. The generated BHK-21 cells with inducible expression of FX were used to investigate the efficacy of the serine transmembrane protease to proteolytic activation of influenza virus hemagglutinin (HA) with monobasic cleavage site. Data showed that the BHK-21/FX stably expressed FX after ten serial passages. The cells could proteolytically cleave the HA of low pathogenic avian influenza virus at multiplicity of infection 0.01. Growth kinetics of the virus on BHK-21/FX, BHK-21, and MDCK cells were evaluated by titrations of virus particles in each culture supernatant. Efficient multicycle viral replication was markedly detected in the cell at subsequent passages. Virus titration demonstrated that BHK-21/FX cell supported high-titer growth of the virus in which the viral titer is comparable to the virus grown in BHK-21 or MDCK cells with TPCK-trypsin. The results indicate potential application for the BHK-21/FX in influenza virus replication procedure and related studies.

INTRODUCTION

Influenza is one of the most economically important viral respiratory diseases of human as well as avian and animal species worldwide. The causative agent is belonging to Orthomyxoviridae, negative-sense, single-stranded RNA, which encodes at least eleven proteins. Type A viruses are subtyped on the basis of the two main surface glycoproteins hemagglutinin (HA) and neuraminidase (NA) and further classified as low pathogenic (LP) or highly pathogenic (HP) on the basis of specific molecular and pathogenesis criteria [1, 2]. Avian influenza viruses replicate mostly in the intestine while human influenza strains replicate in the upper respiratory tract [3, 4]. The infection cycle is initiated by the specific binding of viral HA to a terminal sialic acid-capped glycosylated molecule present on the surface of the host cells. Upon attachment to the cell, receptor-mediated endocytosis occurs either by α2–6Gal or by α2-3Gal linkage [5, 6]. Cellular tropism and the infectivity of influenza viruses are primarily determined by the distribution of these receptors in the cell surface. Also the presence of specific host cellular protease (s) for posttranslational cleavage of HA0 precursor protein is essential for viral infectivity, pathogenicity, and tissue tropism [7, 8]. Influenza A viruses target a wide spectrum of tissues so replication of the viruses has been examined in a variety of cells [9–13]. The results demonstrated that susceptibility to the viruses varied significantly between the cells; and in particular the tracheal epithelial, MDCK, and A549 have been described as suitable permissive cells for the replication of both human and avian influenza viruses. Replication of influenza viruses can be attributed to expression of both 2–6 and 2-3 linked sialic acid receptors on the surfaces of the cells. Beside the viral tropism determined by virus-receptor interactions, local density of receptors, lipid raft microdomains, and host cell proteases activating the viral surface glycoproteins play major roles in influenza infectivity [14, 15]. The viruses have the ability to exploit a host virus-activating protease system to support own replication. Cellular host proteases such as transmembrane serine proteases (TMPRSS), an analogous protease from chicken allantoic fluid to the blood clotting factor 10 (FX), and plasmin were involved in the postentry stages of influenza A virus infection [8, 16]. In particular, there are growing interests in the role of hemostasis during influenza virus infection lately. The fact that factor X might play a role in viral replication suggests that indeed hemostasis and coagulation might be deleterious for the host [17]. Avian influenza viruses reach high titers when grown within chicken-origin cells; however, the efficient replication and infectivity of LP viruses are achieved in the presence of supplemental trypsin [4]. The enzyme enhances the internalization of influenza virus into cells by cleavage of HA but did not improve the ability of the host cells to internalize the virus [18]. It is well documented that the LP viruses cannot be cleaved by ubiquitous intracellular proteases while they replicate efficiently in eggs because of the presence of a protease in allantoic fluid that can cleave HA [19]. Previous studies reveal that FX, a vitamin K-dependent serine protease in the prothrombin family, was induced upon virus infection. The viral activating protein cleaves the fusion proteins of Sendai virus, Newcastle disease virus, and influenza virus at a specific single arginine-containing site and plays a key role in the viral spreading in the allantoic sac [16, 17]. In this study, we cloned the FX mRNA in BHK-21 cell and evaluated the impact of the established BHK-21/FX cell on susceptibility and virus replication kinetics of a LP influenza virus strain to provide insights into the development of future influenza virus diagnostic approaches.

MATERIALS AND METHODS

Cells and Virus Infection

MDCK and BHK-21 cells were grown in Dulbecco's modified Eagle's medium (DMEM) (Invitrogen) supplemented with 10% fetal bovine serum (FBS; Sigma Aldrich), 100 U/mL penicillin, and 100 mg/mL streptomycin, at 37°C with 5% CO_2. With the use of a dose-response test, the optimal L-(tosylamide-2-phenyl)ethyl chloromethyl ketone (TPCK) trypsin concentration was determined for each culture. The BHK-21 cell line had greater susceptibility to trypsin toxicity and received 0.1 mg/mL, while MDCK cell line was the most resistant to toxicity caused by trypsin and received 0.45 mg/mL. Monolayers of the cells at a concentration of

1×10^6 cells/mL were infected with a local isolate influenza H9N2 virus A/chicken/Iran SS8 (2011) at a multiplicity of infection (MOI) of 0.01 in the presence of supplemental trypsin. Following adsorption for 1 h at 37°C, the inoculum was removed and washed before DMEM was replaced. The cultures were incubated up to 72 hours post infection (hpi) and controlled by inverted light microscopy for cytopathic effect (CPE). For each cell, four different sets of tissue culture flasks were infected. Mock virus infected cells served as controls.

Expression Vector pcDNA3.1-FX and Transfection of BHK-21 Cells

The cDNA clone of FX in the pcDNA3.1 expression vector (Invitrogen) was constructed. The open reading frame of FX located in peptidase S1 domain (241–473 nt) isolated from chorioallantoic membrane of embryonated eggs was amplified with primers 5-GGATCCGATGAGTGTCGTCCTGGTGA-3 and 5-AAGCTTAGCCACGCCACTACTACTTT-3, containing restriction sites for BamHI and HindIII (underlines), respectively. BHK-21 cells were transfected with pcDNA3.1-FX plasmid using Lipofectamine 2000 (Life Technologies) according to the manufacturer's instructions. The transfected cells were selected and cultivated in the presence of 10% FBS and 800 g/mL Geneticin (G418; Invitrogen). After two weeks, cells surviving the selection were pooled, passaged three times in 1.5 mg of G418/mL, and frozen in aliquots. The G418-resistant colonies were isolated and subjected to RT-PCR to verify the expression of FX at 232 bp lengths using the extracted RNA (High Pure RNA Extraction Kit, Roche, Germany) and the one-step RT-PCR (iNtRON Biotechnology, Korea).

BHK-21/FX Cell Screening during Multiple Cell Passages

The growth property and plating condition for the BHK-21/FX cells were assessed prior to virus infection. The cells were seeded into two 48-well cell culture plates and incubated overnight. When the monolayers were between 90% and 95% confluent, they were inoculated with the virus at MOI 0.01. One plate received trypsin-supplemented media and one plate received plain media. The cells were incubated for 4 days at 37°C with 5% CO_2 and checked microscopically for the presence of CPE. Mock cells were included in each experiment as controls. On days 1 and 4 post infection (pi), eight wells in each plate were immunostained using influenza NP specific antibody to assess the expression of the viral internal protein. The cells were examined for virus replication kinetics quantification at different times pi. Afterwards, infectious viral particles were quantified and expressed in $TCID_{50}$. Finally, the remaining harvested supernatant was used as inoculum for ten subsequent passages. The infectivity of influenza virus in each passage was determined using immunoassay procedure. Briefly, cells were washed once with phosphate buffered saline (PBS) and fixed with 4% paraformaldehyde in PBS for 15 min at room temperature. The fixed cells were washed and immunostained with mouse anti-influenza NP monoclonal antibody followed by FITC-labeled goat anti-mouse IgG (Dako, Glostrup, Denmark). Immunostained cells were examined under Nikon Eclipse E600 fluorescence microscope.

To evaluate the impact of producing influenza virus in BHK-21/FX cell, we have assessed the sensitivity assay by infecting an amount of 10^5 cells with the H9N2 virus for ten subsequent viral passages. At each passage, virus titer was estimated and the HA and NA nucleotide sequences were determined. In this case, the extracted virus RNA was amplified using the one-step RT-PCR for HA and NA segments at full-lengths [12]. The PCR products of viral genes at different passages were cloned and sequenced subsequently in both directions. Sequences were analyzed using the CLUSTAL W alignment method of the BioEdit sequence alignment editor version 7.0.9 software.

Statistical Analysis

The data are expressed as mean ± SD. Statistical correlation of data was checked for significance by ANOVA and Student's *t*-test. Differences with $P<0.05$ were considered significant.

RESULTS

Cell Lines Validation

The ability of H9N2 influenza virus to infect the MDCK and BHK-21 cell lines was assessed in the absence and presence of supplemental trypsin. The H9N2 virus replicated in both cells and moderate CPEs were evident 48 hpi only in the presence of trypsin. The virus failed to produce CPE in the absence of trypsin because the HA remains uncleaved and virus replication did not occur.

BHK-21/FX Cell Establishment and Screening

BHK-21/FX was established by transfection of plasmid encoding FX. Two weeks continuously under antibiotic selection, the surviving cultivated BHK-21/FX cell was assessed systematically for the sensitivity to primary influenza virus infection and permittivity for virus replication and spread. Specific band amplified from total RNA of BHK-21/FX cell in RT-PCR has confirmed the presence of FX gene in the cell. The morphology of BHK-21/FX cells was not different from BHK-21 cells. Following infection of BHK-21/FX cells with influenza virus, visible CPE was observed by 24 hpi with giant cell formation and massive detaching from the culture flask compared to BHK-21 cell (Figure 1). The BHK-21 cells infected by the virus developed a very light CPE at 72 hpi. Both cells showed an increase in virus titer from $10^{3.0}$ $TCID_{50}$ on the first passage which remained constant during serial passages for BHK-21 cell ($10^{5.5}$ $TCID_{50}$), while the H9N2 virus infected-BHK-21/FX cell exhibited the highest viral titer. The amount of infectious virus yield at the first passage in BHK-21/FX cells was increased ~2500 times ($10^{7.5}$ $TCID_{50}$) in fifth passage which was maintained up to seventh passage. This indicates clearly that the cell permits the production of infectious influenza virus particles (Figure 2). Production and spread of the virus were also monitored by detecting NP expression in immunofluorescence assay up to 72 hpi (Figure 3). From 8 hpi, the numbers of NP-expressed cells increased with time elapses. The virus infected-BHK-21 cells have exhibited the same panel in the presence of supplemental trypsin, while virus replication was not observed in the absence of trypsin. It may be due to protease activation of viral HA cleavage site which supports virus entry and replication.

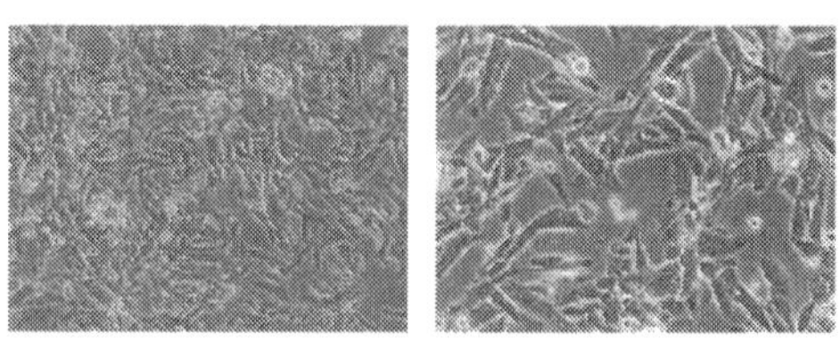

BHK-21/FX Mock   BHK-21/FX 24 hours pi

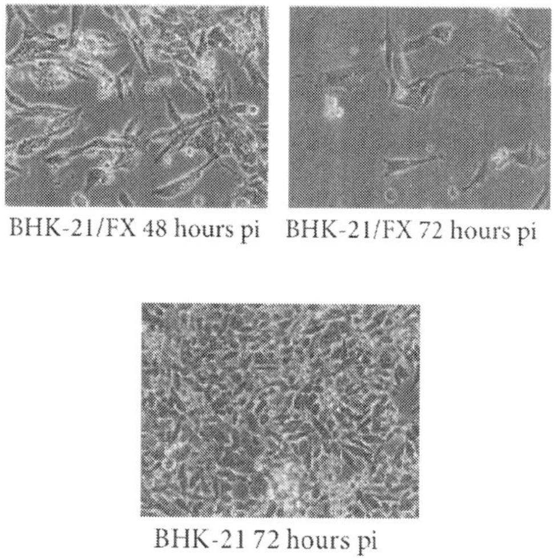

Figure 1: Cytopathogenicity of BHK-21/FX cells to influenza virus infections (MOI 0.01) at interval hours post infection (200x magnification). The BHK-21 cells infected with influenza virus did not manifest cytopathic effects.

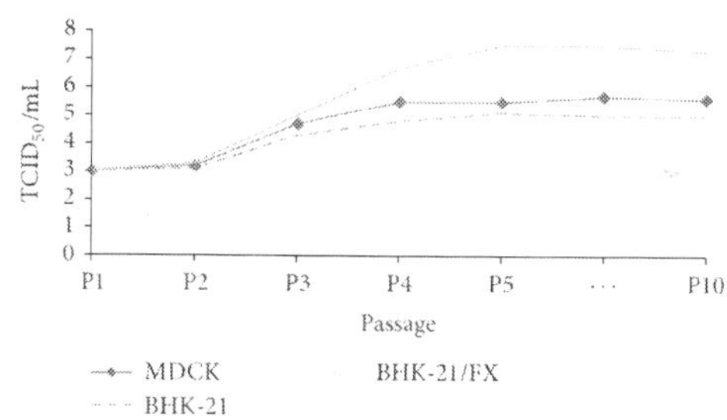

Figure 2: Replication of influenza virus in BHK-21/FX cells. The titer of virus in BHK-21/FX cell supernatants was assayed by $TCID_{50}$ in ten subsequent passages compared to the virus infected MDCK and BHK-21 cells supplemented with trypsin.

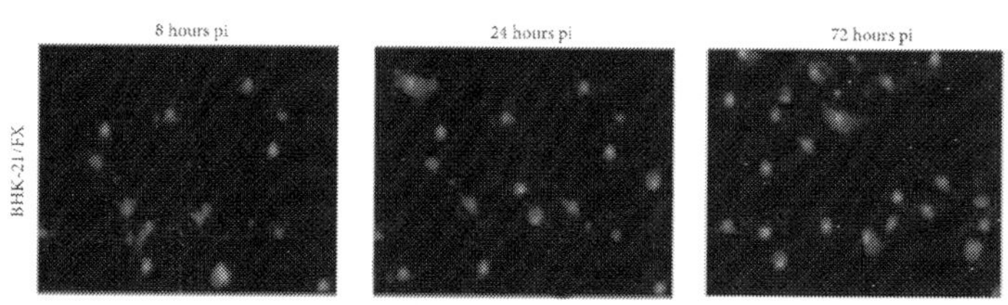

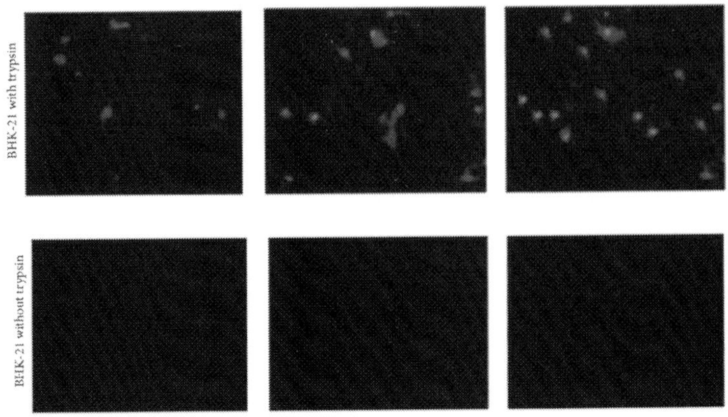

Figure 3: Immunofluorescent detection of influenza virus NP in infected-BHK-21/FX cells (×100). Fluorescence emission was detected in the infected cells at 24 and 72 hours post infection compared to the virus infected-BHK-21 cells supplemented with trypsin and without trypsin.

Nucleotide sequences of the HA and NA genes of the virus passaged seven times in BHK-21/FX cells were analyzed and compared with the parental virus genome sequences. No amino acid change was observed at the cleavage site, in the receptor binding pocket, and within the N-glycosylation sites of HA protein. All of them showed conservation of residues H^{183}, L^{190}, L^{226}, Q^{227}, and G^{228} in the receptor binding pocket and RSSR motif in cleavage site. The comparative sequence analysis of NA indicated that the amino acid sequences at the active sites of NA protein at positions ^{366} IRKDSRAG ^{373}, ^{399} DSDNRSGY ^{406}, and ^{431} PQE ^{433} were conserved. Three simultaneously nucleotide substitutions were found at the fifth passage which did not lead to amino acid changes. In the consequent passages of the virus on BHK-21/FX cell, mutation in the nucleotide sequences of viral genes that resulted in change in amino acid codon was not detected.

DISCUSSION

Since 1997, several cases of human infections with different subtypes of avian influenza viruses have been identified and raised the pandemic potential of avian influenza virus in human population. Thus, early detection is very important to initiate efficient control programs. Virus isolation in embryonated eggs or in cell culture followed by subtyping is considered as the standard protocol for detecting avian influenza virus. Since previous decade some isolates have low growth properties in MDCK cells line [20]. Avian influenza viruses generally grow efficiently on embryonated eggs or primary chick embryo-originated cells due to the specific receptor distribution and host cell proteases. It has been demonstrated that blood-derived proteases promote influenza A virus replication outside the respiratory tract.

Here, we developed BHK-21 cell line that expresses FX, which could affect influenza virus HA activation in absence of supplemental trypsin. The main focus of our project was the characterization of LP avian virus propagation in the genetically manipulated BHK-21 cell and potential use of BHK-21/FX cells for the isolation and replication of the viruses. RT-PCR confirmed the presence of the target gene expression in ten serially passaged cells. Growth kinetics of H9N2 virus on BHK-21/FX, BHK-21, and MDCK cells were evaluated by titrations of virus particles in each culture supernatant. The growth kinetic of the virus in the manipulated cell was comparable to those with MDCK or BHK-21 plus TPCK-trypsin ($P<0.01$). Monitoring of virus titers and release of progeny viruses during multiple cycle infections is of great importance for process characterization and optimization. Based on the length of the interval between the virus inoculation and increasing viral infective titer, the duration of reproductive cycle of influenza virus was estimated as 5-6 hours. Other studies have shown that a single virus progeny production cycle of influenza virus requires 8–10 hours in MDCK cell, the same time in A549 cell, compared to 20 hours in Chang's conjunctival cell [10, 12, 21]. We detected progeny virus release in BHK-21/FX at 8, 18, and 24 hours pi, respectively. The overexpression of 2,6 sialic acid receptors in MDCK cells transfected by type II transmembrane serine proteases members showed sufficient titer recommended for cell-based human influenza vaccine production [7]. Enhancing the virus receptor interaction leads to increasing the membrane fusion, the virion production, and binding rates. Viral entry and initiation infection procedure were also confirmed by detection of viral NP expression in BHK-21/FX cells in absence of exogenous trypsin. The internal viral protein potentially plays an important role in the early stages of the virus life cycle including encapsidating the segmented viral genome into ribonucleoproteins, vRNA synthesis, and interactions with the viral polymerase. The NP protein is an indicator for switching from transcription to replication. Thus, according to the immunofluorescence assay of the protein, the BHK-21/FX cell could cleave the HA protein of H9 subtype in absence of exogenous trypsin. The cleavage is a cell-associated process that leads to virus replication and production of high titer comparable to those in either MDCK or BHK-21 cells plus trypsin. Full-length amplification and sequencing patterns of the HA and NA genes at intervals passages demonstrated that the H9N2 virus at ten passages remained similar to the parent virus. These data indicate that BHK-21/FX is a permissive cell which is capable of optimizing avian influenza virus replication to obtain high titers.

Characterization of a modified cell line during growth and infection will provide the basis for cell culture-based influenza vaccine development, virus isolation, and diagnosis. In this study, we use FX, a Ca^{2+}-dependent serine protease to explore host factor involved in the influenza virus replication. FX is synthesized in liver, circulated in body as plasma protein, and activated into factor Xa via intrinsic and extrinsic pathways. The active site of factor Xa is divided into four subpockets as S1–S4. The S1 subpocket determines the major component of selectivity and binding [16]. Activation and binding of the protease at the surface of infected cells are an alternative mechanism for the proteolytic cleavage of HA which is perquisite for viral replication and pathogenicity [22]. Recent progress in understanding virus replication at the molecular level has revealed that many enveloped viruses, including influenza A viruses, incorporate cellular surface proteins into viral particles during virus-cell fusion and bud from the plasma membrane of their host cells following replication [23]. Among the host-encoded proteins in influenza virus particles, the annexin family is well represented. Annexins are a family of Ca^{2+}/lipid-binding proteins and act as membrane-membrane or membrane-cytoskeleton linkers. The proteins have been implicated in Ca^{2+}-regulated exocytotic events and certain aspects of endocytosis. Tissue specific expression of annexins suggests highly specialized function which correlates well with the ability of the proteins to interact with cellular membranes [24]. The role of annexin members as cellular interactant viral proteins in replication of influenza virus has been studied. As a virus-host interaction, expression of cellular protein annexin 2 (Anx2) is increased in response to influenza infection due to dysregulation of Ca^{2+} homeostasis. The upregulation of cell-surface Anx2 allows the recruitment of plasminogen which is activated by virion. Subsequently, HA cleaves through Anx2 which is incorporated into virus particles [25]. Anx6 expression significantly increased virus production, while its overexpression could reduce the titer of virus progeny, suggesting a negative regulatory role for Anx6 during influenza A virus infection [26]. Anx13 stimulates apical transport of influenza virus HA in MDCK cells [25]. It is assumed that Ann2 and Anx13 promote the Ca^{2+}-dependent association of lipid raft [24]. The rafts as platforms for intracellular sorting and signal transduction events [27] play a decisive role at several steps during virus replication including intracellular transport of viral proteins, assembly and budding of progeny virus at the plasma membrane, environmental stability of the virus particles, and fusion of viral and host cell endosomal membrane upon virus entry [28–30]. The lipid raft is used as a platform to concentrate HA binding receptors; thus, the virus exploits the signalling capacity of raft domains by mediating efficient fusion. Viral NA is concentrated in rafts for its normal incorporation into virions and budding [23, 28]. The raft association is also mediated by the binding of Anx5 to the plasma membrane. Anx5 is incorporated into influenza envelope during the budding process and a substantial proportion of the protein is present in lipid rafts, the site of virus budding [30].

The interactions between viral and cellular factors determine host susceptibility to influenza infection so the absence of remarkable virus replication in some cells could obviously be due to the absence of specific proteases. On the bases of host-pathogen interactions the BHK-21/FX was developed using a cellular protease factor to support high-titer LP influenza virus replication. The impact of the cell on viral titers and sensitivity at subsequent viral passages at defined cell density and virus MOI were analyzed. Distribution of the specific influenza virus receptor, specificity to other subtypes, sensitivity to higher MOIs, cell stability, and the annexin and lipid raft cellular proteins interactions need to be investigated in future studies to establish influenza virus diagnosis or vaccine manufacturing platforms based on the cellular serine protease and engineered BHK-21 cells.

DISCLOSURE

Shahla Shahsavandi as corresponding author of the paper would like to assure that neither the submitted materials nor portions have been published previously or are under consideration for publication elsewhere.

AUTHORS' CONTRIBUTION

Shahla Shahsavandi and other authors participated meaningfully in the study and have approved the final paper.

ACKNOWLEDGMENTS

The authors would like to thank all members of our laboratory for helpful assistances. This study is funded by Grant 2-18-18-91128 from Razi Vaccine and Serum Research Institute.

REFERENCES

1. P. F. Wright, G. Neumann, Y. Kawaoka et al., "Orthomyxoviruses," in Fields Virology, B. N. Fields, D. M. Knipe, and P. M. Howley, Eds., Lippincott-Raven, Philadelphia, Pa, USA, 6th edition, 2013.
2. D. E. Swayne and D. A. Helvorson, "Influenza," in Diseases of Poultry, Y. M. Saif, A. M. Fadly, J. R. Glisson, L. R. McDougald, L. K. Nolan, and D. E. Swayne, Eds., Iowa State University Press, Ames, Iowa, USA, 12th edition, 2008.
3. C.-I. Shen, C.-H. Wang, S.-C. Shen, H.-C. Lee, J.-W. Liao, and H.-L. Su, "The infection of chicken tracheal epithelial cells with a H6N1 avian influenza virus," PLoS ONE, vol. 6, no. 5, Article ID e18894, 2011.
4. S. Shahsavandi, M. M. Ebrahimi, A. Mohammadi, and N. Zarrin Lebas, "Impact of chicken-origin cells on adaptation of a low pathogenic influenza virus," Cytotechnology, vol. 65, no. 3, pp. 419–424, 2013.
5. S. Bertram, I. Glowacka, I. Steffen, A. Kühl, and S. Pöhlmann, "Novel insights into proteolytic cleavage of influenza virus hemagglutinin," Reviews in Medical Virology, vol. 20, no. 5, pp. 298–310, 2010.

6. M. N. Matrosovich, T. Y. Matrosovich, T. Gray, N. A. Roberts, and H.-D. Klenk, "Human and avian influenza viruses target different cell types in cultures of human airway epithelium," Proceedings of the National Academy of Sciences of the United States of America, vol. 101, no. 13, pp. 4620–4624, 2004.
7. E. Böttcher, T. Matrosovich, M. Beyerle, H.-D. Klenk, W. Garten, and M. Matrosovich, "Proteolytic activation of influenza viruses by serine proteases TMPRSS2 and HAT from human airway epithelium," Journal of Virology, vol. 80, no. 19, pp. 9896–9898, 2006.
8. T. H. Bugge, T. M. Antalis, and Q. Wu, "Type II transmembrane serine proteases," The Journal of Biological Chemistry, vol. 284, no. 35, pp. 23177–23181, 2009.
9. K. A. Moresco, D. E. Stallknecht, and D. E. Swayne, "Evaluation and attempted optimization of avian embryos and cell culture methods for efficient isolation and propagation of low pathogenicity avian influenza viruses," Avian Diseases, vol. 54, no. 1, pp. 622–626, 2010.
10. S. Shahsavandi, M. M. Ebrahimi, K. Sadeghi, S. Z. Mosavi, and A. Mohammadi, "Dose- and time-dependent apoptosis induced by avian H9N2 influenza virus in human cells," BioMed Research International, vol. 2013, Article ID 524165, 7 pages, 2013.
11. L. Ollier, A. Caramella, V. Giordanengo, and J.-C. Lefebvre, "High permissivity of human HepG2 hepatoma cells for influenza viruses," Journal of Clinical Microbiology, vol. 42, no. 12, pp. 5861–5865, 2004.
12. N. Z. Lebas, S. Shahsavandi, A. Mohammadi, M. M. Ebrahimi, and M. Bakhshesh, "Replication efficiency of influenza A virus H9N2: a comparative analysis between different origin cell types," Jundishapur Journal of Microbiology, vol. 6, no. 9, Article ID e8584, 2013.
13. H. Ozaki, E. A. Govorkova, C. Li, X. Xiong, R. G. Webster, and R. J. Webby, "Generation of high-yielding influenza A viruses in African green monkey kidney (Vero) cells by reverse genetics," Journal of Virology, vol. 78, no. 4, pp. 1851–1857, 2004.

14. H. Zhang, "Tissue and host tropism of influenza viruses: importance of quantitative analysis," Science in China, Series C: Life Sciences, vol. 52, no. 12, pp. 1101–1110, 2009.
15. T. Suzuki and Y. Suzuki, "Virus infection and lipid rafts," Biological and Pharmaceutical Bulletin, vol. 29, no. 8, pp. 1538–1541, 2006.
16. B. Gotoh, T. Ogasawara, T. Toyoda, N. M. Inocencio, M. Hamaguchi, and Y. Nagai, "An endoprotease homologous to the blood clotting factor X as a determinant of viral tropism in chick embryo," The EMBO Journal, vol. 9, no. 12, pp. 4189–4195, 1990.
17. F. Berri, G. F. Rimmelzwaan, M. Hanss et al., "Plasminogen controls inflammation and pathogenesis of influenza virus infections via fibrinolysis," PLoS Pathogens, vol. 9, no. 3, Article ID e1003229, 2013.
18. F. R. Zuhairi, Maharani, and M. I. Tan, "The role of trypsin in the internalization process of influenza H1N1 virus into vero and MDCK cells," ITB Journal of Science, vol. 44, no. 4, pp. 297–307, 2012.
19. C. Chaipan, D. Kobasa, S. Bertram et al., "Proteolytic activation of the 1918 influenza virus hemagglutinin," Journal of Virology, vol. 83, no. 7, pp. 3200–3211, 2009.
20. N. Asaoka, Y. Tanaka, T. Sakai, Y. Fujii, R. Ohuchi, and M. Ohuchi, "Low growth ability of recent influenza clinical isolates in MDCK cells is due to their low receptor binding affinities," Microbes and Infection, vol. 8, no. 2, pp. 511–519, 2006.
21. A. Abdoli, H. Soleimanjahi, M. T. Kheiri, A. Jamali, and A. Jamaati, "Determining influenza virus shedding at different time points in Madin-Darby canine kidney cell line," Cell Journal, vol. 15, no. 2, pp. 130–135, 2013.
22. K. Nagata, A. Kawaguchi, and T. Naito, "Host factors for replication and transcription of the influenza virus genome," Reviews in Medical Virology, vol. 18, no. 4, pp. 247–260, 2008.
23. N. Chazal and D. Gerlier, "Virus entry, assembly, budding, and membrane rafts," Microbiology and Molecular Biology Reviews, vol. 67, no. 2, pp. 226–237, 2003.
24. U. Rescher and V. Gerke, "Annexins—unique membrane binding

proteins with diverse functions," Journal of Cell Science, vol. 117, no. 13, pp. 2631–2639, 2004.

25. F. LeBouder, B. Lina, G. F. Rimmelzwaan, and B. Riteau, "Plasminogen promotes influenza A virus replication through an annexin 2-dependent pathway in the absence of neuraminidase," Journal of General Virology, vol. 91, no. 11, pp. 2753–2761, 2010.
26. S. Lecat, P. Verkade, C. Thiele, K. Fiedler, K. Simons, and F. Lafont, "Different properties of two isoforms of annexin XIII in MDCK cells," Journal of Cell Science, vol. 113, no. 14, pp. 2607–2618, 2000.
27. P. Scheiffele, A. Rietveld, T. Wilk, and K. Simons, "Influenza viruses select ordered lipid domains during budding from the plasma membrane," The Journal of Biological Chemistry, vol. 274, no. 4, pp. 2038–2044, 1999.
28. M. Takeda, G. P. Leser, C. J. Russell, and R. A. Lamb, "Influenza virus hemagglutinin concentrates in lipid raft microdomains for efficient viral fusion," Proceedings of the National Academy of Sciences of the United States of America, vol. 100, no. 25, pp. 14610–14617, 2003.
29. G. P. Leser and R. A. Lamb, "Influenza virus assembly and budding in raft-derived microdomains: a quantitative analysis of the surface distribution of HA, NA and M2 proteins," Virology, vol. 342, no. 2, pp. 215–227, 2005.
30. T. Takahashi and T. Suzuki, "Role of membrane rafts in viral infection," The Open Dermatology Journal, vol. 3, pp. 178–194, 2009.

CHAPTER

4

OPTIMIZATION OF HETEROLOGOUS PROTEIN PRODUCTION IN CHINESE HAMSTER OVARY CELLS UNDER OVEREXPRESSION OF SPLICED FORM OF HUMAN X-BOX BINDING PROTEIN

Galina Gulis, Kelly Cristina Rodrigues Simi, Renata Rodrigues de Toledo, Andrea Queiroz Maranhao and Marcelo Macedo Brigido

Institute of Biological Sciences, Department of Cell Biology, University of Brasilia, Campus Universitário Darcy Ribeiro

ABSTRACT

Background

The optimization of protein production is a complex and challenging problem in biotechnology. Different techniques for transcription, translation engineering and the optimization of cell culture conditions have been used to improve protein secretion, but there remain many open problems involving post-translational modifications of the secreted protein and cell line stability.

Results

In this work, we focus on the regulation of secreted protein specific productivity (using a recombinant human immunoglobulin G (IgG)) by controlling the expression of the spliced form of human X-box binding protein (XBP-(s)) in Chinese hamster ovary cells (CHO-K1) under doxycycline (DOX) induction at different temperatures. We observed a four-fold increase in specific IgG productivity by CHO cells under elevated concentrations of DOX at 30°C compared to 37°C, without detectable differences in binding activity *in vitro* or changes in the structural integrity of IgG. In addition, we found a correlation between the overexpression of human XBP-1(s) (and, as a consequence, endoplasmic reticulum (ER) size expansion) and the specific IgG productivity under DOX induction.

Conclusions

Our data suggest the T-REx system overexpressing human XBP-1(s) can be successfully used in CHO-K1 cells for human immunoglobulin production.

Keywords- CHO cells Heterologous protein production X-box binding protein T-REx™ system Doxycycline

BACKGROUND

The optimization of the production of secreted proteins, such as therapeutic monoclonal antibodies (mAbs), is still a challenging problem in pharmaceutical biotechnology. Although biopharmaceutical products can be produced by many host cell systems, eukaryotic cells are preferred due to their ability to correctly process and modify human proteins. The primary goal is to establish the ideal combination of a rapid accumulation of productive biomass and the maintenance of cell viability for as long as possible. Many different strategies have been considered for improving both cell viability and the productivity of recombinant proteins, including mAbs. These strategies include physiological optimization and genetic and metabolic engineering [1, 2].

The most common problem during the optimization of protein production is an error in protein folding in the endoplasmic reticulum (ER). The inhibition of protein folding activates the unfolded protein response (UPR), which is a signal transduction network. Overcoming UPR is one of the many strategies for optimizing protein productivity. For instance, protein production has been tested under the expression of survival proteins that play important roles in UPR, including B-cell lymphoma protein 2 (bcl-2), B-cell lymphoma-extra-large protein (bcl-XL) [3, 4, 5], caspase inhibitors [6] and molecular chaperones/heat shock proteins (HSP70) [7]. The role of the spliced form of X-box binding protein (XBP-1(s)) (which plays an important role in regulation processes, such as physical expansion of the ER, increasing the mitochondrial mass and function, increasing the cell size and enhancing total protein synthesis) in optimizing protein production has also been studied [8]. This approach to increasing the secretion capacity of mammalian cells by overexpressing the transcription factor XBP-1(s) was successful in CHO cells; the production of the secreted proteins alkaline phosphatase (SEAP) and alpha-amylase (SAMY) was enhanced upon XBP-1(s) overexpression [9], as was the production of antibody [10]. However, these studies have shown that using overexpression systems without regulation leads to cell apoptosis due to the accumulation of the produced proteins [11].

To overcome accumulation-induced apoptosis, other strategies have been applied to regulate the protein production, such as the use of induction systems. For instance, tetracycline has been used to optimize the overexpression of glycosyltransferases under the control of the Tet on/off system in CHO cells, but unfortunately, a high expression of glycosyltransferases still led to growth inhibition [12]. Furthermore, interesting work using the same expression system has been conducted to control the overexpression of human transferrin (hTf) in human embryonic kidney (HEK-293) cells. That study found favorable concentrations of tetracycline at which the overexpression of hTf was optimal, but again, the high levels of expression limited the cell viability. Such impairment might have been a consequence of the overexpression of the protein of interest, which might have altered the quality of this cell product or even been toxic to the cells [13]. Some studies have attempted to investigate the effect of the expression of an ER-resident molecular chaperone, protein disulfide isomerase (PDI), on the specific production levels of thrombopoietin (TPO) and antibody (Ab) in Chinese hamster ovary cells. Mohan and colleagues used the Tet-off system (in the absence of tetracycline) to regulate PDI, TPO and Ab expression in CHO cells under doxycycline (DOX; a chemical analogue of tetracycline) induction. However, only a small increase in antibody production was observed, and the production of TPO was not affected by PDI expression [14].

Moreover, the optimization of protein production in CHO cells cultured at different temperatures has been addressed. For instance, lowering the temperature from 37°C to 33°C increased the production of erythropoietin (EPO) by approximately four-fold, but at the same time, a low cultivation temperature suppressed cell growth [15]. In addition, a temperature reduction from 37°C to 33°C in the culture of a CHO cell line producing recombinant human granulocyte/macrophage colony-stimulating factor (CHO-K1-hGM-CSF) led to a reduced growth rate, increased cell viability, improved cellular protein production and decreased cell metabolism [16]. One study on the optimization of protein production at 32°C also demonstrated that the specific growth rate of CHO cells producing human mAb decreased by 30–63% at 32°C compared to 37°C. However, the specific antibody productivity of these cells was significantly enhanced at 32°C [17]. Lowering the cultivation temperature even more, from 37°C to 30°C, caused growth arrest associated with a 1.7-fold increase in the specific production of secreted alkaline phosphatase (SEAP) in CHO cells [18].

In this context, we attempted to optimize the specific IgG productivity under different culture temperatures and by regulating the overexpression of apoptotic human protein XBP-1(s) using the T-REx™ system (Invitrogen, Carlsbad, CA, USA). The applied T-REx™ system contains a regulatory plasmid (pcDNA6/TR), which encodes the tetracycline repressor, and an inducible expression plasmid (pcDNA™4/TO/myc-His A) with a tetracycline inductor for expression of the gene of the interest (*xbp-1(s)*). Co-transfected together, these plasmids created a network to regulate XBP-1(s) expression under DOX induction. We cloned *xbp-1(s)* into the T-REx™ system to control its expression with DOX. Then, we transfected the obtained T-REx™-XBP-1(s) system into stably IgG-producing CHO cells and selected stable clones of this system expressing IgG-T-REx-XBP-1(s) to control specific IgG productivity under DOX induction (Figure 1). We determined the optimal concentration of DOX and the temperature at which IgG-T-REx-XBP-1(s) cells produced the maximal amount of IgG without a significant inhibition of cell growth. Moreover, cells treated with DOX for seven days recovered viable cell density to the level of non-treated cells after DOX was washed out from the cell system, and their specific IgG productivity dropped to the basal level. Furthermore, we studied the dependence of specific IgG productivity and viable cell density on the overexpression of XBP-1(s) and ER size expansion.

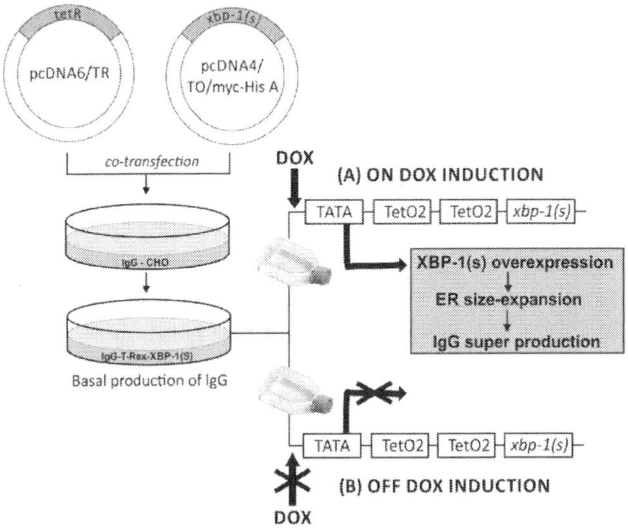

Figure 1: Schematic representation of the DOX-regulated T-Rex™ overexpression XBP-1(s) system. The overproduction of IgG as a result of the XBP-1(s) overexpression and ER size expansion under DOX induction (on DOX induction) **(A)**. The repression of XBP-1(s) overexpression and ER size expansion resulted in the repression of overproduction of IgG in the absence of DOX (off DOX induction) **(B)**.

METHODS

Cell lines and media

The CHO-K1 (ATCC®CCL-61™) and Raji (ATCC®CCL-86™) cell lines were purchased from American Type Culture Collection (ATCC, Manassas, VA, USA). CHO-K1 cells were grown and maintained at 37°C or 30°C with 70% humidity and 5% CO_2 in HAM F12 media (Gibco, Big Cabin, OK, USA) supplemented with 2% fetal bovine serum (FBS, Gibco, Big Cabin, OK, USA) and were used in experiments on protein production. Raji cells were grown and maintained at 37°C with70%

humidity and 5% CO_2 in RAMP media (Gibco, Big Cabin, OK, USA) supplemented with 10% FBS and were used in FACS direct ligation experiments.

Plasmids and cloning

pCOMIRES HIL anti-CD20 is a tricistronic vector that encodes both the heavy and the light chains of an anti-CD20 antibody along with a neomycin resistance gene under the control of a synthetic CMV promoter. This vector was transfected into CHO-K1 cells to obtain IgG (anti-CD20)-producing cells. The human *xbp-1(s)* coding sequence was chemically synthesized by GeneScript (Piscataway, NJ, USA). The restriction enzymes *Hind* III and *Bam*H I (Fermentas, Ontario, Canada) were used to obtain the *xbp-1(s)* insert and then clone it into the inducible expression plasmid pcDNA™4/TO/myc-His A from the Invitrogen T-REx™ system (Invitrogen, Carlsbad, CA, USA). This plasmid was used to co-transfect IgG-producing stable clones of CHO cells along with the regulatory plasmid pcDNA6/TR (Invitrogen, Carlsbad, CA, USA). To confirm *xbp-1(s)* cloning, XL1-blue bacterial cells (Stratagene, La Jolla, CA, USA) were transformed with ligated DNA. Ampicillin (Sigma, Ronkonkoma, NY, USA)-selected colonies were isolated and processed for DNA extraction and purification, which was performed using a QIAprep Miniprep Kit (Qiagen, Valencia, CA, USA). Restriction analysis and sequencing (using CMV forward primer 5′-CGCAAATGGGCGGTAGGCGTG-3′ and BGH reverse primer 5′-TAGAAGGCACAGTCGAGG-3′) confirmed the cloning of the *xbp-1(s)* insert.

Transfection with pCOMIRES anti-CD20 DNA (IgG-encoding plasmid) into CHO cells and generation of stable IgG-producing cells

The transfection of pCOMIRES HIL anti-CD20 plasmid (encoding an anti-CD 20 (IgG) antibody, a secretable protein with molecular weight 150 kDa (two light chains, each with molecular weight 25 kDa, and two heavy chains, each with molecular weight 50 kDa)) into CHO cells was performed using a PolyPlus (JetPrime, New York, NY, USA) kit in six-well test plates (TPP, San Diego, CA, USA) according to the manufacturer's instructions. The clones harboring the pCOMIRES HIL anti-CD20 transgene were selected from a mixed population by the single-cell dilution method. Geneticin (Roche, Gaillard, France) was used for selection at 800 μg/mL.

Transfection with the T-REx™ -XBP-1(s) system into stable IgG-producing clones of CHO cells and generation of stable double clones (IgG-T-REx-XBP-1(s) cells)

The co-transfection of T-REx-*xbp-1(s)* plasmid (encoding a spliced form of human apoptotic XBP-1 protein with predicted molecular weight 40 kDa) along with regulatory plasmid pcDNA6/TR into one of the stable IgG-producing clones was performed using a PolyPlus (JetPrime, New York, NY, USA) kit according to the manufacturer's instructions in six-well test plates (TPP, San Diego, CA, USA). Blasticidin (Sigma, Ronkonkoma, NY, USA) and Zeocin (Sigma, Ronkonkoma, NY, USA) were added to a final concentration of 0.5 μg/mL and 50 μg/mL, respectively. The selective markers encoded by regulatory plasmid pcDNA6/TR and expression plasmid pcDNA™4/TO/myc-His A are against blasticidin and Zeocin, respectively.

Doxycycline induction

Selected IgG-T-REx-XBP-1(s) cells (after the first transfection, IgG clones; after the second, co-transfection, T-REx-XBP-1(s) clones) were induced by DOX at different concentrations: 0 μg/mL for control, 0.1 μg/mL, 0.5 μg/mL and 1 μg/mL. We used these concentrations because we found out that 5 μg/ml and 7.5 μg/ml of doxycycline completely inhibits cells growth for clones and wild type CHO-K1 cells. DOX induction was performed 24 hr after IgG-T-REx-XBP-1(s) cells seeding at a uniform cell density (0.5×10^5 cells/mL) in tissue culture flasks (75 cm^2, TPP, San Diego, CA, USA) and then incubated for seven days at 37°C or 30°C. All cultures reached at least 80% under these conditions. Samples were collected for viable cell density, Ab detection by ELISA, nuclear extract isolation and ER staining. Half of the cells in each group continued to grow for seven more days in DOX-free medium after DOX wash-out. Independently, IgG-T-REx-XBP-1(s) cells were incubated for 42 days at 30°C (150 cm^2 flasks, TPP, San Diego, CA, USA) with or without 1 μg/mL DOX. In all DOX induction experiments, DOX was added (at an appropriate concentration) every three days to the cell culture. Induction experiments were performed twice in duplicate (four independent culture samples per group).

Viability assay

The viable cell density of the IgG-T-REx-XBP-1(s) cells were tested under different DOX concentrations (0 μg/mL, 0.1 μg/mL, 0.5 μg/mL or 1 μg/mL) every day during seven days of cell growth at 37°C and 30°C. Seeding was performed at a uniform cell density (0.06×10^5 cells/mL) in six-well tissue culture plates (TPP, San Diego, CA, USA). At the seventh day all cultures reached at least 80% under these conditions. In addition, the viable cell density of IgG-T-REx-XBP-1(s) cells was tested on the seventh day of growth with DOX and on the seventh day after wash-out in DOX-free medium. In addition, IgG-T-REx-XBP-1(s) cells were tested every seventh day during 42 days of cell growth under 1 μg/mL DOX (or 0 μg/mL as control) at 30°C. The viable cell density was measured using the trypan blue (Sigma, Ronkonkoma, NY, USA) exclusion method with a hemocytometer and light microscope for manual cell counting. Every viable cell density experiment was performed twice in duplicate (single determination from each of two independent culture samples per group in two independent experiments).

ELISA

The supernatants of IgG-T-REx-XBP-1(s) cells in the presence or absence of DOX were collected every seventh day of 37°C or 30°C growth for two weeks or every seventh day for six weeks and processed for analysis by enzyme-linked immunosorbent assay (ELISA) (duplicate determination from each of two independent culture samples per group in two independent experiments). The Lunc/Maxisorp Immunoplate (Thermo Scientific, Waltham, MA, USA) was incubated with primary antibody (goat anti-human IgG (H+L), 1:3000 dilution; Thermo Scientific, Waltham, MA, USA) and blocked with 3% fat-free dehydrated milk solution. After blocking and washing the plate, the supernatants were applied to the plate and incubated for 2 hr. The plate was washed again, and secondary antibody (anti-human IgG Fc-specific, alkaline phosphatase-conjugated, produced in goat, 1:3000 dilution; Sigma, Ronkonkoma, NY, USA) was applied for 1 hr. The plate was washed again, and at the end of the procedure, the signal of absorbance was read at 405 nm by a microplate reader (ELx800 96-well Microplate Reader, MTX Lab Systems, Inc., Vienna, VA, USA) after 4-Nitrophenyl phosphate disodium salt solution (pNPP) (Invitrogen, Carlsbad, CA, USA) addition. In addition, human IgG (whole molecule; Thermo Scientific, Waltham, MA, USA) was used in different concentrations as a control on the same plate.

Isolation and purification of produced proteins

The IgG produced under different temperature conditions by IgG-T-REx-XBP-1(s) cells was purified on the HiTrap™ Protein A HP 1 mL (GE Life Sciences, Pittsburgh, PA, USA) column. The column was first equilibrated with 10 mL Protein A IgG Binding Buffer (Thermo Scientific, Waltham, MA, USA) at a rate of 1 mL/min. Then, the supernatant from IgG-T-REx-XBP-1(s) cells was applied to the equilibrated column. The column was washed with 30 mL Protein A IgG Binding Buffer (Thermo Scientific, Waltham, MA, USA). Then, the protein was eluted with 50 mL IgG Elution Buffer (Thermo Scientific, Waltham, MA, USA), and 2 mL per fraction was collected. Fractions were neutralized with Tris–HCl pH 9.0. The Ab present in the fractions was immunodetected in a dot blot assay. Five microliters of each fraction was directly pipetted onto a nitrocellulose Hybond-C Extra membrane (Amersham® Bioscience, Piscataway, NJ, USA). The membrane was blocked with 3% fat-free milk solution and incubated with anti-human IgG (Fc-specific, alkaline phosphatase-conjugated, produced in goat, 1:3000 dilution) (Sigma, Ronkonkoma, NY, USA), and the proteins were revealed using a BCIP/NBT substrate Kit (Invitrogen, Carlsbad, CA, USA). The Ab-containing fractions were selected for dialysis, which was performed using a Centricon YM-50 (Amicon Bioseparations, Billerica, MA, USA) in PBS buffer (10 mM NaH_2PO_4, 137 mM NaCl, 2.7 mM KCl, pH 7.4).

Western blotting

Anti-CD20 antibody was also detected by western blotting. Five hundred nanograms of IgG sample was loaded in each well of a Bis-Tris gel (NuPAGE® Novex 4-12% Bis-Tris Gel, Invitrogen, Carlsbad, CA, USA) and separated by sodium dodecyl sulfate–polyacrylamide gel electrophoresis (SDS-PAGE) according to the manufacturer's instructions.

The proteins were transferred to the Hybond-C Extra nitrocellulose membrane (Amersham® Bioscience, Piscataway, NJ, USA) and blocked in 3% fat-free milk PBS solution. The immunodetection was performed using anti-human IgG (Fc-specific, alkaline phosphatase-conjugated, produced in goat (1:3000 dilution) (Sigma, Ronkonkoma, NY, USA) with a BCIP/NBT substrate Kit™ (Invitrogen, Carlsbad, CA, USA).

XBP-1(s) was also probed by western blotting. The nuclear extracts from the IgG-T-REx-XBP-1(s) cells were prepared as described by Becker and colleagues [10]. Briefly, the nuclear extracts were prepared from 5×10^6cells/per sample and equal volumes of nuclear extracts were loaded into a Bis-Tris gel (NuPAGE® Novex 4-12% Bis-Tris Gel, Invitrogen, Carlsbad, CA, USA), and SDS-PAGE was performed according to the manufacturer's instructions. Samples were transferred to the Hybond-C Extra nitrocellulose membrane (Amersham® Bioscience, Piscataway, NJ, USA), and after blocking with 3% fat-free milk PBS solution, rabbit anti-human-XBP-1(s) (1:1000 dilution; Sigma, Ronkonkoma, NY, USA) was added, followed by alkaline phosphatase-conjugated anti-rabbit IgG incubation (1:1000 dilution; Sigma, Ronkonkoma, NY, USA). The proteins were revealed using a BCIP/NBT substrate Kit™ (Invitrogen, Carlsbad, CA, USA).

Fluorescence-activated cell sorting (FACS)-ER staining

The IgG-T-REx-XBP-1(s) cells that were grown for seven days under DOX induction and those that were grown for one more week after wash-out were collected at 3×10^5 cells/per staining and washed with HBSS buffer (140 mM NaCl, 4.7 mM KCl, 1 mM $MgCl_2$, 1.5 mM $CaCl_2$, 10 mM glucose, 10 mM HEPES, pH 7.4). After washing with HBSS buffer, the cells were labeled with 250 nM of ER-Tracker™ Green Dye (ER-Tracker™ Green Dye for Live-Cell Endoplasmic Reticulum, Molecular Probes, Invitrogen, Carlsbad, CA, USA) according to the manufacturer's manual. The samples were washed again with HBSS buffer and analyzed using a BD FACS Verse flow cytometer (BD Bioscience, San Jose, CA, USA). Ten thousand events were collected per sample using no gate for acquisition. The dead cells were not excluded in the analysis. We used BD FACSuite to data acquisition. The experiment was performed twice in duplicate.

FACS direct ligation assay

Raji cells were grown for five passages as described above, collected and resuspended in 1 part RAMP media with 10% FBS and 1 part FACS buffer (PBS supplemented with 2% FBS) at 3×10^6 cells/well in a 96-well plate (TPP, San Diego, CA, USA). After centrifugation, the cells were blocked with FcR blocking reagent (MACS, Biotec, Bergisch Gladbach, Germany) on ice for 30 minutes according to the manufacturer's instructions. Purified and dialyzed IgG samples, which were produced by IgG-T-REx-XBP-1(s) cells at 37°C and 30°C, and commercial IgG (rituximab, MabThera, Genetech Inc., South San Francisco, CA, USA) as a positive control were added at 100 ng per well. Samples were incubated on ice for 1 hr and centrifuged after the addition of FACS buffer. The cells were washed twice with FACS buffer and incubated with mouse FITC anti-human IgG (BD Pharmingen™, BD Biosciences, San Jose, CA, USA) according to the manufacturer's manual. The cells were incubated on ice for 30 minutes in the dark, washed again twice with FACS buffer and processed for fluorescence intensity measurements using a BD FACS Verse flow cytometer (BD Bioscience, San Jose, CA, USA). Each experiment was performed twice in duplicate.

RESULTS AND DISCUSSION

Viability and IgG production under induction with DOX in IgG-T-REx-XBP-1(s) cells cultivated at 37°C and 30°C

To establish a DOX-regulated XBP-1(s) cell line, we first created stably IgG-producing clones of CHO cells by transfecting the pCOMIRES HIL anti-CD20 plasmid into CHO-K1 cells. Then, IgG-CHO clones selected with 800 μg/mL of Geneticin were co-transfected with the T-REx-XBP-1(s) system and processed for second selection using 0.5 μg/mL of blasticidin and 50 μg/mL of Zeocin. From these double clones harboring both pCOMIRES HIL anti-CD20 and T-REx-XBP-1(s) system plasmids, we chose one out of 20 for DOX induction at different concentrations (0 μg/mL (control), 0.1 μg/mL, 0.5 μg/mL or 1 μg/mL) and grew them at 37°C or 30°C. Every day, cells were collected to monitor viable cell density. We found that the viable cell density of the IgG-T-REx-XBP-1(s) cells grown at 37°C under 1 μg/mL or 0.5 μg/mL DOX was slightly lower compared to the control and to the cells with 0.1 μg/mL DOX induction (Figure 2A). Moreover, the viable cell density of IgG-T-REx-XBP-1(s) cells grown at 37°C without DOX or with 0.1 μg/mL DOX was 1.25-fold higher compared to the cells grown at 30°C (Figure 2A and 2B). These data agree with several previous studies [15, 16, 17, 18]. Moreover, the viable cell density of IgG-T-REx-XBP-1(s) cells grown at 37°C under 1 μg/mL DOX induction was increased by 20.5% compared to cells incubated at 30°C under the same DOX concentration (Figure 2A and 2B). Thus, IgG-T-REx-XBP-1(s) cells grew more slowly at 30°C with or without DOX compared to those at 37°C.

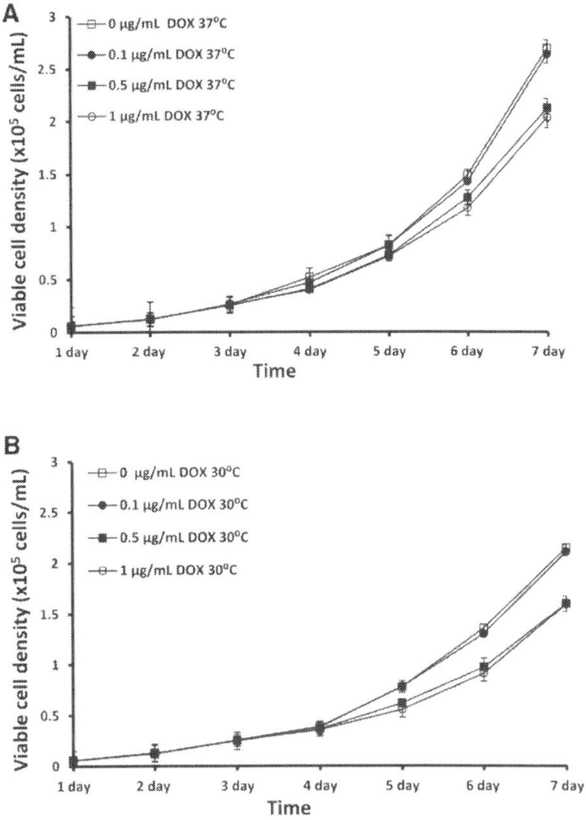

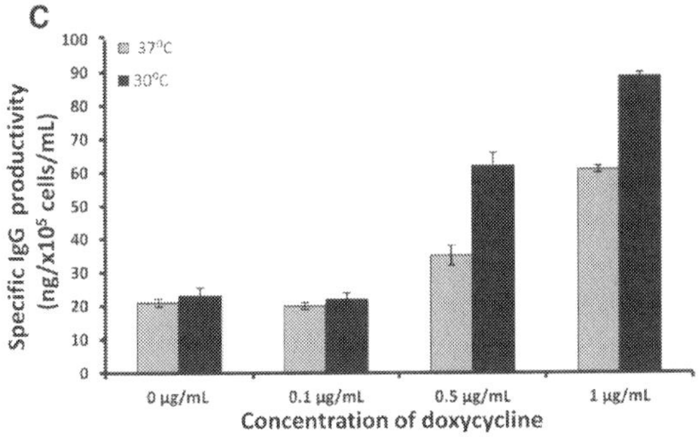

Figure 2: Viable cell density of IgG-T-Rex-XBP-1(s) cells cultured at 37°C (A) or 30°C (B) and their specific IgG productivity at 37°C and 30°C (C) under induction with 0 **µg**/mL (control), 0.1 **µg**/mL, 0.5 **µg**/mL and 1 **µg**/mL DOX. Error bars represent the standard deviation of the mean of two readings from each of two independent culture samples per group in two independent experiments, n=4.

The supernatants from all cells were collected after seven days of induction and tested by ELISA to determine their IgG yields. The specific IgG productivity depended on the concentration of DOX: under 0.5 µg/mL and 1 µg/mL DOX, the increase of specific IgG productivity was 40% and 66%, respectively, compared to the basal level of specific IgG productivity (0 or 0.1 µg/mL DOX) at 37°C (Figure 2C). These data demonstrate that IgG-T-REx-XBP-1(s) cells produced three-fold more IgG compared to untreated cells, even at low viable cell density. At 0.1 µg/mL DOX, there was no improvement in specific IgG productivity at 37°C. Moreover, the data from ELISA indicate that protein production in the cells incubated at 30°C increased four-fold and three-fold under 1 µg/mL and 0.5 µg/mL DOX, respectively. Once more, induction at a low concentration of DOX (0.1 µg/mL) did not increase specific IgG productivity at 30°C, as above at 37°C. In contrast, the specific IgG productivity by IgG-T-REx-XBP-1(s) cells at 30°C increased by 31.5% and 43.5% compared to induction at 37°C under 1 µg/mL and 0.5 µg/mL DOX concentrations, respectively (Figure 2C). However, we did not detect any effect of low temperature on specific IgG productivity *per se* (without the induction of DOX). Tigges and Fussenegger [9] reported the same lack of effect in CHO cells expressing SEAP, whereas other authors reported an increase in the production of different proteins at low temperature and with no inductor [15, 16, 17, 18]. These deviations in experimental results may be due to differences in the proteins and cell lines used in these studies. In conclusion, our data demonstrate a successful improvement of specific IgG productivity using 1 μg/mL DOX in IgG-T-REx-XBP-1(s) cells at 30°C.

Effect of XBP-1(s) expression and ER size expansion on protein production in IgG-T-REx-XBP-1(s) cells

To test the hypothesis that XBP-1(s) expression and ER size expansion indirectly regulate protein production, IgG-T-REx-XBP-1(s) cells were incubated under different concentrations of DOX for seven days at 30°C, and then the same cells were washed with DOX-free medium and incubated for seven more days at the same temperature in fresh DOX-free medium. The supernatant was collected before wash-out and after seven days of incubation in DOX-free medium and processed for ELISA. ELISA showed that specific IgG productivity by cells with DOX (first seven days) increased in a DOX concentration-dependent manner. The specific IgG productivity by cells incubated with 1 μg/mL DOX and 0.5 μg/mL DOX reached four-fold and three-fold that of the untreated cells (treated (t), Figure 3A). In addition, ELISA demonstrated that after DOX removal, the specific IgG productivity returned to the basal level (washed, (w) Figure 3A). Moreover, viability analysis indicated that the growth of the IgG-T-REx-XBP-1(s) cells under different concentrations of DOX was slightly inhibited (first seven days, treated (t), Figure 3B) and then restored to the same level (washed (w), Figure 3B) as cells that had never been exposed to DOX (0 μg/mL (t or w), Figure 3B). In addition, IgG-T-REx-XBP-1(s) cells were used to prepare nuclear extracts, which were analyzed by western blotting for the immune detection of XBP (s). Human XBP-1 (s) was overexpressed in a DOX concentration-dependent manner (first seven days, on DOX induction, treated (t) Figure 3C), but it was absent in cells that were washed and incubated in DOX-free medium for seven days (last seven days, off DOX induction, washed (w), Figure 3C).

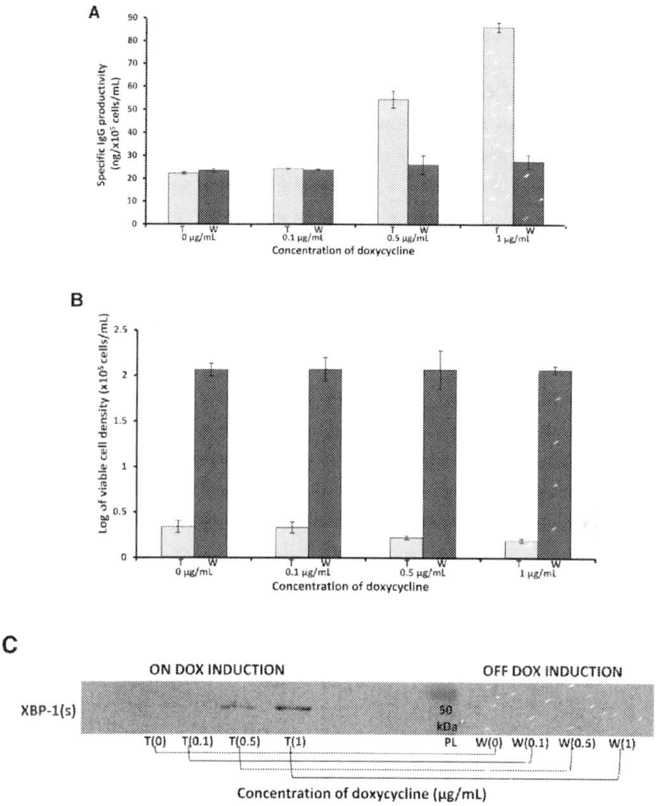

Figure 3: The specific IgG productivity (A), log of viable cell density (B) and western blot analysis of nuclear extracts (C) from the IgG-T-Rex-XBP-1(s) cells grown at 30°C under 0 **µg**/mL, 0.1 **µg**/mL, 0.5 **µg**/mL or 1 **µg**/mL DOX for seven days (treated, t) and from the same cells seven days after DOX wash-out (washed, w). PL, protein ladder. Error bars represent the standard deviation of the mean of double determination from each of two independent culture samples per group in two independent experiments, n=4.

ER expansion was also observed by flow cytometry. The fluorescence change of ER-Tracker™ was used as a measure of ER size expansion. In this analysis, cell staining was more intense in samples treated with 1 µg/mL or 0.5 µg/mL DOX for seven days (on DOX induction, treated (t), Figure 4A) than in those treated with 0.1 µg/mL (on DOX induction, treated (t), Figure 4A) or 0 µg/mL or those that were washed out (w), Figure 4B). Moreover, the signal from the washed out cells was equal among different conditions (off DOX induction, washed (w), Figure 4B), which indicated that DOX was responsible for a cascade of processes leading to ER size expansion. In addition, measurements of the median fluorescence intensity (MFI, Figure 4C) obtained from FACS analysis showed that the MFI of the cells treated with 1 μg/mL or 0.5 μg/mL DOX was 2.7-fold or 1.85-fold higher, respectively, than the MFI of non-treated cells or washed out cells (Figure 4C). Thus, cells under DOX induction and low temperature grew more slowly but, at the same time, exhibited a greater increase in specific IgG productivity. Our data also demonstrate that the wash-out of DOX from the cells restored their viable cell density but reduced their specific IgG productivity to basal levels. Taken together, our results indicate the optimal conditions for specific IgG productivity under DOX induction. XBP-1(s) was overexpressed under induction with DOX, which led to the ER size expansion, and this resulted in an increase of specific IgG productivity/secretion. Our findings corroborate the data obtained by Tigges and Fussenegger, who reported an increased production of SEAP and SAMY under the expression of XBP-1(s) and the expansion of ER and Golgi [9].

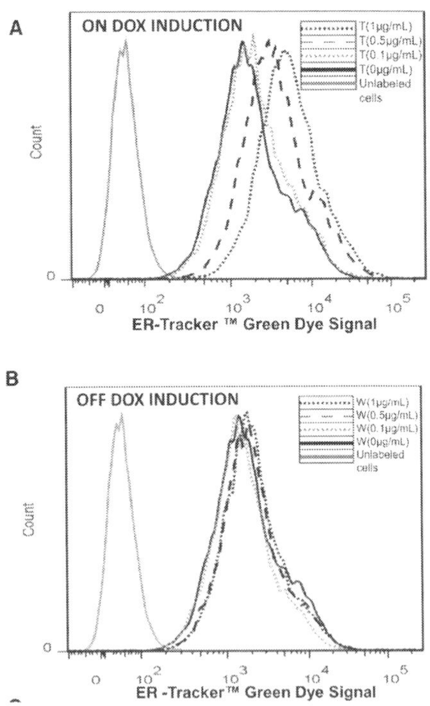

Figure 4: ACS analysis. The samples were run through the flow cytometer until 10.000 events were collected using no gate. The IgG-T-Rex-XBP-1(s) cells

were grown under induction with 0 μg/mL, 0.1 μg/mL, 0.5 μg/mL or 1 μg/mL DOX (on DOX induction, treated (t)) for seven days at 30°C. Then, cells from each group were washed with DOX-free media and grown in DOX-free media for seven more days at 30°C (off DOX induction, washed (w)). The cells from each group were stained with ER-tracker™ green dye, and cell counts vs. ER-Tracker™ signal from IgG-T-REx-XBP-1(s) cells were measured (on DOX induction, treated (t) **(A)** and off DOX induction, washed (w) **(B)**). Median fluorescence intensity (MFI) of ER-Tracker™ Green Dye from IgG-T-REx-XBP-1(s) cells (on DOX induction, treated (t) and off DOX induction, washed (w) **(C)**).

Binding activity of the recombinant proteins produced at different temperatures

To assess its binding activity, the recombinant IgG produced by cells at different temperatures (37°C and 30°C) was purified and tested for its ability to recognize CD20 at the cell surface of Raji cells that were subjected to flow cytometry analysis. FACS analysis indicated no significant difference in the binding activity of IgG produced by IgG-T-REx-XBP-1(s) cells at different temperatures under DOX induction from commercial IgG (rituximab) (Figure 5A). Moreover, protein samples obtained from the IgG-T-REx-XBP-1(s) cells at different temperatures under DOX induction were submitted to western blotting analysis, and the results did not suggest any differences in structural integrity of IgG produced at different temperatures (Figure 5B). These data support the use of low-temperature culture conditions under induction by DOX to increase protein production without eliminating the binding activity and structural integrity of the protein of interest.

A

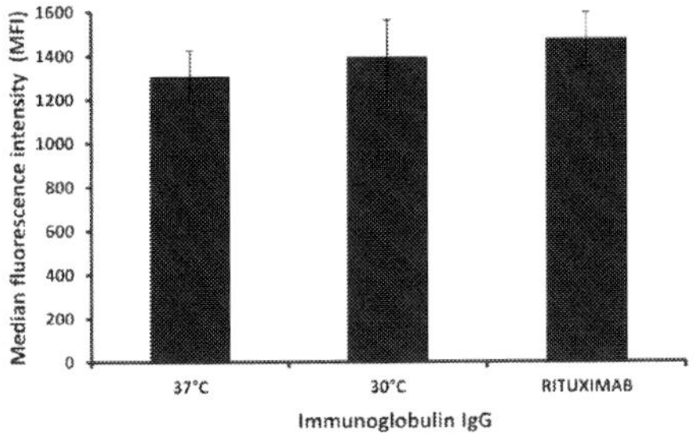

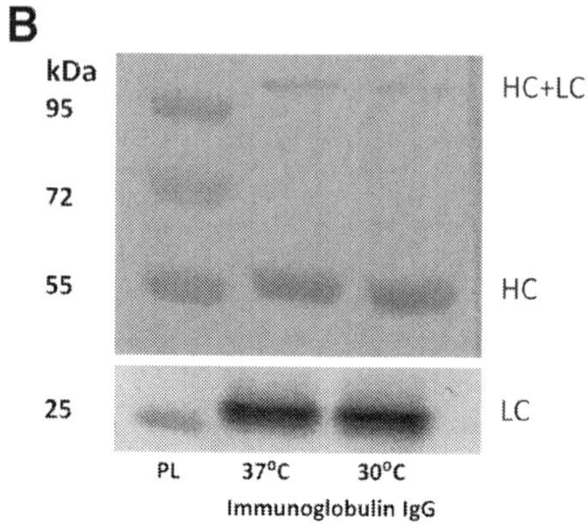

Figure 5: FACS and Western analyses. Median fluorescence intensity measurements of mouse FITC anti-human IgG ligated to the tested IgG (produced by IgG-T-REx-XBP-1(s) cells under DOX induction at 37°C and 30°C (and rituximab, control) that were previously incubated with Raji cells) **(A)**. Western blot analysis of IgG produced by IgG-T-REx-XBP-1(s) cells at 37°C and 30°C under DOX induction (PL, protein ladder) **(B)**.

Establishing a stable protein-producing cell line

To establish a stable cell line, IgG-T-REx-XBP-1(s) cells were seeded and grown at 30°C under 1 μg/mL DOX (or 0 μg/mL DOX as control) for 42 days. The supernatant was collected every seventh day and submitted to ELISA, which showed stable specific IgG productivity around an average of 80.2 ng/10^5 cells/mL by cells incubated with DOX compared to the basal level (25.7 ng/10^5 cells/mL in cells without DOX) (Figure 6A). Moreover, the viable cell density of IgG-T-REx-XBP-1(s) cells under 1 μg/mL DOX was lower than cells without DOX for 42 days (Figure 6B), which allowed cells to be kept in the same culture flask without passaging to new flasks. The described approach might be useful for the production of recombinant secreted proteins, as cells growing under the combination of DOX induction and low temperature multiply more slowly but are more productive.

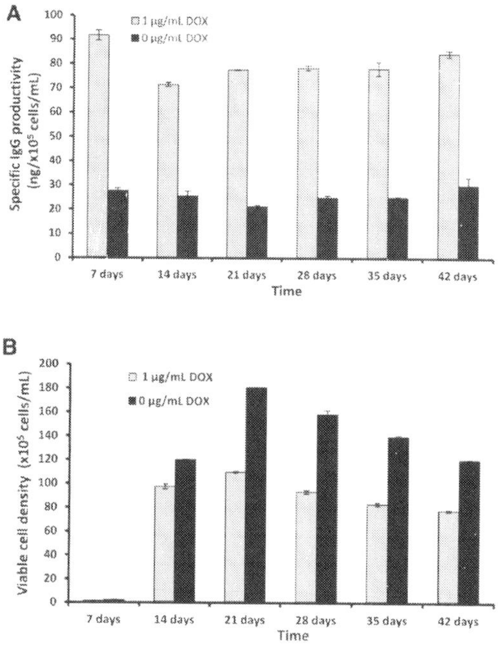

Figure 6: Specific IgG productivity (A) and viable cell density (B) of the IgG-T-REx-XBP-1(s) cells under 1 **µg**/mL DOX induction (0 **µg**/mL DOX, control) for 42 days at 30°C . Error bars represent the standard deviation of the mean of double determination from each of two independent culture samples per group in two independent experiments, n=4.

CONCLUSION

Many studies have been published on improving recombinant protein production. In general, the published data suggest that the optimization of the production of specific target proteins requires specifically adjusted growth conditions and a carefully chosen cell line. In the present study, we optimized the conditions for IgG (human anti-CD20) specific productivity in CHO-K1 cells. We showed that the combination of low temperature (30°C) and XBP-1(s) overexpression regulated by DOX induction significantly improved anti-CD20 specific productivity: under 1 µg/mL DOX treatment, specific IgG productivity was increased by 32% compared to the cells grown under the same concentration at 37°C and 74% compared to the cells grown without DOX induction at 37°C or 30°C. Moreover, the results of our study indicate the direct dependence of specific IgG productivity on the concentration of DOX (under 0.5 μg/mL, the increase was 2.7-fold, and under 1 μg/mL, the increase was 3.9-fold), which allows for the precise regulation of specific IgG productivity in CHO-K1 cells. In addition to the concentration dependence, we demonstrated the possibility of returning the specific IgG over productivity to the basal level of specific productivity by removing DOX. This step also restored the viable cell density, which permitted the cells to overcome the problem of accumulation of the target protein. In the production of proteins, it may be possible to use the T-Rex-XBP-1(s) system to turn up and down the production of protein, repeating this cycle several times to accumulate higher amounts of target protein without a loss of cell viability. We also observed a DOX concentration-dependent relationship involving XBP-1(s) overexpression (western analysis), ER size expansion (FACS measurements) and specific IgG productivity (ELISA). Finally, our data demonstrate that it is possible, under DOX induction at low temperature, to produce a target protein for an extended period of time. Taken together, our data suggest the T-REx-XBP-1(s) system can be used in CHO-K1 cells for human immunoglobulin production.

ABBREVIATIONS

Ab: Antibody

bcl-2: B-cell lymphoma protein 2

bcl-XL: B-cell lymphoma-extra-large protein

CHO-K1: Chinese hamster ovary cells

CHO-K1-hGM-CSF: CHO cell line producing recombinant human granulocyte/macrophage colony-stimulating factor

DOX: Doxycycline

ELISA: Enzyme-linked immunosorbent assay

EPO: Erythropoietin

ER: Endoplasmic reticulum

FACS: Fluorescence-activated cell sorting

HEK-293: Human embryonic kidney cells 293

HSP70:

Heat shock proteins 70

hTf: Human transferrin

IgG: Human immunoglobulin G

mAbs: Monoclonal antibodies

MFI: Median fluorescence intensity

PDI: Protein disulfide isomerase

pNPP: 4-Nitrophenyl phosphate disodium salt solution

SAMY: alpha-amylase

SEAP: Secreted alkaline phosphatase proteins

SDS-PAGE: Sodium dodecyl sulfate–polyacrylamide gel electrophoresis

Tet: Tetracycline

TPO: Thrombopoietin

UPR: Unfolded protein response

XBP-(s): Spliced form of human X-box binding protein.

REFERENCES

1. Costa AR, Rodrigues ME, Henriques M, Azeredo J, Oliveira R: Guidelines to cell engineering for monoclonal antibody production. Eur J Pharm Biopharm. 2010, 74: 127-138. 10.1016/j.ejpb.2009.10.002.
2. Frenzel A, Hust M, Schirrmann T: Expression of recombinant antibodies. Front Immunol. 2013, 4 (217): 1-20.
3. Fussenegger M, Fassnacht D, Schwartz R, Zanghi JA, Graf M, Bailey JE, Pörtner R: Regulated overexpression of the survival factor bcl-2 in CHO cells increases viable cell density in batch culture and decreases DNA release in extended fixed-bed cultivation. Cytotechnology. 2000, 32 (1): 45-61. 10.1023/A:1008168522385.
4. Itoh Y, Ueda H, Suzuki E: Overexpression of bcl-2, apoptosis suppressing gene: Prolonged viable culture period of hybridoma and enhanced antibody production. Biotechnol Bioeng. 1995, 48 (2): 118-122. 10.1002/bit.260480205.
5. Meents H, Enenkel B, Eppenberger HM, Werner RG, Fussenegger M: Impact of coexpression and coamplification of sICAM and

antiapoptosis determinants bcl-2/bcl-x (L) on productivity, cell survival, and mitochondria number in CHO-DG44 grown in suspension and serum-free media. Biotechnol Bioeng. 2002, 80 (6): 706-716. 10.1002/bit.10449.

6. Sauerwald TM, Betenbaugh MJ, Oyler GA: Inhibiting apoptosis in mammalian cell culture using the caspase inhibitor XIAP and deletion mutants. Biotechnol Bioeng. 2002, 77 (6): 704-716. 10.1002/bit.10154.
7. Lasunskaia EB, Fridlianskaia II, Darieva ZA, da Silva MS, Kanashiro MM, Margulis BA: Transfection of NS0 myeloma fusion partner cells with HSP70 gene results in higher hybridoma yield by improving cellular resistance to apoptosis. Biotechnol Bioeng. 2003, 81 (4): 496-504. 10.1002/bit.10493.
8. Shaffer AL, Shapiro-Shelef M, Iwakoshi NN, Lee AH, Qian SB, Zhao H, Yu X, Yang L, Tan BK, Rosenwald A, Hurt EM, Petroulakis E, Sonenberg N, Yewdell JW, Calame K, Glimcher LH, Staudt LM: XBP1, downstream of Blimp-1, expands the secretory apparatus and other organelles, and increases protein synthesis in plasma cell differentiation. Immunity. 2004, 21 (1): 81-93. 10.1016/j.immuni.2004.06.010.
9. Tigges M, Fussenegger M: Xbp1-based engineering of secretory capacity enhances the productivity of Chinese hamster ovary cells. Metab Eng. 2006, 8 (3): 264-272. 10.1016/j.ymben.2006.01.006.
10. Becker E, Florin L, Pfizenmaier K, Kaufmann H: An XBP-1(s) dependent bottle-neck in production of IgG subtype antibodies in chemically defined serum-free Chinese hamster ovary (CHO) fed-batch processes. J Biotechnol. 2008, 135 (2): 217-223. 10.1016/j.jbiotec.2008.03.008.
11. Becker E, Florin L, Pfizenmaier K, Kaufmann H: Evaluation of a combinatorial cell engineering approach to overcome apoptotic effects in XBP-1(s) expressing cells. J Biotechnol. 2010, 146 (4): 198-206. 10.1016/j.jbiotec.2009.11.018.
12. Umaña P, Jean-Mairet J, Bailey JE: Tetracycline-regulated overexpression of glycosyltransferases in Chinese hamster ovary cells. Biotechnol Bioeng. 1999, 65 (5): 542-549. 10.1002/(SICI)1097-0290(19991205)65:5<542::AID-BIT7>3.0.CO;2-Z.

13. Jones J, Nivitchanyong T, Giblin C, Ciccarone V, Judd D, Gorfien S, Krag SS, Betenbaugh MJ: Optimization of tetracycline-responsive recombinant protein production and effect on cell growth and ER stress in mammalian cells. Biotechnol Bioeng. 2005, 91 (6): 722-732. 10.1002/bit.20566.

14. Mohan C, Park SH, Chung JY, Lee GM: Effect of doxycycline-regulated protein disulfide isomerase expression on the specific productivity of recombinant CHO cells: thrombopoietin and antibody. Biotechnol Bioeng. 2007, 98 (3): 611-615. 10.1002/bit.21453.

15. Yoon SK, Song JY, Lee GM: Effect of low culture temperature on specific productivity, transcription level, and heterogeneity of erythropoietin in Chinese hamster ovary cells. Biotechnol Bioeng. 2003, 82 (3): 289-298. 10.1002/bit.10566.

16. Bollati-Fogolín M, Forno G, Nimtz M, Conradt H, Etcheverrigaray M, Kratje R: Temperature reduction in cultures of hGM-CSF-expressing CHO cells: effect on productivity and product quality. Biotechnol Prog. 2005, 21 (1): 17-21.

17. Yoon SK, Hwang SO, Lee GM: Enhancing effect of low culture temperature on specific antibody productivity of recombinant Chinese hamster ovary cells: clonal variation. Biotechnol Prog. 2004, 20 (6): 1683-1688. 10.1021/bp049847f.

18. Kaufmann H, Mazur X, Fussenegger M, Bailey J: Influence of low temperature on productivity, proteome and protein phosphorylation of CHO cells. Biotechnol Bioeng. 1999, 63 (5): 573-582. 10.1002/(SICI)1097-0290(19990605)63:5<573::AID-BIT7>3.0.CO;2-Y.

CHAPTER

5

ENHANCING PROTEIN EXPRESSION IN HEK-293 CELLS BY LOWERING CULTURE TEMPERATURE

Chi-Yen Lin, Zhen Huang, Wei Wen, Andrew Wu, Congzhou Wang, Li Niu

Department of Chemistry, and Center for Neuroscience Research, University at Albany, SUNY, Albany, New York, 12222, United States of America

ABSTRACT

Animal cells and cell lines, such as HEK-293 cells, are commonly cultured at 37°C. These cells are often used to express recombinant proteins. Having a higher expression level or a higher protein yield is generally desirable. As we demonstrate in this study, dropping culture temperature to 33°C, but not lower, 24 hours after transient transfection in HEK-293S cells will give rise to ~1.5-fold higher expression of green fluorescent protein (GFP) and α-amino-3-hydroxy-5-methyl-4-isoxazolepropionic acid (AMPA) receptors. By following the time course of the GFP-expressing cells growing at 37°C and 33°C from 24 hours after transfection (including 19 hours recovery at 37°C in the normal growth medium), we found that a mild hypothermia (i.e., 33°C) reduces the growth rate of HEK-293S cells, while increasing cellular productivity of recombinant proteins. As a result, green cells remain undivided in a longer period of time. Not surprisingly, the property of a recombinant protein expressed in the cells grown at 33°C is unaffected, as shown by the use of AMPA receptors. We further demonstrate with the use of PC12 cells that this method may be especially useful when a recombinant protein is difficult to express using a chemical-based, transient transfection method.

INTRODUCTION

Human embryonic kidney 293 (HEK-293) cells [1] are a popular mammalian, heterologous expression system for producing recombinant proteins [2]. These cells can be also used either intact or in lipid fragments to study structure and function relationships and pharmacological properties of the membrane proteins that are expressed in these cells. The benefit of using HEK-293 cells for expressing recombinant proteins includes an efficient transfection of plasmid DNAs, faithful translation and processing of proteins [2]. However, in using these cells for expressing membrane proteins, such as ion channels, a low signal is sometimes observed (e.g., a signal can be current amplitude from electrophysiology or radioactivity from binding experiments). Low signal is generally related to a low copy number of a receptor protein expressed in a cell or precisely on the surface of a cell. Consequently, when either intact cells or cellular membrane fragments that harbor the expressed proteins or receptors have to be used for assays of a membrane protein, the concentration of that membrane protein is further "diluted". The presence of inhibitors in a measurement further exacerbates the low signal problem and may even prevent a signal from being reliably detected and determined. In addition, the use of HEK-293 cells for a large-scale production of membrane proteins for structural determination is still quite a challenge [3,4,5]. Therefore, it will be useful to find new ways of improving the efficiency of protein expression in HEK-293 cells.

For enhancing expression efficiency, one way is to optimize culture condition. Culture condition (e.g., medium and culture temperature), vector, and host are three major factors that affect the expression of recombinant proteins [6]. Changing culture temperature can be beneficial, because temperature affects cell growth, viability, protein synthesis and metabolism. In this context, lowering temperature from 37°C is known to slow cell growth rate [7,8]. However, early studies showed that lowering temperature had no effect [9,10,11] or even a negative effect on protein expression [12,13,14]. In contrast, some later studies using Chinese hamster ovary (CHO) cells have demonstrated that lowering culture temperature to 30–32°C resulted in higher expression of a variety of recombinant proteins [15,16,17,18,19,20]. These studies suggest that the effect of low temperature on protein expression is cell-line specific, and the enhancement of protein expression is mainly linked to the cold-induced growth arrest within the S or G1 phase of the cell cycle [17,18,21,22]. However, more recent work [23,24] suggests that growth arrest under mild hypothermia conditions (30–35°C) and the associated higher yield of protein expression in mammalian cells may be two independent responses. It should be noted, however, almost of all previous studies of the effect of temperature on production of recombinant proteins have been conducted with CHO cells in suspension culture [6,25]. Very few studies have been attempted with HEK-293 cells, though they are widely used in studies of membrane proteins, especially in static cultures for electrophysiological studies of ion channel proteins.

In the present study, we asked whether lowering culture temperature could enhance protein expression yield in HEK-293 cells (specifically the HEK-293S cell line) and if so, how low temperature can be lowered. We chose two proteins in our study: green fluorescent protein (GFP) [26] and the α-amino-3-hydroxy-5-methyl-4-isoxazolepropionic acid (AMPA) receptors [27,28]. By monitoring green cells with transiently expressed GFP, we followed the time course and the fluorescence intensity of GFP expression in cultures that were subject to mild hypothermia conditions as compared with cultures at 37°C. On the other hand, AMPA receptors are one of the three subtypes of the glutamate ion channel receptor family, and mediate fast synaptic neurotransmission in the central nervous system [29]. Using whole-cell recording, we measured the channel activities of AMPA receptors in the absence and presence of an inhibitor with cells growing under a mild hypothermic condition as compared with those growing at 37°C. In addition, we have tested several, commonly used transfection methods for delivery of plasmid DNAs to explore the temperature effect on the same cell line, including calcium phosphate [30] and Lipofectamine 2000. Based on these results, we describe a protocol in which at 33°C, but not lower, the expression of both GFP and AMPA receptors in HEK-293S cells is increased by ~1.5-fold.

MATERIALS AND METHODS

Plasmid DNAs

The cDNA plasmid that contained the $GluA2Q_{flip}$ gene sequence (unedited at the Q/R site, and flip isoform) was used. Another cDNA construct that contained the $GluA3_{flip}$ gene sequence was also used in this study.

Both DNA plasmids also contained the SV40 replication origin (8.6 kb) [31,32]. A GFP construct [33] was expressed either alone or together with an AMPA receptor; in the latter case, GFP was co-expressed as a marker for cell selection for whole-cell recording. To enhance AMPA receptor expression, we also co-expressed a plasmid DNA encoding large T-antigen (TAg) [34]. We previously reported the use of simian virus (SV) 40 TAg to enhance receptor expression in HEK-293 cells [33]. All the plasmids were propagated in an *E. coli* host (DH5α) and purified using a kit from QIAGEN (Valencia, CA).

Cell line, cell culture and transient transfection

The HEK-293S cell line was used in all studies [33,35]. However, the HEK-293S cells were grown in static culture because we were interested in expressing GFP for cell count and intensity measurement as well as channel proteins for electrophysiological measurements. As we reported previously, in static culture, the majority of 293S cells were indeed attached to Petri dishes, albeit more loosely than regular 293 or 293T cells [33]. The cells were maintained in Dulbecco's modified Eagle's medium (DMEM, Lonza BioWhittaker, Cat. No. 12-604F, Walkersville, MD) supplemented with 10% fetal bovine serum (FBS, Invitrogen, Cat. No. 10082–147), and 100 U/ml penicillin and 100 μg/ml streptomycin (CORNING Cellgro, Cat. No. 30-002-CI, Manassas, VA) in a 6% CO_2, humidified incubator. Two incubators (Forma Series II water-jacked) were used simultaneously. The temperature of one incubator was set at 37°C, whereas the other was set at 33°C or 30°C (see below). A thermometer was placed inside an incubator to provide independent readout of the culture temperature.

To examine the effect of a mild hypothermic culture condition on protein expression, we followed a biphasic temperature culture protocol. The HEK-293 cells were maintained at 37°C normally. After each passage, we continued to grow cells at this temperature to obtain a reasonably high cell density for transfection (see details for transfection below). After transfection, Petri dishes were brought back to 37°C for a minimal five hours before they were transferred to a different incubator with lower culture temperature for a period of time before assays (see Results).

In all of the experiments, transient transfection was performed. A number of transfection reagents were used: calcium phosphate [30], Lipofectamine 2000 (Invitrogen, Cat. No. 18324–111, Carlsbad, CA), Lipofectamine LTX & PLUS (Invitrogen, Cat. No. 15338030) and Metafectene EASY (Biontex Laboratories GmbH, Cat. No. T090-10, Munich, Germany). We also used Opti-MEM I Reduced Serum Medium (Invitrogen, Cat. No. 31985), and the complete medium, which referred to DMEM supplemented with 10% FBS and penicillin/streptomycin. For the PC12 cell line (ATCC, Cat. No. CRL-1721, Manassas, VA), we used the RPMI-1640 Medium (ATCC, Cat. No. 30–2001), supplemented with 10% horse serum (Invitrogen, Cat. No. 26050–088) and 5% FBS with penicillin/streptomycin. The Petri dishes were coated with poly-D-lysine hydrobromide (Sigma-Aldrich, Cat. No. Sigma-P6407, St. Louis, MO).

Whole-cell current recording

Whole-cell recording [36] was used to assess the activity of AMPA receptor channels expressed in HEK-293S cells under mild hypothermic conditions as compared with those grown at 37°C. In this study, we tested two AMPA receptors, i.e., $GluA2Q_{flip}$ and $GluA3_{flip}$; both can form homomeric receptor channels when expressed individually in HEK-293 cells [31,32]. We also measured the ratio of the whole-cell current amplitude in the absence and presence of an inhibitor with the receptor expressed in a lower temperature, and compared that with the one obtained at 37°C.

The procedure for recording AMPA receptor channel activity was previously described [31,32]. In brief, the recording electrodes were pulled from glass capillaries (World Precision Instruments, Sarasota, FL). The electrode resistance was ~3 MΩ when filled with the electrode solution or internal buffer; the electrode solution contained (in mM) 110 CsF, 30 CsCl, 4 NaCl, 0.5 $CaCl_2$, 5 EGTA, and 10 HEPES (pH 7.4 adjusted by CsOH). The external bath solution contained (in mM) 150 NaCl, 3 KCl, 1 $CaCl_2$, 1 $MgCl_2$, 10 HEPES (pH 7.4 adjusted by HCl). Whole-cell recordings were at—60 mV, and 22°C. Specifically, glutamate was applied from a solution flow device to the cell [31,32,37], and the glutamate-induced whole-cell current was recorded using an Axopatch-200B amplifier at cutoff frequency of 2 kHz by a built-in, 4-pole low-pass Bessel filter, and digitized at 5 kHz sampling frequency using a

Digidata 1322A from Axon Instruments (Union City, CA). The data were acquired using pCLAMP 9 (also from Axon). Each data point was an average of at least three measurements collected from at least three cells unless otherwise noted. OriginPro 7 (Origin Lab, Northampton, MA) was used for data plotting. Uncertainties refer to standard deviation of the fits unless noted otherwise.

Fluorescence imaging and quantification

The green fluorescence due to expression of GFP in HEK-293S cells was imaged on a Carl Zeiss Axiovert S200 microscope and captured using a Sony NEX 3 digital camera and an adapter (Model NY1S-EA, MeCan Imaging Inc., Saitama, Japan). The adapter has 1.74× magnification.

Two sizes of cell images were collected in this study. For each 35 mm Petri dish used to express GFP, we took three bright-field images from three randomly chosen locations under a 20× objective lens (therefore, the total magnification was 20 × 1.74 = 34.8×). For the same viewing field, we also took three fluorescence images. On average, a fluorescence image contained several hundred green cells. These 34.8× images were used to count the total green cells from the fluorescence images and the total cells from the bright-field images for calculating the percentage of green cells or the transfection efficiency. Each fluorescence image contained 2,288 × 1,520 pixels, and each pixel was scored to a value possibly ranging from 0 to 255 on 8-bit digital scale using Image J (version 1.46r from http://rsb.info.nih.gov/ij/). The maximum background intensity was 29 on this scale; below it, no green color could be visually recognized. In this case, we assumed that such a cell did not express GFP, at least not appreciably. On the other hand, a typical green cell contained about 3,000 ± 800 pixels, and the fluorescence intensity per pixel was more than 29 but smaller than 255 (i.e., 255 was the maximum intensity value). We found empirically that the fluorescence intensity of any green cell was roughly even. In other words, all of ~3000 pixels used to digitize the fluorescence intensity of a single green cell in fact had similar intensity scale. As an approximation, therefore, we used a single fluorescence scale to reflect the fluorescence intensity of that single green cell. Furthermore, a green cell could possibly be assigned to one of the three fluorescence intensity scales.

We also took three bright-field/fluorescence image sets from three randomly chosen locations under a 5× objective lens (therefore, the total magnification was 5 × 1.74 = 8.7×). These images on average contained thousands of cells and a typical green cell contained ~600 pixels. These images were used to calculate the overall green intensity, an indication for overall protein expression. It was assumed that the fluorescence intensity in a single GFP-expressing cell was linearly proportional to the amount of GFP that cell expressed.

RESULTS

HEK-293S cell growth at 37°C, 33°C and 30°C

To explore whether a mild hypothermic condition would elevate protein expression in HEK-293S cells, we began by expressing and using GFP to establish several parameters that were involved in transient expression. The first experiment was to seed three sets of 35 mm culture dishes with HEK-239S cells at 37°C, 33°C and 30°C. All of the dishes had the same seeding density (~5.5 x10^5 cells/dish). We observed the following phenomena. First, HEK-293S cells grew significantly slower when incubated at either 30°C (Fig 1A and 1B) or 33°C (Fig 1C and 1D). In contrast, the cells maintained at 37°C grew faster, as evidenced by the increase of cell density of >50% at 48 hours (Fig 1). The cell growth was even slower at 30°C as compared with 33°C. Second, immediately after either passage or transfection, incubating HEK-293S cells at a lower temperature led to a significantly poor cell attachment or a significant cell loss. For example, at 30°C, we observed >60% of cell loss as compared with the 37°C growth temperature. From these experiments, we concluded that 33°C was a reasonable starting point to explore enhancing protein expression under a growth arrest condition. Choosing 30°C would seriously reduce biomass (in addition, lowering culture temperature to 30°C actually had a detrimental effect on protein expression, which will be discussed in detail in "Time Course of GFP Expression).

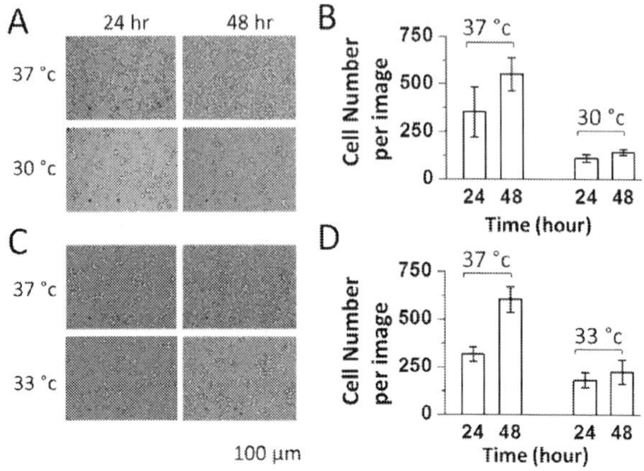

Figure 1: Bright-field images (34.8x magnification) of HEK-293S cells growing at 37°C (upper A and C), 30°C (lower A), and 33°C (lower C).

HEK-293S cells were seeded at a density of 5.5 x 10^5 cells/35 mm Petri dish. The images were taken at 24 and 48 hours using a digital camera mounted on a Carl Zeiss Axiovert S200 microscope (see Methods). Cell number counts per image or 0.22 mm^2were plotted in (B) and (D) based on the images shown in (A) and (C), respectively. For each dish, 3 viewing areas were randomly chosen and their images were taken.

Timing of shifting culture temperature from 37°C to 33°C after transient transfection

Normally, HEK-293S cells were maintained at 37°C. After transient transfection with GFP (using Lipofectamine 2000 as an example), we immediately returned the cells to 37°C. Five hours later, we replaced the transfection solution with a normal growth medium in order to minimize cell toxicity. Medium replacement was especially required for calcium-phosphate precipitation [30]. After the medium replacement, we carried out two experiments. In one experiment, we immediately transferred a set of dishes to 33°C (or five hours after transfection); in the other experiment, we returned another set of dishes to 37°C for the cells to recover from transfection. At the 19th hour (or 24 hours after the transfection), we then transferred the second set of dishes from 37°C to 33°C. In addition, we had one set of control dishes maintained at 37°C throughout the experiment. Our results (Fig 2) showed that the shift of the culture temperature from 37°C to 33°C immediately after the medium replacement resulted in an immediate cell growth arrest or a much lower cell density as compared with the control culture at 37°C. In contrast, a 19th hour delay of shifting the culture temperature to 33°C or an overnight recovery at 37°C from transfection prior to lowering temperature to 33°C led to an increase of green fluorescent intensity in the green cells (a quantitative measure will be presented in the next section below). In other words, a delayed temperature shift from 37°C to 33°C by a day resulted in a higher level of GFP expression, as compared with the 37°C control. In contrast, shifting culture to 33°C without overnight delay resulted in lower green cell counts and lower GFP expression (images are shown in the left column of Fig 2). We arrived at this conclusion based on the assumption that GFP fluorescence intensity in HEK-293S cells was linearly correlated to the amount of GFP expressed in those cells.

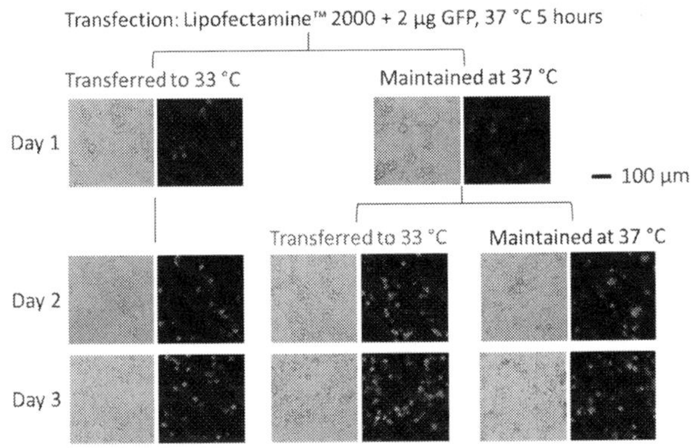

Figure 2: right-field and fluorescence images (34.8x magnification) of GFP-expressing HEK-293S cells.

The cells were transfected with the GFP plasmid using Lipofectamine 2000. Five hours after transfection (the cells were maintained in a 37°C incubator during this 5-hour period), the medium was replaced. Three sets of dishes were subject to three different ways of maintaining culture temperature, as shown. One set of dishes was transferred to a 33°C incubator. Another set was returned to the 37°C incubator; 19 hours later or 24 hours after transfection, this set of dishes was brought to the 33°C incubator and remained there. The control dishes were maintained at 37°C throughout the experiment. The images were taken from the first day (24 hours after transfection) to the third day.

Comparison of different methods and the amount of the GFP plasmid used for transfection

Next, we tested different transfection reagents and methods, and varied the amount of the GFP plasmid in an attempt to find the optimal condition of lowering culture temperature to enhance GFP expression. We first chose the Lipofectamine 2000 reagent and calcium phosphate in HEPES-buffered saline solution [38]. Lipofectamine 2000 is a popular transfection reagent (cationic lipid formulation), whereas the calcium phosphate method is the most inexpensive means to deliver gene to cells and is perhaps the benchmark for evaluating transfection efficiency of chemical-based transfection methods [30]. Using the 19-hour delayed shift of temperature from 37°C to 33°C, we carried out two sets of transfections of GFP, i.e., one with Lipofectamine 2000 and the other with calcium phosphate, with respect to its own control at 37°C. In each set of the experiments, the amount of the GFP plasmid for transfection was varied. Shown in Fig 3A are representative green fluorescence images from different cultures with varying plasmid amounts, whereas in Fig 3B, the results of these two sets of experiments were quantified using these images with respect to individual controls.

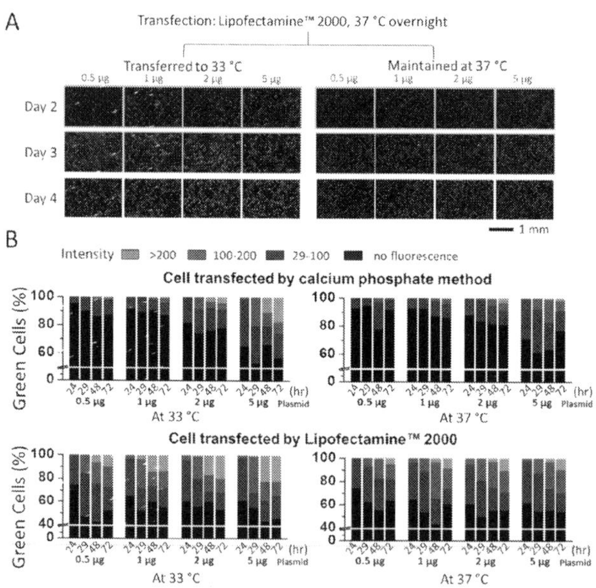

Figure 3: Comparison of different transfection methods and the amount of the GFP plasmid used for transfection.

(A) Enhancement of GFP expression in HEK-293S cells that grew at 33°C (on the left), as compared with 37°C (on the right). HEK-293S cells were transfected with varying amounts of the GFP plasmid, as labeled, using Lipofectamine 2000 or by the calcium phosphate method (photos not shown). A set of dishes (on the left) were transferred to a 33°C incubator, 24 hours after transfection and were maintained at 33°C. Shown are green fluorescence images from day 2 to day 4 (8.7x magnification, 1/5 second exposure time). (B) Percentage of the green cells from each of the dishes with varying amounts of the GFP plasmid and the fluorescence intensity from each of three intensity scales. The upper and lower panels are images from cells transfected using calcium phosphate and Lipofectamine 2000, respectively. The images on the right are controls (at 37°C) whereas the images on the left are cells maintained at 33°C (after overnight culture at 37°C). For determining the percentage of green cells, bright-field images for the same dish were also taken (images not shown). In both cases, 34.8x magnification images were used for data analysis. The white colored line represents the line-break. The green cells were further categorized into three levels of brightness. A typical HEK-293S cell in 34.8x magnification contained 3,000 ± 800 pixels. Based on the green fluorescence image, the pixels from the darkest green cell have an intensity of ≥29. Note the green intensity for each pixel in a JPEG file was recorded in an 8-bit byte, which was converted to a scale from 0 to 255. A cell with the average intensity under 29 indicated no GFP fluorescence, at least not appreciably, which we assigned as black color. A cell with an average brightness between 29 and 100 was labeled low green intensity, colored as dark green bar. A cell with an average brightness between 100 and 200 was considered medium intensity, colored as green. Bright green was the color used to label any cell with an average brightness between 200 and 255. The percentage of each green group was calculated based on the total cell count, and the total number of cells was determined from the bright-field view of the same viewing area (images not shown).

Four conclusions could be drawn from these data. First, when we compared the data horizontally in Fig 3B (or the same transfection method), we found that regardless of the amount of the plasmid we used, shifting the culture from 37°C to 33°C 24 hours after transfection resulted in higher GFP expression. Second, at 5 μg GFP plasmid culture, transfection efficiency seemed to reach the maximum (the transfection efficiency can be read out from the Y-axis in any of the panels in Fig 3B by either from the bottom of the darker green bar or 100%—the percentage value of the top of the black column: the black column represents the percentage of non-transfected cells). At this plasmid amount, the transfection efficiency between the two methods did not differ significantly. Specifically, with Lipofectamine 2000 the transfection efficiency reached ~60%, whereas with the calcium phosphate method the transfection efficiency was ~50%. In contrast, when we used a lower plasmid amount, we found a major difference in transfection efficiency. For example, in the 1 μg plasmid culture, the transfection efficiency was ~50% with the use of Lipofectamine 2000, as compared to <20% with the use of calcium phosphate method. We also observed a similar result even from 37°C control cultures (see the two corresponding columns on the two vertical panels in Fig 3B). Third, the exposure to and maintenance of the culture under the mild hypothermic condition appeared to alter only green fluorescence intensity without changing the percentage of the total fluorescence (Fig 3A and 3B), at least not appreciably. In other words, shifting the culture to 33°C did not enhance the transfection efficiency but did turn green cells greener—this is evident, for instance, from a taller, the bright green bar (with >200 intensity scale) from the 33°C experiment as compared with the same experiment at the 37°C control. This conclusion was true from the transfection experiment regardless of transfection reagent and the amount of the plasmid used, as well as the time in the culture (after transfection). Fourth, by using the average fluorescence intensity per pixel, we found that the fluorescence intensity increased by 1.2-fold (0.5 μg GFP experiment) and 1.6-fold (5 μg GFP experiment) from cells that were maintained at 33°C as compared with those that grew at 37°C, respectively. These values were calculated from the images taken at the 72 hour after transfection using calcium phosphate (Fig 3B, the 72 hr bars in the two upper panels). For the Lipofectamine 2000 method (the two lower panels at the 72 hr in Fig 3B), the fluorescence intensity increased by 1.5-fold (0.5 μg GFP experiment) and 1.2-fold (5 μg GFP experiment) from 33°C to 37°C, respectively. On average, therefore, GFP expression level was ~1.4-fold higher from cells that were subject to the mild hypothermic treatment.

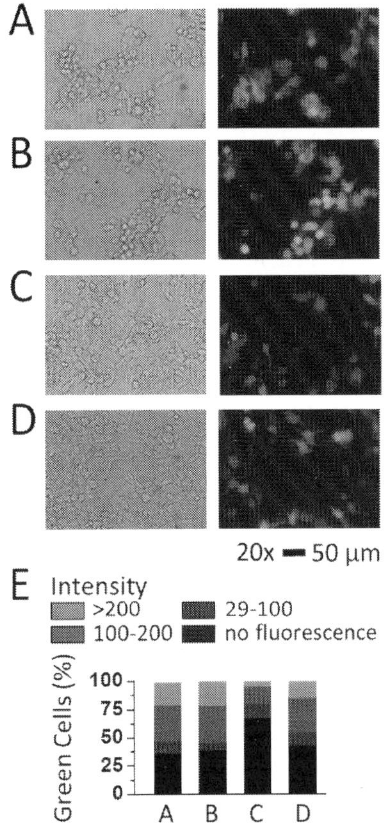

Figure 4: ransfection of the GFP plasmid in HEK-293S cells with different reagents and the expression of GFP at 33°C.

Using the same protocol (i.e., 19 hour after the medium replacement or 24 hour after transfection, lowering the culture temperature from 37°C to 33°C), we also tested additional transfection reagents. Shown in Fig 4 are three more experiments that included Lipofectamine LTX Plus, Metafectene EASY with opti-MEM medium and separately with complete medium. We found that the transfection efficiency for GFP in HEK-293S cells was ~61% for Lipofectamine LTX Plus, ~32% for Metafectene EASY with opti-MEM medium and ~57% for Metafectene EASY with complete medium. As the control for this set of experiment, Lipofectamine 2000 reached over 60% transfection efficiency. Although Lipofectamine LTX Plus and Metafectene EASY were supposed to provide better gene expression in commonly used cell lines, our data indicated that Lipofectaime 2000 was perhaps the preferred reagent for the use of HEK-293S cell line. It should be noted that Metafectene EASY did not work well in the Opti-MEM medium or a reduced serum medium in HEK-293S cells, whereas the same reagent with the complete medium almost doubled the transfection efficiency (Fig 4E).

Bright-field and fluorescence images (34.8x magnification, scale bar = 50 μm) of HEK-293S cells taken at the 48th hour after transfection. Transfection of HEK-293S cells as in images (A) to (D) was carried out using Lipofectamine 2000, Lipofectamine LTX & PLUS, Metafectene EASY with Opti-MEM and Metafectene EASY with DMEM with 10% FBS, respectively. In each of the transfections, 2 μg GFP plasmid for a 35 mm Petri dish was used. On day 1, all transfected dishes were incubated at 37°C; on day 2, these dishes were transferred to a 33°C incubator. Based on these images, the transfection efficiency for transient expression of GFP was determined to be ~64%, ~61%, ~32% and ~57% from (A) to (D), respectively. (E) The green cells were categorized into high (intensity > 200), middle (intensity 100–200), and low (intensity 29–100) three groups. The black color indicates cells visible in the bright view but can't been observed under UV (intensity < 29). The percentage of each group was plotted in stacked columns.

Time course of GFP expression under the mild hypothermic culture condition

We also followed the time course (days) of GFP expression and the enhancement of expression. Shown in Fig 5 as an example (where calcium phosphate transfection method was used), maintaining the culture at 33°C 24 hours after transfection, which is represented by the red line (Fig 5A), inhibited the growth rate (Fig 5B), as compared with the 37°C control culture (black line). Although the total cell count from the delayed (red line) and the immediate temperature shift to 33°C (blue line) after transfection was roughly similar at day 1 (Fig 5B), the delay temperature shift protocol resulted more green cells at the peak time (i.e., days 2 and 3) than the culture shifted to 33°C immediately after the medium replacement (red vs. blue line in Fig 5C). Those green cells in fact remained longer than those maintained at 37°C (red vs. black line in Fig 5C). Furthermore, at 33°C, green cells remained essentially undivided even at the end of the 6 day period, as compared with the 37°C culture (red line vs. black line in Fig 5D). Taken together, the delayed temperature shift protocol led to the enhancement of GFP expression with almost identical biomass on day 2 (Fig 5B and 5C, the black and red line at the 48th hour). Yet a mild hypothermic culture condition (i.e., 33°C) promoted a healthy temperature environment where a higher percentage of green cells survived longer (Fig 5C).

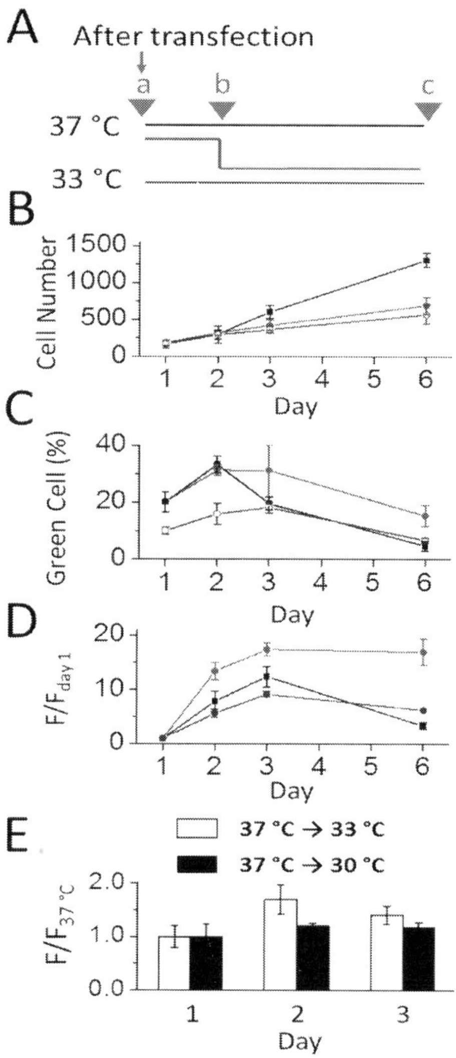

Figure 5: ime course of GFP expression in HEK-293S cells at 30°C, 33°C and 37°C.

(A) Point 'a' represents the 5th hour after transfection at which the medium was replaced (all the data were collected from cells transfected with 2 μg GFP plasmid/35 mm dish using calcium phosphate). Three sets of experiments began as reflected by three color-coded lines as in panels A-D, and in each set, triplicate dishes were used in our experiments. Specifically, blue line represents maintaining a culture at a 33°C incubator. Red line refers to the culture that first stayed at a 37°C incubator overnight (or 19 hours precisely) and then transferred to the 33°C incubator; this was labeled as point 'b' (note that point b correlates to day 1 in panels B-D). Black line represents the control (37°C culture). Point 'c' was the time (day 6) that the experiment ended. (B) The number of cells was counted from 34.8x bright-field images taken at days 1, 2, 3, and 6. Each data point was the average cell number counts of three 34.8x images taken at randomly selected locations of a dish. (C) The percentage of green cells in each experiment. Note that all data points were collected from the 34.8x images as in B. Both bright-field and fluorescence images were used for the determination of the transfection efficiency for each experiment. (D) The ratio of the green fluorescence intensity ($F/F_{day\ 1}$). $F_{day\ 1}$ is the fluorescence intensity for each culture at the 24th hour. F stands for the fluorescence intensity at any particular time (day). All of the $F/F_{day\ 1}$ values were collected from the fluorescence images as in (C). (E) The plot of the $F/F_{37°C}$ values for days 1–3. All of the cells used for these experiments were transfected with GFP and maintained at 37°C for one day. Then, one set of the dishes were transferred to 33°C (hollow square), whereas another set dishes were transferred to 30°C (solid square). Therefore, this plot displaces the differential GFP expression on two different temperature scales, i.e., 33°C and 30°C, relative to 37°C.

A further examination of the time course of green cell counts and fluorescence intensity (Fig 5) showed that a 19-hour delay of lowering culture temperature or the overnight recovery of the transfected culture at 37°C was essential in the use of this biphasic temperature culture protocol to enhance protein expression. This point becomes apparent when the time course of a delay temperature shift is compared with the immediate temperature drop from 37 to 33°C after the medium replacement (5 hours after transfection). As seen in Fig 5B, there was little difference in the total cell counts in both temperature change processes (red and blue lines) in the 6-day period. However, an immediate temperature drop led to the reduction of the transfection efficiency by roughly 50% (blue and red lines in Fig 5C). Furthermore, among the green cells expressed, the overall green fluorescence intensity did not even reach to the peak level of the control (blue line vs. black line for up to day 3 in Fig 5D). In fact, these results were consistent with the fluorescence images taken at 48 hours in Fig 2. Together, these results demonstrated that a 19-hour delay after medium replacement or a 24-hour delay after transfection before lowering the culture temperature from 37 to 33°C was critical for an increase in protein production. Immediate temperature drop after medium replacement or 5 hours after transfection adversely affected both transfection efficiency and protein production in HEK-293S cells. These results are consistent with the notion that growth arrest is unrelated to a higher protein expression at a sub-physiological culture temperature [23,24]. However, we do not yet understand the molecular mechanism of this delayed shift of culture temperature to 33°C in enhancing protein expression.

Because lowering culture temperature to 33°C led to a higher GFP expression, we asked whether dropping temperature even lower, i.e., to 30°C, would further increase GFP expression. Surprisingly, however, we found the opposite. Fig 5E is a comparison of the enhancement of GFP production in HEK-293S cells, represented by the ratio of the fluorescence intensity normalized to the control or 37°C culture. As seen, using the same mild hypothermic protocol, i.e., a 19-hr delayed shift to the respective temperature, green cells from a culture that was subject to 30°C essentially had a similar level of overall fluorescence intensity as the control. In contrast, at 33°C, HEK-293S cells produced more GFP than the control. These data suggested that lowering the culture temperature to 33°C, but not lower, led to a higher GFP expression. Furthermore, 30°C caused a dramatic reduction of cell counts, as we have described earlier (Fig 1).

When cells grew at 37°C, the number of green cells dropped more dramatically from day 2 to day 6 than cells growing either at 33°C or 30°C (Fig 5C, black line vs. either red or blue line). We note, however, that this reduction of green cell numbers was not due to cell death; in fact all cells continued to grow (Fig 5B), as the total cell counts continued to increase. Rather, the reduction of the green cells was due to that green cells began to divide as well, thereby "diluting" the percentage of green cells. Within the first three days, we did not observe any difference in cell viability. However, because we were interested in following the time course longer than 3 days for establishing the optimal condition, we replaced the cell culture media on day 4. From our data (Fig 5C and 5D), it is now clear, however, that within three days after transfection, the cells should be used, because the protein expression is peaked.

Testing the mild hypothermic culture condition with PC12 cells

To show the utility of the biphasic temperature culture protocol, which we established using HEK-293S cells, we further tested the same protocol with PC12 cells. PC12 cell line is derived from pheochromocytoma in the rat adrenal medulla and grows in culture as undifferentiated neuroblasts [39]. This cell line is a commonly used model system for studying neuronal development and function [40]. However, PC12 cells are known to be difficult to transfect by various transfection techniques, including chemical-based transfection methods [41]. Here we expressed GFP in PC12 cells, using Lipofectamine 2000, and one day later we shifted the culture to 33°C (Fig 6). By measuring the green fluorescence intensity, we first achieved ~6% transfection efficiency under the mild hypothermic culture condition (Fig 6B). In contrast, the transfection efficiency of the control culture, which was maintained at 37°C, was barely measurable (i.e., ~2%) (Fig 6A). When we used trypsin to treat the culture during the passage (or one day prior to transfection), which presumably helped dissociate cell clumps into single cells, the transfection efficiency reached ~18% under the mild hypothermic culture condition (Fig 6C) [41]. These results demonstrated that a delayed shift of culture temperature to 33°C could be useful for other types of cells and/or cells that are difficult to transfect and/or the yield of transfection efficiency is very low.

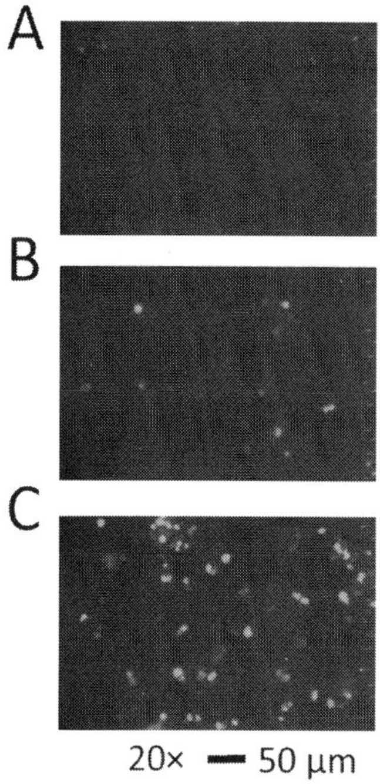

Figure 6: Fluorescence images of GFP-expressing PC12 cells.

PC12 cells were transfected with 1 μg enhanced GFP plasmid DNA in 35 mm dishes using Lipofectamine 2000. Green fluorescence images were taken (34.8x magnification, 1/3 second exposure time, scale bars = 50 μm). In (A), the cells were maintained at 37°C, whereas in (B) the cells were transferred to 33°C on day 2. The images were taken at 96 hours after transfection. The percentage of the green cells was ~2% and ~6% for (A) and (B), respectively. In (C), when the cells were trypsinized during the passage or one day before transfection, the percentage of green cells was ~18%. As the comparison, the cells in (B) were not trypsinized. Like the cells in (B), the cells in (C) were also transferred to 33°C one day later.

Expression of AMPA receptors under the mild hypothermic culture condition and assay of the receptor property using whole-cell recording

We next tested if the same mild hypothermic culture condition could enhance membrane protein expression. Here we separately expressed two AMPA receptor subunits, GluA2 and GluA3 [27,42]. Transient expression of each subunit in HEK-293S cells leads to the formation of a functional, homomeric channel, and the channel property can be characterized using whole-cell recording [31,32] (note that the GluA2 channel we expressed was the unedited isoform in the Q/R editing site). Furthermore, we co-expressed GFP using a separate plasmid as a *cell marker* for whole-cell recording assay, because a green-colored cell would most likely express an AMPA receptor (the ratio of the plasmid used for GFP to an AMPA receptor was 1:5). The purpose of this experiment was to determine whether we could see an enhancement of AMPA receptors at a single-cell level.

Fig 7A is a pair of representative whole-cell current traces from two randomly chosen cells cultured on two different temperature scales. Based on the recording of ~30 cells from each culture for each receptor, we found that the whole-cell current amplitude increased by 1.4 ± 0.2 fold for GluA2Q (Fig 7B) and 1.8 ± 0.4 for GluA3 (Fig 7C) from the cells that were subject to the biphasic temperature culture, respectively, as compared to the control at 37°C for each receptor. Assuming the current amplitude was linearly proportional to the copy number of the receptors expressed on the cell surface, the range of the current amplitude increase suggested the HEK-293S cells produced on average ~1.6-fold more receptors when these cells were cultured at 33°C 24 hours after transfection than those cells maintained at 37°C. This value is consistent with what we found from the GFP experiment in Fig 3B: GFP expression level was ~1.4-fold higher from cells that were subject to the mild hypothermic treatment (the summary of these data is shown in Table 1). That AMPA receptor current amplitude measured at the single-cell level was higher from the cells growing at 33°C was also consistent with the conclusion from the GFP experiment, where the increase of green fluorescence intensity was attributed to the increase of GFP expression from "green" cells, rather than the increase of transfection efficiency. Therefore, ~1.4-fold higher GFP intensity reflected actually the increase of GFP expression at the single-cell (green cell) level.

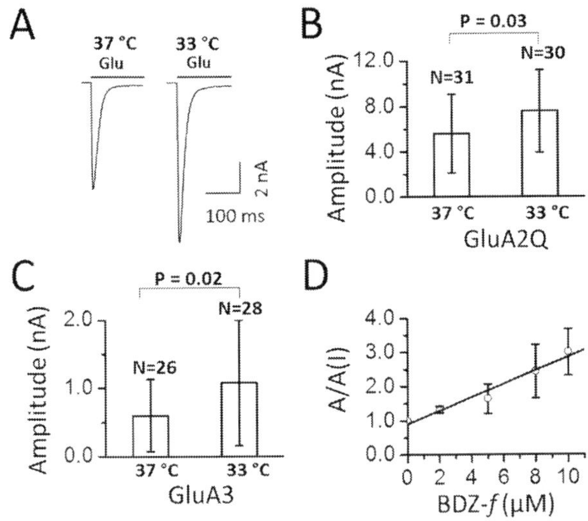

Figure 7: hole-cell recording assay of AMPA receptors expressed in HEK-293S cells grown at 37°C and 33°C.

All of the cells used for the assays were from 48 hours after transfection. (A) Representative traces of 3 mM glutamate induced whole-cell current from HEK-293S cells that transiently transfected with GluA2Q AMPA receptors. The cells cultured at 33°C were maintained at 37°C overnight first. (B) The average current amplitude from GluA2Q channels induced by 3 mM glutamate was shown for each culture. The number of the cells used for the recording was listed on top of each column. The error bar represents standard deviation from the mean. A two sample *t*-test on the whole-cell current data showed a significant difference between the cells cultured at 33°C and those cultured at 37°C (*P* value = 0.03). (C) The average whole-cell current amplitude from the GluA3 AMPA receptor channels induced by 3 mM glutamate. The number of cells used for the recording was listed on top of each column. Similarly, a two sample *t*-test also showed a significant difference between the data collected on the cells cultured at 33°C and those cultured at 37°C (*P* = 0.02). (D) The effect of BDZ-*f* on the GluA2Q AMPA receptor whole-cell current amplitude. The cells that expressed the channel were from the 33°C culture. The K_I value was found to be 5.1 ± 0.5 μM, determined from the slope (see detail in text).

Table 1: AMPA receptor properties.

	Amplitude (nA)		k_{des} (s^{-1})		t_{rise} (ms)	
	37°C	33°C	37°C	33°C	37°C	33°C
GluA2[a]	5.6 ± 0.6	7.6 ± 0.7	108 ± 5	112 ± 5	1.2 ± 0.1	1.1 ± 0.1
GluA3	0.6 ± 0.1	1.1 ± 0.2	188 ± 7	192 ± 7	1.9 ± 0.1	1.6 ± 0.1

[a]For GluA2, the data were the average from 31 and 30 cells cultured at 37°C and 33°C, respectively. For GluA3, the data were the average from 26 and 28 cells cultured at 37°C and 33°C, respectively.

Also found in Table 1 is the comparison of the first-order rate constant of channel desensitization or k_{des} and the current rise time or the time it took for the whole-cell current to reach the peak amplitude (10–90%). No difference was observed in both parameters for the same receptor but from the cells grown normally at 37°C and those grown at 33°C for one day (or 48 hours after transfection). We also measured the inhibition constant for (-)-1-(4-aminophenyl)-4-methyl-7,8-methylenedioxy-4,5-dihydro-3-methylcarbamoyl-2,3-benzodiazepine or BDZ-*f* with GluA2Q [43]. BDZ-*f* is a known noncompetitive inhibitor of AMPA receptors [43]. The inhibition constant (K_I) was found to be 5.1 ± 0.5 μM for the open-channel receptor conformation (Fig 7D), from the plot of the current amplitude in the absence and presence of an inhibitor or A/A(I) [43]. The K_I value was in good agreement with the inhibition constant of 5.4 ± 0.8 μM, which we reported before when we used the same receptor transiently expressed and cultured at 37°C [43]. All of these data (Fig 7 and Table 1) therefore illustrated that the HEK-293S cells expressed the same receptor with the same property regardless of whether these cells were grown at 37°C or the cells were shifted to a 33°C temperature environment 24 hours after transfection, except that the cells that were subject to 33°C treatment exhibited higher whole-cell current amplitude.

DISCUSSION

Animal cells and cell lines, such as HEK-293 cells, are commonly cultured at 37°C. However, as we have demonstrated in this study, lowering the culture temperature to 33°C 24 hours after transient transfection will give rise to ~1.5-fold higher expression of a recombinant protein, such as GFP and AMPA receptors. By following the time course of the GFP-expressing cells growing at 37°C normally and 33°C from 24 hours after transfection (including 19 hours recovery at 37°C in the normal growth medium), we found that a mild hypothermia (i.e., 33°C) increases cellular productivity of recombinant proteins while it reduces cell growth rate and suppresses cell division. As a result, more green cells remain undivided in a longer period of time. The property of a recombinant protein is unaffected, when the HEK-293S cells that express the protein are treated with a mild hypothermia, as shown by the use of two different AMPA receptors. Using PC 12 cells, we have further demonstrated that this method may be especially useful when a recombinant protein is difficult to express using a chemical-based transfection approach.

One important finding from our study is that the transfected culture after medium replacement must return to 37°C for overnight incubation or recovery (or 24 hours after transfection) before the culture temperature can be lowered to 33°C. Without this delayed temperature shift, both the transfection efficiency in HEK-293S cells and the total protein production (or the total intensity of GFP green fluorescence among all green cells) will be significantly reduced (Fig 5Cand 5D), while the number of cells remains essentially the same (Fig 5B). While the molecular mechanism of this phenomenon is unclear, it appears that the initial translation of transiently transfected proteins is sensitive to temperature. Lowering it to 33°C too early or without an overnight incubation at 37°C will result in repression of cellular translation. Furthermore, even with an overnight delay but with the culture temperature lowered to 30°C, the cellular translation is similarly suppressed (Fig 5E). These data demonstrate that lowering temperature to 33°C, but not lower, can augment protein expression in HEK-293S cells, only when the transfected culture is allowed to "recover" at 37°C overnight. These phenomena are consistent with the hypothesis that growth arrest under mild hypothermic conditions (30–35°C) and the associated higher yield of protein expression in mammalian cells are two independent cellular responses [23,24].

As we have shown, HEK-293S cells that are subject to a delayed shift of culture temperature to 33°C one day after transfection produce ~1.5-fold higher expression of a recombinant protein. Such an enhancement is independent of transfection reagents, based on the test of calcium phosphate, Lipofectamine 2000, Lipofectamine LTX Plus, and Metafectene EASY. That said, Lipofectamine 2000 seems to be a better reagent when a lower amount of the cDNA is used for transfection (for GFP this amount can be as low as 0.5 μg/35 mm Petri dish). However, when a sufficient quantity of the DNA plasmid is used (e.g., 5 μg GFP/35 mm Petri dish), the calcium phosphate method is as efficient as Lipofectamine 2000. Because lower temperature does not affect the transfection efficiency, but reduces significantly the rate of cell growth, the increase in protein expression should be attributed to the increase of protein expression at the single cell level. This is consistent with our observation in that (a) green cells become greener when GFP is expressed in HEK-293S cell, and (b) cells expressing AMPA receptors at lower temperature exhibit higher whole-cell current amplitude. The phenomenon can be best explained by the notion that low temperature induces cell growth suppression due to cell cycle arrest at the G(0)/G(1) phase [17,18,21,22]. In fact, cell cycle arrest, particularly in G1 phase, typically using a low temperature culture, has been a well-practiced means to enhance protein yield in production cultures, such as in CHO cells [6].

The method we have described should be simple to implement. For static culture or a small-scale culture, two incubators with two culture temperatures (i.e., one is at 37°C and the other is at 33°C) may be the best way. We have shown previously that using HEK-293S cells in static culture is suitable for electrophysiological measurements, where only the dish-attached cells are selected to make gigaohm seals (in static culture, the majority of 293S cells are attached to Petri dishes, albeit more loosely than either regular 293 or 293T cells) [33]. Furthermore, that the mild hypothermic treatment of a static culture transfected with AMPA receptors can enhance receptor expression on a single cell level by up to ~1.5-fold is also consistent with the level of GFP increase when green fluorescence is used as a "reporter signal". Therefore, it appears that between 37°C and 33°C, the biosynthesis of AMPA receptors, their processing, surface delivery and turnover are not affected or they are not affected differentially. It is known, for example, the endocytosis can be affected by lowering temperature, but temperature will have to be lowered to a single digit [44]. Therefore, our method should be applicable for other membrane proteins, such as other channel proteins. Although we did not test the suspension culture using HEK-293S cells, this cell line is suitable for suspension culture in shake flasks and/or bioreactors for a large-scale production of proteins per unit culture volume [35,45,46]. In suspension culture, changing temperature without changing culture vessels is feasible, because the response of cells to temperature is independent of culture size. Finally, the use of 293S cells may be especially advantageous for preparing membrane proteins, because this cell line is known to express proteins whose activities are modulated by complex post-translation, a feature often observed for membrane proteins [47].

ACKNOWLEDGMENTS

This work was supported in part by grants from the National Institutes of Health and Muscular Dystrophy Association.

AUTHOR CONTRIBUTIONS

Conceived and designed the experiments: CYL ZH LN. Performed the experiments: CYL ZH WW AW CZW. Analyzed the data: CYL ZH WW AW CZW. Contributed reagents/materials/analysis tools: CYL ZH. Wrote the paper: CYL ZH LN.

REFERENCES

1. Graham FL, Smiley J, Russell WC, Nairn R (1977) Characteristics of a human cell line transformed by DNA from human adenovirus type 5. J Gen Virol 36: 59–74. pmid:886304
2. Thomas P, Smart TG (2005) HEK293 cell line: a vehicle for the expression of recombinant proteins. Journal of pharmacological and toxicological methods 51: 187–200. pmid:15862464
3. Standfuss J, Xie G, Edwards PC, Burghammer M, Oprian DD, Schertler GF. (2007) Crystal structure of a thermally stable rhodopsin mutant. Journal of molecular biology 372: 1179–1188. pmid:17825322
4. Gruswitz F, Chaudhary S, Ho JD, Schlessinger A, Pezeshki B, Ho CM, et al (2010) Function of human Rh based on structure of RhCG at 2.1 A. Proceedings of the National Academy of Sciences of the United States of America 107: 9638–9643. pmid:20457942
5. Penmatsa A, Wang KH, Gouaux E (2013) X-ray structure of dopamine transporter elucidates antidepressant mechanism. Nature 503: 85–90. pmid:24037379
6. Kim JY, Kim YG, Lee GM (2012) CHO cells in biotechnology for production of recombinant proteins: current state and further

potential. Applied microbiology and biotechnology 93: 917–930. pmid:22159888

7. Rao PN, Engelberg J (1965) Hela Cells: Effects of Temperature on the Life Cycle. Science 148: 1092–1094. pmid:14289609
8. Watanabe I, Okada S (1967) Effects of temperature on growth rate of cultured mammalian cells (L5178Y). The Journal of cell biology 32: 309–323. pmid:10976224
9. Weidemann R, Ludwig A, Kretzmer G (1994) Low temperature cultivation—a step towards process optimisation. Cytotechnology 15: 111–116. pmid:7765923
10. Borth N, Heider R, Assadian A, Katinger H (1992) Growth and production kinetics of human x mouse and mouse hybridoma cells at reduced temperature and serum content. Journal of biotechnology 25: 319–331. pmid:1382458
11. .Bloemkolk JW, Gray MR, Merchant F, Mosmann TR (1992) Effect of temperature on hybridoma cell cycle and MAb production. Biotechnology and bioengineering 40: 427–431. pmid:18601133
12. Reuveny S, Velez D, Macmillan JD, Miller L (1986) Factors affecting cell growth and monoclonal antibody production in stirred reactors. Journal of immunological methods 86: 53–59. pmid:3944469
13. Sureshkumar GK, Mutharasan R (1991) The influence of temperature on a mouse-mouse hybridoma growth and monoclonal antibody production. Biotechnology and bioengineering 37: 292–295. pmid:18597368
14. Barnabe N, Butler M (1994) Effect of temperature on nucleotide pools and monoclonal antibody production in a mouse hybridoma. Biotechnology and bioengineering 44: 1235–1245. pmid:18618550
15. Ducommun P, Ruffieux P, Kadouri A, von Stockar U, Marison IW (2002) Monitoring of temperature effects on animal cell metabolism in a packed bed process. Biotechnology and bioengineering 77: 838–842. pmid:11835145
16. Furukawa K, Ohsuye K (1998) Effect of culture temperature on a recombinant CHO cell line producing a C-terminal alpha-amidating enzyme. Cytotechnology 26: 153–164. pmid:22358553
17. Kaufmann H, Mazur X, Fussenegger M, Bailey JE (1999)

Influence of low temperature on productivity, proteome and protein phosphorylation of CHO cells. Biotechnology and bioengineering 63: 573–582. pmid:10397813

18. Yoon SK, Song JY, Lee GM (2003) Effect of low culture temperature on specific productivity, transcription level, and heterogeneity of erythropoietin in Chinese hamster ovary cells. Biotechnology and bioengineering 82: 289–298. pmid:12599255
19. Fox SR, Patel UA, Yap MG, Wang DI (2004) Maximizing interferon-gamma production by Chinese hamster ovary cells through temperature shift optimization: experimental and modeling. Biotechnology and bioengineering 85: 177–184. pmid:14705000
20. Trummer E, Fauland K, Seidinger S, Schriebl K, Lattenmayer C, Schertler GF, et al. (2006) Process parameter shifting: Part II. Biphasic cultivation-A tool for enhancing the volumetric productivity of batch processes using Epo-Fc expressing CHO cells. Biotechnology and bioengineering 94: 1045–1052. pmid:16736532
21. Rieder CL, Cole RW (2002) Cold-shock and the Mammalian cell cycle. Cell cycle 1: 169–175. pmid:12429927
22. al-Rubeai M, Emery AN (1990) Mechanisms and kinetics of monoclonal antibody synthesis and secretion in synchronous and asynchronous hybridoma cell cultures. Journal of biotechnology 16: 67–85. pmid:1366816
23. Fox SR, Yap MX, Yap MG, Wang DI (2005) Active hypothermic growth: a novel means for increasing total interferon-gamma production by Chinese-hamster ovary cells. Biotechnology and applied biochemistry 41: 265–272. pmid:15504104
24. Sunley K, Tharmalingam T, Butler M (2008) CHO cells adapted to hypothermic growth produce high yields of recombinant beta-interferon. Biotechnology progress 24: 898–906. pmid:19194899
25. Sunley K, Butler M (2010) Strategies for the enhancement of recombinant protein production from mammalian cells by growth arrest. Biotechnology advances 28: 385–394. pmid:20156545
26. Chalfie M, Tu Y, Euskirchen G, Ward WW, Prasher DC (1994) Green fluorescent protein as a marker for gene expression. Science 263: 802–805. pmid:8303295

27. Hollmann M, Heinemann S (1994) Cloned glutamate receptors. Annu Rev Neurosci 17: 31–108. pmid:8210177
28. Seeburg PH (1993) The TINS/TiPS Lecture. The molecular biology of mammalian glutamate receptor channels. Trends Neurosci 16: 359–365. pmid:7694406
29. Traynelis SF, Wollmuth LP, McBain CJ, Menniti FS, Vance KM, Ogden KK, et al. (2010) Glutamate receptor ion channels: structure, regulation, and function. Pharmacological reviews 62: 405–496. pmid:20716669
30. Chen C, Okayama H (1987) High-efficiency transformation of mammalian cells by plasmid DNA. Mol Cell Biol 7: 2745–2752. pmid:3670292
31. Li G, Pei W, Niu L (2003) Channel-opening kinetics of GluR2Q(flip) AMPA receptor: a laser-pulse photolysis study. Biochemistry 42: 12358–12366. pmid:14567697
32. Pei W, Huang Z, Niu L (2007) GluR3 flip and flop: differences in channel opening kinetics. Biochemistry 46: 2027–2036. pmid:17256974
33. Huang Z, Li G, Pei W, Sosa LA, Niu L (2005) Enhancing protein expression in single HEK 293 cells. J Neurosci Methods 142: 159–166. pmid:15652630
34. Ali SH, DeCaprio JA (2001) Cellular transformation by SV40 large T antigen: interaction with host proteins. Semin Cancer Biol 11: 15–23. pmid:11243895
35. Stillman BW, Gluzman Y (1985) Replication and supercoiling of simian virus 40 DNA in cell extracts from human cells. Mol Cell Biol 5: 2051–2060. pmid:3018548
36. Hamill OP, Marty A, Neher E, Sakmann B, Sigworth FJ (1981) Improved patch-clamp techniques for high-resolution current recording from cells and cell-free membrane patches. Pflugers Arch 391: 85–100. pmid:6270629
37. Udgaonkar JB, Hess GP (1987) Chemical kinetic measurements of a mammalian acetylcholine receptor by a fast-reaction technique. Proc Natl Acad Sci U S A 84: 8758–8762. pmid:2447583
38. Graham FL, van der Eb AJ (1973) A new technique for the assay

of infectivity of human adenovirus 5 DNA. Virology 52: 456–467. pmid:4705382

39. Greene LA, Tischler AS (1976) Establishment of a noradrenergic clonal line of rat adrenal pheochromocytoma cells which respond to nerve growth factor. Proceedings of the National Academy of Sciences of the United States of America 73: 2424–2428. pmid:1065897
40. Grau CM, Greene LA (2012) Use of PC12 cells and rat superior cervical ganglion sympathetic neurons as models for neuroprotective assays relevant to Parkinson's disease. Methods in molecular biology 846: 201–211. pmid:22367813
41. Lee JH, Ahn HH, Kim KS, Lee JY, Kim MS, Lee B, et al. (2008) Polyethyleneimine-mediated gene delivery into rat pheochromocytoma PC-12 cells. Journal of tissue engineering and regenerative medicine 2: 288–295. pmid:18512266
42. Dingledine R, Borges K, Bowie D, Traynelis SF (1999) The glutamate receptor ion channels. Pharmacol Rev 51: 7–61. pmid:10049997
43. Wang C, Sheng Z, Niu L (2011) Mechanism of inhibition of the GluA2 AMPA receptor channel opening: consequences of adding an N-3 methylcarbamoyl group to the diazepine ring of 2,3-benzodiazepine derivatives. Biochemistry 50: 7284–7293. pmid:21751782
44. Steinman RM, Mellman IS, Muller WA, Cohn ZA (1983) Endocytosis and the recycling of plasma membrane. The Journal of cell biology 96: 1–27. pmid:6298247
45. Standfuss J, Edwards PC, D'Antona A, Fransen M, Xie G, Oprian DD, et al. (2011) The structural basis of agonist-induced activation in constitutively active rhodopsin. Nature 471: 656–660. pmid:21389983
46. Reeves PJ, Kim JM, Khorana HG (2002) Structure and function in rhodopsin: a tetracycline-inducible system in stable mammalian cell lines for high-level expression of opsin mutants. Proceedings of the National Academy of Sciences of the United States of America 99: 13413–13418. pmid:12370422
47. Berg DT, McClure DB, Grinnell BW (1993) High-level expression of secreted proteins from cells adapted to serum-free suspension culture. Biotechniques 14: 972–978. pmid:8333965

CHAPTER

6

AN EFFICIENT 3D CELL CULTURE METHOD ON BIOMIMETIC NANOSTRUCTURED GRIDS

Maria Wolun-Cholewa[1], Krzysztof Langer[2], Krzysztof Szymanowski[3], Aleksandra Glodek[1], Anna Jankowska[1], Wojciech Warchol[4], Jerzy Langer[2]

[1]Department of Cell Biology, Poznan University of Medical Science, Poznan, Poland,

[2]Laboratory for Materials Physicochemistry and Nanotechnology, Adam Mickiewicz University, Srem, Poland,

[3]Department of Mother's and Child's Health, Poznan University of Medical Science, Poznan, Poland,

[4]Department of Biophysics, Poznan University of Medical Science, Poznan, Poland

ABSTRACT

Current techniques of *in vitro* cell cultures are able to mimic the *in vivo* environment only to a limited extent, as they enable cells to grow only in two dimensions. Therefore cell culture approaches should rely on scaffolds that provide support comparable to the extracellular matrix. Here we demonstrate the advantages of novel nanostructured three-dimensional grids fabricated using electro-spinning technique, as scaffolds for cultures of neoplastic cells. The results of the study show that the fibers allow for a dynamic growth of HeLa cells, which form multi-layer structures of symmetrical and spherical character. This indicates that the applied scaffolds are nontoxic and allow proper flow of oxygen, nutrients, and growth factors. In addition, grids have been proven to be useful in *in situ* examination of cells ultrastructure.

INTRODUCTION

Cultures of human or animal cells in *in vitro* conditions are usually performed for their identification, proliferation or cell death assessment. Frequently, initial evaluations are carried out in two-dimensional (2D) cell culture systems. Despite their widespread use, observations made in 2D systems do not correspond to the results of *in vivo* studies [1]. Thus, currently, researchers are using three dimensional (3D) cultures that mimic the *in vivo* environment more accurately [2]. Cells grown in three-dimensional cultures are more valid targets for discovering and testing of new drugs for cancer treatment [3]. Contrary to 2D culture systems, 3D *in vitro*models have the potential to provide insights into cellular functions such as: differentiation, migration, and gene expression in a controlled and well-defined manner [4]–[6]. 3D cell culture main feature is the ability to mimic the extracellular matrix (ECM) conditions. Materials used for 3D cell culture systems production include both natural and synthetic biopolymers. Lately also biodegradable materials were introduced in scaffold construction, however it was proven that their stability in liquid environment is limited [5], [7]–[14].

One of the methods for fiber-based 3D scaffolds production is electro-spinning [9]. An advantage of those nanostructured grids, which differentiates them from other types of scaffolds, is the reduced diameter of pore size [15]–[16]. Moreover, electrospun scaffolds are built from very small fibers which create large surface areas [1], [9], [13], [15]. This allows for more accurate evaluation of cell proliferation [9], [11], [17].

In this study, electrospun nanostructured fibers in the form of spatial nanostructured 3D grids, fabricated from a polymer mixture, including polyaniline, were examined as a potential tool for 3D culture *in vitro*.

MATERIALS AND METHODS

Synthesis of Polyaniline

6.9 g of aniline hydrochloride was dissolved in 300 ml of 1 M HCl. The solution of aniline hydrochloride was combined with a solution of 11.4 g ammonium persulphate dissolved in 200 ml of 1 M HCl. The mixture was left at room temperature for 12 hours with continuous mixing. Next the product was filtered, washed with water, methanol and chloroform. Polyaniline (PANI) was transferred into 200 ml of chloroform and dispersed to a suspension using ultrasound bath.

Fabrication of Nanofibers using Electro-spinning

150 mg of polystyrene was dissolved in 2 ml of chloroform, gradually mixed with 50 mg of poly (ethylene oxide) – PEO with continuous mixing. Subsequently, the polystyrene and PEO were mixed with each other and supplemented with 8 ml of the previously obtained polyaniline – PANI suspension (protects against microbiological contamination), with mixing in order to obtain a uniform suspension of polyaniline in the solution of polymers. The process of electro-spinning was conducted in a non-uniform electric field under the voltage of 4.5 kV, inter-electrode distance of 15 cm and polymer solution outflow rate of 0.2 mL/min. Obtained nano- and microfibre random networks were transferred from the electrode into plastic carrier frames, forming final grids.

All reagents used for the synthesis of polyaniline and fabrication of nanofibers using electro-spinning were purchased from Sigma-Aldrich, St Louis, MO, USA.

Cell Culture

Human HeLa cervical epithelial cells (ATCC CCL-2) were purchased from the American Type Cell Culture (ATCC, Manassas, MA). HeLa cells were cultured in standard conditions: RPMI medium (PAN Biotech-Gmbh, Aidenbach, Germany) supplemented with 10% fetal calf serum (FCS; Sigma, St Louis, MO, USA), 2 mmol/L L-glutamine (Cambrex, Charles City, IA, USA), 100 IU/mL penicillin and streptomycin solution (Sigma-Aldrich, St Louis, MO, USA) in a 5% CO2-humidified atmosphere at 37°C.

In order to examine the usefulness of nanostructural grids as a tool for 3D cultures, the cells were seeded on grids at a density of 10 000 cells. Subsequently, the grids with cells were placed into fresh culture medium and after 24, 48 or 72 hours cells morphology and their viability were examined. Every experiment was repeated ten times.

Cell Viability Assessment

The viability of HeLa cells was analyzed using the XTT colorimetric test, based on the reduction of XTT compound (tetrazoline-2,3-bis(2-methoxy-4-nitro-5-sulphophenyl)-2H-5-carboxyanilide, Sigma-Aldrich, St Louis, MO, USA) by living cells. Optical density of colour product was measured at the wavelength of 450 nm.

The percentage of mitochondrial activity was calculated according to the following equation: (OD of nanostructured grid with cells−OD of medium alone)/(OD of nanostructured grid only−OD medium alone)×100; where OD is optical density [18]. Statistical analysis of the results was done with the Kruskal-Wallis test with Dunn's post-test using STATISTICA ver.5 software (Statsoft, Krakow, Poland). *P* value less than 0.05 was considered statistically significant.

Microscopy

Fluorescence microscopy.

After 24, 48 or 72 hours of culture on the grids HeLa cells were stained with 0.1 μg/ml of Hoechst 33342 and 0.125 μg/ml of propidium iodide (Sigma-Aldrich, St Louis, MO, USA). The presence of intact, apoptotic and/or necrotic cells was evaluated using Nicon Diaphot Eclipse TE 200 fluorescence microscope, equipped with UV-2A and FITC/FLUO-3 filters [19].

Confocal microscopy.

In order to visualize the 3D construct the cells were pre-fixed for 10 minutes in 4% paraformaldehyde (Sigma-Aldrich, St Louis, MO, USA) and stained with 0.125 μg/ml of propidium iodide (Sigma-Aldrich, St Louis, MO, USA). Signals were excited at 543 nm wavelength and fluorescence emission was selected with 560 nm bandpass filter. Images of cells were represented as orthogonal projections of 20–60 optical sections in 0.3 μm increments using Zeiss LSM 510 confocal microscope.

Scanning electron microscopy.

After 48 hours of culture on the grids the cells were fixed in a standard way using 2.5% glutaraldehyde (at room temperature for 60 min and, then at 4°C for 24 hours). The grids with cells were washed with PBS for 30 min and the cells were gradually dehydrated in alcohols. The grids with cells were finally coated with palladium and visualized with a scanning electron microscope (Philips SEM 515).

Transmission electron microscopy.

After 48 hours the nanostructured grids with the cells were put into a fixative solution containing 4% glutaraldehyde (Taab, Berkshire, UK), buffered with 0.1 M phosphate buffer, pH 7.3 (Merck). After overnight incubation the nanostructured grids were rinsed for 60 min in 0.1 M phosphate buffer, pH 7.3, and then postfixed in 2% OsO_4 in 0.1 M phosphate buffer for 2 hours (Merck). The specimens were dehydrated in ethanol, block-stained with alcoholic uranyl acetate and embedded in Spurr's medium (Merck). After contrasting, the ultrastructure of HeLa cells was examined using JEOL 100 Transmission Electron Microscope.

RESULTS

Morphology of Polyaniline Fibers

Light microscopy confirmed that the analyzed grids are transparent (Fig. 1A). Moreover, scanning electron microscopy showed that the thickness of electrospun fibers was 2000–2500 nm, and the diameter of pores was 50–300 nm (shorter dimension). The average thickness of a grid is less than 0.5 mm (Fig. 1B).

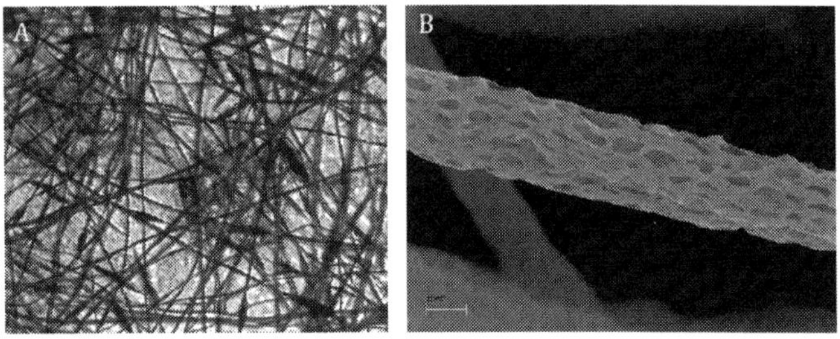

Figure 1. Light microscopy image of nanostructured grid.

Original magnification x200 (A). Scanning electron microscopy image of nanostructured microfiber. Original magnification x500 (B).

Cell Viability

To determine whether the tested electrospun scaffolds can influence the viability of uterine cervix carcinoma cells cultured in 3D, their viability was examined using the XTT assay. Due to XTT-reducing properties of the nanostructured grids alone, the viability of cells grown on the grids were related to the results obtained for measurements of the grids alone.

As shown in Fig. 2 the grids have no cytotoxic effect on HeLa cells. After 48 hours of culture on the nanostructured grids cell viability increased compared to the evaluation made after 24 hours. However, the change was insignificant ($p>0.05$). Subsequent incubation, up to 72 hours, resulted in a significant increase of cell viability, compared to the viability noted after either 24 hour ($p<0.001$) or 48 hour of culture ($p<0.05$) (Fig. 2).

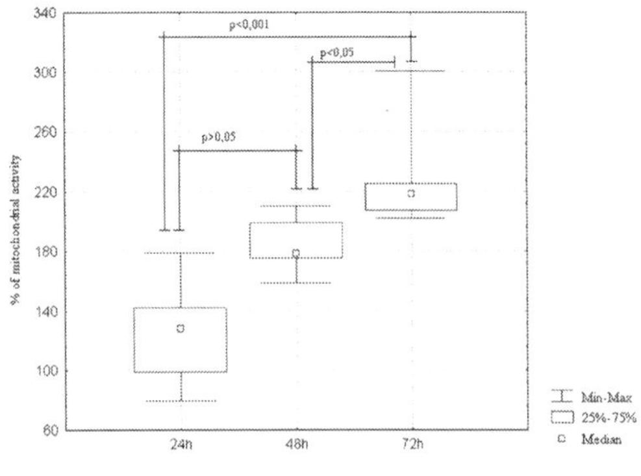

Figure 2. The viability of HeLa cells cultured on the nanostructured grid.

HeLa cells were incubated for 24, 48 72 hours on the nanostructured grids and the percentage of mitochondrial activity was calculated according to the following formula: (OD of nanostructured grid with cells−OD of medium alone)/(OD of nanostructured grid only−OD medium alone)×100; where OD is optical density.

Cell Morphology

The double staining using Hoechst 33342 and propidium iodide allows distinguishing living and necrotic cells as propidium iodide penetrates only cells with a disrupted cell membrane, while Hoechst 33342 penetrates also intact cell membranes. As it was shown in Fig. 3 the nuclei of cells cultured on the nanostructured grids 24 (Fig. 3A), 48 (Fig. 3B) and 72 (Fig. 3C) hours emitted bright blue fluorescence. The staining proved that in the analyzed time intervals cells showed high viability with very low level of necrosis. Moreover, confocal microscopy confirmed that the cells cultured on the grids formed a homogenous 3D constructs (Fig. 3D).

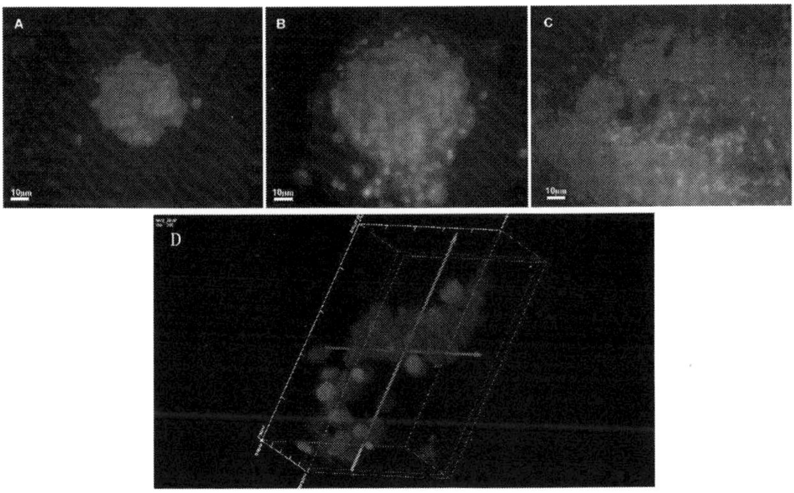

Figure 3: HeLa cells stained with Hoechst 33342 and propidium iodide after 24 (A), 48 (B) and 72 (C) hours of 3D culture on the nanostructure grids.

Original magnification x200. A fragment of the spherical 3D structure of cell growing on nanostructured grids visualized after 48 hours. Original magnification x400 (D).

The results of the light microscopy and scanning electron microscopy observation demonstrated that after 48 hours of culture on the nanostructured grids HeLa cells formed symmetrical and spherical aggregates. The size of cells ranged between 6 and 7 micrometers (Fig. 4).

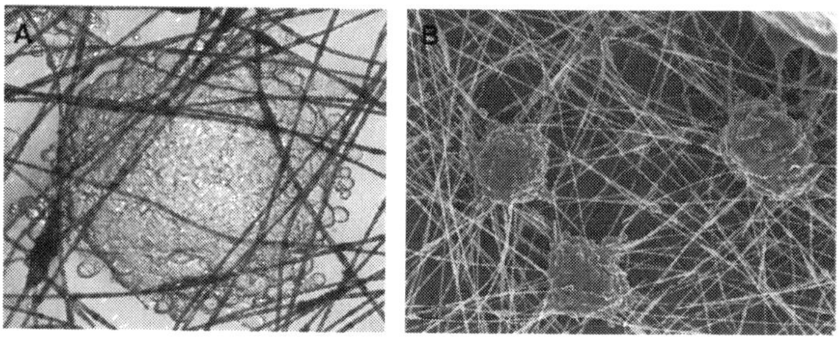

Figure 4: Cells growing on the scaffold visualized using light microscopy after 48 hours of incubation.

Original magnification x200 (A). Scanning electron microscopy image of HeLa cells on the nanostructured grid fibers after 48 hours of incubation. Original magnification x100 (B).

Detailed analysis of the cells grown for 48 hours on the nanostructured grids showed that there were no changes in the ultrastructure of HeLa cells. The cells were oval or elongated in shape with abundant microvilli and mostly euchromatic nuclei. The cytoplasm showed numerous intact organelles. Cisterns of the Golgi apparatus frequently contained electron-dense material in the form of laminae. The cells possessed numerous mitochondria with partially obliterated inner structure, sometimes presenting cristae. Some cells contained phagolysosomes with electron-dense content (Fig. 5).

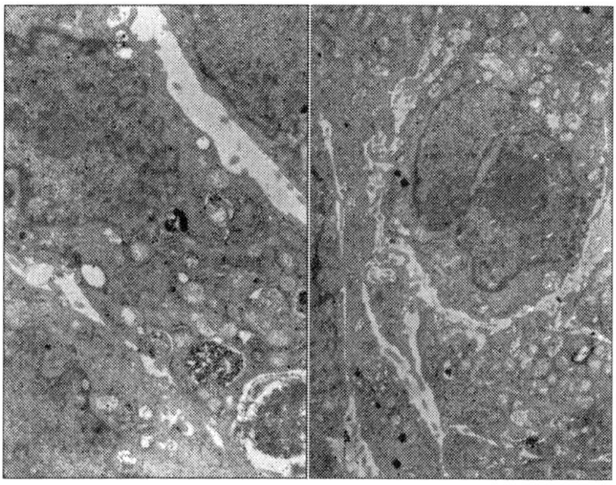

Figure 5: Transmission electron microscopy images of HeLa cells on the nano-structure grid fibers after 48 h culture.

DISCUSSION

Three-dimensional culture scaffolds assure a clearly better growth and proliferation of cells as they mimic the *in vivo* conditions. Moreover the cells cultured on 3D scaffolds have been found to develop a self-specific microenvironment [2],[13],[24]. Culture scaffolds of such properties can be used, for instance in regenerative medicine [2], [7], [9], [12], [22]–[23]. However, not all three-dimensional scaffolds are adequate for cell culture, as the building material of a scaffold must fulfill certain conditions [7], [10], [21], [25]. In particular it should allow for attachment of cells and their undisturbed growth and proliferation.

One of the most common among technologies used in production of nano- and microfibres from polymer solutions is electro-spinning. Studies confirm that this technique can be applied to obtain spatial scaffolds for cell cultures in *in vitro* conditions [20]–[21].

The electrospun three-dimensional nanostructured grids presented in this study were not only proven to be biocompatible but also suitable for cell culture. One of the advantages of these nanostructured grids is in their architecture in the form of nano- and microtubes. What is more in the process of electro-spinning these hollow fibers are being arranged randomly, forming an irregular network with micrometric distance between the tubes preserved. The electrospun capillaries possess nanopores which probably enhance the free flow of the culture medium.

These features have favorable effects on cell attachment and their viability. It has also been shown that these properties may promote migration of the proliferating cells along the fibers or between the overlaying fiber layers [1].

XTT test results have demonstrated that the described scaffold is suitable for cell culture in 3D conditions. The intense and significant increase in cell mitochondrial activity was observed particularly 48 hours after seeding of cells on the grid. The observation has been additionally supported by results from fluorescent microscopy with the use of Hoechst 33342 and propidium iodide. It was demonstrated that within 24 hours of seeding, the cells attached to nanostructured grid fibers and formed spatial agglomerates, which in the course of culture formed multilayered structures.

It can be assumed that the process of cell seeding was accompanied by the selection of particular types of cells, which were capable of proliferation on the fibers and thereby accept the new supporting material. The problem which remains to be resolved involves our inability to calculate the precise number of cells deposited on the nanostructured grid in the first day of the culture. For this reason, the obtained values of mitochondrial activity have been related to results obtained for an empty nanostructured grid and then to each other.

The analysis of results obtained using a scanning electron microscope has shown that the cells deposited on nanostructured grids form protrusions, which may either anchor the cells at the site of deposition or allow their migration along the fibers. It may be expected that the anchored cells may form a barrier preventing further amoeba-like movements of the adjacent cells. This would explain the reason of the inward growth of cells and formation of the spherical spatial structures. The observed spherical shapes of cell agglomerates formed following 24 hours after cell deposition allow for local development of high density cultures on a limited area. Thus, the presented scaffold represents not only an excellent support for growing cells in 3D conditions but may also stabilize the environment of proliferating cells thanks to the specific microenvironment developed. Nevertheless, it should be noted that the above hypothesis requires verification using studies on interaction of cells with individual fibers.

The presence of pores in the described fibers not only promotes cell growth but also makes the grid transparent. The potential for cell observation and their growth monitoring using light microscope is an important advantage of the suggested method of 3D cultures. Such a feature eliminates the need for staining of cells with fluorescent markers or for their transfection using e.g. GFP. Moreover, the nanostructured grid-deposited cells may be used in studies on ultrastructure of cells in *in situ* conditions and thus the simple technique of preparing cultures for Spurr embedding may be here applied.

CONCLUSIONS

A new solution for 3D cultures has been suggested, which is suitable for cell culture in conditions mimicking *in vivo* environment. The relatively high surface to volume ratio permits to load such a nanostructured grid with high number of cells and to grow them in conditions of high density or even overcrowding. Moreover, the suggested novel 3D support in many ways surpasses other similar culture approaches: the scaffolds are transparent and thus allow observation of viable cells in time; standard fluorescence or colorimetric methods such can be used for evaluation of cells. Finally, the cell nanostructured grid can be embedded in *in situ*conditions for ultrastructural studies.

AUTHOR CONTRIBUTIONS

Conceived and designed the experiments: MWC KL KS AG AJ WW JL. Performed the experiments: MWC KL AG. Analyzed the data: MWC KL WW. Contributed reagents/materials/analysis tools: MWC. Wrote the paper: MWC AJ JL.

REFERENCES

1. Yamada KM, Cukierman E (2007) Modeling tissue morphogenesis and cancer in 3D. Cell 130: 601–10 10.1016/j.cell.2007.08.006.
2. Ma PX (2008) Biomimetic materials for tissue engineering. Adv Drug Deliver Rev 60: 184–98 10.1016/j.addr.2007.08.041.
3. Fischbach C, Chen R, Matsumoto T, Schmelzle T, Brugge JS, et al. (2007) Engineering tumors with 3D scaffolds. Nat Methods 4: 855–60 10.1038/nmeth1085.

4. Jang JH, Castano O, Kim HW (2009) Electrospun materials as potential platforms for bone tissue engineering. Adv Drug Deliver Rev 61: 1065–83 10.1016/j.addr.2009.07.008.
5. Zhang S (2008) Designer self-assembling peptide nanofiber scaffolds for study of 3-D cell biology and beyond. Adv Cancer Res 99: 335–62 10.1016/S0065-230X(07)99005-3.
6. Kraehenbuehl TP, Langer R, Ferreira LS (2011) Three-dimensional biomaterials for the study of human pluripotent stem cells. Nat Methods 8: 731–6 10.1038/nmeth.1671.
7. Shin H, Jo S, Mikos AG (2003) Biomimetic materials for tissue engineering. Biomaterials 24: 4353–4364 10.1016/S0142-9612(03)00339-9.
8. Kim JH, Park CH, Lee OJ, Lee JM, Kim JW, et al. (2012) Preparation and in vivo degradation of controlled biodegradability of electrospun silk fibroin nanofiber mats. J Biomed Mater Res A 100: 3287–95 10.1002/jbm.a.34274.
9. Jeong SI, Krebs MD, Bonino CA, Khan SA, Alsberg E (2010) Electrospun alginate nanofibers with controlled cell adhesion for tissue engineering. Macromol Biosci 10(8): 934–43 10.1002/mabi.201000046.
10. Jo JH, Lee EJ, Shin DS, Kim HE, Kim HW, et al. (2009) In vitro/in vivo biocompatibility and mechanical properties of bioactive glass nanofiber and poly(epsilon-caprolactone) composite materials. J Biomed Mater Res B Appl Biomater 91(1): 213–20 10.1002/jbm.b.31392.
11. Lee JJ, Yu HS, Hong SJ, Jeong I, Jang JH, et al. (2009) Nanofibrous membrane of collagen-polycaprolactone for cell growth and tissue regeneration. J Mater Sci Mater Med 20: 1927–35 10.1007/s10856-009-3743-z.
12. Cai YZ, Zhang GR, Wang LL, Jiang YZ, Ouyang HW, et al. (2012) Novel biodegradable three-dimensional macroporous scaffold using aligned electrospun nanofibrous yarns for bone tissue engineering. J Biomed Mater Res A 100: 1187–94 10.1002/jbm.a.34063.
13. Davidenko N, Campbell JJ, Thian ES, Watson CJ, Cameron RE (2010) Collagen-hyaluronic acid scaffolds for adipose tissue engineering. Acta Biomater 6: 3957–68 10.1016/j.actbio.2010.05.005.

14. Ifkovits JL, Burdick JA (2007) Review: photopolymerizable and degradable biomaterials for tissue engineering applications. Tissue Eng 13: 2369–85 10.1089/ten.2007.0093.
15. Rnjak-Kovacina J, Weiss AS (2011) Increasing the pore size of electrospun scaffolds. Tissue Eng Part B Rev 17: 365–72 10.1089/ten.teb.2011.0235.
16. Nam J, Huang Y, Agarwal S, Lannutti J (2007) Improved cellular infiltration in electrospun fiber via engineered porosity. Tissue Eng 2007 13: 2249–57 10.1089/ten.2006.0306.
17. Kim TG, Park TG (2006) Biomimicking extracellular matrix: cell adhesive RGD peptide modified electrospun poly(D,L-lactic-co-glycolic acid) nanofiber mesh. Tissue Eng 12: 221–33
18. Wołuń-Cholewa M, Szymanowski K, Nowak-Markwitz E, Warchoł W (2010) Photodiagnosis and photodynamic therapy of endometriotic epithelial cells using 5-aminolevulinic acid and steroids. Photodiagnosis Photodyn Ther 8: 58–63 10.1016/j.pdpdt.2010.12.003.
19. Fik E, Wołuń-Cholewa M, Kistowska M, Warchoł JB, Goździcka-Józefiak A (2001) Effect of lectin from Chelidonium majus L. on normal and cancer cells in culture. Folia Histochem Cytobiol 39: 215–216.
20. Ahmed I, Ponery AS, Nur-E-Kamal A, Kamal J, Meshel AS, et al. (2007) Morphology, cytoskeletal organization, and myosin dynamics of mouse embryonic fibroblasts cultured on nanofibrillar surfaces. Mol Cel Biochem 301: 241–9 10.1007/s11010-007-9417-6.
21. Rodrigues MT, Martins A, Dias IR, Viegas CA, Neves NM, et al. (2012) Synergistic effect of scaffold composition and dynamic culturing environment in multilayered systems for bone tissue engineering. J Tissue Eng Regen Med 6: e24–30 10.1002/term.499.
22. Garg K, Bowlin GL (2011) Electrospinning jets and nanofibrous structures. Biomicrofluidics 5: 13403 10.1063/1.3567097.
23. Sill TJ, Von Recum HA (2008) Electrospinning: applications in drug delivery and tissue engineering. Biomaterials 29: 1989–2006 10.1016/j.biomaterials.2008.01.011.

24. Bhattarai SR, Bhattarai N, Yi HK, Hwang PH, Cha DI, et al. (2004) Novel biodegradable electrospun membrane: scaffold for tissue engineering. Biomaterials 25: 2595–2602 10.1016/j.biomaterials.2003.09.043.

25. Tuzlakoglu K, Bolgen N, Salgado AJ, Gomes ME, Piskin E, et al. (2005) Nano- and micro-fiber combined scaffolds: a new architecture for bone tissue engineering. J Mater Sci Mater Med 16: 1099–104 10.1007/s10856-005-4713-8.

CHAPTER

7

EXPRESSION OF MURINE 5-AMINOLEVULINATE SYNTHASE VARIANTS CAUSES PROTOPORPHYRIN IX ACCUMULATION AND LIGHT-INDUCED MAMMALIAN CELL DEATH

Erica J. Fratz[1], Gregory A. Hunter[1], Gloria C. Ferreira[1, 2]

[1]Department of Molecular Medicine, Morsani College of Medicine, Tampa, Florida, United States of America,

[2]Department of Chemistry, University of South Florida, Tampa, Florida, United States of America

ABSTRACT

5-Aminolevulinate synthase (ALAS; EC 2.3.1.37) catalyzes the first committed step of heme biosynthesis in animals. The erythroid-specific ALAS isozyme (ALAS2) is negatively regulated by heme at the level of mitochondrial import and, in its mature form, certain mutations of the murine ALAS2 active site loop result in increased production of protoporphyrin IX (PPIX), the precursor for heme. Importantly, generation of PPIX is a crucial component in the widely used photodynamic therapies (PDT) of cancer and other dysplasias. ALAS2 variants that cause high levels of PPIX accumulation provide a new means of targeted, and potentially enhanced, photosensitization. In order to assess the prospective utility of ALAS2 variants in PPIX production for PDT, K562 human erythroleukemia cells and HeLa human cervical carcinoma cells were transfected with expression plasmids for ALAS2 variants with greater enzymatic activity than the wild-type enzyme. The levels of accumulated PPIX in ALAS2-expressing cells were analyzed using flow cytometry with fluorescence detection. Further, cells expressing ALAS2 variants were subjected to white light treatments (21–22 kLux) for 10 minutes after which cell viability was determined. Transfection of HeLa cells with expression plasmids for murine ALAS2 variants, specifically for those with mutated mitochondrial presequences and a mutation in the active site loop, caused significant cellular accumulation of PPIX, particularly in the membrane. Light treatments revealed that ALAS2 expression results in an increase in cell death in comparison to aminolevulinic acid (ALA) treatment producing a similar amount of PPIX. The delivery of stable and highly active ALAS2 variants has the potential to expand and improve upon current PDT regimes

INTRODUCTION

The first committed step of heme biosynthesis in non-plant eukaryotes and some prokaryotes, the pyridoxal 5′-phosphate (PLP)-dependent condensation of glycine and succinyl-coenzyme A to generate 5-aminolevulinate (ALA), coenzyme A (CoA), and CO_2, is catalyzed by 5-aminolevulinate synthase (ALAS) [1], [2]. This reaction is directly coupled to the citric acid cycle via the substrate succinyl-CoA and is the key regulatory step of heme biosynthesis [3]. In mammals, two chromosomally distinct genes each encode an ALAS isoenzyme, and the two isoenzymes are differentially expressed in a tissue specific manner [4]. The human gene for the non-specific or housekeeping isoform, ALAS1, is located on chromosome 3 [5], [6], and is expressed ubiquitously in all tissues [7]. The gene encoding the erythroid specific isoform, ALAS2, is located on the x-chromosome [6], [8] and is expressed only in developing erythroblasts [7].

The two ALAS isoenzymes are translated as precursor proteins with N-terminal mitochondrial matrix import signal sequences that are proteolytically cleaved following importation to yield the mature enzymes [9]–[12]. The activity of the enzyme is only manifested upon localization to the mitochondrial matrix, as this is where the substrate succinyl-CoA is produced [13]–[16]. An important aspect of the import sequences in both ALAS1 [17] and ALAS2 [18] are the presence of heme-regulatory motifs (HRMs), which consist of short amino acid sequences characterized in part by adjacent cysteine-proline (CP) residues [19]. HRMs confer heme-binding properties and have been shown to function as heme-oxygen sensors in bacteria [20], yeast [21] and mammals [18], [22], [23]. In *in vitro* translocation experiments with isolated mitochondria and ALAS2 precursor protein, the two heme-binding motifs in the leader sequence, corresponding to C11 and C38 in murine ALAS2 (mALAS2), were reported to bind heme and prevent translocation of precursor ALAS2 into the mitochondrion [18]. Structural and biochemical data have also demonstrated that a heme-peptide interaction occurs between hemin and the presequence of ALAS2 [24], further indicating the potential of heme to act as a feedback inhibitor of the pathway by preventing the mitochondrial import of precursor ALAS2 when heme levels are sufficient for cellular requirements.

Much of what we know about the chemical and kinetic mechanisms of ALAS2 comes from *in vitro* enzymatic assays that have helped establish and define the microscopic steps of the ALAS-catalyzed reaction, including the rates of glycine and succinyl-CoA binding, formation of the quinonoid intermediates, and product release [2], [25]–[28]. These studies, performed using mALAS2 purified from *E. coli* cells expressing the recombinant mature enzyme, have led to an understanding of the importance of specific regions and single amino acid residues in the intrinsic activity of ALAS2 [29]–[35]. Generally, a mutation made to an amino acid predicted to be of functional importance causes a decrease in activity of the enzyme. For example, K313 of mALAS2 was identified as the amino acid involved in the Schiff base linkage with the PLP cofactor [36], and mutations in K313 completely abolish measurable activity of mALAS2 under standard assay conditions [29], [37]. However, some mutations in mALAS2 cause significantly increased activity of the enzyme, as demonstrated both using purified enzyme [31], [35], and in bacteria when expressing plasmids encoding the variant enzymes [35]. These mutations are all located in an extended conformation region termed the active site loop (Y422-R439), which is predicted to act as a "lid" over the active site following substrate binding [35]. Thus, ALAS is thought to undergo a conformational change from an open conformation, in which the substrates glycine and succinyl-CoA can bind in the active site, to a closed conformation, during which the products ALA, CoA, and CO_2 are formed [25], [28]. When the reaction is complete, the active site loop reopens, and the products are released. It is the opening of the active site loop to allow product release that limits the overall rate of the enzymatic reaction [2], [25], [27]. Based on a combination of kinetic [25], [28] and structural modeling studies [38], it was proposed that mutations in the active site loop can lead to hyperactive forms of ALAS, defined as those with at least a 10-fold increase in catalytic efficiency toward one or both substrates, by accelerating reversion to the open loop conformation upon product formation [35].

Since ALAS catalyzes the rate-determining step of tetrapyrrole biosynthesis in mammals [2], [25], overexpression of ALAS in prokaryotic [35] and eukaryotic cells [39] results in accumulation of the photosensitizing heme precursor, protoporphyrin IX (PPIX). This property has potential for applications of ALAS or ALAS variants in photodynamic therapy of tumors and other non-malignant dermatological indications, such as acne vulgaris, psoriasis, and scleroderma [40], [41]. In this study, we transfected mammalian cells with mALAS2 variants and measured PPIX accumulation using fluorescence activated cell sorting (FACS). We identified the R433K variant with additional mutations of the HRMs in the presequence as the variant causing the most cellular PPIX accumulation. Subsequently, we used the variants causing the most PPIX accumulation to study the cell death caused by the PPIX toxicity and photosensitization.

MATERIALS AND METHODS

Materials

5-Aminolevulinic acid hydrochloride (ALA) was purchased from Acros Organics (Morris Plains, NJ), and dissolved in distilled water at a concentration of 10 mg/mL. Glycine, purchased from Fisher Chemical (Fairlawn, NJ), was dissolved in phenol red-free culture medium purchased from Mediatech, Inc. (Manassas, VA) to give a stock concentration of 1 M. Paclitaxel (6 mg/mL stock in DMSO) was graciously provided by the laboratory of Dr. Scott Antonia (H. Lee Moffitt Cancer Center and Research Institute, Tampa, FL), and was diluted in culture medium directly before use. Deferoxamine mesylate, purchased from Sigma (St. Louis, MO), was dissolved in distilled water to create a 10 mM stock solution. Propidium iodide, 4′,6-diamidino-2-phenylindole (DAPI) and kanamycin sulfate were purchased from Acros Organics (Morris Plains, NJ), and 3-(4,5-dimethylthiazol-2-yl)-2,5-diphenyltetrazolium bromide (MTT) was purchased from Sigma-Aldrich (St. Louis, MO). BmtI and BamHI restriction enzymes were obtained from New England BioLabs, Inc. G418 sulfate was procured from Mediatech, Inc. (Manassas, VA).

Plasmids

The precursor ALAS2 cDNAs were individually subcloned into the multiple cloning site of the pIRES2-ZsGreen1 vector (purchased from Clontech Laboratories, Inc. Mountain View, CA) using the BmtI and BamHI restriction sites. Digested ALAS2-encoding fragments were ligated into the digested pIRES2-ZsGreen1 vector using T4 DNA ligase and ligase buffer (Thermo Scientific Fermentas (Waltham, MA)). Electrocompetent BL21(DE3) cells were transformed by electroporation with the ligated plasmid DNA and selected by spreading them on LB agar medium containing 10 μg/mL kanamycin sulfate. Plasmid DNA was purified from a single colony using a QiaPrep Spin Miniprep kit (Qiagen Inc.; Germantown, MD), and the sequence of the cloned DNA was verified by Genewiz, Inc. (New Brunswick, NJ).

Cell Culture

K562 human immortalized myelogenous leukemia cells (ATCC) were maintained in RPMI-1640 culture medium, purchased from Mediatech, Inc. (Manassas, VA), with 10% fetal bovine serum (FBS; purchased from Thermo Scientific Waltham, MA), gentamicin (50 μg/mL), penicillin (60 μg/mL) and streptomycin (100 μg/mL) at 37°C in a humidified incubator with 5% CO_2. HeLa human cervical carcinoma cells (ATCC) were maintained in DMEM culture medium with 4.5 g/L glucose, L-glutamine, and sodium pyruvate (Mediatech, Inc.; Manassas, VA), with 10% fetal bovine serum (FBS), gentamicin (50 μg/mL), penicillin (60 μg/mL) and streptomycin (100 μg/mL) at 37°C in a humidified incubator with 5% CO_2.

Stable Transfection of K562 Human Erythroleukemia Cells

K562 cells were transfected with Lipofectamine LTX and PLUS Reagent, purchased from Invitrogen (San Jose, CA), according to the supplier's optimized protocol for K562 cells. On the day of transfection, a hemocytometer and trypan blue staining were used to count the cells and determine culture density and viability. In a 6-well plate, K562 cells (5×10^5 cells per well) were seeded in a 6-well plate at a volume of 2 mL of RPMI-1640 growth medium with 10% FBS 30 minutes prior to transfection. For each transfection, 2.5 μg of DNA was added into 500 μL of RPMI medium without serum. 2.5 μL of PLUS reagent (at a 1:1 ratio to DNA) was then added directly to the diluted DNA. After gentle mixing and a 10 minute incubation at room temperature, 10 μL of Lipofectamine LTX was added into the diluted DNA solution, mixed gently and incubated for 35 minutes at room temperature to form DNA-Lipofectamine LTX complexes. The DNA-Lipofectamine LTX complexes were added dropwise to each well containing cells and mixed gently by manually rocking the plate back and forth. 24 hours after transfection, the cells were pelleted by centrifugation at 400×g and resuspended in RPMI-1640 culture medium with 10% FBS, gentamicin (50 μg/mL) and the selection antibiotic, G418 sulfate (500 μg/mL). Every 2–3 days, the cells were pelleted and resuspended in fresh medium with 10% FBS, gentamicin (50 μg/mL), and G418 sulfate (500 μg/mL). Assays were performed 2–6 weeks after the start of selection.

Transient Transfection of Hela Cells

On the day prior to transfection, HeLa cells were trypsinized and counted. Approximately 2×10^4cells were seeded into each well of a 24-well plate in 0.5 mL of DMEM. Cell density was ~30–50% confluent on the day of transfection. For each transfection, 250 ng of DNA was diluted into 100 μl of DMEM without serum. 1 μl of Lipofectamine LTX was added into the diluted DNA solution, mixed gently and incubated for 30–45 minutes at room temperature to form DNA-Lipofectamine LTX complexes. The DNA-Lipofectamine LTX complexes were added dropwise to each well containing cells and mixed gently by manually rocking the plate back and forth for a few seconds. After 4 hours of incubation with the DNA-Lipofectamine LTX complexes, the medium was aspirated out of each well and fresh DMEM with 10% FBS, gentamicin (50 μg/mL), penicillin (60 μg/mL) and streptomycin (100 μg/mL) was added to each well of cells. Cells were incubated at 37°C in a CO_2 incubator for 24 or 48 hours post-transfection before assaying.

Preparation of Cells for FACS and Quantitation of PPIX

While K562 cells were suspended in phenol red-free medium, HeLa cells were washed, scraped and resuspended in phosphate-buffered saline (PBS; 80 mM disodium hydrogen orthophosphate, 20 mM sodium dihydrogen orthophosphate, 100 mM sodium chloride, pH 7.5) before pipetting into BD Falcon tubes with cell strainer caps. Preparation of either K562 or HeLa cells for FACS was done under very low light conditions (1–2 Lux as measured by a Pyle PLMT68 light meter) in order to minimize phototoxicity caused by PPIX accumulation. FACS analyses were performed using a BD LSR II Analyzer (Becton, Dickinson, and Company) and FACSDiva Version 6.1.3 software. ZsGreen1 emission was measured between 515 nm and 545 nm (530/30BP filter) when cells were excited using the 488 nm laser. In order to eliminate any background red fluorescence, the 633 nm-red laser was blocked during the collection of the PPIX emission data. PPIX emission was determined in the 619 nm and 641 nm range (630/22BP filter) when cells were excited with the 405 nm laser. Forward-scatter (FSC) *versus*side-scatter (SSC) dot plots were used to gate the whole cells and thus remove the contribution of the cell debris from the population being examined. A minimum of 10,000 of the gated whole cells were then depicted in dot plots of SSC *versus* ZsGreen1 fluorescence, and the "green-fluorescent population" gate was defined based on untransfected HeLa cells as negative controls. Dot plots of SSC *versus* PPIX fluorescence were used to define the PPIX-accumulating cells for both the "green-fluorescent" and the "non-green fluorescent" populations. The PPIX gating was based on the negative control for PPIX, the pIRES2-ZsGreen1 vector-expressing cells. PPIX fluorescence values were normalized for transfection efficiency using the corresponding ZsGreen1 fluorescence value.

Light Exposure Assays

HeLa cells in 24-well plates were transfected 48 hours prior to FACS analysis. Cells were washed twice with PBS and placed on ice (to prevent overheating) underneath a Sylvania incandescent flood lamp (150W, 120V) for 10 minutes. Light intensity at the samples was measured before each light exposure experiment using a Pyle PLMT68 light meter, and it was between 21–22 kLux for all experiments. The samples were resuspended and pipetted into tubes with cell strainer caps and analyzed first for ZsGreen1 and PPIX fluorescence (as described above), and then for cell viability. In order to determine the statistical significance of cell death directly attributable to mALAS2-induced PPIX accumulation and phototoxicity, samples were compared to the ZsGreen1 controls during t-test evaluations.

Cell Viability Assays

Cell death was assessed by measuring incorporation of either propidium iodide or DAPI into nuclear DNA, as these fluorescent dyes cross the plasma membrane in dying cells much more efficiently than in live cells. HeLa cell cultures were independently incubated with the fluorescent DNA-binding dyes propidium iodide (10 μg/mL) and DAPI (10 μg/mL) for 5 min at 22°C. The cells were then analyzed by determining the fluorescence intensities using FACS. Fluorescence emission of nucleic acid-bound propidium iodide was measured between 585 nm and 625 nm (605/40BP filter) upon excitation of the cells using the 488 nm laser, while DNA-bound DAPI emission was measured between 440 nm and 460 nm (450/20BP filter) following excitation of the cells at 375 nm. Dot plots of SSC *versus* propidium iodide or DAPI fluorescence were used to define the respective gates, which were based on the control samples in which no propidium iodide or DAPI were added. Cell viability was independently determined using the MTT dye reduction assay as described elsewhere [42], [43]. HeLa cells (5×10^3 cells per well) were seeded in 96-well plates, and on the following day were transfected with expression plasmids for mALAS2 variants as described above. Twenty-four hours after plasmid transfection, or 4 hours after addition of 100 μM ALA, cells were exposed to light for 10 minutes. The culture medium was pipetted out, 50 μL of 2 mg/mL MTT was added to each well, and cells were incubated for an additional 4 hours at 37°C. Upon solubilization of the cells with DMSO (100 μL/well) during a 10 minute incubation, the solubilized, MTT-treated cells were thoroughly mixed by pipetting several times, and absorbance was measured at 540 nm using a μQuant plate reader (Bio-Tek Instruments, Inc.).

Confocal Fluorescence Microscopy

HeLa cells were grown in Thermoscientific Nunc Labtek sterile 4-well chambered coverglass until 50% confluent, and then transfected as described above. Six hours later, the culture medium was supplemented with glycine to yield a final concentration of 100 mM, and immediately before obtaining confocal fluorescence microscopy images of the cells, the medium was removed from the wells and the cells were washed with PBS three times. Live cell imaging was performed using a 3i-Olympus spinning disk confocal microscope operated by Slidebook 5 software and equipped with a Photometrics Evolve EMCCD camera. The filter block used consisted of a 350/50 nm excitation filter, a BS400 beamsplitter, and a 630/75 nm emission filter.

RESULTS

Transient Expression of Malas2 Variants Causes Accumulation of PPIX in HeLa Cells

HeLa cells were transfected with mALAS2-encoding plasmids with or without mutations in the mALAS2 presequence and/or mature enzyme sequence (Table 1, Figure 1). Where indicated, the mALAS2 presequence was mutated at three cysteine residues, C11, C38, and C70, in order to yield nonfunctional HRMs and thus eliminate heme feedback inhibition of mitochondrial import [18], [40]. The plasmids were designed such that a single bicistronic mRNA, encoding both mALAS2 and the fluorescent protein ZsGreen1, separated by an internal ribosomal entry site (IRES), would be produced. Transcriptional expression of mALAS2 and ZsGreen1 was under control of the constitutively active cytomegalovirus (CMV) promoter [44]. Twenty-four hours post-transfection, wild-type mALAS2 with a wild-type presequence (WT^{Ψ}) expression caused a slight, but statistically significant as defined by Student's t-test ($p<0.05$), increase in PPIX accumulation (Figure 2A). The K313A mutation in ALAS2 leads to undetectable enzymatic activity values *in vitro* [45], and under *ex vivo* conditions this also appears to be the case, as no PPIX accumulated in K313A-expressing cells (K313A$^{\Psi}$), a result similar to that of the mammalian cells harboring the pIRES2-ZsGreen1 vector (ZsGreen1) alone (Figure 2A). In comparison to WT$^{\Psi}$, the heptavariant mutations in the active site loop alone (HPVT$^{\Psi}$) and mutations in the presequence alone (WT) caused a statistically significant ($p<0.05$) increase in PPIX production in expressing cells (Figure 2A). The combination of the heptavariant mutations in the active site loop and mutations in the presequence (HPVT) resulted in variable PPIX production that was not statistically significant. Somewhat unexpectedly, it was expression of the mALAS2 variant with a mutated presequence containing the R433K mutation (R433K) that caused the largest accumulation of PPIX. Mutation of R433 residue to a lysine in the purified, mature enzyme results in an increase in activity to twice that of the wild-type enzyme (*i.e.*, a 2-fold increase in the k_{cat} value and a 1.65 to 1.85-fold enhancement in the specificity constants for glycine and succinyl-CoA over those of wild type, mature mALAS2) [31]. Similarly, the increase in PPIX accumulation in HeLa cells expressing the R433K precursor with a mutated presequence was approximately 2.5-fold compared to the cells expressing WT$^{\Psi}$ and 2-fold in comparison to cells expressing WT (Figure 2A).

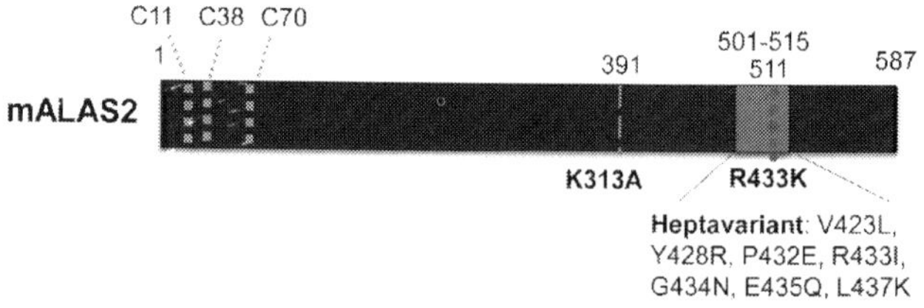

Figure 1: Murine ALAS2 protein schematic indicating mutated amino acid residues.

The green dotted lines represent the relative locations of the three cysteines in the HRMs of ALAS2. C11 and C38 are in the ALAS2 presequence. K313A, R433K, and the heptavariant mutations and their respective positions are indicated according to the amino acid numbering previously described for mature mALAS2 [31], [35], [36]. The amino acid positions according to the numbering for the precursor enzyme are written above the diagram.

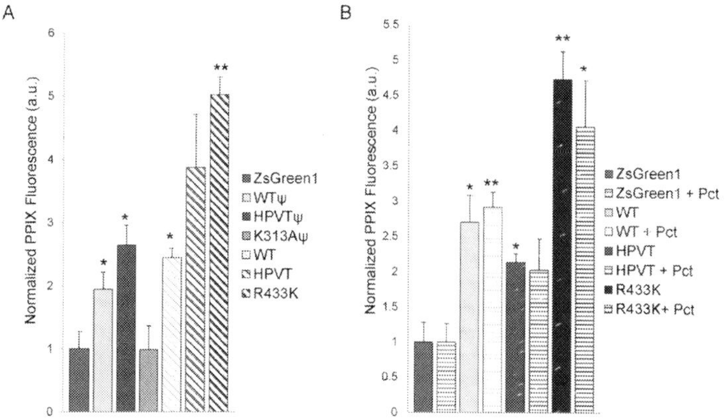

Figure 2: Porphyrin accumulation in transiently transfected HeLa cells with mALAS2 WT- and variant-expressing plasmids.

(A) HeLa cells transfected with mALAS2 variants with and without ($^{\Psi}$) mutated presequences analyzed 24 hours after transfection. (B) HeLa cells transfected with mALAS2 variants with mutated presequences, in cells cultured in the absence or presence of 10 nM paclitaxel (Pct), analyzed 48 hours after transfection. Normalized PPIX fluorescence values were obtained by first dividing the mean PPIX fluorescence by the mean ZsGreen1 fluorescence for each cell population and secondly dividing by the mean PPIX fluorescence/ZsGreen1 fluorescence ratio of the pIRES2-ZsGreen1-transfected cells. Each data set represents three separate experiments ± standard deviation (*p<0.05 and **p<0.01, Student's *t*-test) [a.u., arbitrary units].

Table 1: Plasmid constructs used to express murine ALAS2 variants in mammalian cell lines.

Plasmid Name	Mutations in	
	mALAS2 Presequence	Mature mALAS2 Sequence
pIRES2-ZsGreen1	N/A[1]	N/A[1]
pEF21	---[2]	---[2]
pEF22	---[2]	V423L,Y428R,P432E, R433I, G434N, E435Q and L437K
pEF23	---[2]	K313A
pEF24	C11S, C38S and C70S	---[2]
pEF25	C11S, C38S and C70S	V423L,Y428R,P432E, R433I, G434N, E435Q and L437K
pEF26	C11S, C38S and C70S	K313A
pEF31	C11S, C38S and C70S	R433K

Because PPIX can be used as a photosensitizing agent for PDT of cancers, we examined the effect of combining mALAS2-induced PPIX accumulation with the chemotherapeutic drug paclitaxel, a well-characterized inhibitor of mitosis [46]–[48]. Only mALAS2 variants with mutated presequences were used in the drug combination experiments, since those variants presented the most potential for PPIX accumulation, and thus, subsequent cell death resulting from light-induced phototoxicity. A dose response curve revealed that the concentration of paclitaxel that killed 25% of HeLa cells in 48 hours (IC25) was 10 nM (data not shown), and this concentration was chosen as the desired concentration for the PDT combination experiments. Paclitaxel had no statistically significant effect, as defined by Student's t-tests, on the amount of accumulated PPIX in the mammalian cells transfected with any of the ALAS2 variants after 48 hours (Figure 2B). The largest increase in PPIX fluorescence was observed with the R433K variant, regardless of whether or not paclitaxel was included in the culture medium (Figure 2B).

Supplementation of Cell Culture Medium with Glycine Leads to Increased PPIX Accumulation in Malas2-expressing HeLa Cells

Given that PPIX is a photosensitizer [49] and the Michaelis-Menten constant (K_m) of mALAS2 for glycine, at 25±4 mM [3], [25], [34], is significantly higher than its intracellular concentration of approximately 2.5 mM [50], it is plausible that supplementation of the cell culture medium with glycine would lead to enhanced synthesis of ALA, and consequently, PPIX, which in turn might increase the efficacy of PPIX-induced phototoxicity. To examine whether increased glycine concentration caused enhanced PPIX accumulation, HeLa cells transfected with mALAS2 (wild type and variants) were grown in medium with different glycine concentrations (Figure 3A). Samples with culture medium supplemented with 100 μM ALA for 4 hours served as a control for PPIX accumulation independent of glycine concentration (Figure 3B). The concentration and treatment times for ALA were chosen based on experiments indicating that the extent of PPIX accumulated in HeLa cells supplemented with 100 μM ALA was similar to that of HeLa cells expressing R433K, the mALAS2 variant that induces the highest levels of PPIX accumulation in these cells. HeLa cells expressing WT, HPVT or R433K in culture medium supplemented for 18 hours with 10 mM or 100 mM glycine exhibited significant increases in PPIX in comparison to the "no glycine" controls, with R433K again demonstrating the highest PPIX accumulation ($p<0.01$) (Figure 4A). Supplementing the culture medium with either 10 mM or 100 mM glycine for cells expressing WT resulted in approximately 4- and 6-fold increases in PPIX accumulation, respectively, as compared to no glycine supplementation. For cells expressing R433K, glycine supplementation more than tripled the PPIX accumulation, representing a fourteen-fold increase over the control HeLa cells supplemented with glycine. Cells expressing HPVT were also affected by addition of either 10 mM or 100 mM glycine, increasing the PPIX by 1.4-fold and 2.5-fold, respectively. Glycine elicited no effect on PPIX production in HeLa cells alone, those treated with 100 μM ALA, or HeLa cells expressing the pIRES2-ZsGreen1 vector (Figure 3B).

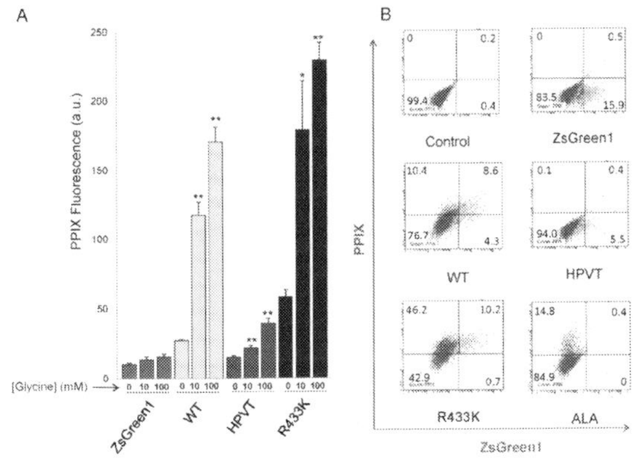

Figure 3: Glycine supplementation of culture medium increases PPIX accumulation in HeLa cells expressing murine wild-type, HPVT, and R433K ALAS2.

(A) PPIX fluorescence measured by flow cytometry 24 hours after transfection and 18 hours after supplementing culture medium with 10 mM glycine or 100 mM glycine. The culture medium of the corresponding control cells and control transfected cells was not supplemented with glycine. Normalized PPIX fluorescence values were obtained by dividing the mean PPIX fluorescence by the mean ZsGreen1 fluorescence for each cell population. Mean PPIX fluorescence values are representative of three separate experiments ± standard deviation (*$p<0.05$ and **$p<0.01$, Student's *t*-test) [a.u., arbitrary units]. (B) ZsGreen1 fluorescence *versus* PPIX fluorescence in HeLa cell populations supplemented with 100 mM glycine as measured by flow cytometry. In each dot plot, cells in the green-fluorescent range are represented in green and to the right of the vertical black line, while cells that are not green-fluorescent are in blue. Cells that are red-fluorescent due to PPIX accumulation are above the horizontal black line, while cells that are not red-fluorescent are below the horizontal black line. The percentage of cells in each quadrant is written in the corners of each dot plot. "Control" refers to non-transfected HeLa cells.

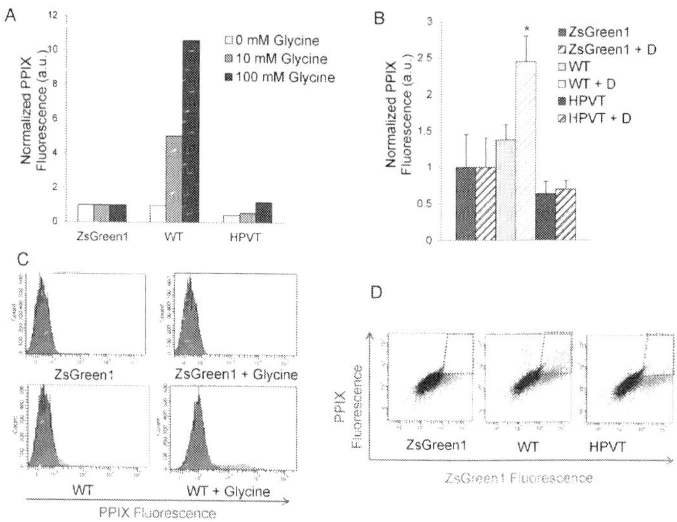

Figure 4: Stable expression of murine ALAS2 variants in K562 cells results in PPIX accumulation when culture medium is supplemented with either glycine or deferoxamine.

(A) PPIX fluorescence 18 hours after supplementation of the culture medium with glycine to yield 10 mM and 100 mM final concentrations. Normalized fluorescence values were obtained by first dividing the mean PPIX fluorescence by the mean ZsGreen1 fluorescence for each cell population and secondly dividing by the mean PPIX fluorescence/ ZsGreen1 fluorescence ratio of the pIRES2-ZsGreen1-transfected cells. (B) PPIX fluorescence 18 hours after supplementation of the culture medium with deferoxamine to a final concentration of 100 μM (+D). (C) PPIX fluorescence of K562 cell populations supplemented with glycine to a 100 mM final concentration. The blue area represents cells that do not express ZsGreen1, and the green area represents the population that expresses ZsGreen1. K562 cells that express ZsGreen1 are not affected by glycine supplementation of the culture medium (upper plots), while cells that express WT exhibit a significant increase in PPIX fluorescence when the culture medium is supplemented with glycine (lower plots). (D) PPIX fluorescence of K562 cell populations grown in culture media containing 100 μM deferoxamine. Cell growth in the presence of deferoxamine significantly increases the PPIX fluorescence in K562 cells expressing WT, but does not affect PPIX fluorescence in cells expressing only ZsGreen1 or HPVT, as indicated by the number of cells within the red dotted-line boxes.

Glycine and Deferoxamine Increase PPIX Accumulation in Malas2-expressing K562 Cells

We generated stable K562 human myelogenous erythroleukemia cell lines expressing mALAS2, expecting that these cells might accumulate high amounts of PPIX, due to their similarity to undifferentiated erythrocytes [51]. We expressed only WT and HPVT in order to test if HPVT could have higher activity in a cell line in which ALAS2 in normally expressed. Initially, the stable cell lines expressing WT did not show an increase in PPIX fluorescence as compared to regular K562 cells (Figure 4A), and we postulated that glycine availability was again a limiting factor, as we had observed in HeLa cells (Figure 3). Another possible limitation upon PPIX accumulation could be a more effective conversion of PPIX into heme in K562 cells, in which case the inclusion of an iron-specific chelator such as deferoxamine should increase the PPIX fluorescence by reducing or even preventing this conversion. To address why the expression of mALAS2 alone did not cause a larger increase in PPIX and what could limit the ALAS2-induced PPIX accumulation, the culture medium of K562 cells expressing (1) ZsGreen1 control, (2) WT, or (3) HPVT was supplemented with glycine to yield final concentrations of 10 mM or 100 mM glycine (Figure 4A and 4C) or deferoxamine mesylate to yield a final concentration of 100 μM deferoxamine mesylate (Figure 4B and 4D). Expression of WT in K562 cells did not increase the cellular PPIX fluorescence, but PPIX fluorescence increased by more than 5- and 10-fold when the culture medium contained 10 mM glycine and 100 mM glycine, respectively (Figure 4A). No significant differences were seen in cells not expressing mALAS2 (Figure 4C) or in cells expressing HPVT when treated with 10 mM glycine (Figure 4A). However, in contrast to HeLa cells, when 100 mM glycine was used, cells expressing HPVT did exhibit an increase in PPIX fluorescence by approximately 3-fold (Figure 4A).

Deferoxamine is a well-characterized iron-specific bacterial siderophore with a long history of clinical use in iron chelation therapy. Deferoxamine has the potential to increase PPIX by decreasing the cellular iron concentration, thereby inhibiting the conversion of PPIX to heme [52], [53]. Treatment of K562 cells with deferoxamine for 18 hours caused no change in PPIX fluorescence in cells not expressing mALAS2 or cells expressing HPVT. Deferoxamine did cause a significant increase in PPIX in cells expressing WT, in which case the mean PPIX fluorescence increased from 1.4-fold over cells expressing ZsGreen1, to 2.4-fold with deferoxamine (Figure 4B and 4D).

PPIX Primarily Accumulates and Localizes at the Plasma Membrane in HeLa Cells Expressing R433K

To evaluate cellular intactness and accumulated PPIX distribution in the transfected HeLa cells, we used confocal microscopy to visualize the fluorescent PPIX in individual HeLa cells (Figure 5A) and HeLa cells transfected with either the pIRES2-ZsGreen1 vector (Figure 5B) or pEF31 (harboring R433K) (Figures 5C and 5D). Transfected cells were grown in medium supplemented with 100 mM glycine for 18 hours in preparation for imaging. The outline of intact cells, and thus their morphological integrity, was evident in the three cases. As expected, green fluorescence was visualized in the transfected (Figures 5B, columns 2, 4 and 5, 5C, columns 2, 4 and 5 and 5D, columns 2, 4 and 5) but not control HeLa cells (Figure 5A, columns 2, 4 and 5). In fact, green fluorescence, arising from the soluble ZsGreen1 green fluorescent protein was observed evenly distributed throughout the cytoplasm of the HeLa cells transfected with either pIRES2-ZsGreen1 (Figure 5B) or the R433K-expression plasmid (Figures 5C and 5D). While the enzyme protoporphyrinogen oxidase catalyzes PPIX production exclusively within the mitochondrion, PPIX accumulated primarily within the plasma membranes of HeLa cells expressing R433K, as visualized by the characteristic red fluorescence; however, PPIX build-up was also apparent within the cells (Figures 5C, columns 3–5 and 5D, columns 3–5). Thus, it appears that much of the PPIX produced in the mitochondria eventually accumulates in the plasma membrane, entirely consistent with the fact that PPIX is a relatively lipophilic molecule. Of note, PPIX accumulated in not only transfected cells, but also in surrounding cells, indicating that PPIX could leave the transfected cells and be taken up by nearby cells (Figures 5, columns 3–5 and 5D, columns 3–5C). It is very likely that, additionally, PPIX accumulated in organelle membranes within the cell, such as in the mitochondrion, but since PPIX photobleached within a few seconds under the conditions utilized here, it was not possible to obtain high-resolution organelle images with our current microscopic parameters. Finally, the cobblestone-like morphology, instead of a slightly elongated shape, of the HeLa cells may result from the cell density and/or the nature of the growth surface. HeLa cells tend to adopt a more cobblestone-like appearance as cultures approach confluence and the HeLa cell morphology varies with the adhesion surface.

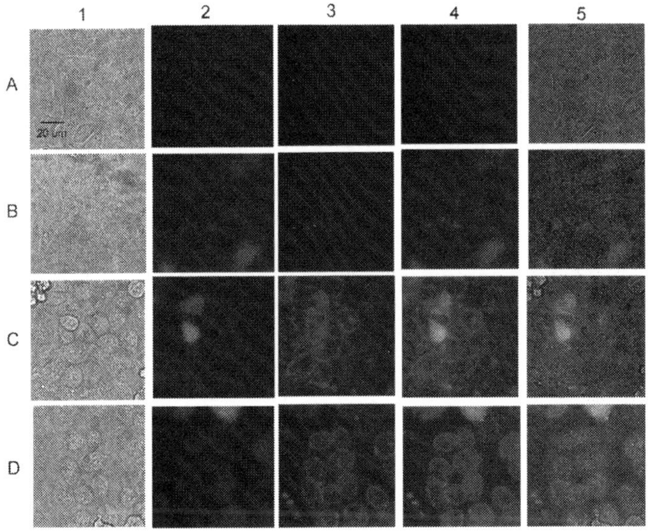

Figure 5: Fluorescence microscopy of HeLa cells expressing GFP and the R433K mALAS2 variant.

From left to right for each row, the panels correspond to (1) brightfield, (2) green fluorescence, (3) red fluorescence, (4) superimposed green and red fluorescence, and (5) superimposed brightfield, green fluorescence, and red fluorescence images. (A) HeLa cells grown to confluence in culture medium with 100 mM glycine. (B) HeLa cells transfected with pIRES2-ZsGreen1 and grown to confluence in culture medium with 100 mM glycine. (C) HeLa cells transfected with pEF31, which expresses the R433K variant from the CMV promoter in the pIRES2-ZsGreen1 vector (Table 1), and grown to confluence in culture medium with 100 mM glycine. (D) HeLa cells transfected with pEF31 and grown to subconfluence (*i.e.*, 80% confluence) in culture medium with 100 mM glycine. Pictures were obtained 24 hours after transfection and 18 hours after glycine supplementation. ZsGreen1 (green) is present throughout the cell, while PPIX (red) accumulates at the plasma membrane of cells expressing R433K and surrounding cells.

ALAS2-induced PPIX Accumulation Followed by Light Exposure Combined with Paclitaxel Treatment Causes Cell Death

Both propidium iodide and DAPI staining were utilized to assess cell viability after mALAS2-induced PPIX accumulation, light exposure, and paclitaxel treatment, based on the specific fluorescence emission of each dye when bound to the DNA of intact cells. Stains were added to cell samples 48 hours after transfection, and the fluorescence of DNA-bound propidium iodide and DAPI were measured by flow cytometric analysis using the respective fluorescence emission maxima of 613 nm and 460 nm. Expression of WT or HPVT did not significantly increase cell death following light exposure. However, expression of R433K caused a statistically significant ($p<0.05$) increase in cell death of up to 30%, as measured both by propidium iodine (Figure 6A) and DAPI (Figure 6B) staining, in comparison to expression of ZsGreen1 alone. Addition of paclitaxel increased cell death in all samples, including controls, by 10–25%. HeLa cells expressing R433K and treated with paclitaxel exhibited the highest percentage of up to 50% cell death (Figure 6A–B). Combination of paclitaxel with mALAS2-induced PPIX accumulation and light treatments exhibited an additive effect in causing death in HeLa cells.

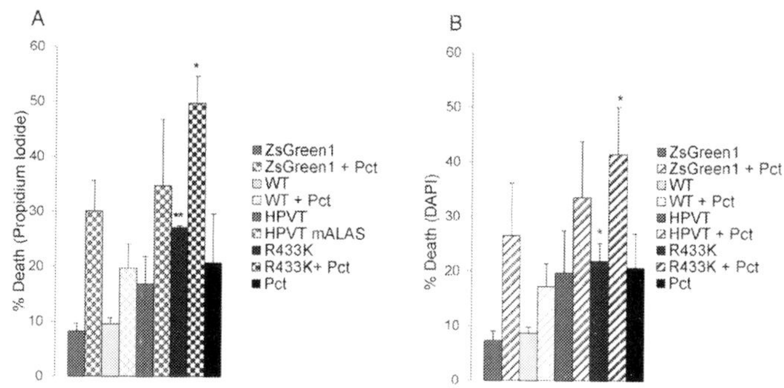

Figure 6: Light-induced cell death of HeLa cells expressing mALAS2 variants.

(A) Cell death measured by changes in DNA-bound propidium iodide fluorescence. (B) Cell death measured by changes in DNA-bound DAPI fluorescence. HeLa cells were transfected with expression plasmids for either wild type (WT) or mALAS2 variants and treated with 10 nM paclitaxel 4-hour post-transfection. 48 hours after transfection, cells were stained and analyzed by flow cytometry. Cell death percentage values were compared to that of the pIRES2-ZsGreen1-transfected cells control. Each data set represents three separate experiments ± standard deviation (*$p<0.05$ and **$p<0.01$, Student's *t*-test).

Supplementation of Cell Culture Medium with Glycine Enhances Phototoxicity and Cell Death in Malas2-expressing HeLa Cells

The phototoxicity and subsequent cell death caused by mALAS2-induced PPIX accumulation, when the culture medium was supplemented with glycine, were also investigated in HeLa cells. Cell viability was measured using propidium iodide (data not shown), DAPI staining, and MTT assays (Figure 7) 24 hours post-transfection, 18 hours after glycine addition, and 4 hours after ALA addition. Significant cell death was ascribed to those samples that exhibited more cell death compared to the pIRES2-ZsGreen1-transfected cells with the same glycine concentration as determined by Student's *t*-tests ($p<0.05$). Transfection of HeLa cells with the pIRES2-ZsGreen1 plasmid caused a decrease in cell viability by an average of 30%, which can be attributed to the mild cytotoxicity of the transfection reagents [54].

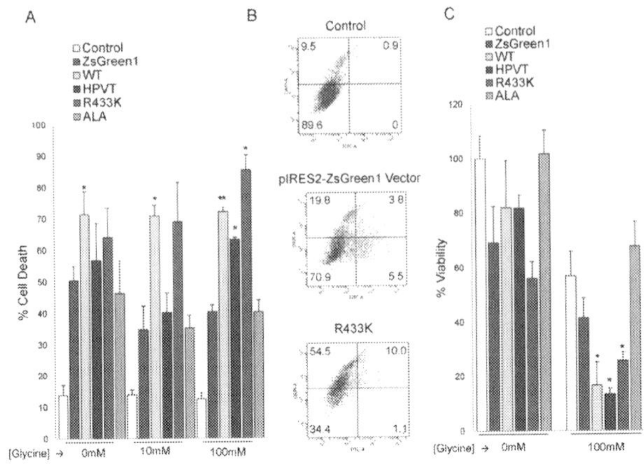

Figure 7: Light-induced cell death in murine ALAS2 variants-expressing HeLa cells cultured in a medium supplemented with glycine.

(A) Cell death as assessed by DAPI fluorescence. Six hours after transfection of the HeLa cells with the different expression plasmids, the cells were grown for an additional 18 hours in medium without or with glycine (10 mM or 100 mM). Cell death percentage values are representative of three separate experiments ± standard deviation (*p<0.05, Student's *t*-test). For control and ALA-treated cells, cell death was calculated by dividing the number of DAPI-stained cells by the total number of cells. For ZsGreen1-, WT-, HPVT-, and R433K-expressing cells, cell death was calculated from the ZsGreen1-positive cell populations by dividing the number of DAPI-stained cells by the total number of ZsGreen1-positive cells. (B) Cell death in HeLa cell populations grown in culture medium supplemented with glycine to a final concentration of 100 mM. The percentage of cells in each quadrant is written in the corners of each dot plot. (C) Cell viability as measured by metabolism of 3-(4,5-dimethylthiazol-2-yl)-2,5-diphenyltetrazolium bromide (MTT). Cells were transfected 24 hours before light treatment, and 100 mM glycine was added 18 hours before light treatment. Cell viability percentage values are normalized to MTT metabolism in healthy, untreated HeLa cells and are representative of at least three separate experiments ± standard deviation (*p<0.05, Student's *t*-test). "Control" refers to non-transfected HeLa cells.

Expressing WT, regardless of the glycine concentration in the culture medium, caused approximately 70% cell death following light exposure, as measured by DAPI staining (Figure 7A–B). In the MTT assays, expression of WT only caused statistically significant cell death ($p<0.05$) with supplementation of 100 mM glycine, in which case only an average of 15% of cells remained viable (Figure 7C). HeLa cells expressing HPVT or R433K did exhibit a glycine-dependent increase in cell death when the glycine concentration reached 100 mM as measured by both DAPI staining and MTT assays. When 100 mM glycine was added, cells expressing HPVT decreased from an average of 81% viable to 13% viable, and cells expressing R433K decreased from 58% viable to 25% viable as measured by MTT assays (Figure 7C). ALA treatment did not cause significant cell death ($p>0.05$), regardless of glycine concentration, when measured by either DAPI staining or MTT assays.

DISCUSSION

Photodynamic therapy (PDT) is a widely utilized clinical procedure for many types of cancer, as well as dermatological conditions such as psoriasis and schleroderma [55]–[57]. PDT is often initiated by application of the pro-drug 5-aminolevulinic acid (ALA) to the patient in order to photosensitize the tissue to be treated [49]. ALA-PDT is very effective, has minimal side effects and is being tested in numerous clinical trials on a wide-spectrum of cancers [49]. There are, however, some limitations to ALA-PDT and other PDTs that provide opportunities for improvement. For example, there is no specific mechanism that targets PPIX to tumor cells after ALA has been applied, other than a slightly preferential uptake of photosensitizers by hyperproliferating cells [58]. Consequently, normal tissue can be damaged from the procedure due to unintentional exposure to light resulting in pain, and conversely, not all hyperproliferating cells necessarily become highly sensitized [55], [59]. In an attempt to restrict photosensitivity by biological, rather than just chemical or physical means, an adenovirus expressing human ALAS2 with mutated HRMs was generated and H1299 lung carcinoma cells were successfully infected [40]. The adenovirus-infected H1299 cells accumulated more PPIX and became more photosensitive than cells supplied with ALA in their media, indicating that delivery of ALAS to tumors could potentially be a useful tool for PDT not only for better targeting, but also for greater photosensitivity [40]. However, the cell death was only 26% even in the presence of deferoxamine [40], leading us to postulate that one or more highly active variants of mALAS2 [31], [35] would increase PPIX accumulation, phototoxicity, and targeted cell death sufficiently to make the approach more clinically attractive.

To explore the ability of mALAS2 variants to cause accumulation of PPIX in mammalian cells, we transfected cells, of both erythroid and non-erythroid lineages, with murine ALAS2-expressing plasmids, and quantified the PPIX fluorescence using FACS analysis. In HeLa cells, we transfected plasmids encoding mALAS2 both with and without mutated presequences and with or without mutations in the mature enzyme sequence (Table 1, Figure 1). As with many mitochondrial proteins, ALAS2 is synthesized in the cytosol and contains a sequence at its N-terminus that targets ALAS2 for mitochondrial import after its synthesis [18]. Within the N-terminus of the ALAS2 precursor, there are three HRMs as recognized by adjacent cysteine-proline residues that have the potential to bind heme. The HRMs located within the presequence of ALAS2 (C11 and C38) have been shown to bind heme [24] and subsequently inhibit the mitochondrial import of ALAS2 [18]. Our experiments support the existing data that the cysteines in the HRMs bind heme, and that mutation of these HRMs relieve the inhibition of mitochondrial import, thus resulting in increased mature, functional ALAS2, as reflected by increased cellular concentrations of PPIX when the HRMs are mutated (Figure 2A). In our study, when the HRMs in the presequences of the mALAS2 constructs were mutated to relieve heme inhibition on mitochondrial import, there were significant increases in PPIX accumulation in HeLa cells expressing WT, HPVT, and R433K (Figure 2A–B).

We tested several mALAS2 variants, covering a range of *in vitro* activity from undetectable to higher than wild-type, for capacity to stimulate PPIX accumulation [31], [35], [36], [60]. Transfection of HeLa cells with the negative control plasmid harboring K313A$^{\Psi}$ resulted in no PPIX increase, as expected (Figure 2A). The mutation of K313 leads to undetectable enzymatic activity *in vitro*, attributable to the role of K313 in formation of a Schiff-base linkage with the PLP cofactor [36], and its additional function as a general base catalyst during the ALAS-catalyzed reaction [3], [28], [36], [61]. In preliminary studies with HeLa cells, it was observed that expression of HPVT$^{\Psi}$, which has seven mutations in its active site loop, yielded significantly more accumulated PPIX than expression of the pIRES2-ZsGreen1 vector control plasmid. HPVT$^{\Psi}$ was chosen for this study as the seven mutations of non-conserved residues in the active site loop resulted in the most active recombinant protein isolated from a variant library at 20°C [35]. Lendrihas *et al.* [35] hypothesized that these mutations in the active site loop, which increase both hydrophilicity and basicity, destabilize the loop by both increasing solubility and eliminating hydrophobic and electrostatic interactions that would typically act to stabilize the loop in its closed confirmation. However, the hyperactivity of HPVT$^{\Psi}$ was temperature-dependent; while at a temperature of 20°C these seven mutations increased the k_{cat} value of the recombinant, mature enzyme to more than 10 times of that of the WT enzyme, at 35°C the k_{cat} values were nearly the same [35]. While the levels of accumulated PPIX in HeLa cells expressing HPVT were significantly affected by supplementation with glycine ($p<0.01$), those in K562 cells expressing HPVT were only modestly increased when the medium was supplemented with 100 mM glycine. However, the HPVT-promoted PPIX concentration enrichments were much lower than those in cells expressing WT or R433K (Figures 3 and 4). The relatively low levels of PPIX associated with HPVT expression are presumably due to the temperature-dependent activity profile of this particular mALAS2 variant. HPVT was isolated from a library of mALAS2 variants engineered to possess greater enzymatic activity than wild type mALAS2 by targeting the active site loop to acquire different degrees of mobility [35]. Since the mutations in HPVT destabilized the active site loop and altered the protein conformation [35], it would not be surprising if this variant had a decreased cellular stability. Additionally, the HPVT mutations may affect protein-protein interactions, specifically the ability of mALAS2 to interact with a known binding partner, succinyl-CoA synthetase [62].

The simple addition of the non-toxic substrate glycine to the cell media increased PPIX production by WT, HPVT, and R433K significantly. When the culture medium was supplemented to a final concentration of 10 mM or 100 mM glycine, PPIX production in HeLa cells expressing WT was increased by more than 4- and 6-fold for each concentration, respectively (Figure 3). In K562 cells expressing WT, the effect was even larger, as PPIX increased by more than 5- and 10-fold for 10 mM and 100 mM glycine, respectively (Figure 4A and 4C). The slightly higher increases in PPIX concentrations in the K562 cells when treated with glycine could be due to the stable expression of ALAS2 *versus* the transient transfection used in the HeLa cells. However, the ability of the K562 cells to tolerate higher expression of WT and production of PPIX, in comparison to HeLa cells, might also be attributable to their erythroid lineage. K562 cells are of the erythroleukemic cell type, and bear some proteomic resemblance to undifferentiated erythrocytes [51], [63] and express endogenous ALAS2. Thus while expressing ALAS2 in HeLa cells introduces an enzyme that does not normally exist in epithelial cells, K562 cells may adapt more easily to the expression and up-regulation of the heme biosynthetic pathway. It seems likely that the capacity of mALAS2 to accumulate PPIX will be found to vary in other mammalian cell types as well.

Supplementing the culture medium of K562 cells expressing WT with 100 μM deferoxamine, an iron-chelator previously shown to increase the amount of PPIX in ALAS adenovirus-infected H1299 cells [40], resulted in increase in the mean PPIX fluorescence per cell (Figure 4B and 4D), but it was much less than with glycine supplementation. However, the modest increase was similar to what Gagnebin *et al.* [40] observed in the ALAS adenovirus-infected cells, suggesting the much larger increases reported here with R433K and glycine supplementation represent substantial advancements in our ability to overproduce PPIX in mammalian cells. R433K is only twice as active as WT *in vitro*, and it is reasonable to believe that significantly more active variants could be readily produced via directed evolution, and that these variants would facilitate even greater levels of PPIX accumulation, perhaps to the point of expanding the clinical applicability of PPIX-based PDT.

The HeLa cells that accumulated the highest amount of PPIX were those expressing R433K. The R433K mutation was analyzed in 1998 by Tan *et al.* [31] in a study that identified the nearby R439 as being a conserved residue in many α-family PLP-dependent enzymes, as well as being involved in binding of the amino acid substrate during catalysis [31]. In that study, the kinetic parameters were defined for R439L, R439K, R433L, and R433K. Although not the primary focus of the article, the kinetic parameters for R433L were found to be comparable to those of the wild-type at 37°C, while for R433K the k_{cat} was increased by two-fold, with no effect on the Michaelis constants, resulting in a doubling of catalytic efficiency for both substrates [31]. In the results detailed here, expression of R433K in HeLa cells produced a 2- to 3-fold increase in PPIX in comparison to WT, and a 4- to 6-fold increase in PPIX fluorescence in comparison to cells transfected with the pIRES2-ZsGreen1 vector control or HeLa cells alone (Figure 2). When the culture medium was supplemented with glycine, PPIX accumulation increased by 13- to 15-fold in comparison to cells transfected with the pIRES2-ZsGreen1 vector control or HeLa cells alone, and represented the conditions for the highest cellular PPIX accumulation. Given that within these samples there were cells fluorescing with PPIX that were not ZsGreen1-expressing (Figure 3B), some of the PPIX produced in the R433K- and WT-expressing HeLa cells appears to have been transported out of the cells into the culture medium, where it was then taken up by non-transfected cells. These populations of red-fluorescing cells were comparable to those cells in culture medium supplemented with ALA (Figure 3B). This is a very important finding in terms of the potential of these constructs for photodynamic therapy, as it indicates that the PPIX produced within a delivery cell could be transported into surrounding diseased tissue and accumulate to clinically relevant levels.

The subcellular distribution of a photosensitizing agent might have important consequences in regards to PDT efficacy [64]. Fluorescence microscopy was successfully utilized to visualize PPIX accumulation in R433K expressing HeLa cells. Due to the extremely short time (a few seconds) it took for PPIX to photobleach, the resolution could not be optimized for visibility of individual organelles such as mitochondria and nuclei. However, it was possible to observe that PPIX had accumulated in the plasma membranes of the R433K expressing cells; and this fluorescence was not seen in the pIRES2-ZsGreen1 vector-expressing cells (Figure 5). It also appears that PPIX can actively exit the transfected cells, potentially via ABC transporters, and enter surrounding cells. Presumably, the lipophilic PPIX accumulated in the organelle membranes as well, but direct microscopic confirmation of PPIX accrual in organelle membranes awaits approaches resulting in better resolution than we were able to achieve here.

In order to study the photosensitivity of cells that accumulated PPIX (caused by expression of WT, HPVT, and R433K), we performed light treatments using an incandescent lamp, which emits light in the visible and infrared spectral regions (400–1000 nm). The cells were exposed for

10 minutes before FACS analysis of PPIX and cell death. Expression of the ZsGreen1 protein alone caused a toxicity that was independent of light treatment, and thus significant cell death was measured against the pIRES2-ZsGreen1 vector-transfected cells ($p<0.05$) as opposed to HeLa cells alone (Figure 7). Regardless of glycine concentration, the expression of WT led to a significant increase in cell death by 1.4- to 2-fold. R433K and HPVT, when the culture medium was supplemented with 100 mM glycine, also caused significant increases in cell death by 2.1- and 1.6-fold, respectively. With no or 10 mM glycine supplementation the error was too high for any of the increases to be significant, although there was a trend towards increased cell death. For WT and R433K, increased PPIX accumulation correlates with increased cell death after light treatment. In spite of the more modest increases in PPIX accumulation, light-induced death also occurred in HPVT-expressing cells. This finding led us to suggest that toxicity inherent to the overproduced HPVT protein is the other contributing factor to cell death. Although purified recombinant, mature HPVT is more active than WT at 37°C (with an enhancement of approximately 1.5 in the k_{cat} value at 37°C *vs.* a 15.5-fold increased k_{cat} at 20°C) [65], when expressed in mammalian cells, HPVT is not as active as WT, as determined by PPIX fluorescence. Since expression of HPVT paired with glycine supplementation caused significant cell death (Figure 7), the potential role of another toxic heme pathway intermediate, such as the pro-oxidant ALA [66]–[69], in causing cell death, is plausible. However, at the present, different levels of protein expression [Quantification of ALAS2 and variant protein levels was not feasible due to the unavailability of a functional ALAS2 antibody.] and distinct modulation of the enzymatic activity between ALAS2 and variants cannot be ruled out as possibilities. Treatment with 100 μM ALA, a positive control for PPIX accumulation, caused a similar cell death after light exposure to expression of the ZsGreen1 protein alone. In summary, the highest cell death was seen in light-treated HeLa cells expressing ZsGreen1 and either WT, HPVT, or R433K, in culture medium supplemented with 100 mM glycine. The total cell death was observed to be as high as 90% in the cells expressing both ZsGreen1 and R433K. This represents a substantial improvement upon the 26% cell death reported previously [40], and indicates that this approach, if carefully developed, may eventually find some clinical utility.

Because ALAS2, especially highly stable and active variants of ALAS2, would be useful in the development of multiplex cancer treatments involving PDT, we experimented with combination PDT and drug treatments of HeLa cells. Paclitaxel (Taxol) is currently approved in the United States for the treatment of AIDS-related Kaposi sarcoma [70], breast cancer [71], non-small lung cell lung cancer [72], and ovarian cancer [73]. Paclitaxel induces apoptosis in cancer cells by binding to tubulin and inhibiting the disassembly of microtubules, thereby resulting in the inhibition of cell division [46], [74]. As expected, paclitaxel did not affect PPIX accumulation, as indicated by the similar mean PPIX fluorescence values between untreated and paclitaxel-treated cells (Figure 2B). However, paclitaxel had an additive effect to PDT and increased cell death in all samples by 10–25%. HeLa cells expressing R433K treated with paclitaxel exhibited the highest percentage of cell death (Figure 6A–B). Ever since PDT was first shown to be able to elicit an immune response [75], many recent advances in PDT are aiming toward creating PDT-generated cancer vaccines [76]. Using highly active and stable ALAS2 variants as part of a vaccine strategy for photosensitization may be useful for this vaccine approach. Further experiments, both in cell culture and live animals, are necessary to test for potential immune response stimulated by ALAS2-PDT.

In this study, we have shown that transfecting mammalian cells, of both erythroid and non-erythroid lineages, with mALAS2 variants, is an effective way to stimulate cellular PPIX accumulation. Furthermore, supplementing the culture medium with glycine vastly increases the intracellular PPIX levels when cells express either WT or R433K. These data offer new ways to accumulate high levels of the photosensitizer PPIX in cancer cells for more targetable and efficient PDT. Human ALAS2 variants with higher than normal activity are now known to occur in nature, and are associated with a form of erythropoietic protoporphyria known as X-linked dominant protoporphyria, which is characterized by a 24-fold increase in erythrocyte PPIX concentrations [39]. It would be appropriate if these variants were eventually utilized in PDT, and could thereby allow clinicians to exploit one disease to treat or cure others.

Porphyrin accumulation in HeLa cells through expression of mALAS2 compared to supplementation of culture medium with ALA. HeLa cells transfected with mALAS2 variants were analyzed 24 hours after transfection, and ALA-treated HeLa cells were analyzed 4 hours after addition of ALA to the culture medium. Mean PPIX fluorescence values were compared to those of the pIRES2-ZsGreen1-transfected cells control; each of PPIX data sets is representative of three separate experiments ± standard deviation [a.u., arbitrary units].

Porphyrin accumulation in HeLa cells through expression of mALAS2 compared to supplementation of culture medium with ALA. HeLa cells transfected with mALAS2 variants were analyzed 24 hours after transfection, and ALA-treated HeLa cells were analyzed 4 hours after addition of ALA to the culture medium. Mean PPIX fluorescence values were compared to those of the pIRES2-ZsGreen1-transfected cells control; each of PPIX data sets is representative of three separate experiments ± standard deviation [a.u., arbitrary units].

ACKNOWLEDGMENTS

We would like to thank Dr. Terri Hunter for her help with experimental design and cell culture work and Dr. Scott Antonia for providing several materials used in the experiments in this study. We would also like to thank Dr. Karoly Szerekes for help and training with the use of the flow cytometer, Dr. Byeong Cha for assistance with microscopy, and Dr. Javier Cuevas, Dr. Peter Medveczky and Shara Pantry for their generosity, teaching, and advice.

AUTHOR CONTRIBUTIONS

Conceived and designed the experiments: EJF GAH GCF. Performed the experiments: EJF. Analyzed the data: EJF GAH GCF. Contributed reagents/materials/analysis tools: GCF. Wrote the paper: EJF GAH GCF.

REFERENCES

1. Jordan PM (1990) In: Dailey HA, editor. Biosynthesis of Heme and Chlorophylls. New York: McGraw-Hill. 55–121.
2. Hunter GA, Ferreira GC (2011) Molecular enzymology of 5-aminolevulinate synthase, the gatekeeper of heme biosynthesis. Biochim Biophys Acta 11: 1467–1473.
3. Ferreira GC, Gong J (1995) 5-Aminolevulinate synthase and the

first step of heme biosynthesis. J Bioenerg Biomembr 27: 151–159.

4. Riddle RD, Yamamoto M, Engel JD (1989) Expression of delta-aminolevulinate synthase in avian cells: separate genes encode erythroid-specific and nonspecific isozymes. Proc Natl Acad Sci USA 86: 792–796.
5. Sutherland GR, Baker E, Callen DF, Hyland VJ, May BK, et al. (1988) 5-Aminolevulinate synthase is at 3p21 and thus not the primary defect in X-linked sideroblastic anemia. Am J Hum Genet 43: 331–335.
6. Bishop DF, Henderson AS, Astrin KH (1990) Human delta-aminolevulinate synthase: assignment of the housekeeping gene to 3p21 and the erythroid-specific gene to the X chromosome. Genomics 7: 207–214.
7. Watanabe N, Hayashi N, Kikuchi G (1983) delta-Aminolevulinate synthase isozymes in the liver and erythroid cells of chicken. Biochem Biophys Res Commun 113: 377–383.
8. Cox TC, Bawden MJ, Abraham NG, Bottomley SS, May BK, et al. (1990) Erythroid 5-aminolevulinate synthase is located on the X chromosome. Am J Hum Genet 46: 107–111.
9. Yamauchi K, Hayashi N, Kikuchi G (1980) Translocation of delta-aminolevulinate synthase from the cytosol to the mitochondria and its regulation by hemin in the rat liver. J Biol Chem 255: 1746–1751.
10. Srivastava G, Borthwick IA, Brooker JD, May BK, Elliott WH (1983) Evidence for a cytosolic precursor of chick embryo liver mitochondrial delta-aminolevulinate synthase. Biochem Biophys Res Commun 110: 23–31.
11. Volland C, Urban-Grimal D (1988) The presequence of yeast 5-aminolevulinate synthase is not required for targeting to mitochondria. J Biol Chem 263: 8294–8299.
12. Neupert W (1997) Protein import into mitochondria. Annu Rev Biochem 66: 863–917.
13. Sadlon TJ, Dell'Oso T, Surinya KH, May BK (1999) Regulation of erythroid 5-aminolevulinate synthase expression during erythropoiesis. Int J Biochem Cell Biol 31: 1153–1167.
14. Ades IZ, Harpe KG (1981) Biogenesis of mitochondrial proteins.

Identification of the mature and precursor forms of the subunit of delta-aminolevulinate synthase from embryonic chick liver. J Biol Chem 256: 9329–9333.

15. McKay R, Druyan R, Getz GS, Rabinowitz M (1969) Intramitochondrial localization of delta-aminolaevulate synthetase and ferrochelatase in rat liver. Biochem J 114: 455–461.
16. Zuyderhoudt FM, Borst P, Huijing F (1969) Intramitochondrial localization of 5-aminolaevulinate synthase induced in rat liver with allylisopropylacetamide. Biochim Biophys Acta 178: 408–411.
17. Dailey TA, Woodruff JH, Dailey HA (2005) Examination of mitochondrial protein targeting of haem synthetic enzymes: in vivo identification of three functional haem-responsive motifs in 5-aminolaevulinate synthase. Biochem J 386: 381–386.
18. Lathrop JT, Timko MP (1993) Regulation by heme of mitochondrial protein transport through a conserved amino acid motif. Science 259: 522–525.
19. Zhang L, Guarente L (1995) Heme binds to a short sequence that serves a regulatory function in diverse proteins. EMBO J 14: 313–320.
20. Qi Z, O'Brian MR (2002) Interaction between the bacterial iron response regulator and ferrochelatase mediates genetic control of heme biosynthesis. Mol Cell 9: 155–162.
21. Gonzalez-Dominguez M, Freire-Picos MA, Cerdan ME (2001) Haem regulation of the mitochondrial import of the Kluyveromyces lactis 5-aminolaevulinate synthase: an organelle approach. Yeast 18: 41–48.
22. Ogawa K, Sun J, Taketani S, Nakajima O, Nishitani C, et al. (2001) Heme mediates derepression of Maf recognition element through direct binding to transcription repressor Bach1. EMBO J 20: 2835–2843.
23. Haro C, Mendez R, Santoyo J (1996) The eIF-2alpha kinases and the control of protein synthesis. FASEB J 10: 1378–1387.
24. Goodfellow BJ, Dias JS, Ferreira GC, Henklein P, Wray V, et al. (2001) The solution structure and heme binding of the presequence of murine 5-aminolevulinate synthase. FEBS Lett 505: 325–331.

25. Hunter GA, Ferreira GC (1999) Pre-steady-state reaction of 5-aminolevulinate synthase. Evidence for a rate-determining product release. J Biol Chem 274: 12222–12228.
26. Hunter GA, Ferreira GC (1995) A continuous spectrophotometric assay for 5-aminolevulinate synthase that utilizes substrate cycling. Anal Biochem 226: 221–224.
27. Zhang J, Ferreira GC (2002) Transient state kinetic investigation of 5-aminolevulinate synthase reaction mechanism. J Biol Chem 277: 44660–44669.
28. Hunter GA, Zhang J, Ferreira GC (2007) Transient kinetic studies support refinements to the chemical and kinetic mechanisms of aminolevulinate synthase. J Biol Chem 282: 23025–23035.
29. Ferreira GC, Vajapey U, Hafez O, Hunter GA, Barber MJ (1995) Aminolevulinate synthase: lysine 313 is not essential for binding the pyridoxal phosphate cofactor but is essential for catalysis. Protein Sci 4: 1001–1006.
30. Tan D, Barber MJ, Ferreira GC (1998) The role of tyrosine 121 in cofactor binding of 5-aminolevulinate synthase. Protein Sci 7: 1208–1213.
31. Tan D, Harrison T, Hunter GA, Ferreira GC (1998) Role of arginine 439 in substrate binding of 5-aminolevulinate synthase. Biochemistry 37: 1478–1484.
32. Gong J, Hunter GA, Ferreira GC (1998) Aspartate-279 in aminolevulinate synthase affects enzyme catalysis through enhancing the function of the pyridoxal 5′-phosphate cofactor. Biochemistry 37: 3509–3517.
33. Lendrihas T, Zhang J, Hunter GA, Ferreira GC (2009) Arg-85 and Thr-430 in murine 5-aminolevulinate synthase coordinate acyl-CoA-binding and contribute to substrate specificity. Protein Sci 18: 1847–1859.
34. Lendrihas T, Hunter GA, Ferreira GC (2010) Serine 254 enhances an induced fit mechanism in murine 5-aminolevulinate synthase. J Biol Chem 285: 3351–3359.
35. Lendrihas T, Hunter GA, Ferreira GC (2010) Targeting the active site gate to yield hyperactive variants of 5-aminolevulinate synthase. J Biol Chem 285: 13704–13711.

36. Ferreira GC, Neame PJ, Dailey HA (1993) Heme biosynthesis in mammalian systems: evidence of a Schiff base linkage between the pyridoxal 5′-phosphate cofactor and a lysine residue in 5-aminolevulinate synthase. Protein Sci 2: 1959–1965.

37. Hunter GA, Ferreira GC (1999) Lysine-313 of 5-Aminolevulinate Synthase Acts as a General Base during Formation of the Quinonoid Reaction Intermediates. Biochemistry 38: 3711–3718.

38. Astner I, Schulze JO, van den Heuvel J, Jahn D, Schubert WD, et al. (2005) Crystal structure of 5-aminolevulinate synthase, the first enzyme of heme biosynthesis, and its link to XLSA in humans. EMBO J 24: 3166–3177.

39. Whatley SD, Ducamp S, Gouya L, Grandchamp B, Beaumont C, et al. (2008) C-terminal deletions in the ALAS2 gene lead to gain of function and cause X-linked dominant protoporphyria without anemia or iron overload. Am J Hum Genet 83: 408–414.

40. Gagnebin J, Brunori M, Otter M, Juillerat-Jeanneret L, Monnier P, et al. (1999) A photosensitising adenovirus for photodynamic therapy. Gene Ther 6: 1742–1750.

41. Issa MC, Manela-Azulay M (2010) Photodynamic therapy: a review of the literature and image documentation. An Bras Dermatol 85: 501–511.

42. Buzzeo R, Enkemann S, Nimmanapalli R, Alsina M, Lichtenheld MG, et al. (2005) Characterization of a R115777-resistant human multiple myeloma cell line with cross-resistance to PS-341. Clin Cancer Res 11: 6057–6064.

43. Yanamandra N, Buzzeo RW, Gabriel M, Hazlehurst LA, Mari Y, et al. (2011) Tipifarnib-induced apoptosis in acute myeloid leukemia and multiple myeloma cells depends on Ca2+ influx through plasma membrane Ca2+ channels. J Pharmacol Exp Ther 337: 636–643.

44. Boshart M, Weber F, Jahn G, Dorsch-Hasler K, Fleckenstein B, et al. (1985) A very strong enhancer is located upstream of an immediate early gene of human cytomegalovirus. Cell 41: 521–530.

45. Tan D, Ferreira GC (1996) Active site of 5-aminolevulinate synthase resides at the subunit interface. Evidence from in vivo heterodimer formation. Biochemistry 35: 8934–8941.

46. Schiff PB, Horwitz SB (1980) Taxol stabilizes microtubules in

mouse fibroblast cells. Proc Natl Acad Sci U S A 77: 1561–1565.

47. Manfredi JJ, Parness J, Horwitz SB (1982) Taxol binds to cellular microtubules. J Cell Biol 94: 688–696.
48. Mole-Bajer J, Bajer AS (1983) Action of taxol on mitosis: modification of microtubule arrangements and function of the mitotic spindle in Haemanthus endosperm. J Cell Biol 96: 527–540.
49. Krammer B, Plaetzer K (2008) ALA and its clinical impact, from bench to bedside. Photochem Photobiol Sci 7: 283–289.
50. Ryan WL, Carver MJ (1966) Free amino acids of human foetal and adult liver. Nature 212: 292–293.
51. Andersson LC, Nilsson K, Gahmberg CG (1979) K562–a human erythroleukemic cell line. Int J Cancer 23: 143–147.
52. Hoffbrand AV, Wonke B (1997) Iron chelation therapy. J Intern Med Suppl 740: 37–41.
53. Roberts DJ, Rees D, Howard J, Hyde C, Alderson P, et al. (2005) Desferrioxamine mesylate for managing transfusional iron overload in people with transfusion-dependent thalassaemia. Cochrane Database Syst Rev: CD004450.
54. Nikcevic G, Kovacevic-Grujicic N, Stevanovic M (2003) Improved transfection efficiency of cultured human cells. Cell Biol Int 27: 735–737.
55. Agostinis P, Berg K, Cengel KA, Foster TH, Girotti AW, et al. (2011) Photodynamic therapy of cancer: an update. CA Cancer J Clin 61: 250–281.
56. Dolmans DE, Fukumura D, Jain RK (2003) Photodynamic therapy for cancer. Nat Rev Cancer 3: 380–387.
57. Darlenski R, Fluhr JW (2013) Photodynamic therapy in dermatology: past, present, and future. J Biomed Opt 18: 061208.
58. Babilas P, Landthaler M, Szeimies RM (2006) Photodynamic therapy in dermatology. Eur J Dermatol 16: 340–348.
59. Warren CB, Karai LJ, Vidimos A, Maytin EV (2009) Pain associated with aminolevulinic acid-photodynamic therapy of skin disease. J Am Acad Dermatol 61: 1033–1043.
60. Hunter GA, Ferreira GC (1999) Lysine-313 of 5-aminolevulinate synthase acts as a general base during formation of the quinonoid

reaction intermediates. Biochemistry 38: 3711–3718.

61. Hunter GA, Ferreira GC (1999) Lysine-313 of 5-aminolevulinate synthase acts as a general base during formation of the quinonoid reaction intermediates. Biochemistry 38: 12526.
62. Bishop DF, Tchaikovskii V, Hoffbrand AV, Fraser ME, Margolis S (2012) X-linked sideroblastic anemia due to carboxyl-terminal ALAS2 mutations that cause loss of binding to the beta-subunit of succinyl-CoA synthetase (SUCLA2). J Biol Chem 287: 28943–28955.
63. Andersson LC, Jokinen M, Klein E, Klein G, Nilsson K (1979) Presence of erythrocytic components in the K562 cell line. Int J Cancer 24: 514.
64. Ji Z, Yang G, Vasovic V, Cunderlikova B, Suo Z, et al. (2006) Subcellular localization pattern of protoporphyrin IX is an important determinant for its photodynamic efficiency of human carcinoma and normal cell lines. J Photochem Photobiol B 84: 213–220.
65. Lendrihas T, Hunter GA, Ferreira GC (2010) Targeting the active site gate to yield hyperactive variants of 5-aminolevulinate synthase. J Biol Chem 285: 13704–13711.
66. Juknat AA, Kotler ML, Quaglino A, Carrillo NM, Hevor T (2003) Necrotic cell death induced by delta-aminolevulinic acid in mouse astrocytes. Protective role of melatonin and other antioxidants. J Pineal Res 35: 1–11.
67. Weiss TS, Pahernik S, Scheruebl I, Jauch KW, Thasler WE (2003) Cellular damage to human hepatocytes through repeated application of 5-aminolevulinic acid. J Hepatol 38: 476–482.
68. Felitsyn N, McLeod C, Shroads AL, Stacpoole PW, Notterpek L (2008) The heme precursor delta-aminolevulinate blocks peripheral myelin formation. J Neurochem 106: 2068–2079.
69. Hunter GA, Rivera E, Ferreira GC (2005) Supraphysiological concentrations of 5-aminolevulinic acid dimerize in solution to produce superoxide radical anions via a protonated dihydropyrazine intermediate. Arch Biochem Biophys 437: 128–137.
70. Cheung MC, Pantanowitz L, Dezube BJ (2005) AIDS-related malignancies: emerging challenges in the era of highly active antiretroviral therapy. Oncologist 10: 412–426.

71. Sparano JA, Wang M, Martino S, Jones V, Perez EA, et al. (2008) Weekly paclitaxel in the adjuvant treatment of breast cancer. N Engl J Med 358: 1663–1671.
72. Ramalingam SS, Khuri FR (2009) The role of the taxanes in the treatment of older patients with advanced stage non-small cell lung cancer. Oncologist 14: 412–424.
73. .Katsumata N, Yasuda M, Takahashi F, Isonishi S, Jobo T, et al. (2009) Dose-dense paclitaxel once a week in combination with carboplatin every 3 weeks for advanced ovarian cancer: a phase 3, open-label, randomised controlled trial. Lancet 374: 1331–1338.
74. Horwitz SB (1992) Mechanism of action of taxol. Trends Pharmacol Sci 13: 134–136.
75. Canti G, Lattuada D, Nicolin A, Taroni P, Valentini G, et al. (1994) Antitumor immunity induced by photodynamic therapy with aluminum disulfonated phthalocyanines and laser light. Anticancer Drugs 5: 443–447.
76. Korbelik M (2010) Photodynamic therapy-generated cancer vaccines. Methods Mol Biol 635: 147–153.

CHAPTER

8

LONG TERM CULTURE OF MDCK STRAINS ALTERS CHROMOSOME CONTENT

Doris Cassio[1, 2]

[1]Inserm UMR-S757, Orsay 91405, France

[2]Université Paris-Sud, Orsay 91405, France

ABSTRACT

Background

The Madin-Darby Canine Kidney (MDCK) cell line and its different strains are widely used as models for studying epithelial simple polarity. Recently Dukes et al. [BMC Cell Biology 12:43, 2011] provided a useful guide to the different MDCK strains, with a directory of where to buy them. The present work focused on chromosome content of MDCK cells, a parameter often disregarded by researchers working with these cells.

Findings

Using a general and reliable method for obtaining high yield of metaphasic preparations, the chromosome content of MDCK, MDCK I, MDCK II obtained from reliable sources was determined after maintaining them in culture for various periods of time. Within two months significant changes were observed in the range and the mean number of chromosomes of MDCK I and MDCK II cells. MDCK II cells routinely cultured in six different laboratories were also examined. In some of these cultures the cells have considerably drifted as shown by a high scattering in their number of chromosomes.

Conclusion

These results entitle me to encourage researchers using MDCK cells obtained from reliable sources, to determine their chromosomal content upon receipt, to check this content after several passages, to use this feature to follow the possible drift of these cells, and finally to avoid working with cells maintained for more than two months in culture.

Keywords: Madin-Darby Canine Kidney (MDCK) cells MDCK I MDCK II Metaphase Chromosome content Cell line drift

FINDINGS

Background

A great variety of cell lines are used to study numerous fields of cell biology. Very few of these lines present a normal karyotype. They are generally aneuploid and often contain rearranged chromosomes. Moreover, depending on the duration and the conditions of culture, populations of these cell lines can change. Because of possible drift, cell lines need to be regularly characterized and checked, one fundamental verification being to test the stability of their chromosome content. Unfortunately, this is often disregarded by researchers working with cell lines, as exemplified here for MDCK cells.

MDCK cells [1] constitute one of the best models for studying protein trafficking, polarity and junctions in epithelial cells (over 6600 references in PubMed [2]). As reported by Dukes et al. [3], researchers often report on "MDCK cells" with no reference to their strain nor to the supplier. Moreover, among the studies on MDCK, very few (4 references in PubMed [4]) have payed attention to the chromosome content of these cells (or that of their subclones [5]) and none had examined the possible drift of this content during the culture.

In the present work the chromosome content of MDCK, MDCK I, MDCK II from different sources was determined at different passages and/or generations. Significant and even large differences were observed in the range and the mean number of chromosomes.

METHODS

The parental MDCK (NBL-2) line and the strains MDCK I and MDCK II, were obtained from the American Type Culture Collection (ATCC; [http://www.atcc.org]), Kai Simons [6] and the European Collection Cell cultures (ECACC; [http://www.hpacultures,org,uk/collections]) (Table 1). MDCK II cells were also obtained from six different laboratories, where these cells are routinely cultured. Cells were cultured as previously described in F-12 Coon's modified medium (Sigma-Aldrich) supplemented with 5% fœtal calf serum [7]. MDCK II cells (from ECACC) were also cultured in DMEM medium. The changes of chromosome content were similar in both media. Cells were plated at 2-4 10^3 cells/cm^2 and passed every 5-7 days; MDCK II was also passed twice a week. At each passage, cells were counted and the number of generations calculated; one passage every 5-7 days corresponded to 5-7 generations. Metaphases were prepared from cultures in exponential phase, using a method giving a high yield in metaphases [8]. This method is easy, reliable and suitable metaphases can be obtained and then counted rapidly. It has been used before to evaluate the mean number and the range of chromosomes of many lines (L(Cl1D), WI38, V79-4, HT29, Fao, BW1-J) [8]. The results were similar to those previously published, showing the reliability of this method. The Giemsa solid staining [9] which gives uniform staining of the chromosomes and makes it easy to count them was used.. The mean number of chromosomes and the range were obtained by analyzing 50 metaphases. In the case of very heterogeneous cultures 85 metaphases were examined. The number of chromosomes of each metaphase was counted twice. Examples of metaphases of MDCK II are shown in Figure 1.

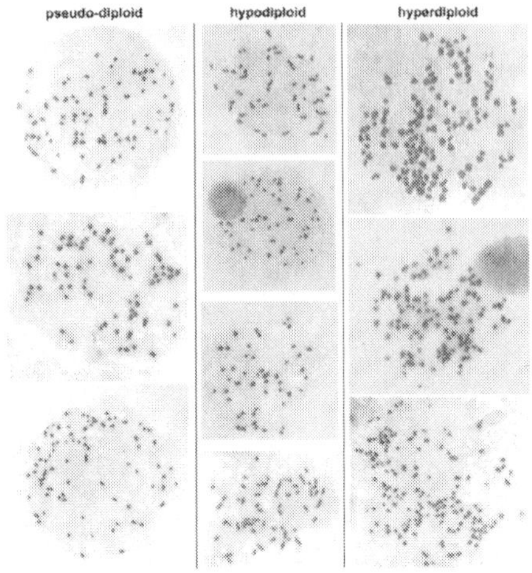

Figure 1: Metaphases of MDCK II cells. Pseudo-diploid metaphases (left) were from cells maintained one month in culture. Hypodiploid metaphases (middle) were from cells maintained 2-3 months in culture. Hyperdiploid metaphases (right) were from top to bottom from the MDCKII cultures shown in Figure 3F, C, E, respectively.

Table 1: Chromosome content of MDCK, MDCK I, and MDCK II cells

Strain	Supplier	Passage, time in culture	Chromo-some mean number	Range	Figure
MDCK	ATCC	passage 11 (upon receipt)	77*	74-81	Figure 1A
		- maintained for two months in cul-ture (10 passages, 60 generations)	75	68-81	Figure 1B
MDCK I	Kai Sim-mons	passage 21 (upon receipt)	79*	76-83	Figure 1C
		- maintained for one month in cul-ture (5 passages, 30 generations)	78	72-82	Figure 1D
		- maintained for two months in cul-ture (10 passages, 60generations)	76	60-80	Figure 1E

MDCK II	ECACC	passage 27 (upon receipt)	75*	69-80	Figure 1F
		- maintained for one month in culture (5 passages, 30 generations)	75	66-82	Figure 1G
		- maintained for two months in culture (10 passages, 60 generations)	73	52-81	Figure 1H
		- maintained for two months in culture (20 passages, 60 generations)	73	51-81	Figure 1I
		- maintained for three months in culture (15 passages, 90 generations)	69	47-82	Figure 1J
MDCK II	from six different laboratories	passage 46	78	74-81	Figure 2A
		passage unknown	75*	67-82	Figure 2B
		passage unknown	75	63-105	Figure 2C
		passage 16	71	51-79	Figure 2D
		passage 26	78 [a]	49-157	Figure 2E
		passage unknown	112[a]	60-144	Figure 2F

* mean number and modal number, [a] 85 metaphases analyzed.

Chromosome content of MDCK, MDCK I, MDCK II and evolution of this content with time

The pseudo-diploid parental MDCK (NBL-2) line and its two subclones MDCK I and MDCK II were obtained at early passages from reliable suppliers (Table 1). The chromosome content was determined upon receipt of the cells, and after several passages (Table 1, Figure 2). For MDCK cells, as for MDCK I and MDCKII, the chromosome content of early cultures was similar to that previously published [5, 6, 8]. The range was narrow and the mean number of chromosomes coincided with the modal number (Table 1). No change was observed for the parental MDCK cells after two months in culture (Table 1, Figure 2A,B). In contrast, changes were noticed as MDCK I and MDCK II cells were maintained in culture for months (Figure 2C-E; Figure 2F-J). The populations became more heterogeneous, the mean number of chromosomes decreased and differed from the modal number, and the range was larger (Table 1). Among MDCK II cells maintained for two months, 20% of metaphases had lost some of their chromosomes (compare Figure 2H,I to Figure 2F) and the phenomenon increased in time, as 34% of metaphases had even fewer chromosomes after three months (Figure 2J). It should be noted that the chromosome content of MDCK II cells maintained for two months and 60 generations in culture was not sensitive to the number of passages (compare 10 passages, Figure 2H to 20 passages, Figure 2I). Moreover, MDCK II cell populations became heterogeneous in morphology with time in culture, as attested by the emergence of small cells forming very tight colonies when cells were cultured for three months (insert in Figure 2J). In parallel, MDCK II cell density at confluency significatively changed from 0.9 10^5 cells/cm^2 (early cultures) to 1.4 10^5 cells/cm^2 (three-month culture) ($p=0.005$). The tendency to select cells with fewer chromosomes as culture lasts longer was also observed for MDCK I (Figure 2C-E).

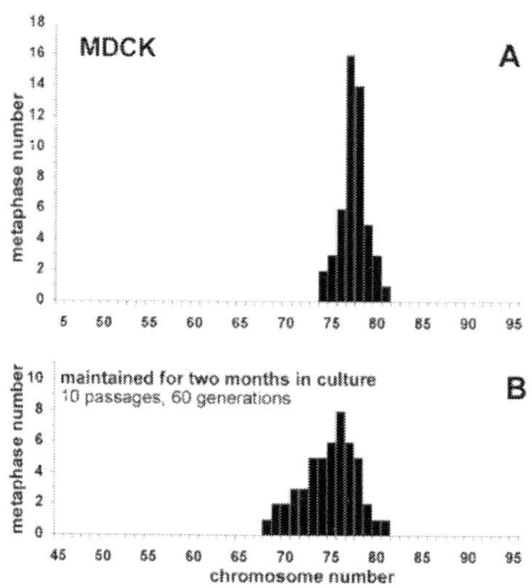

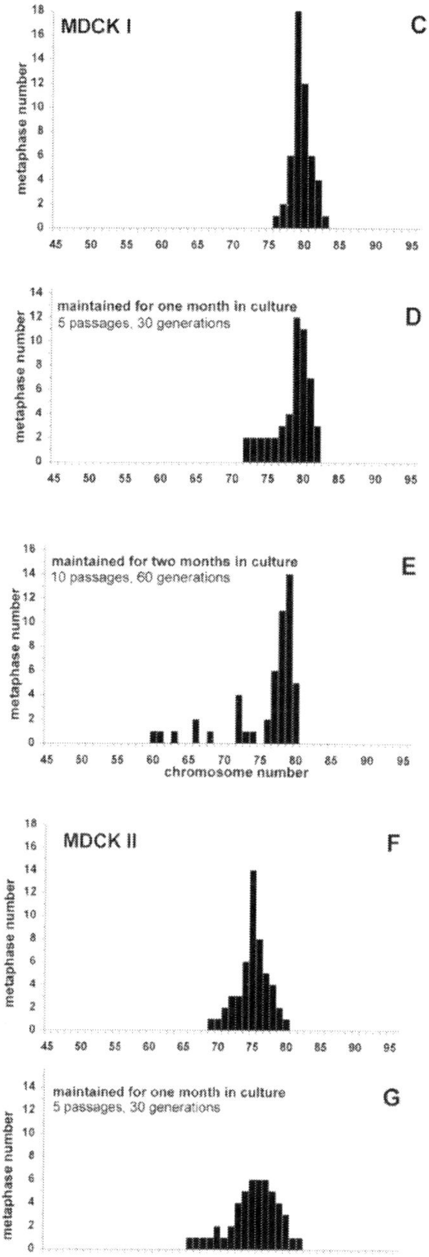
MDCK I
C
metaphase number
maintained for one month in culture
5 passages, 30 generations
D
maintained for two months in culture
10 passages, 60 generations
E
chromosome number
MDCK II
F
maintained for one month in culture
5 passages, 30 generations
G
45 50 55 60 65 70 75 80 85 90 95

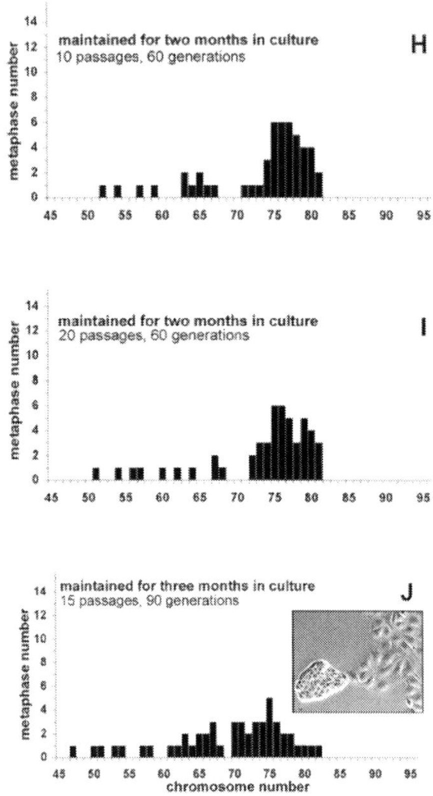

Figure 2: Analysis, and evolution with time in culture, of metaphases of MDCK (A,B), MDCK I (C-E) and MDCK II (F-J) cells. Cells obtained from reliable sources (Table 1) were maintained continuously in culture for 2-3 months. The number of chromosomes of 50 metaphases was determined upon their receipt and after various periods of time in culture. Insert in J: phase contrast image of cells maintained for three months in culture. Note the heterogeneity of the cell population, composed of two types of cells: large and flat cells on the right and small and tight cells on the left. This latter type was observed only after two months in culture. Bar, 10 μm.

Chromosome content of MDCK II cell populations from six different laboratories

MDCK II is by far the most used MDCK strain. Therefore the chromosome content of MDCK II routinely cultured in six different laboratories was examined (Table 1, Figure 3). These cells were kindly given for the purpose of this study, but for most of them, no information on the supplier and/or on passage number was available. Moreover when the passage number was known, it was difficult to determine the real age of the culture, because the conditions such as the plating density at each passage, the time interval between two passages, and more importantly, the number of generations per passage, differed from one laboratory to another.

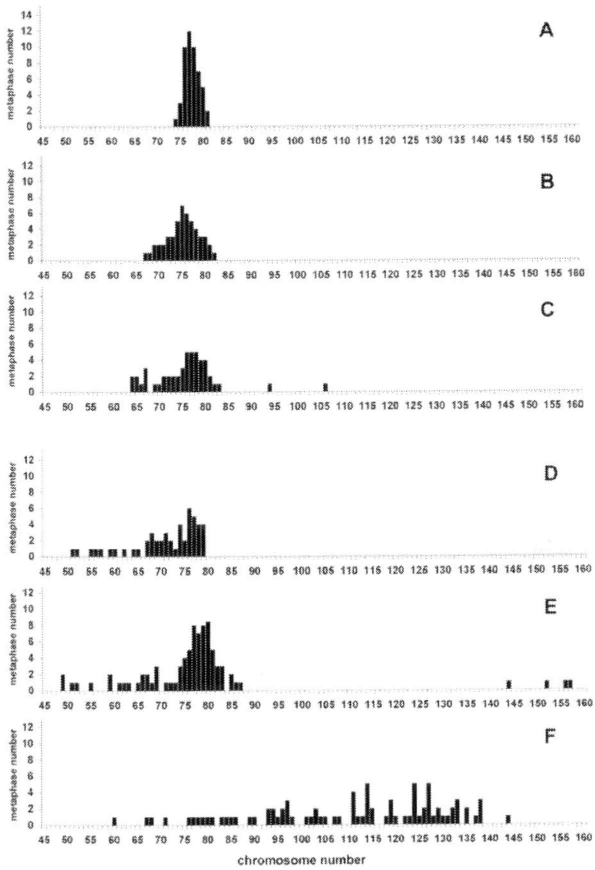

Figure 3: Analysis of metaphases of MDCK II cells obtained from six different laboratories (A-F). Cells were provided by laboratories working routinely with MDCK II cells. The number of chromosomes of 50-85 metaphases (Table 1) of these different MDCK cell populations was determined upon their receipt. The results are presented according to the drift of the population, starting from the most homogeneous one.

Striking differences in the chromosome content were observed between the six populations examined. Two of them (Figure 3A, B) were similar to the culture of MDCK II provided by ECACC (Figure 1F) (similar mean number and range). The four others had become more heterogeneous (Figure 3C-F), and even though the mean number of chromosomes was not far from the expected number (Table 1), the range was wider, reflecting the presence of metaphases with a low number of chromosomes (Figure 3C-F). In some cases, few metaphases with a high number of chromosomes (Figure 3E-F) were present. One population was very different as it was particularly heterogeneous, displaying a wide range of number of chromosomes (60-144) and 81% of metaphases containing more than 90 chromosomes (Figure 3F).

DISCUSSION

The present work aimed at studying the chromosome content of MDCK, MDCK I and MDCK II strains. To ensure that the metaphase preparations were representative of the whole cell populations, several precautions were taken. First, I used a method giving a high yield of metaphases [8]. Second, the fixation steps and the spreading of metaphases were carefully carried out to avoid broken and therefore incomplete metaphases. Finally chromosome counting was performed on round metaphases with no or few overlapping chromosomes and that were still embedded with a little of cytoplasm (Figure 1). The embedding decreased the risk of chromosome loss during the spreading.

The data presented here illustrate the fact that cell populations can considerably drift. The main change observed for MDCK I and MDCK II cells maintained for a long time in culture, was the emergence of an increasing fraction of cells with fewer chromosomes. This phenomenon has been previously reported for other cell lines, in particular for HeLa cells [8]. It is very likely due to the fact that cells with a lower chromosomal content grow more rapidly and prevail on cells with more chromosomes. Another change observed in some MDCK II cell populations (Figure 2E,F) was the presence of few metaphases with a higher number of chromosomes. Such metaphases could result from endoreplication events [10], leading to the generation of cells, initially containing twice as much chromosomes, but eventually losing some over generations.

The number of passages is a parameter often used for characterizing a cell line. Unfortunately in most cases information on the passage itself is missing, in particular the number of generations per passage. Therefore, even if cells are examined at the same passage, they could differ by far. For instance MDCK cells examined at a given passage did not have the same efficiency in forming tumors [11]. Moreover, as reported here, the chromosome contents of MDCK II cells examined at similar passages were very different (compare Figure 2F to Figure 3E). This is the reason why it is highly preferable to evaluate the age of a culture in generations and to compare the results obtained at the same generation, rather than at the same passage.

CONCLUSIONS

The present study illustrates for MDCK strains, one fundamental issue encountered in maintaining cells in culture, namely the potential drift of the cell population over time. This point is most often disregarded, or even ignored. Even though it is impossible to prevent such a drift, precautions can be taken to limit the consequences of the drift. First it is highly recommended to obtain cell lines from reliable sources rather than from laboratories working with these lines, except if these laboratories can give information on the source, the number of generations (or passages), the chromosome content and the homogeneity of their cell populations. Second, upon receipt of the cells, a frozen stock of several vials has to be constituted. The chromosome content of these cells needs to be established and further checked after several generations (or passages). Finally, even if there are no important changes in the number and the range of chromosomes over time, I strongly recommend to avoid working with cells maintained for more than two months in culture. One should rather go back to the original frozen stock of cells created upon their receipt. As illustrated in the review of Hughes et al. [12], the use of over-passed cell cultures could lead to erroneous and spurious results.

ACKNOWLEDGEMENTS

I am grateful to Chantal Hamon Benais for letting me share her interest in chromosomes and karyotypes. I thank Jean-Pierre Mauger for discussion about the various MDCK strains. I acknowledge Kai Simons for MDCK I and colleagues from several laboratories in different countries, for their

MDCK II cells. I thank Pascale Dupuis-Williams, Pascal Colosetti and Isabelle Doignon for careful reading the manuscript. Finally, I warmly thank Elisabeth Mintz for examination of the manuscript, linguistic corrections and helpful comments.

REFERENCES

1. Gaush CR, Hard WL, Smith TF: Characterization of an established line of canine kidney cells (MDCK). Proc Soc Exp Biol Med. 1966, 122 (3): 931-935.
2. Dukes JD, Whitley P, Chalmers AD: The MDCK variety pack: choosing the right strain. BMC Cell Biology. 2011, 12: 43-10.1186/1471-2121-12-43.
3. Wünsch S, Gekle M, Kersting U, Schuricht B, Oberleithner H: Phenotypically and karyotypically distinct Madin-Darby canine kidney cell clones respond differently to alkaline stress. J Cell Physiol. 1995, 164 (1): 164-171. 10.1002/jcp.1041640121. Hansson GC, Simons K, van Meer G: Two strains of the Madin-Darby canine kidney (MDCK) cell line have distinct glycosphingolipid compositions. EMBO J. 1986, 5 (3): 483-489.
4. Colosetti P, Tunwell RE, Cruttwell C, Arsanto JP, Mauger JP, Cassio D: The type 3 inositol 1,4,5-trisphosphate receptor is concentrated at the tight junction level in polarized MDCK cells. J Cell Sci. 2003, 116 (Pt 13): 2791-2803.
5. Cassio D: A General and Reliable Method for Obtaining High Yield Metaphasic Preparations from Adherent Cell Lines: Rapid Verification of Cell Chromosomal Content. Cell Biology, A Laboratory Handbook. Edited by: Celis J. 2005, Waltham: Academic Press, vol III, chapter 39, 3
6. Worton RG, Duff C: Karyotyping. Methods Enzymol. 1979, 58: 322-344.
7. Lee HO, Davidson JM, Duronio RJ: Endoreplication: polyploidy with purpose. Genes Dev. 2009, 23 (21): 2461-2477. 10.1101/gad.1829209.
8. Omeir RL, Teferedegne B, Foseh GS, Beren JJ, Snoy PJ, Brinster LR, Cook JL, Peden K, Lewis AM: Heterogeneity of the tumorigenic

phenotype expressed by Madin-Darby canine kidney cells. Comp Med. 2011, 61 (3): 243-250.

9. Hughes P, Marshall D, Reid Y, Parkes H, Gelber C: The costs of using unauthenticated, over-passaged cell lines: how much more data do we need?. Biotechniques. 2007, 43 (5): 575-584. 10.2144/000112598.

CHAPTER

9

CYCLIN AND DNA DISTRIBUTED CELL CYCLE MODEL FOR GS-NS0 CELLS

David G. García Münzer[1], Margaritis Kostoglou[2], Michael C. Georgiadis[3], Efstratios N. Pistikopoulos[1], Athanasios Mantalaris[1]

[1]Biological Systems Engineering Laboratory, Centre for Process Systems Engineering, Department of Chemical Engineering, Imperial College London, London, United Kingdom,

[2]Department of Chemistry, Aristotle University of Thessaloniki, Thessaloniki, Greece,

[3]Department of Chemical Engineering, Aristotle University of Thessaloniki, Thessaloniki, Greece

ABSTRACT

Mammalian cell cultures are intrinsically heterogeneous at different scales (molecular to bioreactor). The cell cycle is at the centre of capturing heterogeneity since it plays a critical role in the growth, death, and productivity of mammalian cell cultures. Current cell cycle models use biological variables (mass/volume/age) that are non-mechanistic, and difficult to experimentally determine, to describe cell cycle transition and capture culture heterogeneity. To address this problem, cyclins—key molecules that regulate cell cycle transition—have been utilized. Herein, a novel integrated experimental-modelling platform is presented whereby experimental quantification of key cell cycle metrics (cell cycle timings, cell cycle fractions, and cyclin expression determined by flow cytometry) is used to develop a cyclin and DNA distributed model for the industrially relevant cell line, GS-NS0. Cyclins/DNA synthesis rates were linked to stimulatory/inhibitory factors in the culture medium, which ultimately affect cell growth. Cell antibody productivity was characterized using cell cycle-specific production rates. The solution method delivered fast computational time that renders the model's use suitable for model-based applications. Model structure was studied by global sensitivity analysis (GSA), which identified parameters with a significant effect on the model output, followed by re-estimation of its significant parameters from a control set of batch experiments. A good model fit to the experimental data, both at the cell cycle and viable cell density levels, was observed. The cell population heterogeneity of disturbed (aftcr cell arrest) and undisturbed cell growth was captured proving the versatility of the modelling approach. Cell cycle models able to capture population heterogeneity facilitate in depth understanding of these complex systems and enable systematic formulation of culture strategies to improve growth and productivity. It is envisaged that this modelling approach will pave the model-based development of industrial cell lines and clinical studies.

INTRODUCTION

Monoclonal antibodies (mAb) represent a key growth section of the high-value bio-pharmaceuticals (biologics) market [1]. These biologics are commonly produced by mammalian cell culture systems due to their ability to perform human-compatible post-translation modification (glycosylation) of proteins. Mammalian cells represent complex production systems whereby a large number of interlinked metabolic reactions control productivity and product quality, which are influenced by culture parameters. Mammalian cell cultures are intrinsically heterogeneous at all scales from the molecular to the bioreactor level [2–4]. The key underlying source of heterogeneity is cell cycle segregation [5–7], which is at the centre of cellular growth, death, and productivity, all of which vary during the different cell cycle phases. Specifically, the cell cycle phase can influence the mAb productivity, both of which have been reported to be cell cycle-, cell line-and promoter-dependent [8, 9]. Therefore, a better understanding and knowledge of the cell cycle timing, transitions, and associated production profiles can aid the development (modelling, control, and optimisation) of these industrially-relevant systems [10]. Recently, metabolic flux analysis (MFA) has become a key tool for the study of mammalian cell cultures aiming at improving productivity and product quality. These studies [11–14] provide valuable insight on cell behaviour and assist in understanding cell metabolism. However, they neglect the intrinsic heterogeneity (e.g. cell cycle, genotypic, and phenotypic variations) [15, 16] of cell culture systems. Moreover, MFA applicability to mammalian cells is limited due to their complexity, pseudo steady-state approximation, unbalanced cell growth behaviour, and adaptation to changing environments [17, 18].

Mathematical models can aid the study of the complex mammalian cell culture systems by capturing the heterogeneity of the cell population, both at the biophase and cellular levels. A number of studies have dealt with the development of cell cycle models and the advancement of modelling techniques to capture the segregation of cell cultures. Alas, the development of relevant segregated cell cycle models has proved challenging, particularly due to difficulties in providing quantitative experimental validation. The vast majority of cell cycle models can be classified as cell ensemble models (CEMs) or population balance models (PBMs) [19–22]. CEMs capture heterogeneity by including a large number of single cell models (SCMs) that differ according to key cellular properties. The approach is based on the assumption that a sufficiently large number of individual cells will represent the behaviour of the population. When CEMs describing a small number of cells were coupled to SCMs computation time was small [23, 24]. However, the number of cell-simulated affects both the resolution of the obtained distributions and the computational burden. An alternative to model segregation is based on the population balance equation (PBE), which describes the number distribution of internal cell states as a function of time. This type of formulation allows the incorporation of information on stage-to-stage transition (e.g. cell cycle) and partition of cell material upon cell division. A detailed review of PBEs is available [25]. A series of studies [26–28] on the solution of PBEs in biological systems showed an increased computational solution time when increasing the number of variables (even with an unstructured cell model). Therefore, most of the available PBMs in cell cultures are coupled to unstructured models in order to reduce the computational burden. The computational time is also significantly increased when increasing model complexity, such as number of stages, variables or structure [21, 26, 29, 30].

Two main features describe the existing PBM formulations: age or mass/volume. Early reports used age as the sole distribution variable [31–33]. Later, more complex formulations were reported using age and an additional intracellular property (e.g. product, RNA) [34, 35]. Similarly, relevant mass/volume based PBMs have been reported [7, 29, 36, 37], such as the coupling a PBE with a SCM [30], which demonstrated the computational intensive nature of the PBM-SCM description. Both PBE formulations for the cell cycle transition in mammalian cell cultures, namely age and mass/volume, are non-mechanistic [38–40] and difficult to experimentally quantify.

In mammalian cells, the cell division cycle occurs in sequential distinct phases: G_1 (first gap), S (DNA synthesis), G_2 (second gap) and M (mitosis). However, cells can enter a non-proliferating or quiescent state (G_0) when conditions (internally or externally) are not favourable for DNA synthesis [41]. Three cell cycle checkpoints are reported [42]: at the G_1-S phase transition, at the end of the S phase, and between the G_2-M phase transition. The cell cycle is controlled mainly by two proteins, cyclin-dependent-kinases (CDKs) and cyclins [43]. The transition checkpoints are regulated by the varying concentration of the different cyclins, whereas the concentration of CDKs remains in constant excess throughout [44]. Specifically, the transition between the G_1-S checkpoint has been reported to be regulated by two cyclins, cyclin D and E. Similarly, cyclin B is reported to regulate the cell's entrance into mitosis [43].

Herein, cell population heterogeneity was captured by formulating a biologically-relevant distributed cell cycle model for the industrial cell line (GS-NS0). The cyclin and DNA distributed cell cycle model utilised cyclin E1 as a marker for the G_1 phase, DNA content for the S phase, and cyclin B1 for the G_2/M phase. Each of these markers can be quantitatively experimentally validated. The model was developed using the mathematical modelling framework previously described [45–47]. Specifically, the parameters' effect on the model output was evaluated by GSA and the identified significant parameters were re-estimated using a set of batch control experiments. The model simulations were compared to independent batch experiment (with varying initial cell cycle distribution and inoculation density) data to test the model's accuracy and predictive power. The model can be used as a tool for cell cycle-specific (productivity) studies and model-based control and optimisation applications due to its low computational requirements.

RESULTS

Model Structure

The proposed model is a multivariable and multistage PBM. The PBM is coupled to an unstructured metabolic model in order to more accurately represent the system. The model is cyclin distributed for the lumped phases—G_1/G_0 (Equation 1) and G_2/M (Equation 3)—and is DNA distributed for the S phase (Equation 2). The model formulation presented does not have boundary conditions since the communication between the stages occurs internally through source and sinks terms [48].

G_1/G_0 phase:

$$\frac{\partial N_{G1}(cycE,t)}{\partial t}+\frac{\partial r_{G1}(cycE,s)N_{G1}(cycE,t)}{\partial cycE}=-\Gamma_{G1}(cycE,s)N_{G1}(cycE,t)$$

$$-k_d(s)N_{G1}(cycE,t)+2\delta(cycE)\int_{cycBmin}^{cycBmax}\Gamma_{G2}(cycB,s)N_{G2}(cycB,t)dcycB \tag{1}$$

S phase:

$$\frac{\partial N_S(DNA,t)}{\partial t}+\frac{\partial r_S(DNA,s)N_S(DNA,t)}{\partial DNA}$$

$$=\delta(DNA-1)\int_{cycEmin}^{cycEmax}\Gamma_{G1}(cycE,s)N_{G1}(cycE,t)dcycE-k_d(s)N_S(DNA,t) \tag{2}$$

G_2/M phase:

$$\frac{\partial N_{G2}(cycB,t)}{\partial t}+\frac{\partial r_{G2}(cycB,s)N_{G2}(cycB,t)}{\partial cycB}=-\Gamma_{G2}(cycB,s)N_{G2}(cycB,t)$$

$$-k_d(s)N_{G2}(cycB,t)+\delta(cycB-cycB_{S-G2/M})r_S(DNA=2,s)N_S(DNA=2,t) \tag{3}$$

The number distribution (N_i=G_1,S,G_2) is the number of cells distributed on a variable for each phase. The PBM is experimentally derived from the cell's cyclin blueprint [49], which is used to formulate the cyclin intensity functions. The model utilises cyclin E1 (*cycE*, key for the transition from G_0/G_1 to S phase), *DNA* (key for the transition between S phase and G_2/M) and cyclin B1 (*cycB*, key for the transition between G_2/M phases to G_0/G_1). The cyclin growth functions ($r_{i = G1,G2}$, Equation 4) denote cyclin (E1 and B1) build-up on each phase (G_0/G_1 and G_2/M) and are responsible for the transition of the cells within each cyclin domain. Similarly, the DNA growth function ($r_{i = S}$, Equation 5) denotes the DNA build-up or synthesis in the S phase and is responsible for the transit of the cells within that phase. The cyclin and DNA growth functions (r_i, Equation 4–5) are substrate/metabolite concentration *(s)* dependent. The substrate/metabolite *(s)* stands in general for the main substrates (glucose, glutamate) and metabolite (lactate). Specifically, the growth functions (r_i) are glutamate-dependent (Equation 7). This dependency was introduced since it has been reported that glutamine, or glutamate in the GS system [50], is the main energy-yielding metabolic route [11, 51, 52] with over 90% flowing into the TCA cycle [53]. In addition, the cyclin growth functions ($r_{i = G1,G2}$) were also formulated as lactate-dependent since it has been reported [54, 55] that its concentration inhibits cell growth. Lactate dependency was not included in the DNA growth function ($r_{i = S}$), since it has been reported to be fairly constant in duration [56]. Therefore, the cells entered the S phase—by means of the Dirac delta function $\delta(DNA\text{-}1)$—with an initial DNA content and after doubling their DNA, transit occurs immediately to the G_2/M phase, i.e. $\Gamma_S(DNA = 2)$ *equals to infinity*. Cyclin B1 content in the S phase is approximated, as it is not biologically relevant for the S to G_2 phase transition [43]. Hence, the cells that transit from the S to the G_2/M phase, enter—by means of the Dirac delta function $\delta(cycB\text{-}cycB_{S\text{-}G2/M})$—with a cyclin B1 content equal to the one observed experimentally between these two phases. It should be noted that in the general case, the S phase must be described by a bivariate probability density function having as internal coordinates the DNA and cyclin B1 content. However, under specific conditions for the rate functions a mathematical dimensionality reduction of the S equation can be achieved as detailed in S1 Text. This dimensionality reduction is essential since it reduces the computational effort of the model by at least 2 orders of magnitude [7, 21, 30], thus permitting its further use for model-based applications (e.g. sensitivity analysis, parameter estimation, control, and optimisation).

The cyclin transition functions are defined as probability of transition between the G_0/G_1 and S phases ($\Gamma_{i=G1}$), as well as between G_2/M and G_0/G_1 phases ($\Gamma_{i=G2}$). The cells complete the cycle by giving birth to two daughter cells and enter—by means of the Dirac delta function $\delta(cycE)$— the G_1 phase with a minimum cyclin E1 content. The cyclin transition ($\Gamma_{i=G1,G2}$, Equation 6) is formulated as a step function, which is zero below the respective cyclin threshold for the phases G_1 (cyclin E1) and G_2 (cyclin B1), and becomes positive above the threshold, though it is glucose dependent. Similar step transition functions have been recently identified [57] in age-structured models from experimental data. The glucose dependency was introduced [13] since even though glucose is not the growth limiting substrate, it does prolong the stationary phase and delays the decline phase.

The substrate/metabolite (Equation 7) dependencies were formulated observing Monod type kinetics [58, 59]; where *Glu, Glc*, and *Lac* represent the extracellular glutamate, glucose, and lactate concentrations, respectively. The $K_{i=Glu,Glc,Lac}$ parameters represent the Monod constant, and the exponential factors (*n, m*, and *p*) capture the sensitivity towards the respective substrate/metabolite.

Cyclins growth functions:

$$r_{i=G1,G2}(cyc,s) = Cycgrow \cdot f_{\lim Glu} \cdot f_{\lim Lac} \qquad \textbf{(4)}$$

DNA growth function:

$$r_S(DNA,s) = DNAgrow \cdot f_{\lim Glu} \qquad \textbf{(5)}$$

Cyclins transition functions:

$$\Gamma_{i=G1,G2}(cyc,s) = \begin{cases} 0 & , \quad below \quad threshold \\ prob_i \cdot f_{\lim Glc} & , \quad above \quad threshold \end{cases} \qquad \textbf{(6)}$$

Limiting functions:

$$f_{limGlu} = \frac{Glu^n}{Glu^n + K_{Glu}^{\ n}} \quad ; \quad f_{limGlc} = \frac{Glc^m}{Glc^m + K_{Glc}^{\ m}} \quad ; \quad f_{limLac} = \frac{K_{Lac}^{\ p}}{K_{Lac}^{\ p} + Lac^p} \tag{7}$$

The different populations are calculated as follows, see (Equation 8).

Phase populations:

$$phase_{i=G1,S,G2} = \int_{phase\_min}^{phase\_max} N_i(x_i)dx_i \tag{8}$$

Total population:

$$total = phase_{G1} + phase_S + phase_{G2} \tag{9}$$

Cell cycle fractions:

$$f_{i=G1,S,G2} = \frac{phase_i}{total} \tag{10}$$

Viable cells:

$$X_V = \frac{total}{V} \tag{11}$$

Dead cells:

$$\frac{d(X_D V)}{dt} = k_d X_V V - k_{lys} X_D V \tag{12}$$

The total population is calculated as the sum of all three distributed domains (Equation 9) and the cell cycle distribution is estimated by Equation 10. The viable cell density (X_V, Equation 11) is expressed by dividing over the reactor working volume (V). The model operates under the standard assumption of ideal mixing and is presented in batch operation mode. The dead cell population (X_D, Equation 12) is calculated by accounting for the specific death rate (k_d) of the viable cells and the lysis specific rate (k_{lys}) of the dead cells. The specific death rate (Equation 13) is linked to the extracellular substrate concentrations, particularly glutamate as this is an essential amino acid for the GS system [50]. The glutamate affinity death parameter (k_{dGlu}) and the exponential factor *(q)*, define the sensitivity of the cell death to glutamate depletion. The growth rate (μ, Equation 14) is calculated based on the cyclin and DNA growth functions. The calculated times of transit for each phase are based on the minimum threshold for transition and the corresponding cyclin/DNA growth functions.

Specific death rate:

$$k_d(s) = k_{d\max}\left(\frac{k_{dGlu}}{k_{dGlu} + Glu}\right)^q \tag{13}$$

Specific growth rate:

$$t_{i=G1,G2} = \frac{cyclin\_threshold}{r_{i=G1,G2}} \quad ; \quad t_S = \frac{1}{r_S} \quad ; \quad \mu = \frac{\ln(2)}{t_{G1} + t_S + t_{G2}} \tag{14}$$

The substrate (Equation 15), metabolite (Equation 16), and antibody (Equation 17) reactor mass balances are specified. The biomass yield on glutamate and glucose is represented by ($Y_{i=Glu,Glc}$, Equation 15). Lactate was assumed to be produced (Y_{Lac}, Equation 16) by both glucose and glutamate consumption, as previously reported [11, 12, 51, 53]. In addition, lactate is also consumed (Q_{Lac}) [14, 55], which in this case was modelled and triggered by the depletion of glutamate. Antibody production (q_i, Equation 17) is specific for each cell cycle phase. A summary of the declared model variables can be found in Table 1.

Table 1: Model variables.

Nomenclature	Description	Units
$r_{i=G1,G2}$	cyclin transition functions	n/a
$r_{E,i}$	cyclin E1 transition function on sub-interval i	n/a
$r_{B,k}$	cyclin B1 transition function on sub-interval k	n/a
r_S	DNA transition function	n/a
μ	cell growth rate	h^{-1}
cycE	cyclin E1 distributed content	% cyclin
cycEmax	maximum cyclin E1 content used for domain truncation	% cyclin
cycB	cyclin B1 distributed content	% cyclin
cycBmax	maximum cyclin B1 content used for domain truncation	% cyclin
cycBmin	minimum cyclin B1 content used for the domain start	% cyclin
DNA	DNA distributed content	DNA relative content
$f_{i=G1,S,G2M}$	cell fraction in phase i	n/a
f_{lim}	Monod kinetics functions	n/a
Glc	extracellular glucose concentration	mM
Glu	extracellular glutamate concentration	mM
$h_{i=E,DNA,B}$	resolution of a sub-interval (bin)	% cyclin or DNA relative content

Lac	extracellular lactate concentration	mM
mAb	antibody concentration	mg/L
$N_{i=G1,S,G2}$	number of cells distributed on a variable for each phase	Cells
$N_{E,i}$	number of cells on sub-interval i of domain cyclin E1	Cells
$N_{DNA,j}$	number of cells on sub-interval j of domain DNA	Cells
$N_{B,k}$	number of cells on sub-interval k of domain cyclin B1	Cells
$n_{i=E,DNA,B}$	number of sub-intervals (or bins) within each domain interval	n/a
$Q_{i=Glu,Glc}$	substrate consumption rate	mM/cells/h
r_i	growth functions of the distributed variable	h^{-1}
$r_{E,i}$	cyclin E1 growth function on sub-interval i	% cyc E/h
$r_{DNA,j}$	DNA growth function on sub-interval j	DNA/h
$r_{B,k}$	cyclin B1 growth function on sub-interval k	% cyc B/h
total	total number of viable cells	cells
$phase_{i=G1,S,G2}$	number of cells in phase i	cells
(s)	substrate/metabolite concentration	mM
V	Volume	mL
X_V	viable cell density	cells/mL
X_D	death cell density	cells/mL

Substrates:

$$Q_{i=Glu,Glc} = \frac{\mu}{Y_{i=Glu,Glc}} \quad ; \quad \frac{ds}{dt} = -Q_{i=Glu,Glc} X_V V \tag{15}$$

Metabolites:

$$\frac{dLac}{dt} = Q_{Glc} Y_{Lac} f_{\lim Glu} + Q_{Glu} \frac{Y_{Lac}}{2} X_V V - Q_{Lac}(1 - f_{\lim Glu}) \tag{16}$$

Antibody:

$$\frac{dmAb}{dt} = (q_{G1} f_{G1} + q_S f_S + q_{G2} f_{G2}) X_V V \tag{17}$$

The population balance model (Equation 1–3) resembles the McKendrick type epidemic model [60] and in principle it admits large type asymptotic solutions. Nevertheless, these solutions are not possible in case of coupling with the metabolic model. The metabolic model coupling is essential for systems of changing substrate environments, which is the case in industrially-relevant systems (batch & fed-batch cultures). In the general case of studying cell distribution evolution and considering external time dependency (i.e. metabolic model) the complete problem must be numerically solved.

The system of the hyperbolic partial integrodifferential equations (1–3) is discretised in the cell content space using conservative first order finite differences [61, 62]. In this way it is transformed to a system of ordinary differential equations with respect to time, as the model structure is ideally suited to this type of discretization. The code for the solution of the underlying model was originally developed and a detailed assessment was made to ensure compliance with the Courant-Friedrichs-Lewy (CFL) stability condition. Furthermore, it was found that the optimal time step was the one that exactly corresponded to the CFL condition (even better than smaller time steps) since it eliminated numerical diffusion errors. Finally, it was decided to employ the method of lines (ODEs) combined to backwards first order finite differences in gPROMS in order to utilise its advanced non-linear optimisation capabilities for model validation (e.g. parameter estimation) and model-based analysis (global sensitivity analysis). The stability is automatically checked by the ODE integrator by adjusting the time step in gPROMS. The temporal accuracy is guaranteed by the ODE integrator while the spatial accuracy by making a detailed comparison of results for different discretizations. The detailed discretization of equations is presented below:

The interval (*0, cycEmax*) is divided in n_E equal size sub-intervals, the interval (*DNA = 1, DNA = 2*) is divided in n_{DNA} equal size sub-intervals, and the interval (*cycBmin, cycBmax*) in n_B equal size sub-intervals. Let us define:

$$h_E = \frac{cycE\max}{n_E} \quad ; \quad h_{DNA} = \frac{(2-1)}{n_{DNA}} \quad ; \quad h_B = \frac{(cycB\max - cycB\min)}{n_B} \tag{18}$$

$$N_{E,i}(t) = N_{G1}(ih_E, t) \;\; ; \;\; N_{DNA,j}(t) = N_S(1 + jh_{DNA}, t) \;\; ; \;\; N_{B,k}(t) = N_{G2}(cycB\min + kh_B, t) \tag{19}$$

$$r_{E,i} = r_{G1}(ih_E) \quad ; \quad r_{DNA,j} = r_S(1 + jh_{DNA}) \quad ; \quad r_{B,k} = r_{G2}(cycB\min + kh_B) \tag{20}$$

$$\Gamma_{E,i} = \Gamma_{G1}(ih_E) \quad ; \quad \Gamma_{B,k} = \Gamma_{G2}(cycB\min + kh_B)$$
$$where \quad i = 0, 1, 2, \ldots, n_E \;\; ; \;\; j = 0, 1, 2, \ldots, n_{DNA} \;\; ; \;\; k = 0, 1, 2, \ldots, n_B \tag{21}$$

The discretization of the equation (1) for G_0/G_1 phase, equation (2) for S phase, and equation (3) for G_2/M phase leads to the following system of ordinary differential equations.

G_0/G_1 phase:

$$\frac{dN_{E,i}}{dt} = \frac{-r_{E,i}N_{E,i} + r_{E,i-1}N_{E,i-1}}{h_E} - \Gamma_{E,i}N_{E,i} - k_d N_{E,i} \, i = 1, 2, 3, \ldots, n_E \tag{22}$$

$$\frac{dN_{E,0}}{dt} = \frac{-r_{E,0}N_{E,0}}{h_E} - \Gamma_{E,0}N_{E,0} - k_d N_{E,0} + 2\frac{h_B}{h_E}\sum_{k=0}^{n_B} w_k \Gamma_{B,k} N_{B,k} \tag{23}$$

S phase:

$$\frac{dN_{DNA,j}}{dt} = \frac{-r_{DNA,j}N_{DNA,j} + r_{DNA,j-1}N_{DNA,j-1}}{h_{DNA}} - k_d N_{DNA,j} \quad j = 1,2,3,\ldots,n_{DNA} \tag{24}$$

$$\frac{dN_{DNA,0}}{dt} = \frac{-r_{DNA,0}N_{DNA,0}}{h_{DNA}} - k_d N_{DNA,0} + \frac{h_E}{h_{DNA}}\sum_{i=0}^{n_E} w_i \Gamma_{E,i} N_{E,i} \tag{25}$$

G_2/M phase:

$$\frac{dN_{B,k}}{dt} = \frac{-r_{B,k}N_{B,k} + r_{B,k-1}N_{B,k-1}}{h_B} - \Gamma_{B,k}N_{B,k} - k_d N_{B,k} + \frac{1}{h_B} r_{DNA,n_{DNA}} N_{DNA,n_{DNA}} \delta\left(k - \mathrm{int}\left(\frac{cycB_{S-G2/M}}{h_B}\right)\right) \quad k = 1,2,3,\ldots,n_B \tag{26}$$

$$\frac{dN_{B,0}}{dt} = \frac{-r_{B,0}N_{B,0}}{h_B} - \Gamma_{B,0}N_{B,0} - k_d N_{B,0} \tag{27}$$

where w_i (Equation 23) and w_k (Equation 25) are the weights of the employed numerical integration scheme (trapezoidal). The cyclin B1 content between S and G_2/M phase ($cycB_{S\text{-}G2\,M}$) is approximated as explained in the S1 Textand the term int(x) denotes the integer part of x. The system of discretized equations (Equation 22–27) is coupled to the metabolic model. A schematic representation of the discretized model is presented (Fig. 1). All simulations were performed using a desktop computer with an Intel Core i7 CPU.

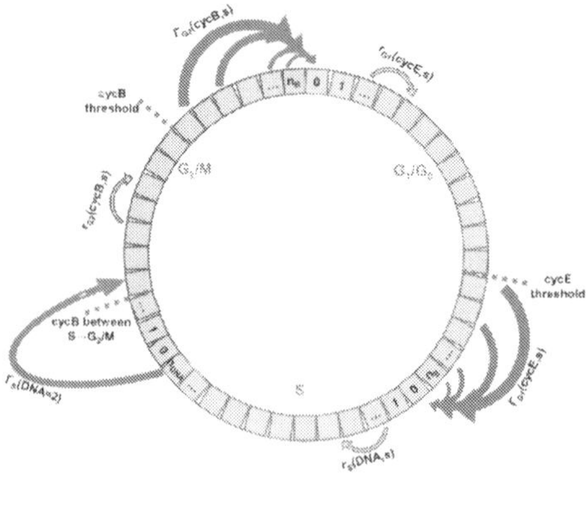

Figure 1: Schematic representation of the discretized cell cycle model.

Cell Proliferation Assay

Cell proliferation assays provide a direct way to estimate the duration of cell cycle phases. A specific population (which was active during DNA synthesis) of the culture is identified by labelling the DNA base analogue (e.g. EdU) introduced into its DNA. In order to estimate the average transit times based on the DNA content (i.e. G_0/G_1, S and G_2/M phases), EdU positive and negative cells were tracked for each time point of the frequent sampling. The shifts in the distribution of EdU positive cells indicate a fast transit of the cells from S to G_2/M (Fig. 2A); after 2h of culture a significant increase in the G_2/M fraction is observed. However, at 2h the fraction of cells at G_1/G_0 remains comparable to the starting level, and only after 4h a significant increase is observed. This indicates that the initial EdU positive population traversed the G_2/M phase in approximately 4h. In order to estimate the G_1/G_0 phase time, the EdU positive fraction in the S phase was monitored (Fig. 2A). The S phase was observed to be minimal between 8–10h of culture and soon after an increase was observed (circa 12h). Such increase indicated the completion of the G_1/G_0 phase (12h) of the initial EdU positive population (which entered that phase at around 4h); therefore, the estimated timing for this phase is of 8h. The S phase timing can be estimated based on the EdU negative population (Fig. 2B). After 2h, there was a significant change in the S phase, which indicated the entrance of cells into this phase. At around 8h, the number of cells in the S phase peaked with a concurrent increase in the G_2/M phase fraction. Therefore, the traversing estimate for the S phase is approximately 6h. An average cell cycle time of 18h can be estimated for the GS-NS0 under the specific culture conditions. The estimated value was in agreement with previous reported maximum growth rates of $0.04h^{-1}$ [63], which corresponds to a doubling time of 17.3h. The cell marker Ki-67 was used to determine the proliferative state of the cell population and discriminate between the G_0 and G_1 fractions. A high proliferative state was observed with a greater than 95% positive Ki-67 fraction over 3 days of culture (Fig. 3). After 80h of culture, cell viability started to decline as well as the Ki-67 positive fraction, indicating that some of the cells were entering the non-proliferative state (G_0). The entrance to the G_0 phase could not be confirmed to be associated to the commitment to apoptosis (as it was not quantified), however the Ki-67 proliferation marker showed a high correlation with the cell viability.

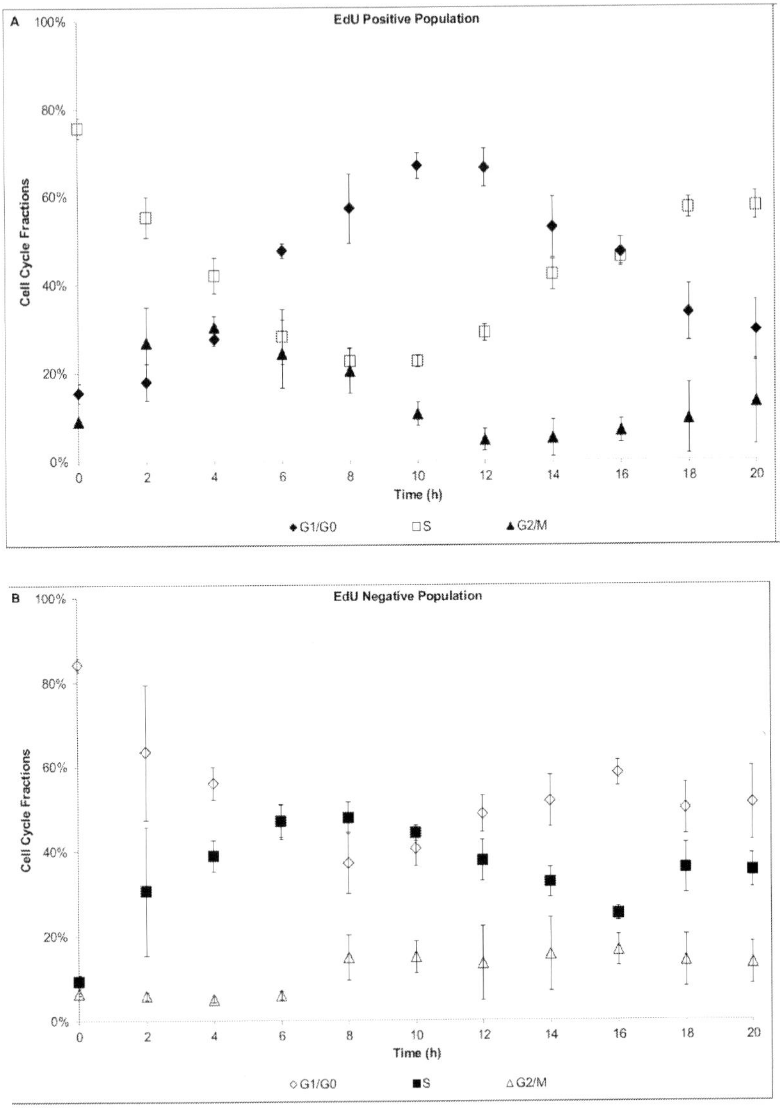

Figure 2: Proliferation assay—Cell cycle distribution.

A) EdU positive population; B) EdU negative population.

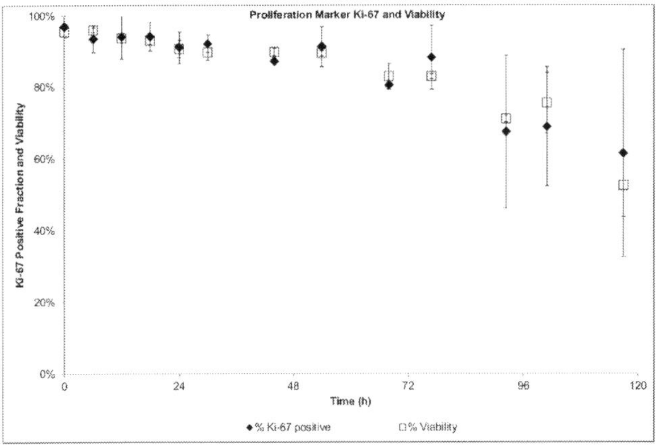

Figure 3: Proliferation assay—Proliferation marker Ki-67 and viability.

Cyclin Profiles

The cyclin profiles for the GS-NS0 cells under similar culture conditions have been recently reported [49], which are in agreement with cyclin expression patterns and thresholds presented herein. By plotting the cyclin content of the EdU positive and negative populations as they traverse the different phases, the cyclin profiles and thresholds could be identified. Cyclin E1 expression profile for the EdU positive population (Fig. 4A) showed the rapid decrease of cyclin E1 expression in the S phase (0–2h), as well as after the G_1 to S phase transition (18–20h). The cyclin E1 threshold level (between 15–20%) could be identified at the transition (18h). Similarly, the EdU negative population confirmed the cyclin E1 transition threshold (18h, Fig. 4B), as well as the rapid decrease in cyclin expression after the transit (as observed between 0–4h). Cyclin B1 expression profile for the EdU positive population (Fig. 4C), identified the rapid cyclin increase following S to G_2/M transition at 2h and 18–20h. A comparable profile was observed for the EdU negative population (Fig. 4D). Cyclin B1 build-up in the S phase started at minimum levels to approximately 15% (at 8h). Soon after transition to G_2/M, cyclin B1 expression builds-up rapidly and peaks between 40–60%, after which cells transit to G_1. The average cyclin E1 and cyclin B1 expression before (i.e. 0 to approximately 68h of culture) and after (i.e. after 68h of culture) glutamate exhaustion is shown in Fig. 4E. Both cyclins had a significantly lower expression ($p<0.05$) after the glutamate was exhausted from the media.

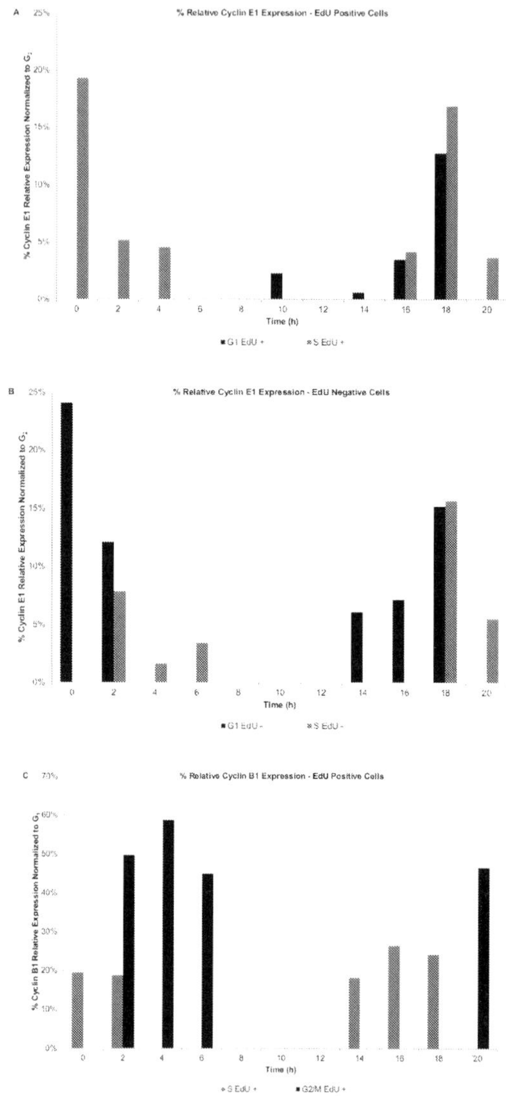

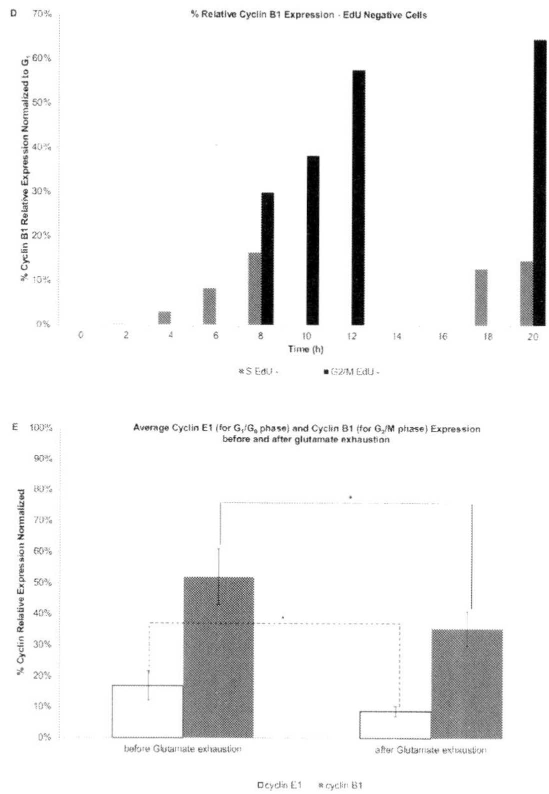

Figure 4: Proliferation assay—Cyclin expression profiles.

A) Cyclin E1 expression—EdU positive cells, B) Cyclin E1 expression—EdU negative cells, C) Cyclin B1 expression—EdU positive cells, D) Cyclin B1 expression—EdU negative cells, E) Average cyclin E1 (for G_1/G_0 phase) and cyclin B1 (for G_2/M) expression before and after glutamate exhaustion.

Model Analysis and Parameter Estimation

As outlined in the mathematical modelling development framework [45–47], following the development of the model type and structure, the influence of parameter uncertainty on model output was evaluated. GSA was applied in order to systematically evaluate output variability due to the nonlinear nature of the model. Prior to GSA, the conservation of entities in the system was verified. The birth term was fixed to 1 (to avoid generation), the death rate was fixed to zero ($k_{dmax}=0$) and the interaction with the nutrients was neglected ($f_{limGlu/Glc/Lac} = 1$). A uniform distribution was applied for all domains (cyclin E1 between 0–95%, DNA between 1–2, and cyclin B1 between 0–95%). The full range of 24 parameters was varied ±50% [30] from their nominal value and evaluated in over 100 scenarios to test the entities conservation. The nominal values were taken from experimental data (e.g. cell cycle blueprint) and when data were not available, approximated order-of-magnitude values were assigned [30, 63]. The number of bins for the cyclin E1 and B1 domains was increased in order to ensure that over 99% of the entities remained in the system after 120h of simulation. The selected number of bins was 200, both for cyclin E1 and cyclin B1. After fixing the number of bins, GSA was performed on the system. In order to evaluate the relative influence of the uncertainty of the parameters in the outputs, a large number of sampling points is typically required. The parameters were varied ±50% from their nominal value (restoring the birth term, death rate, and limiting functions) and the sampling points were generated using the sobolset function in Matlab. Each scenario was evaluated using the go:MATLAB tool of the gPROMs modelling software. A total of 40,000 scenarios were evaluated and the sensitivity indexes (SIs) were calculated using the GUI-HDMR software [64]. Five outputs were evaluated (X_v, f_{G1}, f_S, f_{G2}, and mAb) at three simulation time points (36, 72, and 120h). The SIs are the result of the GSA and vary between 0 and 1, with 0 being not significant. To evaluate the results of the GSA a threshold of 0.1 was selected, representing an experimental error of 10% [65]. Therefore, if the SI is above the threshold, the parameter is identified as significant and re-estimated. The non-significant parameters can be fixed at their nominal value.

The entities conservation test facilitated the selection of an appropriate number of bins to minimize particle loses during GSA. A trade-off between the number of bins and the solution time is generally observed. Despite selecting a considerable number of bins (i.e. 200) for each cyclin domain and 20 for the DNA domain, the computational time was low (<4s). This enabled completion of the sensitivity analysis (40,000 simulations) in less than 30h using a desktop pc. Both the time for a single run (<4s, two fold faster [7]) and to complete the GSA (<30h) were fast compared to previous reports [7, 21, 26, 30]. Out of the initial 24 parameters, 9 were identified as significant (Fig. 5). The viable cell population output (X_v)

was found to be sensitive to the cyclin and DNA growth rates (at early culture time 36h). Also it showed increasing sensitivity (from mid stage 72h towards late stage 120h) to the biomass yield on glutamate (Y_{Glu}) and maximum death rate (k_{dmax}). The G_1/G_0 cell fraction output (f_{G1}) showed recurrently (at all stages) sensitivity towards the cyclin E1 growth rate (*CycEgrow*) and DNA growth rate (*DNAgrow*). However, the cyclin E1 growth rate proved to be the most influential parameter for this cell fraction. At mid/late stages also it was found to be sensitive to the cyclin B1 growth rate (*CycBgrow*) stressing the link to the previous cell cycle phase (G_2/M). The S phase fraction output (f_S) showed a constant and recurrent sensitivity to the cyclin E growth rate for all stages. However, the DNA growth rate was more significant throughout. The G_2/M cell fraction output (f_{G2}) showed sensitivity towards the cyclin E1 growth rate at early stage, whereas a constant sensitivity throughout all stages for the DNA growth rate (preceding cell cycle phase) was observed. Also an increasing sensitivity (from early to late stages) towards the cyclin B1 growth rate (*CycBgrow*) was observed. Finally, the antibody titre output (*mAb*) was sensitive to all the cell cycle specific productivities ($q_{i\,=\,G1,S,G2}$) at early stage (36h). At mid stage it was sensitive to only S and G1 phase specific productivity (q_S and q_{G1}) whereas at late stage (120h) only to the G_1 phase specific productivity (q_{G1}). The identified 9 significant parameters were then re-estimated in order to minimize the uncertainty in the model outputs. Specifically, the control runs for thymidine and dimethyl sulfoxide (DMSO) experiments [49] were used to re-estimate the identified significant set of parameters, while setting the remaining parameters to their nominal value. The model showed good agreement between simulation and experimental results after parameter re-estimation for the cell growth kinetics, peak and decline, as well as the decline in viability (Fig. 6A) for both control runs. The glucose/lactate (Fig. 6B) and glutamate/mAb (Fig. 6C) profiles were also satisfactorily captured. In addition, the model simulated correctly the cell cycle distribution (Fig. 6D) trends, particularly at early stages of the culture. The simulation results were expected to be comparable between the two sets of experiments as the experimental differences are minimal and linked mainly to the initial conditions (cell cycle distribution, centrifugation steps prior to start, glucose concentration [49]).

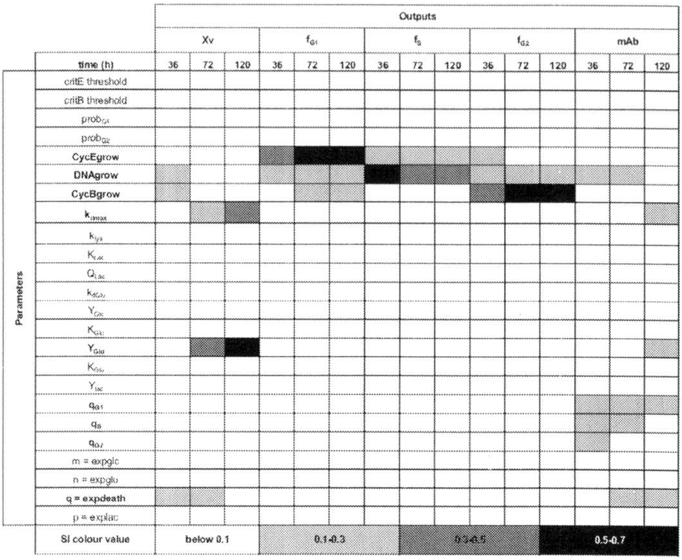

Figure 5: GSA indexes.

Outputs: X_V = viable cell density, $f_{G1} = G_1/G_0$ cell fraction, f_S = S cell fraction, $f_{G2} = G_2/M$ cell fraction, and mAb = antibody titre.

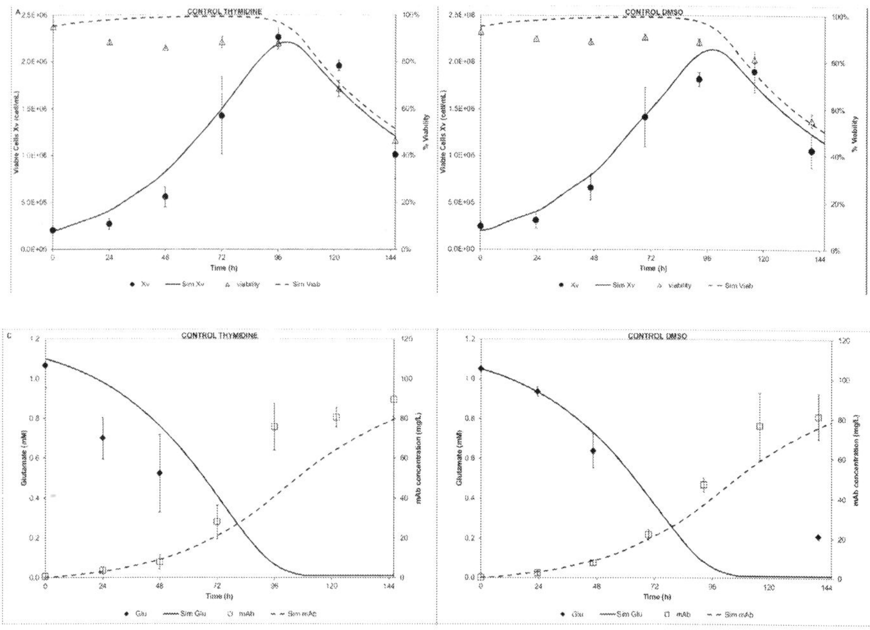

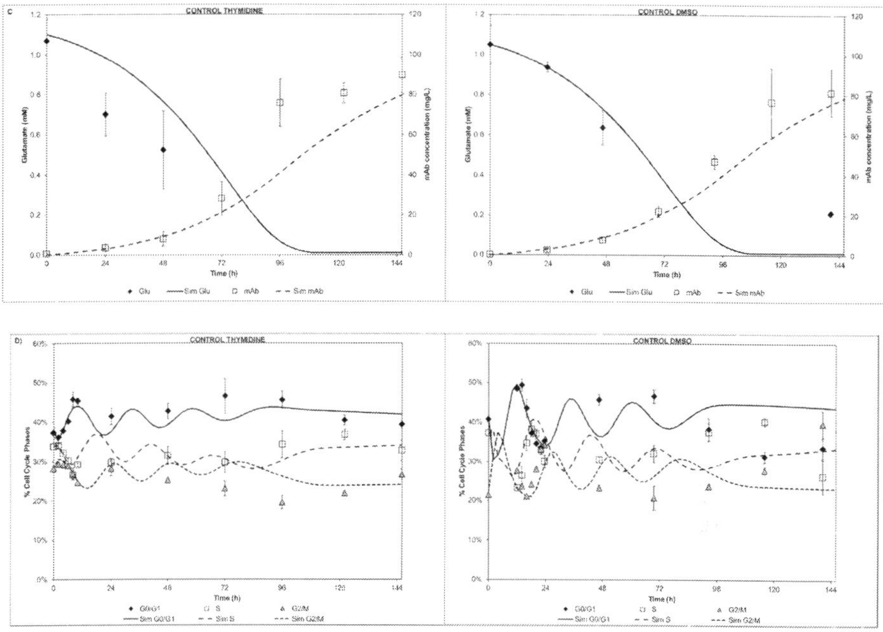

Figure 6: Modelling of control experiments.

A) Cell growth and viability, B) Glucose and lactate concentration profiles, C) Glutamate and mAb concentration profiles, D) Cell cycle distribution.

Modelling of Arrest Release Experiments

The model was then tested against the experiments performed after an arrest release [49]. Although the initial substrates/metabolites conditions were similar to the control runs, the arrest release experiments provided significant different initial cell cycle distribution scenarios. Following thymidine arrest, the cell growth was comparable to the control run, which was reflected by the good model prediction (Fig. 7A). Furthermore, the glucose (Fig. 7B), glutamate, and mAb (Fig. 7C) model simulations showed a good fit. However, the lactate profile (Fig. 7B) showed some discrepancies at early stage (24h) whereas after 48h good agreement was obtained. A plausible explanation to the experimental high lactate levels at 24h has been reported [49]. Nonetheless, the cell cycle distribution (Fig. 7D) showed good agreement with the experimental data. The sharp changes at the early stages of the culture (first 10h), right after the thymidine release, were captured by the model. The good cell cycle fit validated the overall modelling approach and stressed the relevance of the cyclin distribution on the cell cycle transitions.

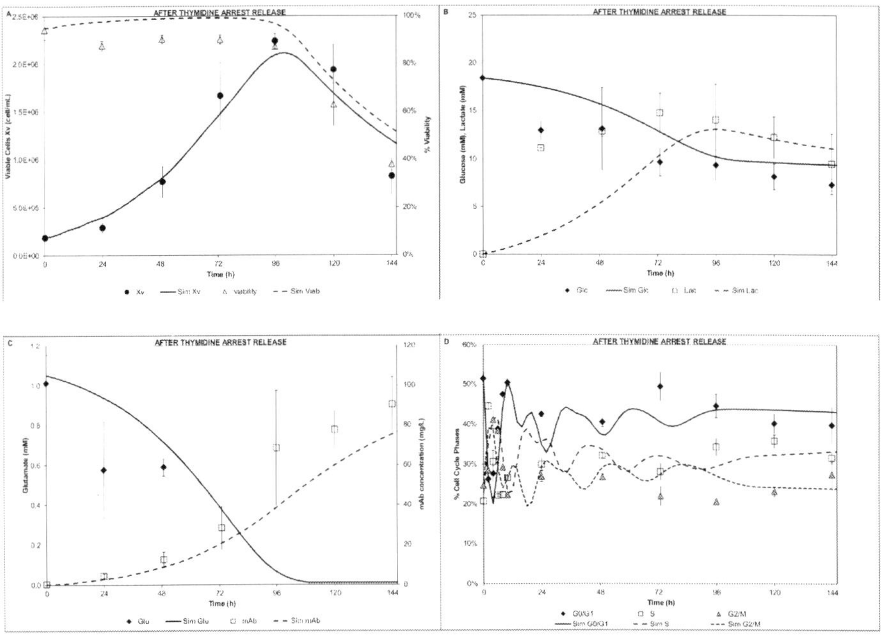

Figure 7: Modelling of after thymidine arrest release experiment.

A) Cell growth and viability, B) Glucose and lactate concentration profiles, C) Glutamate and mAb concentration profiles, D) Cell cycle distribution.

The DMSO arrest release experiment was also modelled. The overall trends for cell growth and viability were in good agreement (Fig. 8A, grey lines), although there was systematic difference between model predictions and experimental data (due to the pronounced experimental lag phase). However, the glucose/lactate Fig. 8B, grey lines) and glutamate (Fig. 8C, grey lines) profiles were satisfactorily predicted. Conversely, the modelled mAb profile (Fig. 8C, grey lines) illustrated a significant deviation from the experimental data (overestimating the final titre). The mAb deviation could be attributed to the higher cell density predicted. Similarly, the cell cycle modelled distributions (Fig. 8D) showed significant differences. Although the trends were correct, the model failed to predict the proper timing to restart proliferation. DMSO has been shown to produce imbalances in the G_1 cyclins [66]. Therefore, the hypothesis of a slower G_1 phase completion during the first 24h due to DMSO's effect was tested in the model. For this purpose, the cyclin E1 growth rate (*cycEgrow*) was decreased. In order to make the simulation comparable to the initial scenario, the biomass yield on glutamate (Y_{Glu}) and the biomass yield on glucose (Y_{Glc}) were adjusted by the same factor. The estimated decreased factor for the cyclin E1 growth rate was found to be ~2.3. The three parameters were modified (DMSO delay simulation values, Table 2) for the first 24h of simulation. After 24h, when the effect of DMSO was considered negligible, the modified parameters were set back to their initial values (nominal or estimated values, Table 2). All other model parameters were kept at the reported values. The delay growth simulations showed a marked improvement in the agreement with the experimental data. The cell growth and viability (Fig. 8A) showed an improved fit. Particularly, cell growth and initial lag phase were satisfactorily captured. The glucose/lactate (Fig. 8B) and glutamate (Fig. 8C) profiles showed comparable results to the initial simulation. The modelled mAb titre showed improved prediction when compared to the experimental data (Fig. 8C), although still over predicted the final mAb titre. The slower growth in the G_1 phase during the first 24h, did seem to improve the cell cycle distribution prediction (Fig. 8E). The model satisfactorily captured the timing for restarting the proliferation and the sharp oscillating cell cycle fractions.

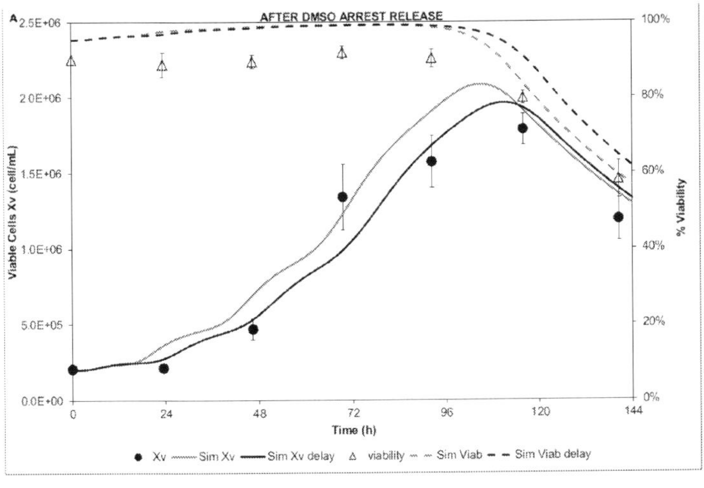

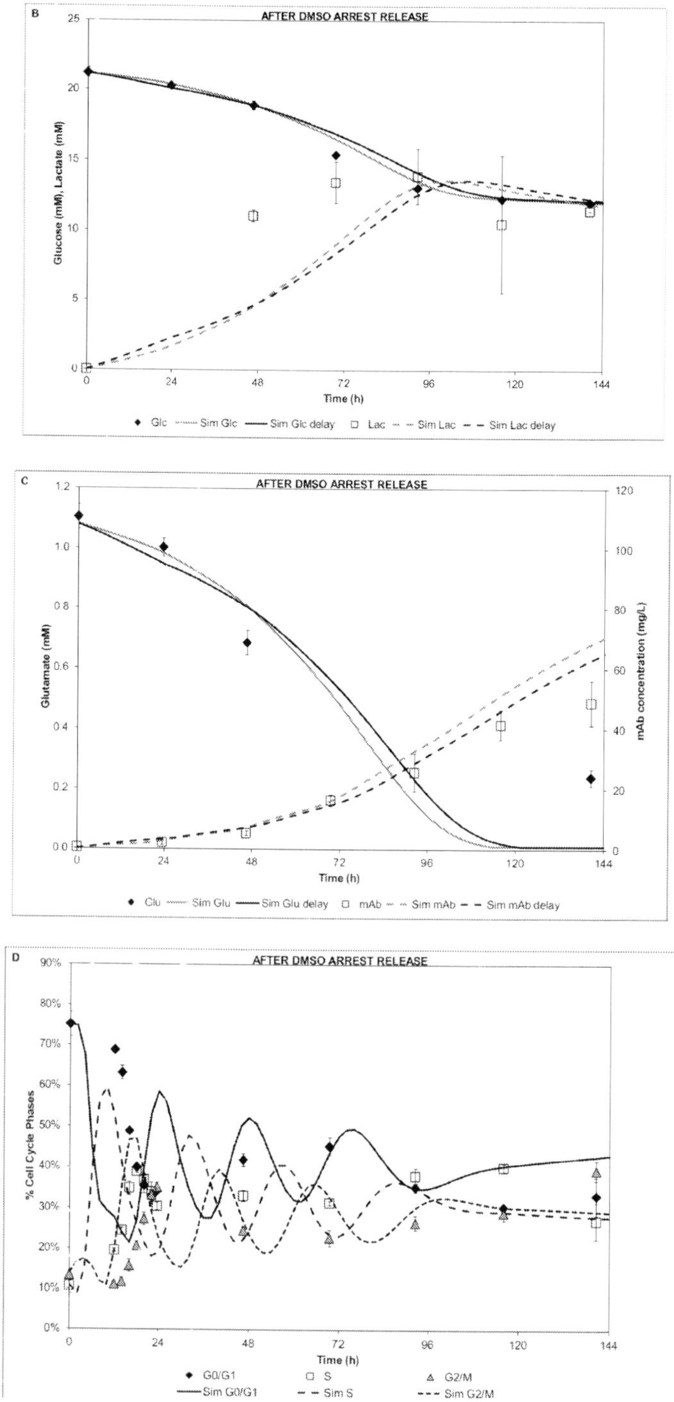
B
AFTER DMSO ARREST RELEASE
Glucose (mM), Lactate (mM)
Time (h)
Glc
Sim Glc
Sim Glc delay
Lac
Sim Lac
Sim Lac delay
C
AFTER DMSO ARREST RELEASE
Glutamate (mM)
mAb concentration (mg/L)
Time (h)
Glu
Sim Glu
Sim Glu delay
mAb
Sim mAb
Sim mAb delay
D
AFTER DMSO ARREST RELEASE
% Cell Cycle Phases
Time (h)
G0/G1
S
G2/M
Sim G0/G1
Sim S
Sim G2/M

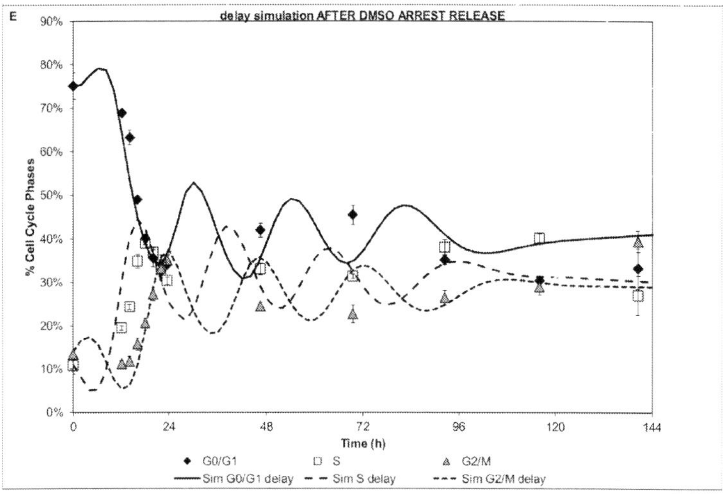

Figure 8: Modelling of after DMSO arrest release experiment.

A) Cell growth and viability, B) Glucose and lactate concentration profiles, C) Glutamate and mAb concentration profiles, D) Cell cycle distribution, E) Cell cycle distribution delay simulation. Grey lines: simulation with estimated parameters; Black lines: simulation with DMSO delay parameters (Table 2)

Table 2: Model parameters values.

Parameter	Nominal Value	Units	Estimated Value	95% Conf. Interval	DMSO delay simulation
critE threshold	15	% cyc E			
critB threshold	37	% cyc B			
$prob_{G1}$	0.6	n/a			
$prob_{G2}$	0.6	n/a			
CycEgrow	3.33	% cyc E/h	3.49	1.46E-01	1.52
DNAgrow	0.15	DNA/h	0.16	4.76E-03	
CycBgrow	8.22	% cyc B/h	8.07	2.99E-01	
$\mathbf{k_{dmax}}$	3.0E-02	h^{-1}	2.31E-02	1.90E-03	
k_{lys}	5.0E-03	h^{-1}			
K_{Lac}	50	mM			
Q_{Lac}	2.5E-11	mM/h			
k_{dGlu}	0.05	mM			
Y_{Glc}	2.1E+08	cell/mM			9.1E+07
K_{Glc}	5	mM			
$\mathbf{Y_{Glu}}$	1.8E+09	cell/mM	1.80E+09	7.20E+07	7.4E+08
K_{Glu}	0.20	mM			
Y_{lac}	2.3	n/a			
$\mathbf{q_{G1}}$	5.0E-10	mg/(cell*h)	9.57E-10	2.10E-09	
$\mathbf{q_{S}}$	5.0E-10	mg/(cell*h)	1.00E-10	4.30E-09	
$\mathbf{q_{G2}}$	5.0E-10	mg/(cell*h)	1.00E-11	2.00E-09	
m = expglc	2	n/a			
n = expglu	1	n/a			
q = expdeath	2	n/a	2		
p = explac	1	n/a			

Model Prediction of an Undisturbed Cell Cycle Experiment

An experiment with different starting cell density (2 fold higher) and undisturbed cell cycle distribution was used to test the predictive power of the validated model. The model was able to capture the cell growth kinetics, peak timing, and decline (Fig. 9A). Similarly, viability was closely predicted. Glucose (Fig. 9B) as well as glutamate and mAb (Fig. 9C) profiles were also satisfactorily predicted. At the late stages of the culture (~96h), glutamate seemed to be either produced by the viable cells or released by cell lysis, both of which the model did not account for. The lactate profile (Fig. 9B) showed a good agreement the first 48h of culture whereas at later stages, although the decreasing trend was captured by the model, the decrease in the experimental data was significantly steeper. The observed lactate decrease between 48 and 72h indicated a shift in the cell metabolism. The modelled cell cycle distribution (Fig. 9D) showed a fair prediction. Although the trends were correct a lack of fit was observed during the first 12h, particularly for the G_1 phase. Despite trying out different initial distribution scenarios, the differences could not be resolved.

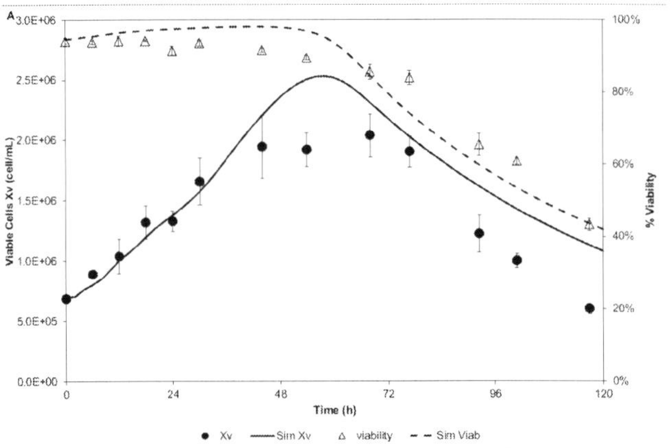

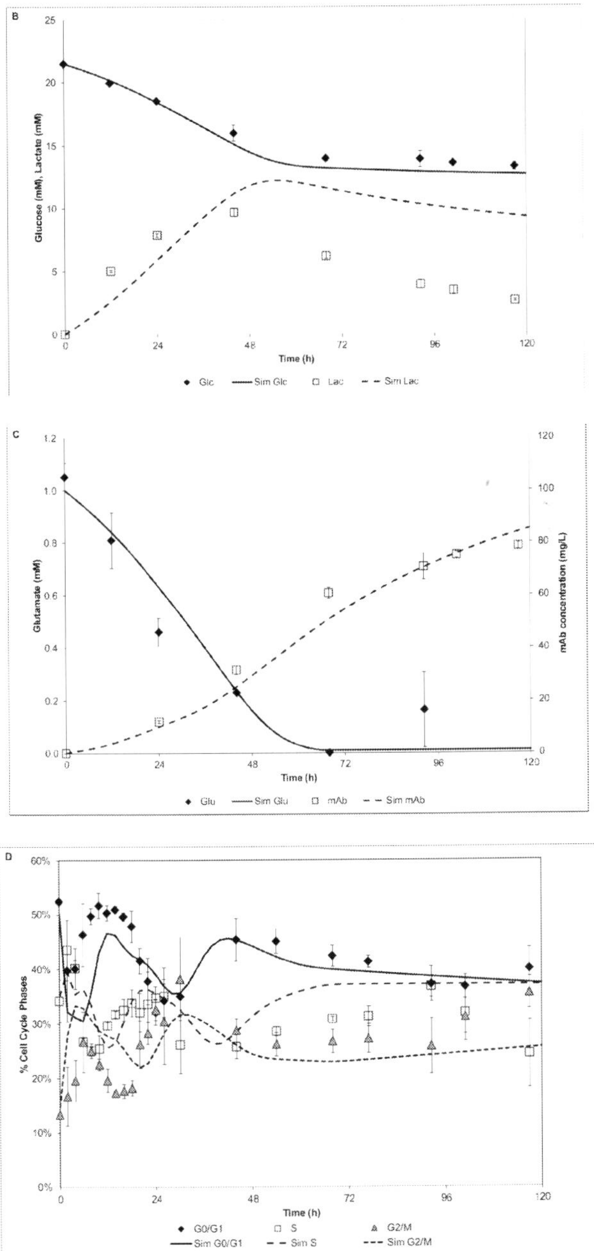

Figure 9: Model prediction of an undisturbed cell cycle experiment.

A) Cell growth and viability, B) Glucose and lactate concentration profiles, C) Glutamate and mAb concentration profiles, D) Cell cycle distribution.

DISCUSSION

The presented model represents a methodical approach to capture the cell culture heterogeneity by including relevant complex cell cycle mechanisms for an industrially relevant cell line. By using the mathematical modelling framework [45, 47, 58, 65, 67] it was possible to evaluate the structure and sensitivity of the model outputs in a systematic manner. Particularly, global sensitivity analysis guided model development and experimentation by identifying key model interactions and parameters, thus reducing experimental costs.

The fact that none of the transition probabilities were identified as significant is not evident. The model uses transition probabilities to determine dispersion around the thresholds. Therefore, as long as there is a positive transition value, the value itself of the parameter is not as significant. It should be noted that if a different model structure for the transition probability was formulated or other outputs were evaluated, it could probably be significant. Similarly, the thresholds for the cyclins were not identified as significant. This does not imply that the transitions are independent of the threshold. The justification is that once a cell reaches the threshold (regardless of its value), the cell becomes a candidate to proceed to the next stage. Therefore, cyclin thresholds and cyclin growth function act as a biologically relevant, mechanistically accurate, and quantifiable substitute to the previously employed variables (e.g. age/mass/volume). Moreover, these markers can be meaningfully coupled with the bioprocessing variables (e.g. substrate availability, chemical arrest agents). In this context, the cyclin blueprint is relevant not for its absolute values of cyclin expression but also in identifying the patterns, timings, and schedule of expression, as this will determine which biomarkers to use in model formulation. Nonetheless, should model validation be extended to cyclin distribution values or considered as an output in the sensitivity analysis, cyclin thresholds might be identified as significant parameters. The relevance of the cyclins and their distribution is evident by their ability to predict cell cycle variations. Finally, the sensitivity of the mAb titre output to the specific cell cycle productivities, although expected, suggests some association. The G_1 phase specific productivity was the only identified consistently throughout the mAb output as significant. This result suggests a significant correlation of the G_1/G_0 phase (compared to the other two phases, i.e. S and G_2/M) to the total productivity. A similar outcome is observed if the productivity is evaluated in terms of the cumulative cell hours [68] and the analysis is broken down per cell cycle phase [10]. Both a partially growth associated productivity and a better correlation towards the G_1/G_0 phase is observed.

Overall, the GSA identified the key parameters that need to be experimentally refined in order to reduce the model's output uncertainty. This systematic approach is in agreement with previous reports [69] that clearly stress the importance of mathematical models and tools for the analysis and understanding of biological systems.

Model formulation of a PBM coupled to an unstructured metabolic model in conjunction with model reduction demonstrated clear benefits. The computational times were at least two fold faster [7, 28] and enabled the completion of model-based applications, such as GSA, in just over 1 day. This is significantly faster when compared to the 3000h (in a 25 PCs cluster) for the completion of the GSA of a PBM-SCM [30]. The identified significant parameters were re-estimated by tailored made experiments such as the proliferation assay. This experiment allows extracting most of the required data for the model formulation, such as estimated cell cycle times, cyclins' thresholds, and cyclin/DNA growth rates. Herein, the reported cyclin levels are in agreement with the levels previously reported [49]. The observed cyclin blueprint is cell line and culture condition dependent. These results confirmed the cyclin blueprint without the need to perform any kind of cell synchronisation, and support the cyclin distributed model development. A key aspect included in the model is the link between cyclin expression levels and the culture environment (i.e. substrate and metabolite concentrations). Several reports [70, 71] have addressed the link of the cyclin D family of proteins with the culture environment (growth factors). Although cyclin D is not explicitly formulated in the model, it is expected that cyclin D expression will cascade on the other cyclins (e.g. cyclin E1, B1) and this is explicitly formulated. The experimental data did suggest that the relative expression levels decrease along the batch culture for both cyclins, particularly after the depletion of key energy/anabolic yielding substrates such as glutamate. Nonetheless, due to the employed media formulation (serum-containing), further studies are required. In particular, use of chemically defined media would enable the refinement of the modelled cyclin-extracellular substrate/metabolite concentrations.

The lumped G_1/G_0 phases assumption in the model development was supported by a high proliferating cell fraction (Ki-67 positive population). As observed, the cells maintained a high proliferative state as long as key substrates are available. Similarly, the lumped G_2/M is justified by the short length of the M phase when compared to the other cell cycle phases [72]. However, it should be noted that the cyclin B1 is responsible for the G_2 to M transition, and herein is approximated to be used for the G_2/M to G_1 transition. However, the model can be further refined by segregating the G_2/M lumped phases (if considered necessary) as M phase markers are also available [73].

The benefits of the presented modelling approach extend to several levels. In first place, the model allows to segregate the population throughout biologically relevant markers. Therefore, the model predictive power is extended to both (un)-disturbed cell cycle conditions, as the cell markers can account for such mechanistic effects. In particular, the impact of the segregation is fully appreciated in the after arrest release experiments. The cyclin segregation provided the right markers to account for the different cell cycle profiles. In particular, after the DMSO arrest release the model did partially capture the experimentally observed delay to restart the proliferation based on the low cyclin E1 expression. More importantly, the model allows us to hypothesize the reason for the delayed start of proliferation and *in silico* postulate a plausible explanation. Herein, by simulating a slower G_1 transit (capturing indirectly the several mechanistic imbalances caused by the DMSO [66]) for the first 24h *in silico*, was possible to validate and explain the experimentally delayed cell growth, lower peak density and partially account for the lower productivity. Previous models would fail to predict such variations as size/volume cannot account for the delays in the cell cycle transition.

Some model simulations discrepancies were observed in the lactate profiles. Although, the model formulation accounts for the lactate being produced both from glucose and glutamate, it is evident that other sources might be available (e.g. pyruvate readily available in the media [74]). In addition, the included lactate consumption triggered by the glutamate exhaustion showed a fair prediction.

Although, there are no literature reports that directly link lactate consumption and glutamate exhaustion, the selection was based on: 1) in-house experiments of the mid-exponential growing GS-NS0 cells (in the presence of glutamate) showing lactate consumption when cultured on spent medium (without glutamate, S2 Text) and 2) the well-characterised central cellular metabolism that includes glutamate and lactate. The adopted metabolic link herein is supported by studies that report a high conversion of glutamine into lactate [74] and the accumulation of glutamate when lactate is consumed [75]. Similarly, lactate consumption has been correlated to the cessation of rapid growth [55], which occurs in batch mode after the exhaustion of key substrates (such as glutamate) [50]. However, lactate consumption has been shown to be independent of the residual glucose and lactate concentrations in industrial fed-batch cultures [76]. Nonetheless, there is a need to perform additional experiments to confirm the events that trigger such metabolic shifts.

Model prediction of the undisturbed cell cycle experiment showed overall good agreement with the experimental results. However, the need to improve the metabolic description was also noticed. Particularly, a significant deviation was observed in the lactate profile after the glutamate exhaustion, when lactate was noticeably consumed. The observed lactate switch has been reported [14, 55], which as observed has an impact on the proliferation and therefore, on the cell cycle. Specifically, the metabolic switch seemed to be responsible for the lower cell density and provides an opportunity to link key metabolites (carbon source dependent growth rates) to the cell cycle progression. This type of links will help to further develop our understanding of the effect of substrate and metabolites complementing the MFA approach.

Ultimately, the development of cell culture models accounting for the population heterogeneity via mechanistically relevant markers is now possible. The proposed model structure captures the cell population heterogeneity, which can easily accommodate further biological detail (e.g. apoptosis, cell metabolism, energy requirements, and proliferation cessation, among others).

The presented model can be extended to fed-batch processes of industrial relevance. Typically, during fed-batch operation viable biomass, viability, and culture longevity are targeted for production optimisation without considering culture segregation (heterogeneity) [77]. However, the presented model can aid in the selection of optimal proliferation control strategies accounting for culture heterogeneity, which is linked to cell growth limitations, cell cycle distribution, and cell cycle-specific productivity. A key aspect of the presented model is its ability to accommodate for various proliferation control strategies, such as temperature shifts (temperature dependent cyclin/DNA growth rates), variation of medium osmolarity (slowing growth and increasing productivity) [78, 79], and chemical approaches for cell cycle arrest (shown herein). The model could be used to optimise different targets, including the time for arrest induction (either by temperature shifts or chemical addition), maximization of the integral of viable cells (temperature, osmotic effects), or maximizing productivity (temperature shifts, osmotic effects, or chemical addition). Cell cycle segregation relevance in fed-batch processes accounts for cessation of proliferation despite the avoidance of substrate exhaustion. In addition, the model can be coupled to a glycosylation model [80] in order to capture product quality concerns. Several studies have reported changes [81–84], even improvements, on the glycosylation patterns linked to the cell culture environment (e.g. temperature, pH). Although the effects have been attributed mainly to the environmental changes rather than the effects on proliferation [85], the glycosylation coupled model would enable a systematic study of any potential associations between cell cycle and glycosylation. Therefore, the model can accommodate for environmental factors as well as proliferation control strategies, which can include mechanistic effects in the segregated model formulation and cell cycle-specific productivity.

Moreover, the low computational requirements enable the use of the model for model-based applications of industrial relevance such as control and optimisation. Similarly, the used model development framework provides a systematic platform for the development of biological models and quantitatively analysis of the experimental data. The presented approach helps to close the gap between the development of experimental techniques and modelling approaches of cell cultures. Particularly, cell culture systems that need to account for their heterogeneity, such as co-cultures and mixed cultures can benefit from the presented approach. Thus, the methodology can be extended to accommodate more than one cell line/type in describing the population dynamics. The significance of this approach is not limited to industrial scenarios (where the cell cycle productivity or cell growth can be optimised) but also to clinical environments (tumours growth models, design and optimisation of cancer treatments, etc.).

For the first time, to our knowledge, a biologically mechanistic and experimentally quantifiable cell cycle distributed model is formulated. The key role of the cyclins is confirmed at different levels accounting for the proliferative state and as a quantifiable marker. The presented model formulation with its low computational requirements and associated to experimental techniques provide a versatile toolbox that can be further refined to capture the complex heterogeneity and multi-scale nature of mammalian cell culture systems. The developed model satisfactorily predicted different experimental scenarios and can be used as a systematic computer-aided tool to derive optimal operating policies to improve product titre while satisfying tight operating constraints and product specifications. Thus, the overall integrated modelling framework ultimately will pave the development of relevant cell cycle model both at the industrial and clinical environments.

MATERIALS AND METHODS

Cell Culture

Batch cultures over 5 days of GS-NS0 cells (kindly provided by Lonza; passage 5) with expression of a chimeric B72.3 IgG4 antibody were performed as previously described [49].

Cell Culture Analysis

Cell counts and viability were determined manually using the dye exclusion method (Erythrosin-B; ATCC) and a haemocytometer on a Leica DM-IL inverted phase microscope (Leica). Samples (1.5–2mL) were centrifuged at 800rpm for 5min. The cell-free supernatant was stored at -20°C for metabolic profiling and antibody (IgG4) titre determination. The cell pellets were used for flow cytometry analysis. Metabolic profiles concentrations were measured from the stored supernatant samples with the Nova BioProfile 400 Analyzer (Nova Biomedical). Monoclonal antibody (IgG4) quantification was performed using a high pressure liquid chromatography (HPLC) Agilent 1260 Infinity system (Agilent Technologies). The supernatant was filtered through a 13mm syringe filter with 0.22μm PTFE membrane (VWR, 28145–491) prior to injection. The affinity column, Bio-Monolith Protein A (Agilent Technologies,

5069–3639) was used to quantify IgG4 antibody through the use of a calibration curve. The samples (50μL) were loaded at a flowrate of 1mL/min in a phosphate buffered saline (PBS) mobile phase (pH 7.4) for 2 minutes. Thereafter, a change in gradient was performed using an acetic acid (0.5M, pH 2.6) mobile phase to elute, which was maintained for 5 minutes. Finally the column was again conditioned with a PBS (pH 7.4) mobile phase for 2 minutes. Elution was monitored with a UV detector at 280nm.

Cell Proliferation Assay

Cells in mid-exponential growth were exposed to 30μM of 5-ethynyl-2′-deoxyuridine (EdU, Invitrogen C10419), which is a nucleoside analogue to thymidine and is incorporated into DNA during active DNA synthesis, for a period of 3 hours. After the exposure, the cells were washed twice with PBS and re-suspended in the standard growth media (in the absence of EdU) and cultured in batch mode for 5 days. A parallel control group was established (including all the centrifugation and washing steps without exposure to EdU). Samples were taken every 2h for the first 24h and every 24h afterwards.

Flow Cytometry

1×10^6 cells were fixed and stored prior to flow cytometric analysis as previously described [49]. Preparation for analysis included removing the fixing agent and washing the cells with a rinsing solution followed by centrifugation, as described [49]. Cell membrane permeabilisation was achieved using 50μL of a saponin-based wash reagent (component E, C10419, Invitrogen) for 15min at room temperature as per the vendor's instructions. The EdU was detected by adding 0.5mL of the click-IT reaction cocktail (prepared as per vendor instructions). Incubation (room temperature in the dark for 30min) was followed by washing (rinsing solution). The rinsing solution was removed after centrifugation followed by addition of 50μL of a saponin-based wash reagent. Then 20μL of a PE mouse anti-human Ki-67 or the isotype (BD, 556087) were added to stain for the proliferation marker. Incubation (room temperature in the dark for 30min) was followed by washing (rinsing solution). After centrifugation, the rinsing solution was removed and 50μL of a saponin-based wash reagent was added. Antibodies were used for cyclin staining according to supplier instructions as previously reported [49]. Briefly, incubation (room temperature in the dark for 30min) was followed by washing (rinsing solution) and centrifugation. After removing the rinsing solution, 1mL of rinsing solution was added and DNA was stained by adding 1μL of FxCycle violet stain (Invitrogen, F10347). Controls (isotypes, single colours, and isotypes) were also prepared. Flow cytometry was performed on a LSRFortessa (BD) collecting 10,000 events/sample. The following flow cytometry configuration was used: 405nm excitation wavelength and 450/50 filter collection for the violet stain; 488nm excitation and 530/30 filter collection for FITC; 561nm excitation and 582/15 filter collection for PE; 640nm excitation and 670/14 filter collection for Alexa647. FlowJo software (Tree Star Inc) was used for data analysis. Analysis included selection of the cell population from the forward versus side scatter plot, followed by doublet discrimination analysis [49]. Apoptotic cells, debris, and aggregates were gated out from the DNA histogram. Bivariate plots (cyclin vs. DNA) were used to extract the cyclin expression profiles as reported [49]. The proliferating population was gated above the Ki-67 isotype control region (bivariate plots Ki-67 vs. DNA); the G_0 population was identified in the isotype region. The EdU positive population is gated above the negative EdU population (bivariate plots EdU vs. DNA, cells treated with click-IT reaction cocktail but that were not exposed to EdU).

Statistical Analysis

All experiments presented were carried out simultaneously in triplicate and the shown values represent the mean ± standard deviation (SD). SigmaStat (Systat Software Inc.) was used for statistical analysis performing analysis of variance (ANOVA) with a level of significance of $p<0.05$ on data drawn from a normally distributed population. The Kruskal-Wallis ANOVA on ranks with a significance of $p<0.05$ (Student-Newman-Keuls method) was used when data were drawn from a non-normally distributed population or with unequal variances in order to make multiple comparison.

Acknowledgments

The authors are thankful to Lonza for generously supplying us the GS-NS0 cell line.

Author Contributions

Conceived and designed the experiments: DGGM ENP AM. Performed the experiments: DGGM. Analyzed the data: DGGM AM. Contributed reagents/materials/analysis tools: DGGM MK MCG ENP AM. Wrote the paper: DGGM MK MCG ENP AM.

REFERENCES

1. Nelson AL, Dhimolea E, Reichert JM (2010) Development trends for human monoclonal antibody therapeutics. Nat Rev Drug Discov 9: 767–774. pmid:20811384
2. Pichler J, Galosy S, Mott J, Borth N (2011) Selection of CHO host cell subclones with increased specific antibody production rates by repeated cycles of transient transfection and cell sorting. Biotechnology and Bioengineering 108: 386–394. pmid:20842656
3. Du ZM, Mujacic M, Le K, Caspary G, Nunn H, et al. (2013) Analysis of heterogeneity and instability of stable mAb-expressing CHO cells. Biotechnology and Bioprocess Engineering 18: 419–429.
4. Davies SL, Lovelady CS, Grainger RK, Racher AJ, Young RJ, et al. (2013) Functional heterogeneity and heritability in CHO cell populations. Biotechnology and Bioengineering 110: 260–274. pmid:22833427
5. Zhang Z, Engineering UoMC (2007) Phenotypical Genomics of Bacteria Escherichia Coli: University of Minnesota.
6. Pilbrough W, Munro TP, Gray P (2009) Intraclonal protein expression heterogeneity in recombinant CHO cells. PLoS One 4. pmid:20037651
7. Karra S, Sager B, Karim MN (2010) Multi-scale modeling of heterogeneities in mammalian cell culture processes. Industrial & Engineering Chemistry Research 49: 7990–8006.
8. Alrubeai M, Emery AN (1990) Mechanisms and kinetics of monoclonal-antibody synthesis and secretion in synchronous and

asynchronous hybridoma cell-cultures. Journal of Biotechnology 16: 67–86. pmid:1366816

9. Alrubeai M, Emery AN, Chalder S, Jan DC (1992) Specific monoclonal-antibody productivity and the cell cycle-comparisons of batch, continuous and perfusion cultures. Cytotechnology 9: 85–97. pmid:136918
10. Dutton RL, Scharer J, Moo-Young M (2006) Cell cycle phase dependent productivity of a recombinant Chinese hamster ovary cell line. Cytotechnology 52: 55–69. pmid:19002865
11. Dorka P, Fischer C, Budman H, Scharer J (2009) Metabolic flux-based modeling of mAb production during batch and fed-batch operations. Bioprocess Biosyst Eng 32: 183–196. pmid:18560901
12. Nolan RP, Lee K (2011) Dynamic model of CHO cell metabolism. Metab Eng 13: 108–124. pmid:20933095
13. Sellick CA, Croxford AS, Maqsood AR, Stephens G, Westerhoff HV, et al. (2011) Metabolite profiling of recombinant CHO cells: designing tailored feeding regimes that enhance recombinant antibody production. Biotechnology and Bioengineering 108: 3025–3031. pmid:21769861
14. Templeton N, Dean J, Reddy P, Young JD (2013) Peak antibody production is associated with increased oxidative metabolism in an industrially relevant fed-batch CHO cell culture. Biotechnology and Bioengineering 110: 2013–+. pmid:23381838
15. Wang D, Bodovitz S (2010) Single cell analysis: the new frontier in 'omics'. Trends in Biotechnology 28: 281–290. pmid:20434785
16. Sitton G, Srienc F (2008) Mammalian cell culture scale-up and fed-batch control using automated flow cytometry. Journal of Biotechnology 135: 174–180. pmid:18490070
17. Niklas J, Heinzle E (2012) Metabolic flux analysis in systems biology of mammalian cells. In: Hu WS, Zeng A-P, editors. Genomics and Systems Biology of Mammalian Cell Culture: Springer Berlin Heidelberg. pp. 109–132.
18. Gòdia F, Cairó J (2002) Metabolic engineering of animal cells. Bioprocess Biosyst Eng 24: 289–298.
19. Fredrickson AG, McGee RD III, Tsuchiya HM (1970) Mathematical

models in fermentation processes. Advanced Applied Microbiology 23.

20. Tziampazis E, Sambanis A (1994) MODELING OF CELL-CULTURE PROCESSES. Cytotechnology 14: 191–204. pmid:7765590
21. Pinto MA, Immanuel CD, Doyle FJ (2007) A feasible solution technique for higher-dimensional population balance models. Computers & Chemical Engineering 31: 1242–1256.
22. Kiparissides A, Koutinas M, Pistikopoulos EN, Mantalaris A (2010) Model development and analysis of mammalian cell culture systems. Process Systems Engineering: Wiley-VCH Verlag GmbH & Co. KGaA. pp. 403–439.
23. Domach MM, Shuler ML (1984) A finite representation model for an asynchronous culture of Escherichia-coli. Biotechnology and Bioengineering 26: 877–884. pmid:18553472
24. Henson MA (2005) Cell ensemble modeling of metabolic oscillations in continuous yeast cultures. Computers & Chemical Engineering 29: 645–661.
25. Ramkrishna D (2000) Population Balances. Theory and Applications to Particulate Systems in Engineering. San Diego: Academic Press.
26. Mantzaris NV, Daoutidis P, Srienc F (2001a) Numerical solution of multi-variable cell population balance models: I. Finite difference methods. Computers & Chemical Engineering 25: 1411–1440.
27. Mantzaris NV, Daoutidis P, Srienc F (2001b) Numerical solution of multi-variable cell population balance models. II. Spectral methods. Computers & Chemical Engineering 25: 1441–1462.
28. Mantzaris NV, Daoutidis P, Srienc F (2001c) Numerical solution of multi-variable cell population balance models. III. Finite element methods. Computers & Chemical Engineering 25: 1463–1481.
29. Mantzaris NV, Srienc F, Daoutidis P (2002) Nonlinear productivity control using a multi-staged cell population balance model. Chemical Engineering Science 57: 1–14.
30. Sidoli FR, Asprey SP, Mantalaris A (2006) A coupled single cell-population-balance model for mammalian cell cultures. Industrial & Engineering Chemistry Research 45: 5801–5811.

31. Cazzador L, Mariani L (1993) Growth and production modeling in hybridoma continuous cultures. Biotechnology and Bioengineering 42: 1322–1330. pmid:18612960
32. Hatzis C, Srienc F, Fredrickson AG (1995) Multistaged corpuscular models of microbial—growth—Monte—Carlo Simulations. Biosystems 36: 19–35. pmid:8527693
33. Martens DE, Sipkema EM, Degooijer CD, Beuvery EC, Tramper J (1995) A combined cell-cycle and metabolic model for the growth of hybridoma cells in steady-state continuous-culture. Biotechnology and Bioengineering 48: 49–65. pmid:18623460
34. Cain SJ, Chau PC (1998) Transition probability cell cycle model with product formation. Biotechnology and Bioengineering 58: 387–399. pmid:10099273
35. Rounseville KJ, Chau PC (2005) Three-dimensional cell cycle model with distributed transcription and translation. Med Biol Eng Comput 43: 155–161. pmid:15742735
36. Mantzaris NV, Liou JJ, Daoutidis P, Srienc F (1999) Numerical solution of a mass structured cell population balance model in an environment of changing substrate concentration. Journal of Biotechnology 71: 157–174.
37. Liu YH, Bi JX, Zeng AP, Yuan JQ (2007) A population balance model describing the cell cycle dynamics of myeloma cell cultivation. Biotechnology Progress 23: 1198–1209. pmid:17691814
38. Conlon IJ, Dunn GA, Mudge AW, Raff MC (2001) Extracellular control of cell size. Nature cell biology 3: 918–921. pmid:11584274
39. Conlon I, Raff M (2003) Differences in the way a mammalian cell and yeast cells coordinate cell growth and cell-cycle progression. Journal of biology 2: 7. pmid:12733998
40. Echave P, Conlon IJ, Lloyd AC (2007) Cell size regulation in mammalian cells. Cell Cycle 6: 218–224. pmid:17245129
41. Pardee AB (1989) G1 events and regulation of cell-proliferation. Science 246: 603–608. pmid:2683075
42. Murray A (1994) Cell cycle checkpoints. Current Opinion in Cell Biology 6: 872–876. pmid:7880536
43. Darzynkiewicz Z, Gong JP, Juan G, Ardelt B, Traganos F (1996)

Cytometry of cyclin proteins. Cytometry 25: 1–13. pmid:8875049

44. Morgan DO (1995) Principles of CDK regulation. Nature 374: 131–134. pmid:7877684
45. Kontoravdi C (2006) Development of a combined mathematical and experimental framework model for modeling mammalian cell cultures. London: Imperial College London.
46. Sidoli FR, Mantalaris A, Asprey SP (2004) Modelling of mammalian cells and cell culture processes. Cytotechnology 44: 27–46. pmid:19003227
47. Kiparissides A, Koutinas M, Kontoravdi C, Mantalaris A, Pistikopoulos EN (2011) 'Closing the loop' in biological systems modeling—From the in silico to the in vitro. Automatica 47: 1147–1155.
48. Georgiadis MC, Kikkinides ES, Kostoglou M (2008) Modeling Frameworks of Complex Separation Systems. Computer Aided Process and Product Engineering: Wiley-VCH Verlag GmbH. pp. 137–169.
49. Munzer DGG, Kostoglou M, Georgiadis MC, Pistikopoulos EN, Mantalaris A (2013) Developing a cyclin blueprint as a tool for mapping the cell cycle in GS-NSO. Biochemical Engineering Journal 81: 97–107.
50. Barnes LM, Bentley CM, Dickson AJ (2000) Advances in animal cell recombinant protein production: GS-NS0 expression system. Cytotechnology 32: 109–123. pmid:19002973
51. Naderi S, Meshram M, Wei C, McConkey B, Ingalls B, et al. (2011) Development of a mathematical model for evaluating the dynamics of normal and apoptotic Chinese hamster ovary cells. Biotechnology Progress 27: 1197–1205. pmid:21618458
52. Zhou F, Bi JX, Zeng AP, Yuan JQ (2007) A macrokinetic model for myeloma cell culture based on stoichiometric balance. Biotechnology and Applied Biochemistry 46: 85–95. pmid:16800813
53. Zhao L, Fan L, Wang JQ, Niu HX, Tan WS (2009) Responses of GS-NS0 myeloma cells to osmolality: cell growth, intracellular mass metabolism, energy metabolism, and antibody production. Biotechnology and Bioprocess Engineering 14: 625–632.

54. Ozturk SS, Riley MR, Palsson BO (1992) Effects of ammonia and lactate on hybridoma growth, metabolism, and antibody production. Biotechnology and Bioengineering 39: 418–431. pmid:18600963

55. Mulukutla BC, Gramer M, Hu W-S (2012) On metabolic shift to lactate consumption in fed-batch culture of mammalian cells. Metab Eng 14: 138–149. pmid:22244936

56. Heath JK (2001) Principles of cell proliferation. Oxford: Blackwell Science.

57. Billy F, Clairambaultt J, Fercoq O, Gaubertt S, Lepoutre T, et al. (2014) Synchronisation and control of proliferation in cycling cell population models with age structure. Mathematics and Computers in Simulation 96: 66–94.

58. Kontoravdi C, Asprey SP, Pistikopoulos EN, Mantalaris A (2005) Application of global sensitivity analysis to determine goals for design of experiments: An example study on antibody-producing cell cultures. Biotechnology Progress 21: 1128–1135. pmid:16080692

59. Kiparissides A, Pistikopoulos EN, Mantalaris A (2010) Towards energy-based dynamic optimization of monoclonal antibody producing GS-NS0 cultures. 20th European Symposium on Computer Aided Process Engineering 28: 589–594.

60. McKendrick AG (1926) Applications of mathematics to medical problems. Proceedings of the Edinburgh Mathematical Society 54: 98–130.

61. Kostoglou A, Karabelas AJ (1995) Evaluation of numerical methods for simulating an evolving particle size distribution in growth processes. Chem Eng Commun 136: 177–199.

62. Lapidus L, Pinder GF (1982) Numerical Solution of Partial Differential Equations in Science and Engineering. New York Wiley-Interscience.

63. Kiparissides A (2012) Development of a combined mathematical and experimental framework for the control and optimisation of mammalian cell culture systems. London: Imperial College London. 247 p.

64. Ziehn T, Tomlin AS (2009) GUI—HDMR—A software tool for global sensitivity analysis of complex models. Environmental Modelling & Software 24: 775–785.

65. Sidoli FR, Mantalaris A, Asprey SP (2005) Toward global parametric estimability of a large-scale kinetic single-cell model for mammalian cell cultures. Industrial & Engineering Chemistry Research 44: 868–878.

66. Ponzio G, Loubat A, Rochet N, Turchi L, Rezzonico R, et al. (1998) Early G(1) growth arrest of hybridoma B cells by DMSO involves cyclin D2 inhibition and p21(CIP1) induction. Oncogene 17: 1159–1166. pmid:9764826

67. Kiparissides A, Kucherenko SS, Mantalaris A, Pistikopoulos EN (2009) Global sensitivity analysis challenges in biological systems modeling. Industrial & Engineering Chemistry Research 48: 7168–7180.

68. Dutton RL, Scharer JM, Moo-Young M (1998) Descriptive parameter evaluation in mammalian cell culture. Cytotechnology 26: 139–152. pmid:22358552

69. Bailey JE (1998) Mathematical modeling and analysis in biochemical engineering: Past accomplishments and future opportunities. Biotechnology Progress 14: 8–20. pmid:9496667

70. Matsushime H, Ewen ME, Strom DK, Kato JY, Hanks SK, et al. (1992) Identification And Properties Of An Atypical Catalytic Subunit (P34(PSK-J3)/CDK4) For Mammalian-D Type-G1 Cyclins. Cell 71: 323–334. pmid:1423597

71. Sherr CJ (1993) Mammalian G(1)-cyclins. Cell 73: 1059–1065. pmid:8513492

72. Alberts B, Johnson A, Lewis J, Raff M, Roberts K, et al. (2008) Molecular Biology of The Cell: Garland Science, Taylor & Francis Group.

73. Darzynkiewicz Z, Bedner E, Smolewski P (2001) Flow cytometry in analysis of cell cycle and apoptosis. Seminars in Hematology 38: 179–193. pmid:11309699

74. Dean J, Reddy P (2013) Metabolic analysis of antibody producing CHO cells in fed-batch production. Biotechnology and Bioengineering 110: 1735–1747. pmid:23296898

75. Zhou WC, Chen CC, Buckland B, Aunins J (1997) Fed-batch culture of recombinant NS0 myeloma cells with high monoclonal antibody production. Biotechnology and Bioengineering 55: 783–

792. pmid:18636588

76. Le H, Kabbur S, Pollastrini L, Sun Z, Mills K, et al. (2012) Multivariate analysis of cell culture bioprocess data--lactate consumption as process indicator. Journal of Biotechnology 162: 210–223. pmid:22974585
77. Kumar N, Gammell P, Clynes M (2007) Proliferation control strategies to improve productivity and survival during CHO based production culture—A summary of recent methods employed and the effects of proliferation control in product secreting CHO cell lines. Cytotechnology 53: 33–46. pmid:19003188
78. Stansfield SH, Allen EE, Dinnis DM, Racher AJ, Birch JR, et al. (2007) Dynamic analysis of GS-NS0 cells producing a recombinant monoclonal antibody during fed-batch culture. Biotechnology and Bioengineering 97: 410–424. pmid:17115445
79. Ho Y, Kiparissides A, Pistikopoulos EN, Mantalaris A (2012) Computational approach for understanding and improving GS-NS0 antibody production under hyperosmotic conditions. Journal of Bioscience and Bioengineering 113: 88–98. pmid:22018734
80. Jimenez del Val I, Nagy JM, Kontoravdi C (2011) A dynamic mathematical model for monoclonal antibody N-linked glycosylation and nucleotide sugar donor transport within a maturing Golgi apparatus. Biotechnology Progress 27: 1730–1743. pmid:21956887
81. Borys MC, Linzer DI, Papoutsakis ET (1993) Culture pH affects expression rates and glycosylation of recombinant mouse placental lactogen proteins by Chinese hamster ovary (CHO) cells. Bio/technology (Nature Publishing Company) 11: 720–724. pmid:7832523
82. Andersen DC, Bridges T, Gawlitzek M, Hoy C (2000) Multiple cell culture factors can affect the glycosylation of Asn-184 in CHO-produced tissue-type plasminogen activator. Biotechnology and Bioengineering 70: 25–31. pmid:10940860
83. Trummer E, Fauland K, Seidinger S, Schriebl K, Lattenmayer C, et al. (2006) Process parameter shifting: Part I. Effect of DOT, pH, and temperature on the performance of Epo-Fc expressing CHO cells cultivated in controlled batch bioreactors. Biotechnology and Bioengineering 94: 1033–1044. pmid:16736530

84. Gawlitzek M, Estacio M, Fürch T, Kiss R (2009) Identification of cell culture conditions to control N-glycosylation site-occupancy of recombinant glycoproteins expressed in CHO cells. Biotechnology and Bioengineering 103: 1164–1175. pmid:19418565

85. Kaufmann H, Mazur X, Marone R, Bailey JE, Fussenegger M (2001) Comparative analysis of two controlled proliferation strategies regarding product quality, influence on tetracycline-regulated gene expression, and productivity. Biotechnology and Bioengineering 72: 592–602. pmid:11460250

CHAPTER

10

IMPROVED POLIOVIRUS D-ANTIGEN YIELDS BY APPLICATION OF DIFFERENT VERO CELL CULTIVATION METHODS

Yvonne E. Thomassen[a], Olaf Rubingh[a], René H. Wijffels[b], Leo A. van der Pol [a], Wilfried A.M. Bakker [a]

[a]Institute for Translational Vaccinology, Process Development, PO BOX 450, Bilthoven 3720 AL, The Netherlands

[b]Wageningen University, Bioprocess Engineering, PO BOX 8129, Wageningen 6700 EV, The Netherlands

ABSTRACT

Vero cells were grown adherent to microcarriers (Cytodex 1; 3 g L^{-1}) using animal component free media in stirred-tank type bioreactors. Different strategies for media refreshment, daily media replacement (semi-batch), continuous media replacement (perfusion) and recirculation of media, were compared with batch cultivation. Cell densities increased using a feed strategy from 1×10^6 cells mL^{-1} during batch cultivation to 1.8, 2.7 and 5.0×10^6 cells mL^{-1} during semi-batch, perfusion and recirculation, respectively.

The effects of these different cell culture strategies on subsequent poliovirus production were investigated. Increased cell densities allowed up to 3 times higher d-antigen levels when compared with that obtained from batch-wise Vero cell culture. However, the cell specific d-antigen production was lower when cells were infected at higher cell densities. This cell density effect is in good agreement with observations for different cell lines and virus types. From the evaluated alternative culture methods, application of a semi-batch mode of operations allowed the highest cell specific d-antigen production.

The increased product yields that can easily be reached using these higher cell density cultivation methods, showed the possibility for better use of bioreactor capacity for the manufacturing of polio vaccines to ultimately reduce vaccine cost per dose. Further, the use of animal-component-free cell- and virus culture media shows opportunities for modernization of human viral vaccine manufacturing.

Keywords: Microcarriers Perfusion Feed Recirculation Batch Adherent

INTRODUCTION

Recently, we have produced Sabin-IPV (inactivated polio vaccine based on attenuated Sabin strains) clinical lots under cGMP for phase I safety (and indicative immunogenicity) studies in human adults and infants 1,2. The applied production process was based on a scale-down model of the Salk-IPV manufacturing process 3. The use of this scale-down model allowed fast development of a first generation Sabin-IPV, for which the specifications are closely related to that for the regular IPV product 2. Parallel to this fast-track development an optimization and modernization research program for the manufacturing of Sabin-IPV was started. Examples of modernization are replacement of the used animal derived components (e.g. bovine serum and porcine trypsin) and antibiotics. These components should preferably be omitted (for the antibiotics primarily to prevent any potential allergic reaction), or respectively replaced by animal component free (ACF) alternatives to minimize the risk of adverse effects (e.g. the potential transfer of viruses and/or prions). Moreover, a better scientific understanding of the process, resulting in improved process control and ability for troubleshooting, can be created. Optimization improvements can possibly be found in the currently used, low cell densities (1×10^6 cells mL^{-1}). Assuming comparable virus quality and yields per cell, the use of increased cell densities can potentially result in more efficient use of bioreactor capacity, and ultimately reduce the cost per dose. The demand for IPV is increasing as in 2012 the WHO SAGE group advised all countries to introduce at least one dose IPV in their routine immunization schedules 4. With the increased IPV demands, which will further increase after oral polio vaccine (OPV) cessation, the production capacity will have to increase by scale-up and optimization causing the current IPV price of $ 3.00 per dose to decrease to $ 0.52–$ 1.95 5.. This is still four to fifteen times the current price of OPV (cost per dose $ 0.14), the vaccine used in most countries. Process optimization for IPV manufacturing will be needed to be able to further reduce manufacturing costs below $ 0.50 to keep polio vaccination economically feasible when switching from OPV to IPV 6.

Here we report initial studies where four different adherent Vero cell cultivation methods were applied using ACF cell culture media: (i) batch, the currently used method for Sabin-IPV preparation; (ii) semi-batch, where daily media refreshments were applied; (iii) perfusion where continuous media refreshment was applied; and (iv) recirculation where media was circulated through the bioreactor and re-used. With these commonly known cell culture methods 7. higher cell densities were obtained and the subsequent virus culture, also using ACF virus culture media, produced higher quantities of infectious and immunogenic poliovirus.

MATERIALS AND METHODS

Cells, virus

Vero cells obtained from WHO (10-87) originally derived from ATCC (CCL-81) were used as host for poliovirus production. Poliovirus seeds 1. Sabin type 1 (LSc 2ab KP_2; SO + 3), Sabin type 2 (P712 Ch2ab-KP_2; SO + 3) and Sabin type 3 (Lot 457-III-Pfizer; RSO3) were used.

Cultivation methods

Pre-cultures

Vero cells were cultured in T-flasks and Hyperflasks (Corning) in VP-SFM (Invitrogen) to expand the cell number. After trypisinization (TrypLE Select; Invitrogen) cells were resuspended in VP-SFM and added to the bioreactor.

Bioreactor Vero cell cultures

Different cultivation methods have been applied where Vero cells were grown adherent to microcarriers (3 g L^{-1} Cytodex 1; GE Healthcare). The cultures were maintained at pH 7.2, 37 °C, 50% dissolved oxygen (DO) by headspace aeration only (1 L min^{-1}) and sampled at least once a day.

Cell cultures were carried out in standard glass stirred-tank type bioreactors, optionally equipped with a spin filter (70 μm) to retain cells on microcarriers in the bioreactor when needed (perfusion and recirculation culture mode). Alternatively, a harvest pipe with a 75 μm sieve was used to remove media while retaining microcarriers. Cultivations were controlled using Sartorius DCU-3 control units and MFCS-win software (Sartorius AG, Melsungen, Germany).

Batch cultivations were carried out at 4 L working volume with inoculation densities of 0.1×10^6 cells mL^{-1}. During cultivation, glucose and glutamine were added by bolus feeding to 10 mM glucose and 2 mM glutamine when concentrations were below 5 mM and 0.5 mM respectively. Semi-batch cultivations were essentially performed as described by Mendonça (1998) 8. at 3 L working volume with an inoculation density of 0.1×10^6 cells mL^{-1}. From day two onwards, daily 1 L culture medium (1/3 culture volume) was replaced with fresh medium. Media replacement was done after sedimentation of the microcarriers without agitation. In addition, bolus feeding of glucose and glutamine was done once 4 days after the start of cultivation to obtain concentrations of 20 mM glucose and 2 mM glutamine.

Perfusion cultivations were carried out using 1.5 L working volume. Cells were inoculated at 0.1×10^6 cells mL^{-1} and retained in the bioreactor. After 2 days of batch cultivation, continuous media feed was started at 1.5 L day^{-1} (1 culture volume per day). Media feed rate was kept constant for the remainder of the perfusion cultures.

Recirculation cultures, where cells are retained in the bioreactor (3 L working volume) while medium (15 L total volume = culture volume + circulated volume) is circulated, were carried out essentially as described previously 9.. Cells were inoculated at a cell density of 0.6×10^6 cells mL^{-1}. After 1 day, a continuous recirculation flow was started with 15 L day^{-1} (5 culture volumes per day), followed by daily increments to 22 L day^{-1} and 30 L day^{-1} (respectively 7.3 and 10 culture volume per day) at days 3 and 4.

Virus culture

Prior to virus infection, using the same bioreactor vessel used for Vero cell culture, the media feed was stopped and pH, DO and temperature settings were adjusted to 7.4, 25% and 32.5 °C, respectively. Media was not refreshed but glucose and glutamine were fed when concentrations were below 5 mM and 0.5 mM, respectively. Cells were infected with poliovirus with an MOI (multiplicity of infection) of 0.01. Virus cultivation was considered finished when 100% CPE (cytopathic effect) was observed microscopically.

Analytics

Cells were counted daily using a Nucleocounter NC-100 (Chemometec).

Cell culture metabolites such as glucose, lactate, glutamine, glutamate and ammonia were monitored using a Bioprofile 100 Plus (Nova Biomedical Waltham, MA). Poliovirus was quantified with a virus titer assay as described previously 10.. The amount of d-antigen was assessed using a d-antigen ELISA 11..

RESULTS

Vero cell growth

Vero cell cultures were performed in four different cultivation modes, batch, semi-batch, perfusion and recirculation. Batch cultivations were performed to obtain a reference growth curve for later comparison with the more sophisticated culture methods where either media is refreshed (semi-batch and perfusion) or circulated (recirculation). After 3–4 days of cultivation, a cell density at 1.0×10^6 cells mL^{-1} was reached in batch cultivation with an average growth rate of 0.036 h^{-1} during exponential growth and a growth rate of 0.022 h^{-1} at the moment of virus infection on day 4 (Fig. 1; Table 1). At this point cells are present as a monolayer on the microcarriers (Fig. 2). Applying a daily partial medium renewal in a semi-batch mode allowed cell growth to continue and after 2 additional days of culture (6 days in total) a cell density of 1.8×10^6 cells mL^{-1} was obtained. Here comparable growth rates to batch cultivation were observed. The growth rate declined during the feed phase from 0.034 h^{-1} at day 3 to 0.006 h^{-1} at day 6. Using a perfusion mode, where medium renewal is continuous, cell growth could be prolonged to yield a cell density of 2.7×10^6 cells mL^{-1} in 7 days. The growth rates of the Vero cells were lower during the feed phase compared to the growth rates observed in semi-batch cultivations and decreased from 0.018 h^{-1} at day 3 to 0.005 h^{-1} at day 7. Cells were present in a multilayer on the microcarriers at these cell concentrations (Fig. 2). In the so-called recirculation method 9. cells were retained in the bioreactor while medium from an external container was circulated. When starting with an inoculation density of 0.6×10^6 cells mL^{-1} a monolayer was already formed after one day of cultivation, and cells started to grow in a multilayer rapidly. Cell concentrations of 5.0×10^6 cells mL^{-1}were found after a culture time of 4 days, while growth rates decreased linearly during the feed phase from 0.025 h^{-1} at day 2 to 0.0004 h^{-1} at day 4. It should be noted, that the latter cultivation method was applied using higher inoculation densities, which implies that a more extensive (or longer) pre-culture was needed.

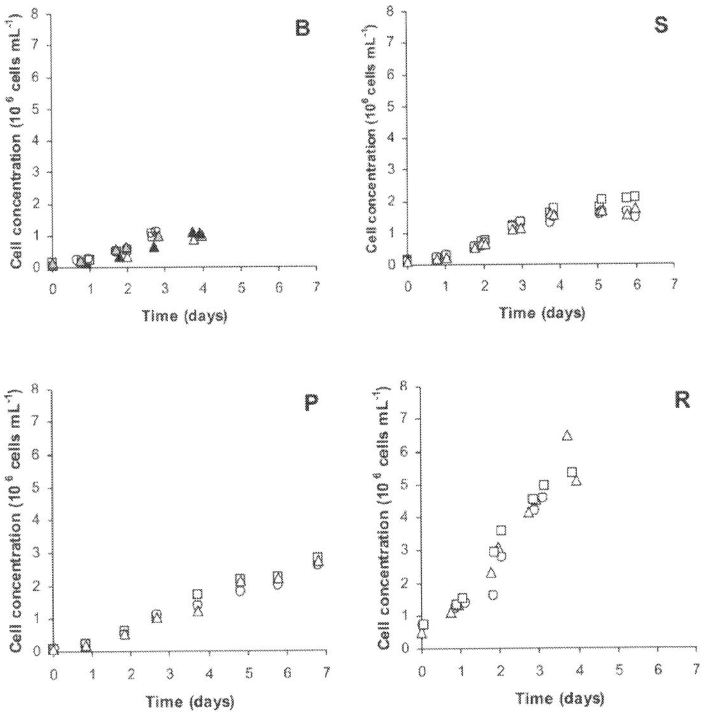

Figure 1: Growth curves of Vero cells using different cultivation modes batch (B), semi-batch (S), perfusion (P) and recirculation (R). Different batches are indicated with different grey shades and symbols, where squares, circles and triangles represent cultures that were infected with poliovirus type 1, 2 and 3 respectively. Infection of cultures was done at the end of the shown growth curve.

Table 1: Cell culture characteristics of different Vero cell cultivation methods.

Cultivation method	Inoculation cell density ($\times 10^6$ cells mL^{-1})	Cell density at time of infection ($\times 10^6$ cells mL^{-1})	Culture time (days)	Growth rate at virus infection (h^{-1})
Batch (*n* = 5)	0.10 ± 0.04	1.02 ± 0.07	4	0.022
Semi-batch (*n* = 3)	0.14 ± 0.01	1.75 ± 0.32	6	0.006
Perfusion (*n* = 3)	0.08 ± 0.01	2.72 ± 0.10	7	0.005
Recirculation (n = 3)	0.64[a] ± 0.11	5.01 ± 0.40	4[a]	0.0004

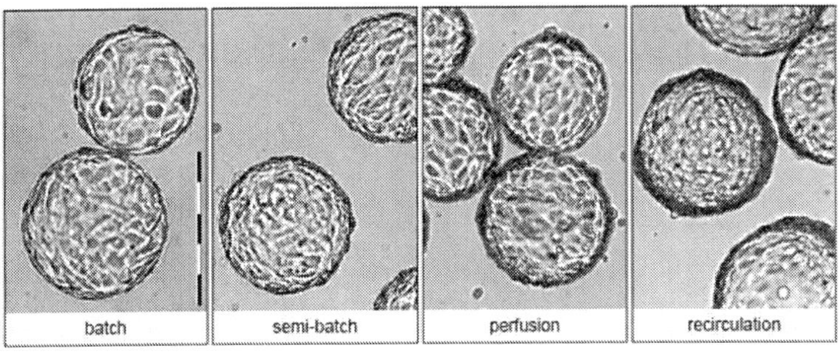

Figure 2: Photographs of Vero cells on microcarriers at the time of virus inoculation. Presence of multilayers at higher cell concentrations (1.0, 1.8, 2.7 and 5.0×10^6 cells mL^{-1} for respectively batch, semi-batch, perfusion and recirculation) is clearly visible.

Glucose and glutamine concentrations during cell culture

The concentrations of glucose and glutamine were analyzed during the Vero cell growth in different cultivation modes. Glucose and glutamine concentrations decreased rapidly when the culture was in batch mode (Fig. 3). When media was refreshed daily (semi-batch) or continuously (perfusion) or when media was circulated (recirculation), sufficient glucose and glutamine were present during the complete cultivation time. During perfusion and recirculation cultivations it is clear that from the moment the feed was started the glucose and glutamine levels remained reasonably constant, whereas during semi-batch cultivations glucose and glutamine concentrations varied more. This was directly correlated to the feeding times. It should be noted that during semi-batch cultivations, an additional bolus feed of glucose and glutamine was given at day 4 (Fig. 3).

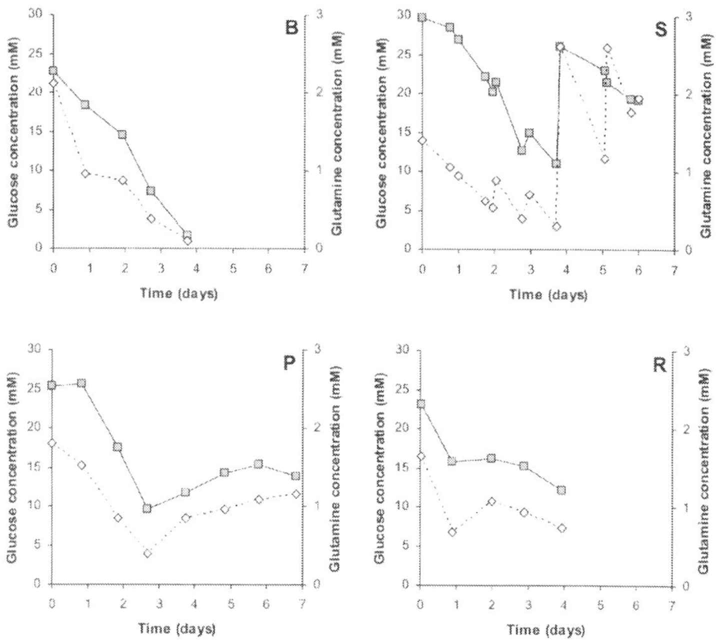

Figure 3: Glucose (squares) and glutamine (diamonds) concentrations present in Vero cell cultivations using batch (B), semi-batch (S), perfusion (P) or recirculation (R). Figures are averaged numbers. Daily media exchange was started at day 2 (S), media feed at day 3 (P) and circulation at day 1 (R).

Lactate and ammonia concentrations

During the batch cultivation lactate and ammonia concentrations increased and within 3 days concentrations up to 30 mM lactate were reached. Daily media replacements allowed to keep lactate concentration below 30 mM whereas continuous media replacement lowered the lactate concentration. Recirculation of media caused a relative constant lactate and ammonia concentration during the cultivation time. Although lactate levels reach high concentrations (above 20 mM), the Vero cell growth continued and therefore it was concluded that this did not inhibit cell growth severely. Ammonia concentrations were below 2 mM under all growth conditions (Fig. 4).

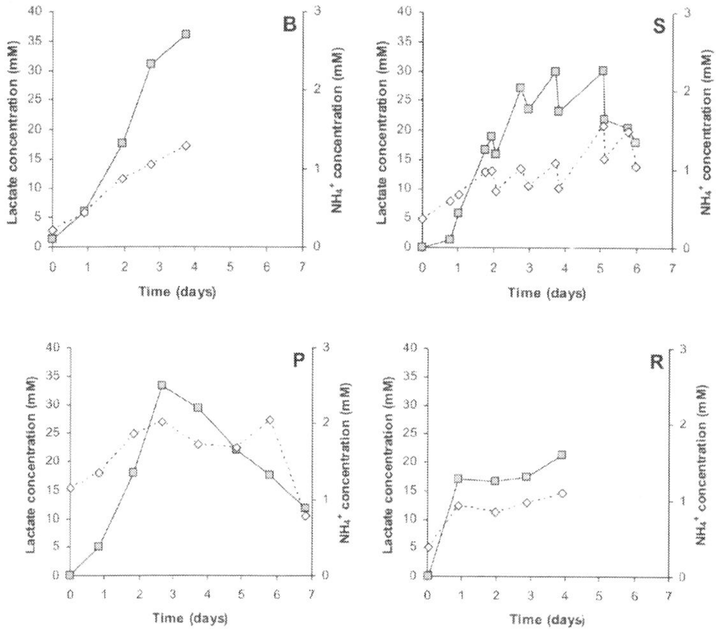

Figure 4: Lactate (squares) and NH_4^+ (diamonds) concentrations present in Vero cell cultivations using batch (B), semi-batch (S), perfusion (P) or recirculation (R). Figures are averaged numbers. Daily media exchange was started at day 2 (S), media feed at day 3 (P) and circulation at day 1 (R).

Virus culture

To determine the variability in poliovirus yields, three cell cultures (in batch mode) were infected with poliovirus type 3. When virus culture was complete, virus titers were measured to determine the amount of infectious poliovirus and d-antigen was measured to quantify the amount of immunogenic poliovirus. The RSD (relative standard deviations) were 9% for the virus titer and 8% for the d-antigen concentration. Both are within 10%, which can be considered comparable. This means that cultures were very comparable as the virus titer assay is valid within 0.5 log (=6%) and the RSD for test reproducibility for the d-antigen ELISA is 10.6% 11. Based on good virus culture reproducibility, it was chosen to compare the effects of different cell culture strategies on the virus yield with $n = 1$ for all three virus types.

Comparable virus titers were found independent of the cell culture method that was applied (Table 2). On the other hand, for all three poliovirus types differences in d-antigen concentrations were more pronounced. In all cases where media refreshments were used during cell cultures an increase of the d-antigen yield was observed, when compared with batch-wise cell culture. These increases ranged from approx. 1.5- to 2-fold when cell cultures were carried out in semi-batch and perfusion mode to approx. 2.4- to 2.9-fold when cell cultures were executed in recirculation mode. Since cell concentrations at the start of virus culture were different in the different settings (Table 1), the cell specific d-antigen yields were calculated and compared (Fig. 5). Cell specific d-antigen yields were the highest when virus culture was carried out based on semi-batch cell cultures for poliovirus type 1 and batch or semi-batch cell cultures for type 2 and 3. When perfusion or recirculation cultures were used prior to virus culture, the cell specific d-antigen yields were a factor 2 lower.

Table 2: Virus titers and d-antigen concentrations observed using different Vero cell culture methods before virus replication.

Cultivation method during cell culture	Poliovirus type 1		Poliovirus type 2		Poliovirus type 3	
	Virus titer (10 LOG $TCID_{50}$ mL^{-1})	d-antigen (DU mL^{-1})	Virus titer (10 LOG $TCID_{50}$ mL^{-1})	d-antigen (DU mL^{-1})	Virus titer (10 LOG $TCID_{50}$ mL^{-1})	d-antigen (DU mL^{-1})
Batch	8.7	175	8.8	44	8.8 ± 0.8 (n = 3)	78 ± 6 (n = 3)
Semi-batch	8.8	356	8.7	72	7.9	137
Perfusion	9.3	320	8.3	65	8.6	115
Recirculation	9.5	508	8.8	115	8.9	185

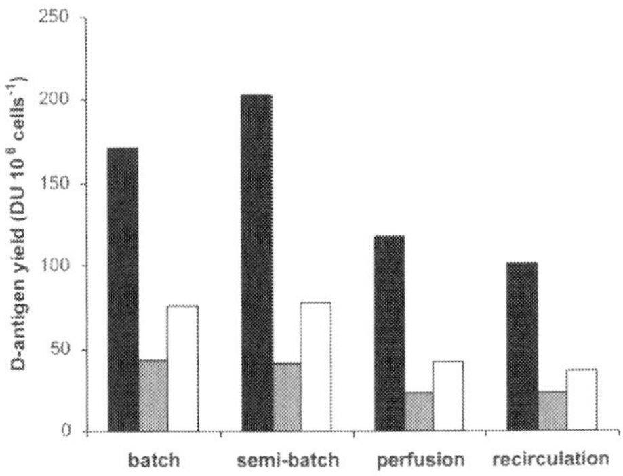

Figure 5: Cell specific d-antigen yields (poliovirus type 1, black; type 2, grey; type 3, white) after different cell culture methods. Similar virus culture conditions were applied.

DISCUSSION

The Vero cell line is one of the commonly used cell lines to produce viral vaccines 12.. Classic cell culture processes used in vaccine manufacturing are often based on batch-wise cell and virus cultivations followed by extensive downstream processing, concentration, purification and inactivation to yield a product 13,14.. While downstream processing is important, the virus of interest is generated during upstream processing, i.e. cell and virus culture. It is also at this stage where the intrinsic product quality is determined. Whereas product yields may be related to both the cell concentration and the metabolic state of the cells, product quality is likely largely influenced by the cells metabolic condition and the virus culture conditions. In other words, the cell culture method may impact product quality. The cell cultures are discussed first, followed by the observed d-antigen levels as indicator of product quality.

The application of different cell culture strategies resulted in higher cell densities, up to 5×10^6 cells mL^{-1} during recirculation cultures. These cell concentrations were at comparable levels to those previously reported for recirculation cultures 15.. In addition, the cell densities reached using perfusion, semi-batch and batch cultures were comparable to those reported by others 8,16,17.

At the higher cell densities, cells were growing in multilayers on the microcarriers. Recently it has been reported that the tumorgenicity of Vero cells is dependent not only on the passage level as reported previously 18., but also on the culture conditions 19.. The growth in dense cultures as well as the adaptation to serum free media may result in the acquisition of a tumorgenic phenotype. Moreover differences in cell morphology, i.e. the compactness of the monolayer, have been reported for Vero cell growth in different serum free media 20.. As such, tumorgenicity of the Vero cells growing in multilayers in a specific ACF medium should be investigated before these cells are used to produce clinical materials.

During all cell cultures, sufficient concentrations of glucose and glutamine were present. At the end of cell culture lactate concentrations were high, up to 36 mM during batch, approx. 20 mM during semi-batch and recirculation and 12 mM during perfusion cultures. Reports of high lactate concentrations during Vero cell growth are abundant when the carbon source is glucose, see for example Petiot et al. (2010) 17., and are caused by the overflow metabolism. High lactate concentrations may be prevented by using other carbon sources like fructose or galactose 8,17.. The ammonia concentration was around 1 mM at the end of the cultivations, which is at an acceptable level that does not inhibit cell growth 21.. Since media was not changed prior to virus culture, these lactate and ammonia concentrations were present at virus infection.

The use of VP-SFM during cell and virus culture appeared beneficial for virus yields when compared to cultivation using serum containing medium during cell culture and M199 during virus culture. In earlier studies 1., using the latter media, d-antigen levels reported for production at 350-L scale were 120, 25 and 56 DU mL^{-1} for respectively Sabin poliovirus type 1, 2 and 3. The use of VP-SFM resulted in a 1.5 times higher level of antigenic product concentration using batch cultivations and 4 fold when using a recirculation culture prior to virus infection. It should be noted that here virus cultures were carried out using spent media. Regarding the nutrient and waste metabolite concentrations it might be even more beneficial to change the media prior to virus culture or to feed possible depleted nutrients during virus culture. This type of optimization may result in a favourable host cell metabolic condition with respect to virus production.

Differences in d-antigen yield per cell between batch or semi-batch and perfusion or recirculation were observed (Fig.5). At higher cell densities the virus yield per cell decreased. This might be an example of the so-called "cell density effect" first described by Wood et al. 22. and observed for different virus cultivation systems 16,20,23,24.. In several cases nutrient limitation or the presence of inhibitory factors may have caused this effect 16,23,24.. In others, the cause remains to be found 20,25.. Here, the concentrations of the main nutrients, glucose and glutamine, and waste products, lactate and ammonia, were at less favourable levels during batch or semi-batch, while the highest specific product yields were observed under these conditions. We thus concluded that these concentrations are less relevant when compared with other phenomena that influence the cells ability to produce virus. These other aspects could be the growth rate at virus infection, the presence of multilayers, or the capacity (surface space) to continue growth after virus infection. Cell growth rates at time of virus infection were lower under all high cell density strategies compared to the growth rates observed in batch cultivations and thus do not explain for the difference in cell specific d-antigen yield observed between semi-batch and perfusion or recirculation cultures. Possibly, the presence of a multilayer has a more important negative effect. It could be that in a multilayer not all cells present on the microcarrier support virus replication in the same way. Moreover, the capacity to continue cell growth at the moment of virus infection may be important as the applied MOI was 0.01 which means 99% of the cells will not be infected during the first virus replication cycle and can potentially grow further. These topics are currently under investigation to be able to further optimize the virus culture at increased cell densities.

The highest virus yields, based on d-antigen concentrations, were observed using the recirculation mode for cell culture. At the first glance, to maximize bioreactor capacity, this seems to be the best choice. However, it should be mentioned that a larger pre-culture needs to be prepared as here the cell culture is started at 0.6×10^6 instead of 0.1×10^6 cells mL^{-1} used for the other cell culture strategies. Hence, extending the overall process throughput time. Further, considering the cell specific d-antigen productivity, the semi-batch cell culture strategy appeared to be a good alternative. In addition, this method can be applied in existing manufacturing equipment without large investments. At present, we are optimizing this method with respect to microcarrier concentration, feed frequency and feed/bioreactor volume ratio. In addition, adaptation of downstream processing to concentrate and purify the poliovirus obtained from increased cell density cultures is studied. Focus points are the filter load with cell debris during clarification and concentration and the removal of the increased concentrations of host cell proteins and host cell DNA during column chromatography. Also, product quality and immunogenicity after purification remains to be assessed. In that way, discrimination between intact virus particles and virus progeny, which may have attributed to the observed increased d-antigen levels, can be made.

CONCLUSIONS

This study shows that adherent Vero cell culture using different methods of medium refreshment allows higher cell densities. Increased cell densities allowed up to 3 times higher d-antigen levels when compared with that obtained from batch-wise Vero cell culture. The cell specific d-antigen production was lower when cells were infected at higher cell densities. Application of a semi-batch mode of operations allowed the highest cell specific d-antigen production, while 2 fold lower cell specific d-antigen yields were found using perfusion or recirculation cultures. This reduction may be related to the presence of multilayers of cells on the microcarriers, which were observed at higher cell densities that were reached using perfusion or recirculation mode. In our view, the most promising concept for polio d-antigen yield optimization would be semi-batch cultivations. This strategy has potential to be further improved and can be implemented in current manufacturing facilities. Using the here presented method for semi-batch cell culture and subsequent virus culture, d-antigen yields per run can be doubled.

Concluding, the use of increased cell densities can result in more efficient use of bioreactor capacity, and may so reduce the vaccine cost per dose.

ACKNOWLEDGEMENTS

This work was supported by the World Health Organization using funds provided by a grant from the Bill and Melinda Gates Foundation.

REFERENCES

1. Bakker WAM, Thomassen YE, van't Oever AG, Westdijk J, van Oijen MGCT, Sundermann LC, et al. Inactivated polio vaccine development for technology transfer using attenuated Sabin poliovirus strains to shift from Salk-IPV to Sabin-IPV. Vaccine 2011;29(41):7188–96.
2. Thomassen YE, van't Oever AG, Van Oijen MGCT, Wijffels RH, Van der Pol LA, Bakker WAM. Next generation inactivated polio vaccine manufacturing to support post polio-eradication biosafety goals. PLoS ONE 2013;8(12):e83374.
3. Thomassen YE, van't Oever AG, Vinke M, Spiekstra A, Wijffels RH, van der Pol LA, et al. Scale-down of the inactivated polio vaccine production process. Biotechnol Bioeng 2013;110(5):1354–65.
4. WHO. Meeting of the strategic advisory group of experts on immunization. November 2012 – conclusions and recommendations. Wkly Epidemiol Rec 2013;88(1):1–16.
5. Venczel L, Landry S, Aylward B, Sutter R, Sabow A, Smith G. Global post-eradication IPV supply and demand assessment: integrated findings commissioned by the Bill & Melinda Gates Foundation and prepared by Oliver Wyman Inc. Available online at: www.polioeradication.org. 2009, accessed 16 September 2013..
6. Sangrujee N, Caceres VM, Cochi SL. Cost analysis of post-polio certification immunization policies. Bull World Health Organ 2004;82(January (1)):9–15.
7. Pirt SJ. Principles of Microbe and Cell Cultivation. Oxford, United Kingdom: Blackwell Scientific Publication; 1975.
8. Mendonc¸a RZ, Pereira CA. Cell metabolism and medium perfusion in VERO cell cultures on microcarriers in a bioreactor. Bioproc Biosyst Eng 1998;18(3):213–8.
9. van Wezel AL. Monolayer growth systems: homogeneous unit processes. In: Spier RE, Griffith JB, editors. Animal Cell Biotechnology. London: Academic Press; 1985. p. 265–82.
10. Thomassen YE, van derWelle JE, van Eikenhorst G, van der Pol LA, BakkerWAM. Transfer of an adherent Vero cell culture method between two different rocking motion type bioreactors

with respect to cell growth and metabolic rates. Process Biochem 2012;47(2):288–96.

11. ten Have R, Thomassen YE, Hamzink MRJ, Bakker WAM, Nijst OEM, Kersten G, et al. Development of a fast ELISA for quantifying polio D-antigen in in-process samples. Biologicals 2012;40(1):84–7.
12. Barrett PN, Mundt W, Kistner O, Howard MK. Vero cell platform in vaccine production: moving towards cell culture-based viral vaccines. Expert Rev Vaccines 2009;8(5):607–18.
13. Montagnon BJ, Fanget B, Vincent-Falquet JC. Industrial-scale production of inactivated poliovirus vaccine prepared by culture ofVero cells on microcarrier. Rev Infect Dis 1984;6(Suppl 2):S341–4.
14. van Wezel AL, van SG, van der MP, Osterhaus AD. Inactivated poliovirus vaccine: current production methods and new developments. Rev Infect Dis 1984;6(Suppl 2):S335–40.
15. Trabelsi K, Rourou S, Loukil H, Majoul S, Kallel H. Comparison of various culture modes for the production of rabies virus by Vero cells grown on microcarriers in a 2-l bioreactor. Enzyme Microb Technol 2005;36(4):514–9.
16. Rourou S, van der Ark A, van der Velden T, Kallel H. A microcarrier cell culture process for propagating rabies virus in Vero cells grown in a stirred bioreactor under fully animal component free conditions. Vaccine 2007;25(19): 3879–89.
17. Petiot E, Guedon E, Blanchard F, Geny C, Pinton H, Marc A. Kinetic characterization of vero cell metabolism in a serum-free batch culture process. Biotechnol Bioeng 2010;107(1):143–53.
18. Levenbook IS, Petricciani JC, Elisberg BL. Tumorigenicity of Vero cells. J Biol Stand Oct 1984;12(4):391–8.
19. Manohar M, Orrison B, Peden K, Lewis Jr AM. Assessing the tumorigenic phenotype of VERO cells in adult and newborn nude mice. Biologicals 2008;36(January (1)):65–72.
20. Chen A, Poh SL, Dietzsch C, Roethl E, Yan ML, Ng SK. Serum-free microcarrier based production of replication deficient influenza vaccine candidate virus lacking NS1 using Vero cells. BMC Biotechnol 2011;11:81.

21. Hassell T, Gleave S, Butler M. Growth inhibition in animal cell culture. The effect of lactate and ammonia. Appl Biochem Biotechnol 1991;30(1):29–41.
22. Wood HA, Johnston LB, Burand JP. Inhibition of Autographa californica nuclear polyhedrosis virus replication in high-density Trichoplusia ni cell cultures. Virology 1982;119(June (2)):245–54.
23. Bock A, Schulze-Horsel J, Schwarzer J, Rapp E, Genzel Y, Reichl U. High-density microcarrier cell cultures for influenza virus production. Biotechnol Progr 2011;27(1):241–50.
24. Kamen A, Henry O. Development and optimization of an adenovirus production process. J Gene Med 2004;6(S1):S184–92.
25. Le Ru A, Jacob D, Transfiguracion J, Ansorge S, Henry O, Kamen AA. Scalable production of influenza virus in HEK-293 cells for efficient vaccine manufacturing. Vaccine 2010 May 7;28(21):3661–71.

CHAPTER

11

AO38, A NEW CELL LINE FROM EGGS OF THE BLACK WITCH MOTH, ASCALAPHA ODORATA (LEPIDOPTERA: NOCTUIDAE), IS PERMISSIVE FOR ACMNPV INFECTION AND PRODUCES HIGH LEVELS OF RECOMBINANT PROTEINS

Yoshifumi Hashimoto, Sheng Zhang and Gary W Blissard

Boyce Thompson Institute at Cornell University, Tower Road, Ithaca NY 14853 USA

ABSTRACT

Background

The insect cell line is a critical component in the production of recombinant proteins in the baculovirus expression system and new cell lines hold the promise of increasing both quantity and quality of protein production.

Results

Seventy cell lines were established by single-cell cloning from a primary culture of cells derived from eggs of the black witch moth (*Ascalapha odorata*; Lepidoptera, Noctuidae). Among 8 rapidly growing lines, cell line 38 (Ao38) was selected for further analysis, based on susceptibility to AcMNPV infection and production of secreted alkaline phosphatase (SEAP) from a baculovirus expression vector. In comparisons with low-passage High Five (BTI-Tn-5B1-4) cells, infected Ao38 cells produced β-galactosidase and SEAP at levels higher (153% and 150%, respectively) than those measured from High Five cells. Analysis of N-glycans of SEAP produced in Ao38 cells revealed two N-glycosylation sites and glycosylation patterns similar to those reported for High Five and Sf9 cells. Glycopeptide isoforms consisted of pauci- or oligomannose, with and without fucose on N-acetylglucosamine(s) linked to asparagine residues. Estimates of Ao38 cell volume suggest that Ao38 cells are approximately 2.5× larger than Sf9 cells but only approximately 74% of the size of High Five cells. Ao38 cells were highly susceptible to AcMNPV infection, similar to infectivity of Sf9 cells. Production of infectious AcMNPV budded virions from Ao38 cells peaked at approximately 4.5×10^7 IU/ml, exceeding that from High Five cells while lower than that from Sf9 cells. Ao38 cells grew rapidly in stationary culture with a population doubling time of 20.2 hr, and Ao38 cells were readily adapted to serum-free medium (Sf-900III) and to a suspension culture system. Analysis of Ao38 and a parental *Ascalapha odorata* cell line indicated that these lines were free of the alphanodavirus that was recently identified as an adventitious agent in High Five cell lines.

Conclusions

Ao38 cells represent a highly productive new insect cell line that will be useful for heterologous protein expression and other applications in biotechnology

BACKGROUND

Insect cell lines are essential for basic studies of insect viruses and arboviruses and represent critical components in the baculovirus expression vector system. In addition insect cell lines have been used for studies of immunological, toxicological, and hormonal responses [1]. For instance, hemocyte-like cell lines have been used to characterize signaling pathways and other processes regulating hemocyte immune responses [2, 3] and an epidermal cell line derived from the integument of *Helicoverpa armigera* showed gene expression responses to

20-hydroxyecdysone [4]. An insect specific scorpion toxin, AaIT, was demonstrated to be highly toxic to Sf9 cells but not to a human MCF-7 cells [5]. New cell lines that are permissive for replication of the *Autographa californica* nucleopolyhedrovirus (AcMNPV) have potential for use in biotechnological applications related to the baculovirus expression vector (BEV) system. Specifically, cell lines that provide improved protein production or post-translational processing are particularly desirable as they have the promise of higher protein yields, and may provide more biologically active or useful recombinant proteins for research, therapeutics, or vaccine production. For applications in biotechnology, desirable characteristics of an insect cell line include rapid growth (cell division times of ≤ 24 h), adaptation to and rapid growth in large-scale suspension cultures, growth in serum-free media, and susceptibility to infection by AcMNPV and/or other expression vectors. An additional factor in the utility of an insect cell line is the possible presence of adventitious agents. In some cases, the presence of such agents may pose no difficulties. In other cases, cell lines that are free of adventitious agents are more highly desirable since industrial proauction of human therapeutic proteins or vaccine candidates must adhere to rigorous specifications related to purity and content.

The baculovirus AcMNPV is one of the most commonly used and well-developed of eukaryotic protein expression vectors. This is due in large part to the so-called hyper expression of genes inserted under the control of the polyhedrin promoter of this virus. While very late genes of AcMNPV are expressed at extremely high levels, the level of expression achieved from heterologous proteins may vary considerably in different cell lines. Lines such as the High Five (BTI-Tn-5B1-4) cell line have been used widely since they were initially shown to produce as much as 7 fold higher levels of heterologous protein when compared with standard Sf9 cell lines [6]. While the High Five cell line likely represents the most highly productive cell line currently in widespread use, there remains enormous potential for development of new cell lines from different insect species, from different tissue sources, and by engineering existing cell lines for specific improvements.

In the current studies, we generated primary cell cultures and individual cell lines from eggs of the black witch moth (*Ascalapha odorata*; Lepidoptera, Noctuidae), a large moth indigenous to Central and

North America. Here we describe the generation and analysis of single-cell derived lines from primary cultures, and the characterization of one line, Ao38. To examine the potential utility of Ao38 cells, we measured the production of secreted human placental alkaline phosphatase (SEAP) and β-galactosidase expressed from recombinant AcMNPV baculoviruses. We also examined cell growth under various conditions and found that Ao38 cells adapted readily to serum-free culture medium and to a suspension culture system. We also found that Ao38 cells appear to be free of an alphanodavirus that was recently reported to latently or persistently infect High Five cells.

RESULTS

Primary *Ascalapha odorata* cell cultures and cell line cloning

Primary cultures of cells were generated from *Ascalapha odorata* egg tissue suspensions in various dilutions of TNMFH medium. Floating and attached cells were separated and maintained separately. Seventy-seven days after primary cultures were initiated, several isolated colonies of cells were observed in culture flasks containing attached cells (cultures 1.0AA, 2.0AA, and 3.0AA) although vigorous cell proliferation occurred only in culture 2.0AA which contained three distinct foci of cells producing proliferating cells at the edge of dark brown flat tissue explants. Two foci producing cells with different shapes and sizes are shown in Fig. 1. Minor foci producing proliferating cells were also present in culture 2.0AA. The parental culture consisting of mixed cells derived from 2.0AA was named AoP (*Ascalapha odorata* Parental) and was first sub-cultured 92 days after the primary cell culture was initiated. AoP was initially passed at 4 - 5 day intervals (by diluting cells 1:10 in fresh TNMFH medium) and later by similar dilution at 3-day intervals. Single cell-derived lines were generated by dilution of cells from the 2.0AA primary culture in 96-well plates, and single cells were confirmed by visual screening. Twenty-three rapidly growing cell lines were obtained from 70 single cell-derived Ao cell lines generated in this manner. Some of the rapidly growing cell lines showed variation in morphology, size, and aggregation (Fig. 2A).

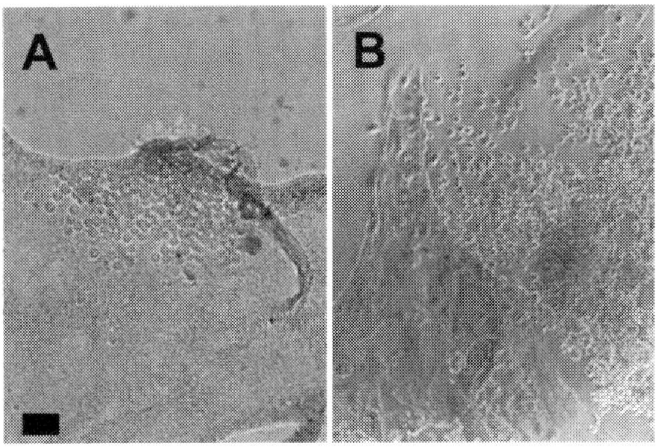

Figure 1: Generation of *Ascalapha odorata* cell lines. Cell line development from a primary culture of *Ascalapha odorata*cells. Seventy-seven days after initiating primary cultures from *A. odorata* eggs, the Ao 2.0AA primary culture showed three cell foci containing proliferating cells. Left and right panels show Foci 1 and 2, respectively, which generated cells with distinct morphologies. The black bar represents 100 μm.

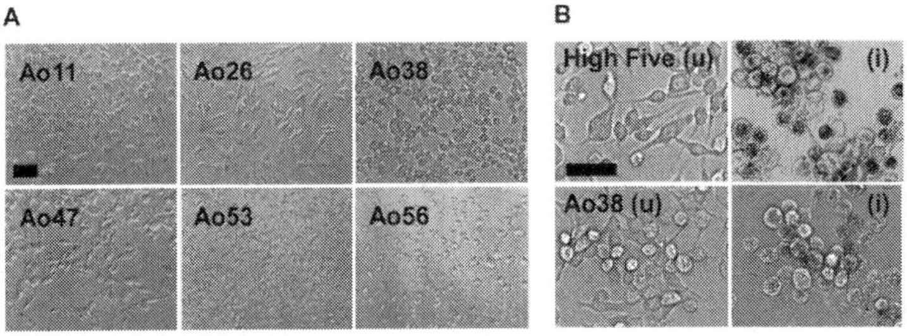

Figure 2: Clonal *Ascalapha odorata* cell lines and AcMNPV infection. (A) Clonal cell lines Ao11, Ao26, Ao38, Ao47, Ao53, and Ao56, are shown in the panels. The black bar represents 50 μm. (B) Comparison of uninfected and infected High Five and Ao38 cell lines. Left panels show uninfected High Five cells and Ao38 cells. Right panels show corresponding AcMNPV-infected High Five and Ao38 cells at 48 h pi. The black bar represents 50 μm.

Susceptibility to AcMNPV infection

In preliminary studies, AoP and AoF1-F3 cells were challenged with WT AcMNPV at 1 IU/cell. AcMNPV virus infection and replication were observed in all cultures. Rapidly growing Ao cell lines (Ao11, Ao24, Ao26, Ao38, Ao45, and Ao56) were initially screened by analyzing protein production from cell lines infected with a recombinant AcMNPV (SEAP-AcMNPV) that expresses secreted alkaline phosphatase (SEAP) [7]. Based on that analysis, cell line Ao38 (a single cell-derived line from AoP) was selected for further studies. A recombinant AcMNPV expressing β-galactosidase (βgal-AcMNPV) was also used to evaluate the Ao38 cell line for susceptibility to viral infection and for analysis of recombinant protein production [6] (see below).

Morphology and growth characteristics

Similar to the morphology of High Five cells, Ao38 cells have a flat spindle-like shape in TNMFH medium (Figs. 2A and 2B). Based on microscopic examination, the average cell size of Ao38 cells appeared similar to that of High Five cells although the general morphology shows slight differences in cell shape, with High Five cells exhibiting slightly longer cell extensions. Some cell aggregates were observed in Ao38 cell culture as cell densities approached confluency and this was observed in both TNMFH and Sf-900III media. Aggregates of cells appeared to consist of approximately 50 to 100 cells.

Adaptation to serum-free medium

Adaptation of Ao38 cells from TNMFH medium into serum free medium (Sf-900III) was performed in two steps, by incrementally lowering the FBS concentration over a 12 day period. Thereafter, the culture was passed by diluting cells 1:10 in fresh medium. The interval between the passages decreased from 8 to 3 days over time (approximately < 3 weeks). During adaptation to Sf-900III medium, the cell morphology changed slightly, becoming more rounded. After adapting Ao38 cells to Sf-900III medium, we also attempted to adapt Ao38 cells to several other commercially available serum-free media formulations, including Sf-900II, Ex-Cell 420, Express V, and ESF921. Ao38 cells cultured in Sf-900II and Ex-Cell 420 media were maintained for several passages without apparent ill effects. However, in those media the cells later began dividing more slowly and showing irregular shapes and sizes, and granules were observed in the cytoplasm. Express V and ESF921 media did not initially support growth of Sf-900III-adapted Ao38 cells (data not shown) and extended efforts were not attempted.

Adaptation to suspension culture

After > 50 passages of Ao38 cells in Sf-900III medium in T-flasks, the cells were used to establish a suspension (shaker) culture. Approximately 2×10^5 cells/ml (in 80 ml of Sf-900III medium) were added to a 250 ml

Erlenmeyer flask (Corning, Cat. No. 431046) and placed on a shaker platform at 100 rpm. After growth for approximately 3-4 days in shaker cultures, cell aggregates were removed by placing the flask on a stationary platform and allowing aggregates to settle for 3-5 min. The single and less aggregated cells from the top of the culture flask were then removed to another culture flask and placed again at 100 rpm. After repeating this process for five passages, the majority of cells in the shaker culture appeared to be single suspended cells and cell morphology did not appear to change during adaptation to the shaker culture system. Passage of cells by diluting 1:10 in fresh medium every 3 days appeared to be optimum for maintaining shaker cultures of Ao38 cells (Fig. 3B).

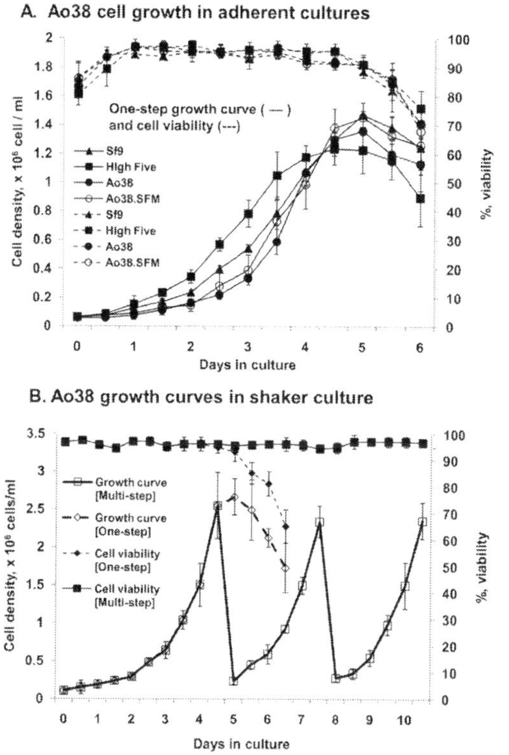

Figure 3: Growth of Ao38 cells in culture. (A) The graph shows a one-step cell culture growth curve (solid line) of adherent cells. In addition, measurements of cell viability (dashed lines) are also included. Cell lines Sf9 (closed triangle), High Five (closed squares), and Ao38 (closed circles) grown in TNMFH medium were also compared with Ao38 cells

grown in Sf-900III medium (open circles). All data points represent averages of triplicate samples and standard deviations are indicated by bars. (B) The graph shows a one step (dashed line) and a multi-step step (solid line) growth curve (open boxes) and viability (closed boxes) of suspension-culture Ao38 cells that were grown in Sf-900III medium in a shaker culture vessel at 100 rpm. Cells were diluted to approximately 0.25×10^6 cells/ml in fresh media on days 5 and 8. Data points represent the average of triplicate experiments and standard deviations are plotted.

Cell Growth in adherent and suspension cultures

Growth rates for attached Ao38 cells grown in T-flasks were determined by one-step growth curves. For these studies Ao38 cells were propagated in either TNMFH or Sf-900III culture media and for comparisons, High Five and Sf9 cells were propagated in TNMFH culture medium (Fig. 3A). During the first four days, Sf9 and High Five cells grew at a higher rate than that of Ao38 cells. However, High Five cells peaked at approximately 1.2×10^6cells/ml at approximately 4.5 days whereas Ao38 and Sf9 cell densities peaked slightly later, at 5 days, with densities of 1.36 - 1.47×10^6 cells/ml. Doubling times for Sf9, High Five, Ao38 (in TNMFH) and Ao38 (in Sf-900III) cells were 22.6, 21.4, 20.2, and 20.9 hr, respectively. Thus growth rates and doubling times were similar for Ao38, High Five, and Sf9 cells. In addition, growth curves of Ao38 cells in serum-containing and serum-free media were similar to those of High Five and Sf9 cells in TNMFH. While the above data represent studies in adherent cultures, we also examined Ao38 cell growth in suspension cultures (Fig. 3B). Growth curves and cell viability assays in both one-step and continuous multi-step shaker cultures showed greater than 90% cell viability and cell densities up to 2.5×10^6 cells/ml.

Cell volume comparisons

To compare the relative size of Ao38 cells with that of the commonly used Sf9 and High Five cells, we measured packed cell volumes (PCV) of cells from each cell line [8]. PCV measurements suggest that the size of Ao38 cells is 2.5 and 0.7 times that of Sf9 and High Five cells, respectively (Fig. 4A). Thus, Ao38 cells were substantially larger than Sf9 cells yet appear to be slightly smaller than High Five cells (which are approximately 3 times the volume of Sf9 cells).

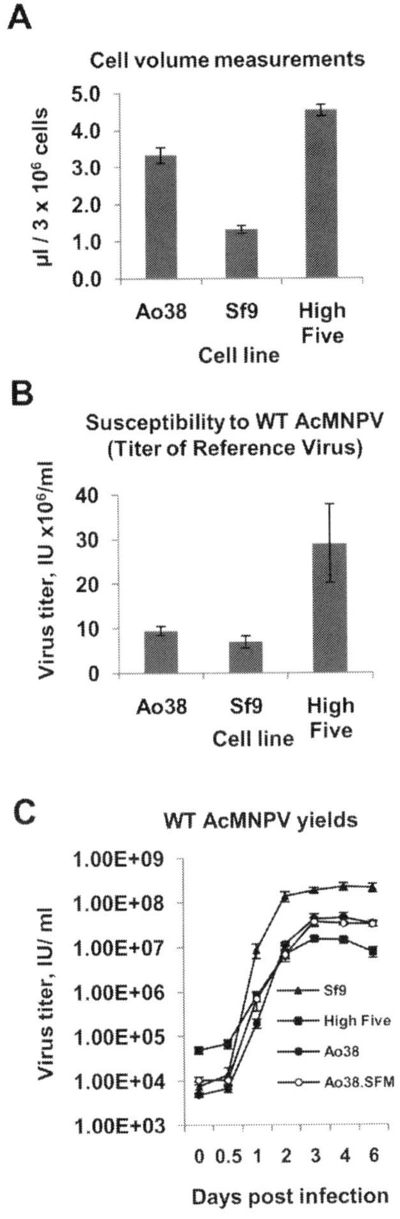

Figure 4: Ao38 cell volumes, AcMNPV susceptibility, and virus yields. (A) Measurements of packed cell volumes (PCV, μl per 3 × 10^6 cells) of Ao38, Sf9 and High Five cells are compared in a bar graph. Gray bars represent averages and standard deviations from 3 replicate measurements are indicated. (B) Susceptibility of

Ao38, Sf9, and High Five cells to AcMNPV infection. The relative susceptibilities of Ao38, Sf9, and High Five cells to AcMNPV infection was measured by infecting each cell line with an AcMNPV virus (βgal-AcMNPV, 1×10^7 IU/ml) that was initially titred on Sf9 cells. The same virus preparation (βgal-AcMNPV) was then titred on each cell line. Gray bars represent averages of virus titers determined on each cell line at 4 days p.i. and standard deviations of triplicate infections are indicated. (C) Comparisons of WT AcMNPV production from Sf9, High Five, and Ao38 cells in TNMFH medium, and from Ao38 cells in Sf-900III serum-free medium. Virus titres were determined from the supernatants of triplicate infections of each cell line with WT AcMNPV at various times post infection (Y-axis).

Virus susceptibility

To examine the relative susceptibility of different cell lines to infection by AcMNPV, we used a standard preparation of the βgal-AcMNPV virus that was previously titred on Sf9 cells (1×10^7 IU/ml), for infection of each cell line. In parallel, the titre of the standard virus preparation was determined on each candidate cell line. Titration of the standard virus inoculum on Sf9 cells produced a titer of 0.70 (+/- 0.13) $\times 10^7$ IU/ml. The titres determined on Ao38 and reference High Five cells were 0.95 (+/- 0.10) $\times 10^7$ and 2.90 (+/- 0.88) $\times 10^7$IU/ml, respectively. These results indicated that Ao38 cells were similar to, or slightly more susceptible to AcMNPV infection than Sf9 cells. In addition and as reported previously [9], High Five cells were approximately three times more susceptible to AcMNPV infection than Sf9 cells (Fig. 4B).

AcMNPV replication in Ao38 cells

To compare AcMNPV production in Ao38 cells to that in Sf9 and High Five cell lines, we infected cells with WT AcMNPV and generated one-step viral growth curves in each cell line (Fig. 4C). Peak virus titres were 1.5×10^7, 4.5×10^7, and 2.8×10^8 IU/ml in High Five, Ao38, and Sf9 cell lines, respectively. Although the kinetics of viral production were generally similar, AcMNPV titres from Sf9 cells appeared to peak earlier that in Ao38 or High Five cells. Thus, Ao38 cells generated virus titers at relatively high levels (4.5×10^7 IU/ml) that exceeded levels measured from High Five cells, but below that from Sf9 cells. We also examined Ao38 cells that were propagated in, and infected in serum-containing TNMFH and in serum-free (Sf-900III) media. Cells propagated and infected in both media resulted in similar AcMNPV yields. Thus the presence of serum did not appear to substantially affect virus production.

Recombinant protein production in Ao38 cells

For maximal utility of new lepidopteran insect cell lines in biotechnology, the capacity for high-level protein production is an important characteristic. To determine whether Ao38 cells were suitable for high-level protein production, Ao38 cells propagated in both TNMFH and Sf-900III media were infected with recombinant AcMNPV viruses that express either an intracellular, or a secreted reporter protein under the control of the AcMNPV polyhedrin promoter. For these studies, Ao38 cells were compared with Sf9 and High Five cells as standards for high-level protein production. The virus βgal-AcMNPV was used for studies of intracellular protein production, and recombinant β-galactosidase expressed in the cell lines was quantified (see Materials and Methods). β-galactosidase production peaked at day 4 for High Five and Sf9 cells, and at day 6 for Ao38 cells. Peak levels for High Five cells were 29.4 $\times$ 10^3 U/10^6 cells whereas those for Ao38 were 44.8 $\times$ 10^3 U/10^6cells (Fig. 5A). Thus, while β-galactosidase production peaked one day later in Ao38 cells, expression levels were 152% of that observed for High Five cells and 231% of that observed for Sf9 cells.

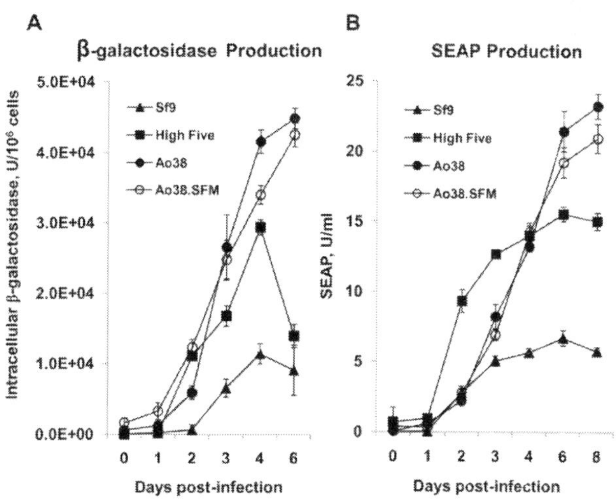

Figure 5: β-gal and SEAP production of Sf9, High Five, and Ao38 cells. (A) Beta-galactosidase production was measured from Ao38, High

Five, and Sf9 cells at various time post infection (Days post infection) by virus βgal-AcMNPV. Ao38.SFM represents Ao38 cells grown in serum-free medium (Sf-900III medium) and all other samples were produced in TNMFH containing 10% FBS. Data points and bars represent averages and standard deviations, respectively, from triplicate infections. (B) Secreted human placental alkaline phosphatase (SEAP) was measured from Ao38, High Five, and Sf9 cells at various time post infection (Days post infection) by virus SEAP-AcMNPV. Ao38.SFM represents Ao38 cells grown in serum-free medium (Sf-900III medium) and all other samples were produced in TNMFH containing 10% FBS. Data points and bars represent averages and standard deviations, respectively, from triplicate infections.

To examine the production of a secreted protein in Ao38 cells, cells were infected with SEAP-AcMNPV (MOI 10) and levels of secreted alkaline phosphatase (SEAP) were measured at various times post infection. SEAP production from infected Sf9 and High Five cells was examined in parallel. SEAP production in Ao38 cells was measured at levels of > 20 U/ml (Fig. 5B). As also observed for β-galactosidase production, SEAP production from infected Ao38 cells continued for a longer period of time and was 300% of that observed from Sf9 cells and 150% of that from High Five cells. Thus, in both cases of intracellular and secreted proteins, recombinant protein production in Ao38 cells peaked slightly later and was substantially higher than that from High Five or Sf9 cells.

Analysis of SEAP N-glycans

To examine protein glycosylation in Ao38 cells, we used LC-MS/MS to examine the N-linked carbohydrate from the 64 kDa SEAP glycoprotein produced from the recombinant virus SEAP-AcMNPV. Using SDS-PAGE analysis of culture medium from SEAP-AcMNPV infected Ao38 cells, we identified an intensely staining band of approximately 64 kDa (Fig. 6). A similar band was absent from culture medium from WT AcMNPV infected Ao38 cells. The 64 kDa band was excised and the tryptic peptide composition was examined by LC analysis and confirmed as SEAP. The SEAP protein has two N-glycosylation motifs (N-X-S/T/C, where X can be any amino acid except proline) located at N139 and N266. The two predicted motifs are located in the tryptic glycopeptides, FNQCN^{139}TTR and YVWN^{266}R. The observed masses of these glycopeptides are summarized in Additional file 1. The Additional file 1 also shows the proposed structures of these N-glycans (based on theoretical sugar masses) and their predicted relative abundance. The mass spectrometry spectra showed the presence of an isoform at the N^{139} glycopeptide with a difference in mass of 14 Da in the amino acid backbone. MS/MS spectra showed the additional 14 Da on the alkylated cysteine residue 138 (data not shown). We suspect that the difference reflects an additional methylene (-CH_2-) group and is caused by residual acrylamide in SDS gel labeling free cysteine to produce propinamide-Cys rather than expected acetamide-Cys or the presence of a mixture of cysteine and homo-cysteine in the tryptic peptide. In both N-glycans at N^{139}, the major sugar composition of the N-glycan was a fucosylated, mono- or bi-antennary, pauci- or oligo-mannose structure [Man2-4GlcNAc2Fuc1,2]. The N-glycan at N^{266} had a molecular mass corresponding to a non-fucosylated pauci- or oligo-mannose structure [Man2-6GalNAc2]. A small portion of the N-glycan corresponds to a bi-antennary tri-mannose structure with a GlcNAc at one branch [GlcNAcMan3GlcNAc2]. In Sf9 cells infected with SEAP-AcMNPV, pauci- or oligo-mannnocidic N-glycans with and without fucose were also produced in the major portion of N-glycan. In High Five cells infected with SEAP-AcMNPV, similar glycan components were also found by PNGase F digestion followed by CE capillary analysis of released glycans [10]. But complex type N-glycans such as Gal3GlucNAc3Man3GucNAc2, Gal3GlucNAc3Man3GucNAc2Fuc, Gal3GlucNAc4Man3GucNAc2Fuc were reported as a substantial proportion (>40%) of the total N-glycans in High Five cells [11]. SEAP produced from CHO cells contains complex and sialylated N-glycan structures in >50% of the N-glycans [12]. A comparison of biochemical features of SEAPs expressed from three insect cell lines is summarized in Table 1.

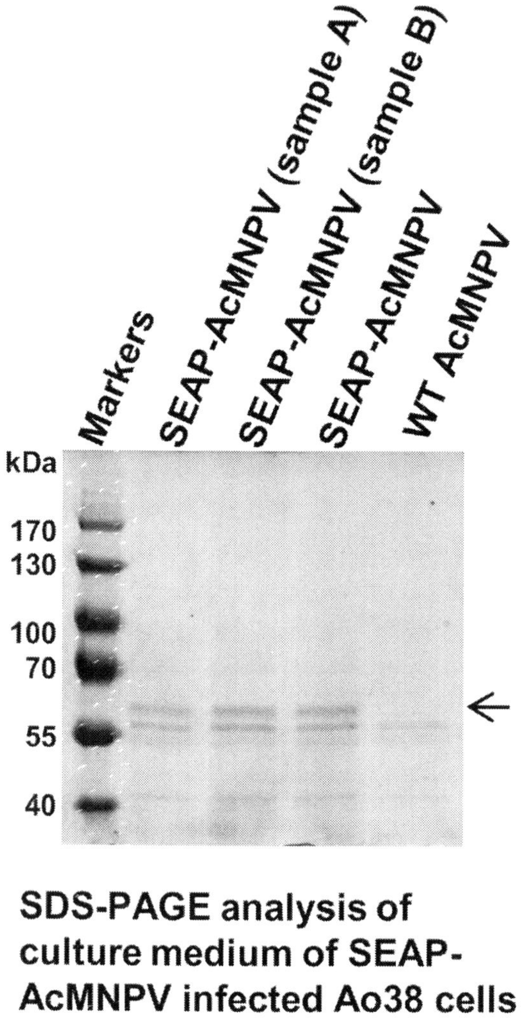

Figure 6: SDS-PAGE analysis of SEAP production in SEAP-AcMNPV infected Ao38 cells. Samples of 20 μl from the supernatant of SEAP-AcMNPV infected Ao38 cells grown in Sf-900III medium were electrophoresed and major protein bands identified by Coomassie blue staining. Lane 1; protein marker sizes in kDa are indicated on the left. Lanes 2-3 represent two independent preparations of SEAP-AcMNPV infected culture medium (20 μl). Lane 4, SEAP-AcMNPV infected culture medium (20 μl) after centrifugation (80,520 × g for 60 min) to remove baculovirus virions. Lane 5, Control from WT AcMNPV infected culture medium (20 μl). An arrow indicates the position of the protein band that was excised and subjected to N-glycan analysis by LC-MS/MS.

Table 1: Comparison of expression levels and N-glycan composition of SEAP in Sf9, High Five, and Ao38 cells

Cell line	Sf9	High Five	Ao38
Expression level (U/ml)	6	11.6	23.2
Identified glycans	Man2, Man3, Man3F, Man4, Man5-8	Man2, Man3, Man4, Man4F, Man5-8	Man2F, Man3F, Man4F, Man3F1,2
			Man2-6, HexNAc-Man3
	Man3F: 30%	Man3: 25%	Man3F: 53% for N^{139}
Major component[a]	Man4: 40%	Man4F: 15%	Man6: 31% for N^{266}
			Man3: 29% for N^{266}

[a] Percent composition of the major component of N-glycans were determined by LC/MS/MS for each glycan (N139 or N266) of Ao38-expressed SEAP, and compared with published data from total glycans of SEAP expressed in Sf9 and High Five cells (see reference [10]).

Absence of TNCL Virus

Recently, an alphanodavirus named Tn5 Cell Line Virus, or TNCL Virus, was described in High Five cells (BTI-Tn-5B1-4). TNCL virus was induced upon infection by WT and recombinant AcMNPV [13]. In order to determine a) if that virus resulted from the initial cell cultures or isolation of BTI-Tn-5B1-4 cells, and b) whether Ao38 or other Ao cell lines also carried this virus, we used RT-PCR to examine AcMNPV infected and uninfected cells from Ao38, Sf9, and High Five. For analysis of High Five cells, we used a relatively early passage of that cell line (passage 98). RT-PCR was performed to detect TNCL virus RNAs in total RNA isolated from WT AcMNPV infected AoP (passage 5) and Ao38 cells (Fig. 7A). Five primer sets were used to sample the bipartite RNA genome of TNCL virus (Table 2, Fig. 7A). The five primer sets produced amplicons with the expected sizes from AcMNPV-infected High Five cell RNA, confirming the presence of the TNCL virus. However, in a parallel analysis no amplicons were detected from cellular RNA preparations from AcMNPV-infected AoP, Ao38 or Sf9 cells. As an internal control, AcMNPV *gp64* transcripts were examined in each RNA preparation to confirm the quality and specificity of each preparation. One-step RT-PCR and PCR was performed with the RNAs, either with or without DNase I treatment to confirm the detection of transcripts. When a DNase I-treated total RNA preparation from AcMNPV infected cells was used as template, the *gp64*-specific product was detected by one-step RT-PCR, but not by PCR alone, confirming the detection of the gp64 transcript (Fig. 7B).

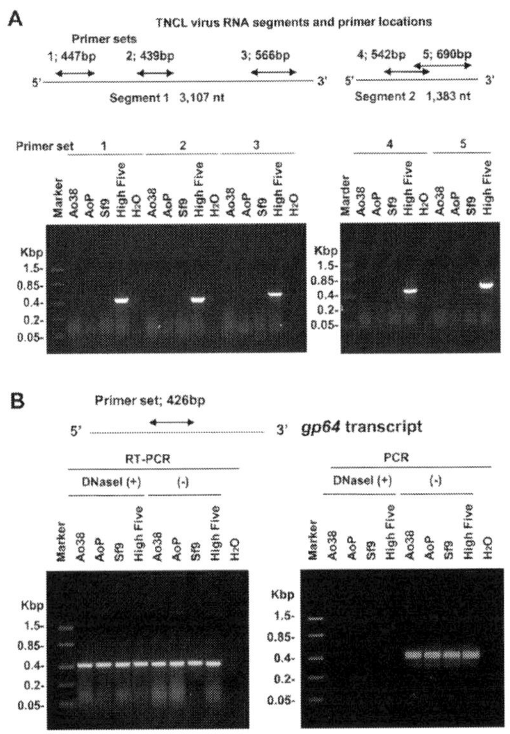

Figure 7: RT-PCR analysis and TNCL virus detection. (A) To examine cell lines for the presence of a TNCL virus, five primer pairs were used for RT-PCR analysis of total RNA from cell lines. Ao38, AoP (passage 5), Sf9, and High Five cells were infected with AcMNPV 48 h prior to RNA isolation. The relative locations of primer pairs on TNCL Virus RNAs 1 and 2 are indicated in the upper diagram, and primer sequences are listed in Table 2. Primer sets were labeled as follows: Set 1, Noda-R1-190F and Noda-R1-636R; Set 2, Noda-R1-1093F and Noda-R1-1531R, Set 3; Noda-R1-2368F and Noda-R1-2933R; Set 4, Noda-R2-269F and Noda-R2-810R; Set 5, Noda-D4 and Noda-U4. Primer sets and cell lines are indicated above each lane, and sizes (Kbp) of marker

DNAs are indicated on the left of each gel.(B). RT-PCR analysis of gp64 transcript RNA control. RT-PCR (left) and PCR (right) were used to analyze total RNA from Ao38, Ao-P (passage 5), Sf9, and High Five cells infected with WT AcMNPV, AcMNPV gp64 primers are listed in Table 2. Samples treated with DNaseI (+) or not treated with DNaseI (-) are indicated above lanes. Sizes (kbp) of marker DNAs are indicated on the left.

Table 2: Primers used in TNCL virus study

Primer	Sequence (5'-3')	TNCL RNA segment	Position on viral genome*	Direction
Noda-R1-190F	gggaaccgagttacacgcg-cattgc	1	190..214	>
Noda-R1-636R	ggtgaatggtgagtcagcatc	1	616..636	<
Noda-R1-1093F	caatctgtcaacgctaggct-tatcgg	1	1093..1188	>
Noda-R1-1531R	ccgccctaagttgtagtt-gttgggacgg	1	1504..1531	<
Noda-R1-2368F	tgtaccgatgcgcttactc-cgttgatatcgg	1	2368..2398	>
Noda-R1-2933R	ccacgctgggtttctccag-cagtgatgttacc	1	2902..2933	<
Noda-R2-269F	ggaatacctgatagattt-gaaggcaaag	2	269..296	>
Noda-R2-810R	ggcaatgtttggataccctc-caatatgtcg	2	781..810	<
Noda-D4	acatccagatccgatcaagt	2	491..510	>
Noda-U4	gccaggaatgttgcttgcaa	2	1161..1180	<
Primer	**Sequence (5'-3')**	***gp64***	**Position***	**direction**
C-177Fs	attttggacgctgagggc		109139..109156	<
L-330Rs	cttgttgatgtgcgcatgcat-cagctc		108731..108757	>

*; position of each primer is based on sequences of deposited in GenBank under accession numbers EF690537 (TNCL virus RNA1), EF690538 (TNCL virus RNA2), and L22858 (AcMNPV)

DISCUSSION

In the current studies, we established a variety of proliferating cell lines from *Ascalapha odorata* egg tissue. After generating primary cultures, proliferating cells were identified as individual colonies at approximately 77 and 90 days. Cells from proliferating colonies were used for single cell cloning by limiting dilution at this relatively early stage (13 days after identifying three foci of proliferating cells in culture 2.0AA) in order to preserve the diversity of potentially useful cells. The 23 rapidly growing cell lines exhibited variations in size, shape, substrate adherence and growth rates. In a preliminary screen, cloned cells were examined based on growth characteristics, AcMNPV infectivity, and production of recombinant secreted alkaline phosphatase (SEAP) from a baculovirus vector. From these, one cell line (Ao38) was selected for more detailed analysis.

Ao38 cells grown in TNMFH are moderately-to-tightly adherent, spindle shaped, and are relatively large in comparison with the more commonly used Sf9 cells. Measurements of packed cell volumes indicated that Ao38 cells are larger than Sf9 cells, but smaller than High Five cells. Ao38 cells have a morphology similar to that of High Five cells (Fig. 2B). Our subsequent analysis showed that Ao38 cells had the following culture characteristics. Ao38 cells can be plated at low densities (approximately 5% confluency) without substantial effects on growth rate, and cells grow rapidly (doubling time of approximately 20.2 h) to relatively high densities of over 1.4×10^6 cells/ml in stationary cultures. As cells grow beyond confluency, some vertical aggregation of cells was observed. The observed clumping may result from a lower level of contact inhibition than observed in other cell lines, and that may contribute to the observation of higher cell densities in adherent and suspension cultures and continued recombinant protein production. Cell clumping was minimized by stepwise adaptation to suspension culture: removing cell aggregates at intervals.

Perhaps the most important and widespread applications of insect cell lines involve the production of secreted proteins that will be later purified from the growth medium. Therefore growth of cells and production of recombinant proteins in serum-free medium is an important characteristic. Ao38 cells were adapted to a commercial serum-free medium, Sf-900III, within a relatively short time period (12 days) and without laborious or stepwise adaptation. However, the Sf-900III-adapted cells did not grow successfully in three other commercial media that have been used for Sf9 and/or High Five cells. Ao38 cells also adapted readily to shaker cultures, using 5 steps to remove aggregates of cells during the first 15 days of growth in the suspension cultures. The procedures for adaptation to suspension cultures were similar to those we have previously used for High Five cells. Cell growth curves in suspension cultures and cell viability assays indicated that Ao38 cells grew optimally in shaker cultures at cell densities ranging from 0.2×10^6 to 2.5×10^6 cells per ml.

We also examined Ao38 cells in terms of viral infectivity, yields, and production of recombinant proteins. Comparisons with Sf9 and High Five cells revealed that Ao38 cells were similar to Sf9 cells in susceptibility to WT AcMNPV infection and both were less susceptible than High Five cells. In contrast to viral susceptibility studies, we found that yields of infectious virus were highest in Sf9 cells. WT AcMNPV yields from Ao38 cells were slightly less than observed from Sf9 cells, but higher than virus yields from High Five cells. Thus, in comparisons of viral susceptibility and virus yield in Sf9 and High Five cells, Ao38 cells compare favorably and are intermediate between these two well-established cell lines. The kinetics of β-galactosidase and SEAP protein production also suggest that progression of the infection cycle is similar in the three cell lines although future studies will examine this in more detail. More importantly though, recombinant protein yields from AcMNPV-infected Ao38 cells were higher those from Sf9 and High Five cells. While the kinetics of β-galactosidase production were generally similar among the three cell lines, peak protein levels were substantially higher from Ao38 cells. Although β-galactosidase production in High Five and Sf9 cells appeared to decrease after 4 days, we found that β-galactosidase continued to accumulate through 6 days post infection in Ao38 cells. A decrease in intracellular β-galactosidase activity in Sf9 and High Five cells after 4 or 5 days p.i. has been previously reported [6, 14]. Production of peak levels of the secreted protein, SEAP, was most rapid in High Five cells. However, peak SEAP levels in Ao38 cells were approximately 50% greater than peak SEAP levels in High Five cells (Fig. 5B). This may be explained by the observation that SEAP production in Ao38 cells appears to continue over a longer time period, permitting accumulation of higher quantities of the secreted protein. It was previously reported [15] that on a culture volume basis, High Five cells produced at least fivefold more SEAP than Sf9 cells. The maximum concentration of SEAP reported for High Five cells was 9.5 U/ml. These results are comparable to SEAP data from the current study (Fig. 5B). We measured peak SEAP production in High Five cells at 15.5 U/ml whereas peak SEAP production in Ao38 cells was measured at 23.2 U/ml in serum-containing medium, and 20.9 U/ml in serum-free medium. Overall our analysis of protein production using two reporter proteins expressed from recombinant baculoviruses indicates that Ao38 cells are capable of producing higher levels of both intracellular and secreted proteins. The reason for higher level protein production by Ao38 cells is unclear although it may result from either more robust protein synthetic machinery, more efficient protein transport and processing, or some combination of these factors. Alternatively, our preliminary observations suggest that when compared with Sf9 and High Five cells, Ao38 cells appear to remain viable for a longer period after infection initiates, possibly resulting in a prolonged period for protein production. Future studies will address these and additional questions.

Insect cells are known to produce glycoproteins with paucimannnosidic or oligomannose N-glycan structures and this is believed to be due to either divergence of N-glycan processing pathways in subcellular organelles, and/or different regulation of the enzymes responsible for generating complex N-glycan structures [16, 17, 18]. To examine and evaluate protein glycosylation in Ao38 cells, we performed LC-MS/MS analysis on the N-glycan of recombinant SEAP protein produced from an AcMNPV expression vector. Based on the consensus sequence for N-linked glycosylation, SEAP has three potential N-glycosylation sites at N^{136}, N^{139} and N^{266}, and prior studies suggested that only the N^{266} site was utilized for glycosylation [12, 19]. In the current study of Ao38 cells, we applied the precursor ion scans monitoring the specific marker ion ($HexNAc^{+}$ at m/z 204) in LC-MS/MS analysis for selective identification of all glycan isoforms associated with each of peptides. We identified N-linked glycosylation of two tryptic peptides at position N^{139} and N^{266} which were eluted at 8.5 min and 20 min, respectively under our LC conditions (Additional file 1). This analysis is complementary with previous approaches for released glycans from SEAP protein, allowing us to determine the glycoisoforms and the difference for each of the N-glycan sites. Sugar composition at both N-glycan sites was paucimannosidic or oligomannose, but more mannose residues were present in the N-glycan at N^{266}. Interestingly, fucosylation at N-acetylglucosamine linked to Asn selectively occurred for the N-glycan at N^{139}. It is possible that the selective fucosylation may have occurred due to the nature of the structure around N^{139}. The nature of the fucose linkage (alpha 1,6 or alpha 1,3) in SEAP from Ao38 cells is not known but insect cell lines have been reported to differ in whether or not their glycoproteins contain core alpha (1,3) fucose (reviewed in reference [17]), and fucosylation may have significance in various applications. While recombinant glycoproteins with a core alpha (1,3) fucose may not be problematic as therapeutics [20], they have resulted in false positives when used in diagnostic assays [21]. It will be of interest in future studies to determine whether this specific linkage is found in glycans from Ao38 cells.

Glycans at the N^{139} residue were predominantly Man3F and Man2F in composition (Additional file 1). In contrast, Man3 and Man6 are the dominant glycan isoforms found at the N^{266} residue. Nevertheless, the identification of the glycans in SEAP expressed in the Ao38 cell line reported in this work is similar to that identified previously from Sf9 and High Five cell lines [10, 11]. It should be noted that the relative quantitation of identified N-linked glycan isoforms should be treated as an estimate since the assumption that all sugar isoforms share a similar ionization efficiency may not be accurate.

Recently an adventitious virus was identified in High Five cell lines. The alphanodavirus, TNCL virus, was identified during production of hepatitis E virus-like particles in High Five cells infected with a recombinant baculovirus vector [13]. Using RT-PCR and primers specific for five regions of the TNCL virus genome, we were unable detect the TNCL Virus in AcMNPV infected Ao38 cells, an uncloned parental cell line (AoP) from *Ascalapha odorata*, or Sf9 cells (a negative control) (Fig. 7A). However, TNCL virus was detected in AcMNPV-infected High Five cells (passage 98) used as a positive control. Persistent viral infections have been observed occasionally in insect cell lines [22, 23] and nodaviruses are thought to occasionally infect wild marine invertebrates either latently or persistently [24]. A betanodavirus was also reported to be latently infecting a cell line derived from the brain tissue of the barramundi fish, *Lates calcarifer* [25].

CONCLUSIONS

We have described the development and characterization of a new cell line, Ao38, derived from cells of the Black Witch Moth, *Ascalapha odorata*. Ao38 cells were found to be highly productive for expression of recombinant protein and exceeded protein expression from the well-characterized Sf9 and High Five cell lines. The data described in this study indicate that Ao38 cells will be a valuable new tool for protein expression using the baculovirus expression system, and in addition, Ao38 cells may be important for cell line engineering by insertion of genes into the Ao38 cell genome.

METHODS

Establishment of cell lines

Primary cultures of cells were prepared on February 19, 2008, from eggs of *Ascalapha odorata* (black witch moth; Lepidoptera, Noctuidae) collected at the Boyce Thompson Institute from adult female moths that were originally collected in Hawaii. In brief, approximately 150 two day old eggs were placed in a 100 μm Nylon cell strainer (BD Falcon) and surface sterilized in the following manner. The eggs were immersed in 2% Clorox for 10 min, rinsed with 70% ethanol for 4 min, then rinsed briefly with autoclaved distilled water. The eggs were then crushed and cells were forced through the nylon mesh strainer with a cell scraper (BD Falcon), into a six well cell culture plate containing 3 ml of TNMFH culture medium [26] supplemented with 10% fetal bovine serum (HyClone, Inc.). The egg cell suspension was diluted to a volume of 10 ml with fresh TNMFH medium. The diluted tissue suspension was then mixed with fresh TNMFH in various ratios, 1:6 to 6:1 (egg cell suspension: TNMFH) to a final volume of 3.5 ml, and each was placed in a 25 cm^2 T-flask (Corning). These primary cultures were incubated for one week at 27°C with no change of medium. The primary cultures were passaged by replacing the medium with fresh TNMFH every 2-3 weeks in order to separate floating and attached cells. Floating cells were retained in separate T-flasks. After approximately 11 weeks, three separate cell foci were observed at the periphery of egg tissue explants in a primary culture, 2.0AA. (Culture 2.0AA was derived from an initial mixture of 2.0 ml of egg tissue suspension and 1.5 ml of TNMFH, a 4:3 dilution as described above.) The attached cells of culture 2.0AA were selected by twice removing floating cells. After each cell focus had grown to a few hundred cells, portions of each focus were carefully removed using a micropipette tip and transferred into new T-flasks. Cell lines derived from foci 1 to 3 are referred to as AoF1, AoF2, and AoF3. The parent culture (that contained the remainder of the original 3 foci) was split into two cultures. One was used for single cell cloning by limiting dilution, and the other was used to develop a parental mixed cell line (AoP). AoF1-AoF3 and AoP cells were frozen at passage 1 and 2. Among seventy cell lines cloned as single cells, 23 rapidly growing cell lines were selected for further analysis based on their rapid growth rate in 96 well plates. After amplifying the cells, each cell line was frozen at passage 1 and 2, and used in the current study after over 10 passages in T-flasks. A second primary culture was generated from *A. odorata* eggs and processed in the same manner on March 16, 2009.

Analysis of cell growth rates

To examine the rate of cell growth, cells (5×10^4 cells) were seeded in a 6-well cell culture plate and cell density and viability were determined at 12 hour intervals. For density calculations, adherent cells were dislodged from one well of a 6-well plate (9.6 cm^2 per well), and suspended in 0.5 ml culture medium. The density and viability of the suspended cells were determined using a hemocytometer and trypan blue staining. A cell growth curve was generated from the results of three independent experiments.

Cell lines and viruses

Cell lines Sf9 (ATCC No: CRL-1711) [27] and BTI-Tn-5B1-4 (High Five) [7, 28] were used as controls in this study. Because protein expression may be reduced in very high passage insect cell lines, we used High Five cells from a relatively low passage stock (84 to 95 passages) for this study. Both cell lines were maintained in TNMFH medium. Wild type (WT) AcMNPV (strain E2) was used in this study. As a reporter for production of intracellular protein, we used a recombinant AcMNPV containing the *Escherichia coli* β-galactosidase gene under the transcriptional control of the polyhedrin gene promoter, a virus that was used in similar studies and previously described [6]. As a reporter for production of secreted protein production, we used a recombinant AcMNPV containing a secreted human placental alkaline phosphatase (SEAP) gene under the control of a polyhedrin promoter [7].

Cell growth in serum-free medium and suspension culture

To examine Ao38 cell adaptation and growth in serum-free medium, Ao38 cell adaptation was initially examined in several commercial serum-free media formulations. Ao38 cells adapted most rapidly to Sf-900III medium (Invitrogen Corp.), which was used for subsequent studies. For adaptation studies, a fifty percent confluent culture of Ao38 cells in T25 flasks, was first incubated in a mixed medium of TNMFH and Sf-900III (1:9) for 6 days at 27°C. The medium was then changed to 100% Sf-900III, and the culture was incubated for another 6 days. During the later 6-day period, the culture became confluent. The culture was passed by diluting cells 1:10 in Sf-900III medium at 3-4 day intervals, over 10 passages before initiating the cell line characterization studies. Sf-900III-adapted Ao38 cells were also examined for growth in other serum-free media. For those studies, cultures of Sf-900III-adapted Ao38 cells (approximately 80% confluent) were transferred directly into four different media formulations: Sf-900II (Invitrogen), Express V (Invitrogen), ESF921 (Expression Systems), and Ex-Cell 420 (SAFC Biosciences).

Growth of Ao38 cells in suspension cultures was examined in the following manner. Sf-900III-adapted Ao38 cells were initially seeded into a 250 ml tissue culture flask (Corning) by adding 80 ml of Ao38 cells (1×10^5cells/ml) in Sf-900III medium. The culture was placed on an orbital shaker at 100 rpm at 27°C. During the first several passages, large cell aggregations of several hundred cells were observed in addition to single cells. To reduce cell aggregation, the culture was removed from the shaker platform and aggregates allowed to settle for 3-5 minutes. The upper phase of the culture (containing single and less aggregated cells) was transferred to another culture flask and returned to the shaker platform as described above. After five repetitions of this process, single cells predominated in the shaker culture. Passage was then carried out by diluting cells 1:10 in fresh medium every three days. Growth curves of cells in the shaker culture were generated by measuring cell density and viability every 12 h. Three replicates of each shaker culture were included in these studies.

Cell volume comparisons

To compare the relative volumes of Ao38, High Five, and Sf9 cells, we examined the packed-cell volumes (PCV) of each cell line using cells grown in TNMFH medium and VoluPAC tubes (Sartorius). We initially examined the optimum g-force for pelleting cells by comparing PCVs at g-forces of 100, 247, 523, 989, 1490, and 2000 × g for 1 min. At 1,490 and 2,000 × g, consistent PCVs were obtained with each cell line. Therefore, cells (3×10^6 cells) from a log-phase culture were pipetted into a VoluPAC tube, centrifuged at 2,000 × g for 1 min, and the volume of the cell pellet was measured to determine the PCV. Averages and standard deviations were calculated based on triplicate sample data.

COMPARISONS OF ACMNPV INFECTIVITY AND YIELD

Viral infectivity

To compare cell lines for susceptibility to AcMNPV infection, we infected Sf9, High Five, and Ao38 cells in parallel with a recombinant AcMNPV virus that was produced and titred on Sf9 cells. βgal-AcMNPV (1×10^7IU/ml, generated and titrated on Sf9 cells) was used to prepare 10^2 to 10^9 fold virus dilutions. Ten μl of each viral dilution was added to 2×10^4 Sf9, High Five, or Ao38 cells in 90 μl TNMFH medium in 12 wells of a 96 well cell culture plate. Four days later, the cells were stained with an X-gal solution and incubated at 27°C for 1 hr to identify infected cells and determine viral titres. The experiment was run in triplicate.

Virus Yields

To examine and compare AcMNPV virus yields from the Ao38 cell line, three cell lines (Ao38, High Five, and Sf9) were infected with WT AcMNPV. Each of the three cell lines was propagated in TNMFH medium, and in addition Ao38 cells were also propagated in Sf-900III medium. Cells were infected by a 1 h incubation with WT AcMNPV at an MOI of 10 IU/cell, followed by rinsing the cells once with fresh medium. After infection, culture supernatant samples were collected at 24 h or 48 h intervals through 6 days post infection. For each supernatant sample, cells or debris were removed by a low speed centrifugation and the supernatant was used for virus titration by $TCID_{50}$ on Sf9 cells.

Analysis of recombinant protein production from recombinant AcMNPV

To compare recombinant protein production in Ao38 cells, we infected Ao38, High Five, and Sf9 cells with either a) a recombinant virus (βgal-AcMNPV) expressing the intracellular protein β-galactosidase, or b) a recombinant virus (SEAP-AcMNPV) expressing the secreted protein, SEAP.

β-galactosidase

Infections with βgal-AcMNPV were carried out in the same manner as that for studies of WT AcMNPV virus yields (as described above). At each time point, infected cells were collected by gently scraping cells from plates and pelleting by low speed centrifugation (700 × g, 5 min). The cell pellet was rinsed once with 1 ml phosphate-buffered saline (pH 7.0), then lysed in 200 μl of lysis buffer (20 mM Tris-HCl, pH 7.5, 5 mM EDTA, 150 mM NaCl, 1% Triton X-100). Debris were removed by centrifugation (12,000 × g for 5 min) and 2 μl of the supernatant was then added to 80 μl of Z buffer (60 mM Na_2HPO_4, 40 mM NaH_2PO_4, 10 mM KCl, 1 mM $MgSO_4$, 50 mM β-mercaptoethanol, pH 7.4) in a 96 well plate. Twenty μl of pre-warmed substrate solution (OPNG, 4 mg/ml in Z buffer) was added to each well and absorbance at 420 nm was monitored every 1 min for 15 min. β-galactosidase activity was calculated according to the equation: U/ml = (OD_{420} · 0.102 ml · dilution factor)/(0.0045 [extinction coefficient] · sample volume).

SEAP

Infection with SEAP-AcMNPV was carried out in a same manner as that for studies of WT AcMNPV virus yields, as described above. Endogenous alkaline phosphatase activity present in the infected culture medium was inactivated by incubation of supernatant samples at 65°C for 5 min. Two μl of each infected-cell supernatant sample (containing SEAP) was mixed with 198 μl of SEAP assay buffer (1.0 M diethanolamine, pH 9.8, 0.5 mM $MgCl_2$, 10 mM homoarginine) in a well of a 96 well plate, and the mixture was incubated at 27°C for 30 min. Twenty μl of freshly prepared and pre-warmed substrate solution (120 mM p-nitrophenylphosphate in SEAP assay buffer) was added to each well. Absorbance at 405 nm was monitored every 1 min for 15 min. SEAP production (units/ml) was calculated according to the equation: U/ml = (change in absorbance × 220 μl × dilution factor)/(sample volume × 18.8 [extinction coefficient] × 1 cm).

Analysis of N-glycan processing

Recombinant SEAP from the culture medium of Sf-900III-adapted Ao38 cells infected with SEAP-AcMNPV (MOI 10) was collected at 96 hr p.i. The culture medium was cleared by centrifugation at 2,000 × g for 10 min. The culture medium of WT AcMNPV-infected cells was prepared in the same manner and used as a SEAP-negative control. Proteins from each culture medium were separated on 8% polyacrylamide gels by SDS-polyacrylamide gel electrophoresis and visualized by Coomassie brilliant blue staining. Because the BV GP64 protein has a similar migration with SEAP, supernatant samples were also subjected to ultracentrifugation at 80,520 × g for 60 min to pellet BV particles, as a control. GP64 did not appear to be detectable in supernatant samples prepared as described above. Bovine serum albumin protein (0.1 to 2.0 μg) was electrophoresed in parallel to generate a standard curve for quantifying SEAP in the stained SDS-PAGE gels. The identity of the SEAP band was also confirmed by in-gel alkaline phosphatase assay after in-gel protein re-folding [29]. Briefly, after the SDS-polyacrylamide gel electrophoresis, proteins were incubated in a transfer buffer (0.01% Triton X-100, 48 mM Tris, 39 mM glycine, 20% methanol, pH 9.2) twice for 15 min. The gel was rinsed in a Western blot detection buffer for 5 min and alkaline phosphatase activity was detected by staining with BCIP/NBT. A band containing approximately 1 μg of the SEAP protein was excised from the gel and subjected to in-gel digestion and tryptic peptide extractions following a previously published protocol [30]. Briefly, gel slices were destained, reduced with 10 mM DTT and alkylated by treatment with 55 mM iodoacetamide in 20 mM ammonium bicarbonate, (in the dark at room temperature for 1 hr). Samples (gel slices) were incubated overnight in

40 μl of 25 mM ammonium bicarbonate containing 0.2 μg trypsin. The resulting tryptic-digested peptides in the solution were collected after centrifugation for 2 min at 4,000 × g. The remaining peptides in the gel were then extracted by sonication in 50 μl of 5% formic acid in 50% acetonitrile and collected similarly. All collected tryptic peptides were combined and evaporated to dryness in a Speedvac SC110 (Thermo Savant, Milford, MA). The digested samples were reconstituted in 15 μl of 2% acetonitrile with 0.5% formic acid, of which 4 μl were injected using a Famous auto sampler onto a C18 column (5 μm, 300 μm × 5 mm, Dionex) for on-line desalting and then separated on a PepMap C-18 RP nano column (3 μm, 75 μm × 15 cm, Dionex), eluted in a 45-minute gradient of 5% to 45% acetonitrile in 0.1% formic acid at 275 nl/min. The nanoLC was connected in-line to a hybrid triple quadrupole linear ion trap mass spectrometer, 4000 Q Trap equipped with a Micro Ion Spray Head II ion source (Applied Biosystems/MDS SCIEX, Framingham, MA).

MS data acquisition was performed using Analyst 1.4.2 software (Applied Biosystems) for precursor ion scan triggered information dependent acquisition (IDA) analysis and enhanced MS-based IDA analysis. The precursor ion scan at m/z 204.08 was monitored at a step size of 0.2 Da across a mass range of m/z 400 to 1500 for detecting glycopeptides containing N-acetylhexoamine unit. The nanospray voltage was 2.0 kV, and was used in positive ion mode for all experiments. The declustering potential was set at 50 eV and nitrogen as collision gas. In IDA analysis, after each precursor ion scan or EMS scan, and enhanced resolution scan, the two to three highest intensity ions with multiple charge states were selected for tandem MS (MS/MS) with rolling collision energy applied for detected ions based on different charge states and m/z values. All acquired MS/MS spectra from EMS-IDA were subjected to Mascot database search against Human database. All acquired MS/MS spectra for detected glycopeptides ions by precursor ion scanning were manually inspected and interpreted with BioAnalysis 1.4 software (Applied Biosystems). The peak areas of detected precursor ions were determined by extracted chromatogram (XIC) at each specific m/z representing glycopeptides isoforms. The relative quantitations of the glycan isoforms of N-linked peptide ions were carried out based on precursor ion peak areas under the assumption that all glycan isoforms linked to the same core peptide have identical or similar ionization efficiency.

RT-PCR analysis of various cell lines for TNCL Virus

Ao, Sf9 and High Five cell lines were examined for the presence of an alphanodavirus (Tn5 cell line virus or TNCL Virus), which was recently shown to persistently infect cultures of High Five cells. TNCL Virus is produced in High Five cells at a significant level when challenged by wild type or recombinant AcMNPV [13]. To detect TNCL Virus, five sets of primers, which cover the bipartite TNCL genome segments (RNA1 and 2) were designed (Table 2) and used for one-step RT-PCR analysis of total RNA from Ao38, AoP (passage 5), High Five (passage 98) and Sf9 cells that were all infected with WT AcMNPV at 10 IU/cell. Total RNA was extracted from the infected cells at 48 hr p.i. using Trizol reagent (Invitrogen) according to the manufacturer's protocol. Each RT-PCR reaction (25 μl) consisted of 100 ng total RNA, 0.5 μM of each primer, 0.25 μl of SuperScript III RT/Platinum Taq mix, 1 × reaction mix (0.2 mM of each dNTP, 1.6 mM $MgSO_4$). RT-PCR was run under the following conditions: 1 cycle of 45°C, 30 min; 1 cycle of 94°C, 2 min; 33 cycles of 94°C, 30 sec/55°C, 15 sec/68°C, 45 sec; 1 cycle of 68°C, 5 min. RT-PCR products were separated on a 1.5% agarose gel.

A portion of the total RNA sample (5 μg) was treated with DNase I (2 units) in a 50 μl reaction containing 1 × DNase I reaction buffer (10 mM Tris-HCl, pH 7.6, 2.5 mM $MgCl_2$, 0.5 mM $CaCl_2$) at 37°C for 30 min, and the reaction was stopped by adding 1 μl of 0.5 M EDTA followed by heating the mixture at 75°C for 10 min. Mock DNase I treated samples were prepared and processed in parallel by replacing DNase I with water. One hundred ng of DNase I-treated total RNA was used in PCR and one-step RT-PCR using AcMNPV *gp64* primers (Table 2). Each PCR reaction (25 μl) consisted of 0.625 unit of Go-Taq DNA polymerase, 1.5 mM $MgCl_2$, 0.2 mM each of dNTP, 0.2 μm each primer, 1 × Colorless Go-Taq Reaction buffer (Promega). PCR was run under the following conditions: 1 cycle of 95°C, 2 min; 33 cycles of 95°C, 30 sec/55°C, 30 sec/72°C, 30 sec; 1 cycle of 72°C, 5 min and the PCR products were separated on a 1.5% agarose gel.

ACKNOWLEDGEMENTS

This research was supported by a DARPA HI-MEMS program and BTI project B00103-R06-01255. We also thank Ms. Abigail McSweeney for reviewing the manuscript. We are grateful to R. Granados and H. A. Wood for providing recombinant AcMNPV viruses containing reporter genes and R. Granados for valuable discussions throughout the course of this work.

REFERENCES

1. Smagghe G, Goodman CL, Stanley D: Insect cell culture and applications to research and pest management. In Vitro Cell Dev Biol Anim. 2009, 45 (3-4): 93-105. 10.1007/s11626-009-9181-x.
2. Dimopoulos G: Insect immunity and its implication in mosquito-malaria interactions. Cell Microbiol. 2003, 5 (1): 3-14. 10.1046/j.1462-5822.2003.00252.x.
3. Strand MR: The insect cellular immune response. Insect Science. 2008, 15 (1): 1-14. 10.1111/j.1744-7917.2008.00183.x.
4. Shao HL, Zheng WW, Liu PC, Wang Q, Wang JX, Zhao XF: Establishment of a new cell line from lepidopteran epidermis and hormonal regulation on the genes. PLoS ONE. 2008, 3 (9): e3127-10.1371/journal.pone.0003127.
5. Ji SJ, Liu F, Li EQ, Zhu YX: Recombinant scorpion insectotoxin AaIT kills specifically insect cells but not human cells. Cell Res. 2002, 12 (2): 143-150. 10.1038/sj.cr.7290120.
6. Wickham TJ, Davis T, Granados RR, Shuler ML, Wood HA: Screening of insect cell lines for the production of recombinant proteins and infectious virus in the baculovirus expression system. Biotechnol Prog. 1992, 8 (5): 391-396. 10.1021/bp00017a003.
7. Davis TR, Trotter KM, Granados RR, Wood HA: Baculovirus expression of alkaline phosphatase as a reporter gene for evaluation of production, glycosylation and secretion. Biotechnology (N Y). 1992, 10 (10): 1148-1150. 10.1038/nbt1092-1148.
8. Stettler M, Jaccard N, Hacker D, De Jesus M, Wurm FM, Jordan M: New disposable tubes for rapid and precise biomass assessment for suspension cultures of mammalian cells. Biotechnol Bioeng. 2006,

95 (6): 1228-1233. 10.1002/bit.21071.

9. Woo SD, Roh JY, Choi JY, Jin BR: Propagation of Bombyx mori Nucleopolyhedrovirus in Nonpermissive Insect Cell Lines. J Microbiol. 2007, 45 (2): 133-138.
10. Zhang F, Murhammer DW, Linhardt RJ: Enzyme kinetics and glycan structural characterization of secreted alkaline phosphatase prepared using the baculovirus expression vector system. Appl Biochem Biotechnol. 2002, 101 (3): 197-210. 10.1385/ABAB:101:3:197.
11. Joshi L, Davis TR, Mattu TS, Rudd PM, Dwek RA, Shuler ML, Wood HA: Influence of baculovirus-host cell interactions on complex N-linked glycosylation of a recombinant human protein. Biotechnol Prog. 2000, 16 (4): 650-656. 10.1021/bp000057p.
12. Nam JH, Zhang F, Ermonval M, Linhardt RJ, Sharfstein ST: The effects of culture conditions on the glycosylation of secreted human placental alkaline phosphatase produced in Chinese hamster ovary cells. Biotechnol Bioeng. 2008, 100 (6): 1178-1192. 10.1002/bit.21853.
13. Li TC, Scotti PD, Miyamura T, Takeda N: Latent Infection of a new Alphanodavirus in an insect cell line. J Virol. 2007, 81 (20): 10890-10896. 10.1128/JVI.00807-07.
14. Wang P, Granados RR, Shuler ML: Studies on serum-free culture of insect cells for virus propagation and recombinant protein production. J Invertebr Pathol. 1992, 59 (1): 46-53. 10.1016/0022-2011(92)90110-P.
15. Davis TR, Wickham TJ, McKenna KA, Granados RR, Shuler ML, Wood HA: Comparative recombinant protein production of eight insect cell lines. In Vitro Cell Dev Biol Anim. 1993, 29A (5): 388-390. 10.1007/BF02633986.
16. Harrison RL, Jarvis DL: Protein N-glycosylation in the baculovirus-insect cell expression system and engineering of insect cells to produce “mammalianized” recombinant glycoproteins. Adv Virus Res. 2006, 68: 159-191. 10.1016/S0065-3527(06)68005-6.
17. Tomiya N, Narang S, Lee YC, Betenbaugh MJ: Comparing N-glycan processing in mammalian cell lines to native and engineered lepidopteran insect cell lines. Glycoconj J. 2004, 21 (6): 343-360. 10.1023/B:GLYC.0000046275.28315.87.

18. Rendic D, Wilson IB, Paschinger K: The glycosylation capacity of insect cells. Croatia Chimica Acta. 2008, 81: 7-21.
19. Endo T, Ohbayashi H, Hayashi Y, Ikehara Y, Kochibe N, Kobata A: Structural study on the carbohydrate moiety of human placental alkaline phosphatase. J Biochem. 1988, 103 (1): 182-187.
20. Altmann F: The Role of Protein Glycosylation in Allergy. International Archives of Allergy and Immunology. 2007, 142 (2): 99-115. 10.1159/000096114.
21. Hancock K, Narang S, Pattabhi S, Yushak ML, Khan A, Lin SC, Plemons R, Betenbaugh MJ, Tsang VC: False positive reactivity of recombinant, diagnostic, glycoproteins produced in High Five insect cells: effect of glycosylation. J Immunol Methods. 2008, 330 (1-2): 130-136. 10.1016/j.jim.2007.08.002.
22. Grace TDC: Tissue Culture for Arthropods. Ann NY Acad Sci. 1958, 77: 275-282. 10.1111/j.1749-6632.1959.tb36908.x.
23. Lery X, Fediere G, Taha A, Salah M, Giannotti J: A new small RNA virus persistently infecting an established cell line of Galleria mellonella, induced by a heterologous infection. J Invertebr Pathol. 1997, 69 (1): 7-13. 10.1006/jipa.1996.4617.
24. Gomez DK, Baeck GW, Kim JH, Choresca CH, Park SC: Molecular detection of betanodaviruses from apparently healthy wild marine invertebrates. Journal of Invertebrate Pathology. 2008, 97 (3): 197-202.
25. Chi SC, Wu YC, Cheng TM: Persistent infection of betanodavirus in a novel cell line derived from the brain tissue of barramundi Laes calcarifer. Dis Aquat Organ. 2005, 65: 91-98. 10.3354/dao065091.
26. Hink WF: Established insect cell line from the cabbage looper, *Trichoplusia ni*. Nature. 1970, 226: 466-467. 10.1038/226466b0.
27. Vaughn JL, Goodwin RH, Tompkins GJ, McCawley P: The establishment of two cell lines from the insect *Spodoptera frugiperda* (Lepidoptera; Noctuidae). In Vitro. 1977, 13 (4): 213-217. 10.1007/BF02615077.
28. Granados RR, Li GX, Derksen ACG, Mckenna KA: A new insect cell line from Trichoplusia ni (BTI-Tn-5B1-4) susceptible to Trichoplusia

ni single enveloped nuclear polyhedrosis virus. J Invertebr Pathol. 1994, 64 (3): 260-266. 10.1016/S0022-2011(94)90400-6.

29. Zhou J, Blissard GW: Mapping the conformational epitope of a neutralizing antibody (AcV1) directed against the AcMNPV GP64 protein. Virology. 2006, 352: 427-437. 10.1016/j.virol.2006.04.041.

30. Shevchenko A, Wilm M, Vorm O, Mann M: Mass spectrometric sequencing of proteins silver-stained polyacrylamide gels. Anal Chem. 1996, 68 (5): 850-858. 10.1021/ac950914h.

CHAPTER

12

TRANSCRIPTOME ANALYSES OF INSECT CELLS TO FACILITATE BACULOVIRUS-INSECT EXPRESSION

Kai Yu[1,2], Yang Yu[1,2], Xiaoyan Tang[1,2], Huimin Chen[1,2], Junyu Xiao[2,3]&, Xiao-Dong Su[1,2]

[1]Biodynamic Optical Imaging Center, School of Life Science, Peking University, Beijing 100871, China

[2]State Key Laboratory of Protein and Plant Gene Research, Peking University, Beijing 100871, China

[3]Peking-Tsinghua Center for Life Sciences, Peking University, Beijing 100871, China

ABSTRACT

The High Five cell line (BTI-TN-5B1-4) isolated from the cabbage looper, *Trichoplusia ni* is an insect cell line widely used for baculovirus-mediated recombinant protein expression. Despite its widespread application in industry and academic laboratories, the genomic background of this cell line remains unclear. Here we sequenced the transcriptome of High Five cells and assembled 25,234 transcripts. Codon usage analysis showed that High Five cells have a robust codon usage capacity and therefore suit for expressing proteins of both eukaryotic- and prokaryotic-origin.

Genes involved in glycosylation were profiled in our study, providing guidance for engineering glycosylated proteins in the insect cells. We also predicted signal peptides for transcripts with high expression abundance in both High Five and Sf21 cell lines, and these results have important implications for optimizing the expression level of some secretory and membrane proteins.

Keywords: High Five cell line baculovirus-insect cell system codon usage glycosylation signal peptide

INTRODUCTION

The baculovirus-insect cell expression system is one of the most popular platforms for recombinant protein expression. It is widely used for protein structure and function studies in academic laboratories, and facilitates massive protein production in industry (Kost et al. 2005). The two common cell lines in this binary system are Sf21 (IPLB-Sf21AE) from *Spodoptera frugiperda* (Vaughn et al. 1977), and High Five (BTI-TN-5B1-4) from ovarian tissues of *Trichoplusia ni* (cabbage looper) (Wickham et al.; Davis et al. 1992).

Protein expression in insect cells has several advantages such as high expression level and easy manipulation. In addition, difficult proteins especially eukaryotic proteins that need posttranslational processing usually fold better in insect cells than in the *E. coli* expression system (Brondyk 2009). However, compared to the mammalian cells, post-translational modifications are still limited in insect cells, with glycosylation as the most significant example (Jarvis 2003; Kost et al. 2005). Due to the defect in glycosylation, functions of some recombinant glycoproteins are impaired (Xu and Ng 2015). For example, the insect cells cannot produce sialylated N-linked glycans. In the past two decades, various efforts were made to import the mammalian glycosylation pathway related genes into the insect cells to engineer the required glycosylation modification (Castilho 2015). For example, Hollister first reported in 1998 that an engineered Sf9 cell line expressing the B4GALT1 gene could produce foreign glycoproteins with terminally galactosylated N-glycans (Hollister et al. 1998). It was also reported in 2001 that both Sf9 and High Five cells were engineered to produce sialylated proteins by adding the ST6GAL1 gene (Hollister and Jarvis 2001; Breitbach and Jarvis 2001). Hollister et al. (Hollister et al. 2002) later transformed a set of other genes to generate the SfSWT-1 cell line which produce biantennary, terminally α-2,6- and α-2,3- sialylated N-glycans. More work has been recently done to obtain more powerful insect cell lines.

Next generation sequencing technology is widely used recently in biological studies. Genomes and transcriptomes of different species are sequenced, which generate high-input information for genomic studies and molecular modifications. We believe similar information can also be explored to provide guidance to engineer new strains of insect cell line for expressing proteins with mammalian type of posttranslational modifications. Genome and transcriptome of Sf21 cell line have already been reported in 2014 and 2015, respectively (Kakumani et al. 2014; Kakumani et al. 2015); however, the global genomic information of the High Five cell line is still unknown.

We constructed and sequenced an mRNA library of the High Five cell line, assembled a reference transcriptome for function and expression studies. We analyzed some protein-expression-related problems by comparing our High Five transcriptome with the reported Sf21 transcriptome (Kakumani et al. 2015). In addition, we extracted codon usage information from their coding sequences and compared it with other expression systems and model species. We also annotated transcripts that may have glycosylation-related functions, and evaluated their expression abundance to generate the global view of glycogenes in High Five and Sf21 cell lines. High expression transcripts, which have predicted signal peptide sequences, were analyzed for predicting highly efficient signal peptide sequences for secretory protein expression.

RESULTS AND DISCUSSION

Reference transcriptome assembly

Considering the genome size of several reference-ready insects, a total 49.5 million 101 bp paired-end reads were sequenced and yield 4.95 Gb bases raw data, 48 million clean reads were kept after low quality reads were filtered. After reads trimming with Trimmomatic, we used Trinity pipeline to do the *de novo* transcriptome assembly and obtained 31,068 transcripts with an N50 value of 2,276 bp (Haas et al. 2013; Bolger et al. 2014).

In total, 39.4 Mb bases are assembled and the average transcript length is 1,269 bp. To reduce the redundancy of the assembly, cd-hit-est was used and transcript number was reduced to 25,234 under 90% sequence identity threshold. These transcripts data sets are the so-called 'unigene'. 13,732 coding peptide sequences are predicted with TransDecoder. Detailed statistics numbers are shown in Table 1.

Table 1: Assembly statistics information

	Raw assembly	**Duplicate removed assembly**
Trinity 'genes'	27,389	24,000
Trinity transcripts	31,068	25,234
GC content (%)	40.84	40.71
Median contig length (bp)	722	622
Average contig length (bp)	1269.6	1160.9
Total assembled bases (bp)	39,444,068	29,294,166

All clean reads have been submitted to NCBI SRA database under accession number SRP068276. Assembly version in this paper has been submitted to NCBI TSA database (GEEM01000000).

Assembly assessment

To evaluate the assembly quality, we employed several strategies for quality assessment. First, we used bwa (v0.7.10) (Li and Durbin 2010) to align all clean reads back to the assembly. 97.4% of the reads could be aligned and 95.3% are properly paired, indicating that the completeness of our assembly is high and very reliable. All transcripts were then aligned to SwissProt to check the proportion of transcripts that may be full-length or near full-length. As shown in Fig. 1A, about 6000 transcripts, with more than 10% of its contig length, could be aligned to a homolog in SwissProt. Among them, about 1/3 of the transcripts are fully aligned and more than 5000 have at least 30% sequence overlapping with known homolog. The result could be underestimated as the homologs of different species' genes have different proportion in the SwissProt database, but we believe our data reached our expectation.

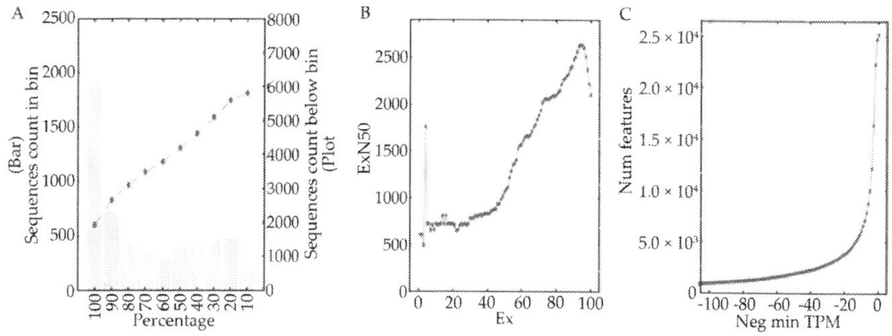

Figure 1: Assembly quality assessment. (A) Full-length transcript assessment. Bin on x-axis represent the percentage of the hit's length included in the alignment to the Trinity transcript. Left y-axis with bar plot is the transcript count in each bin and right y-axis with point plot is the accumulate count below that bin. (B) N50 of subset of transcript by decreasing the expression level. Ex is the top most expressed transcripts that represent x% of the data. ExN50 is the length of a transcript while the total length of transcripts shorter that it reached 50% of total length of all transcripts in this dataset. (C) Transcript count with a threshold of negative minimum TPM value

Taking expression values into consideration, we recalculated the N50 value after low expression contigs were eliminated (Fig. 1B) and plotted expression value distribution pattern in Fig. 1C. Ex in Fig. 1B means a subset of top x% highly expressed transcript, the ExN50 reached the max length at E95, showing that 8,147 transcripts are in the top 95% expression subset with the minimum TPM of 6.1. From this, we can conclude from these data that most of the extremely high expression transcripts in High Five cells are in the range from 600 to 1000 bp. Longer transcripts have more regular expression level. As shown in Fig. 1C, the transcript number was reduced to 10,047 after transcripts of low TPM values (below 5) were eliminated.

Function annotation

To annotate functions of transcripts and coding peptides, we searched homologous genes in SwissProt, TrEMBL90 and NCBI nr databases with blast. Among the 25,234 transcripts, 13,492 got blast hits in nr database, 9,639 and 13,767 have similar sequences in SwissProt and TrEMBL90 database, respectively. With 13,732 coding peptides, 9,427 and 12,198 got alignments in SwissProt and TrEMBL90 database, respectively.

While executing GO annotation and EggNOG annotation, we also used protein sequence predicted from published Sf21 transcriptome data with the same analysis procedural. In High Five transcriptome, 13,447 transcripts have been annotated for GO terms and 13,459 have been annotated in EggNOG database. More specific function classification data are shown in Fig. 2and Fig. 3, and detailed annotations for each transcript are described in supplemental file S1. From the global pattern of these figures, we can tell that the differences in GO terms and EggNOG categories between High Five and Sf21 cells are quite similar. Transcript numbers are higher for some function classes, including those related to intracellular trafficking, secretion, vesicular transport, posttranslational modification, protein turnover, transcription, translation, etc. Possessing plentiful genes with these functions make these insect cell lines an ideal host for protein production.

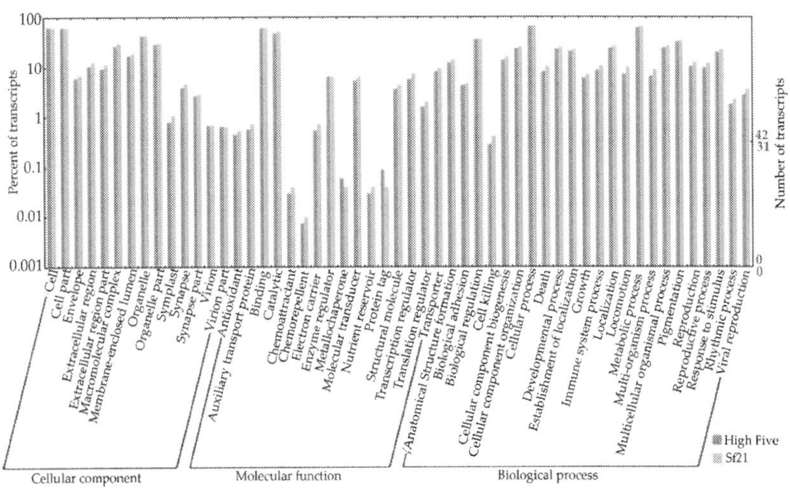

Figure 2: Gene Ontology of High Five and Sf21 transcriptome. Summarized in three main GO categories: Cellular component, Molecular function and Biological process. Right y-axis is the transcript count in that function item, left y-axis is the corresponding percentage of transcripts number

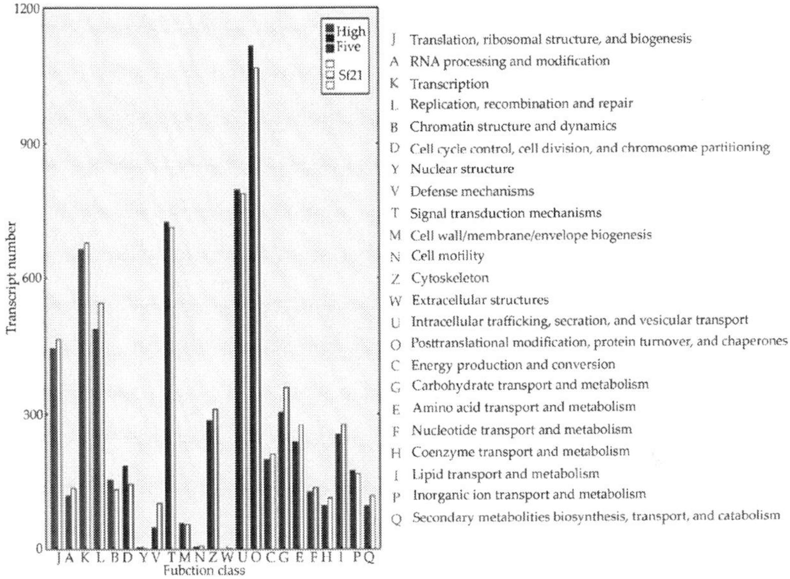

Figure 3: Transcript number in each EggNOG function classes. Divided into 3 parts by colors. (Red) Information storage and processing; (Blue) Cellular processes and signaling; (Purple) Metabolism

Codon usage among recombinant protein expression systems

Among different organisms, synonymous codons are usually utilized with different frequencies; a phenomenon generally referred to as the 'codon usage bias'. Codon usage bias is a major factor that affects the level of recombinant protein expression (Holm 1986). We collected CDS sequences from the Sf21 and High Five cells and compared their codon usage preference to other commonly used expression hosts, including the prokaryotic system *E. coli* BL21, the eukaryotic expression system *S. cerevisiae* and mammalian system CHO. We also compared the codon preference of insect cells with five other species including human, mouse, drosophila, zebrafish and arabidopsis. RSCU (relative synonymous codon usage) values are used to compare the use of synonymous codons. RSCU values of all amino acids in 10 species' CDS sequence were calculated separately for downstream studies as shown in Fig. 4.

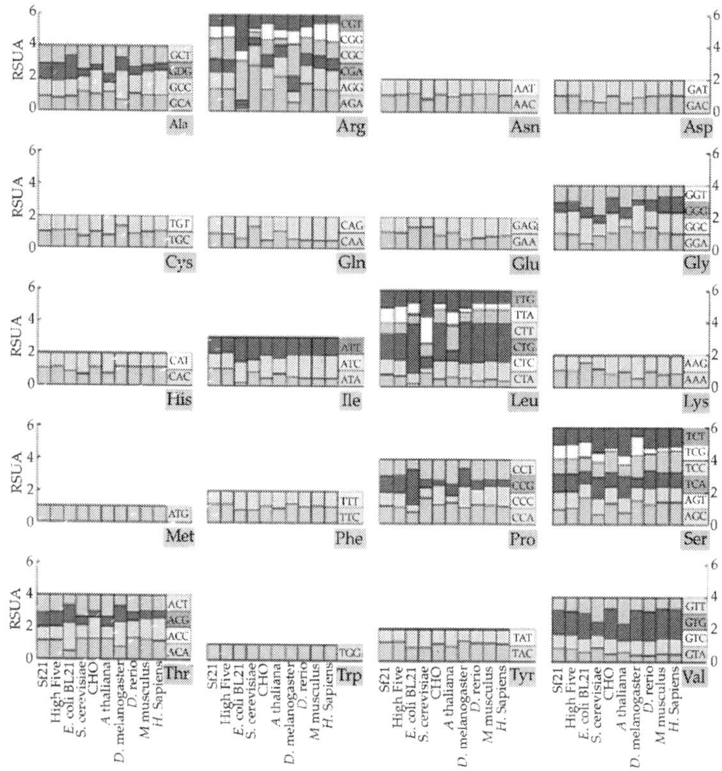

Figure 4: Codon usage of 20 amino acids across 10 different species. Each sub-plot is an amino acid, x-axis is different species and y-axis is the RSCU value of codons coding that amino acid

With the characteristic of RSCU value, all x synonymous codon RSCU values of an amino acid always get a sum up value equals to x. For amino acids only encoded by one or two codons, there is no extreme distribution in it, but others are quite different. We calculated the RSCU range value of a set of codon related to one amino acid. By comparing this value, we can tell in which species this set of codon have greater bias. For example, the range value of arginine, isoleucine, leucine and proline in BL21 reached 2.29, 1.32, 2.81 and 1.63, while the minimum codon's RSCU is only 0.13, 0.21, 0.21 and 0.49. This situation exactly indicates that optimization of codon usage is of great significance. For instance, if BL21 is used to produce recombinant protein, failure to avoid these minimum codons may dramatically reduce the expression level.

In comparison with the range and standard deviation values in all species, no matter which amino acid you are using, both High Five and Sf21's range value are at a relatively low level. The homogeneity of codon usage in baculovirus-insect system could be an advantage for protein expression. Coding sequence cloned from most species could be normally expressed in insect cells without codon optimization. This robust property of codon usage in baculovirus-insect system made it a good platform for both eukaryotic and prokaryotic recombinant protein expression. But some previous publications claim that codon with lower RSCU value is intended to slow down the translation speed in order to produce well-folded proteins (Chaney and Clark 2015). More experimental evidence is required to show whether the usage of codon with relative higher RSCU values could become a disadvantage for expression of proteins of complex folding.

Glycogene profile in protein expression insect cells

Post-translational modification is one of the most important characteristics of baculovirus-insect expression system. But the truncated N-glycosylation pathways in insect cells limit its application on some glycoprotein expression (Jarvis 2003). Several glyco-engineering modifications have been reported in the past two decades. Some modifications require importation of glycogenes into baculovirus-insect system. Glycosylation is mediated with complicated pathways and a number of genes are involved. Without a global gene map of the insect cells, we cannot thoroughly understand glycosylation related problems. Since GGDB and CAZy databases included genes associated with glycan synthesis procedural, we used them as references to find homologs of glycogenes in our High Five transcriptome and previous Sf21 transcriptome.

Here we identified 69 glycogenes in the High Five transcriptome and 72 in Sf21, with an overlap of 66 genes. Those genes are marked in blue with their expression value in Fig. 5A, and detailed informations are described in supplemental file S2. Glycogenes can be classified into several types according to their functions. Total gene counts of each type in High Five and Sf21 are shown in Fig. 5B. Insect cells have more or less homologs among most types. But for sialyltransferases and N-acetylgalactosaminyltransferases, no similar transcript was found in these two cell lines. That is the main reason why baculovirus-insect cannot produce complete mammalian N-glycosylation proteins. O-glycan modification is more complicated and not well studied. From the gene matrix, we found that some O-glycosylation required enzymes are detectable, such as OGT, POFUT1/2, XYLT1, POMT2, etc. Previous study suggests that O-mannosylation in insect species may occur more frequently than what is currently believed (Vandenborre et al. 2011). Understanding the profile of glycogenes in insect cells would be helpful for more detailed research on glycosylation.

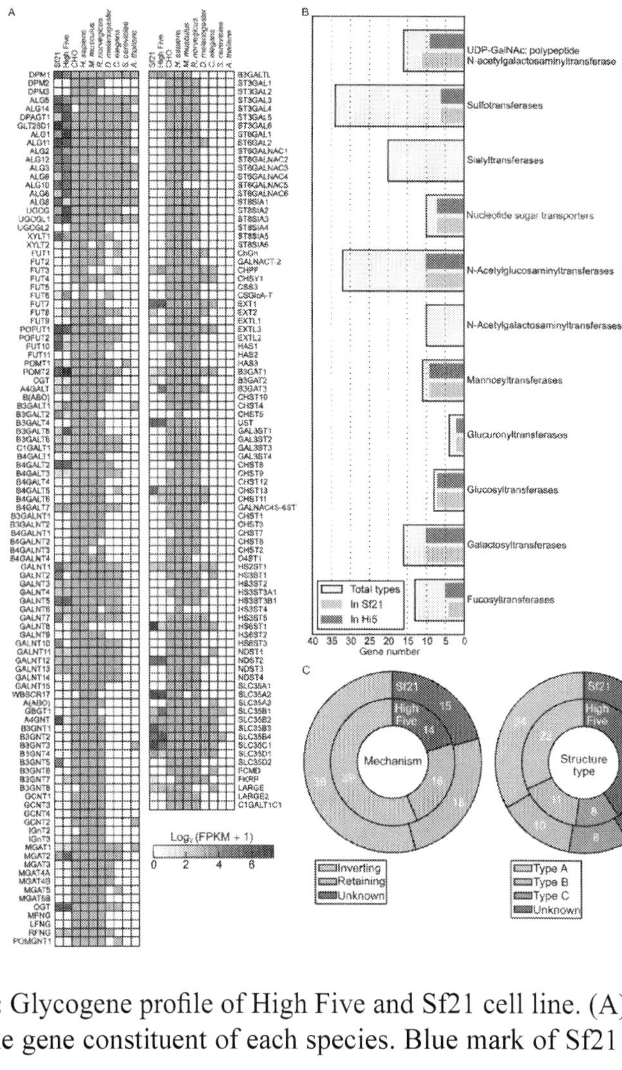

Figure 5: Glycogene profile of High Five and Sf21 cell line. (A) Heatmap represents the gene constituent of each species. Blue mark of Sf21 and High Five

showed the expression value of each gene. Red mark only represent they have this gene. (B) Bar plot of glycogene categories. (C) Ring plot representing the properties of glycogenes in High Five and Sf21 cell lines

We also compared the glycogenes' functioning mechanism and structure status in High Five and Sf21 (Fig. 5C). More than 50% of the glycosyltransferases function as inverting mechanism by catalyzing group transfer with inversion at anomeric reaction center of substrate, and less than 30% are for retentions. About 30% glycosyltransferase consist of two closely abutting β/α/β Rossmann domains, 20% consist of two β/α/β Rossmann domains that face each other and is flexibly linked. Remaining part have not yet been experimentally determined or studied (Lairson et al. 2008).

Here we complemented the glycogene database with data from baculovirus-insect expression system related cell lines. Our data would be valuable for introducing supplemental mammalian glycogesnes into insect cell lines and more efficiently modifying their glycosylation properties.

Highly expressed signal peptide containing genes

Baculovirus-insect system is a good platform for secretory protein expression. When there is a signal peptide fused to the recombinant protein, product usually secreted to the outside environment through the secretory pathway. Optimization of signal peptide would be helpful to get a better yield (Olczak and Olczak 2006). Here we used SignalP software (Petersen et al. 2011) to predict all possible signal peptide in all protein sequences, and then sorted them with their transcript expression value. Fig. 6 showed all predicted transcripts that may have signal peptide sequence in High Five and Sf21 transcriptome. We identified signal peptide sequences from top 100 expressed transcripts and believe they are good candidates for higher protein production. Related peptide sequence, CDS sequence and functional annotation are described in supplementary file S3. Because higher expression value usually means higher protein amount, secretion efficiency could be closely linked with the amount of the signal peptide containing protein. Moreover, signal peptide sequence from insect itself is of the best choice because endogenous secretory signal peptide is more efficient than exogenous signal peptides (von Heijne and Abrahmsén 1989; Soejima et al. 2013).

Combined with expression value and corresponding function of these proteins, we believe this approach would be useful and simplify the complexity of finding good signal peptides for protein expression.

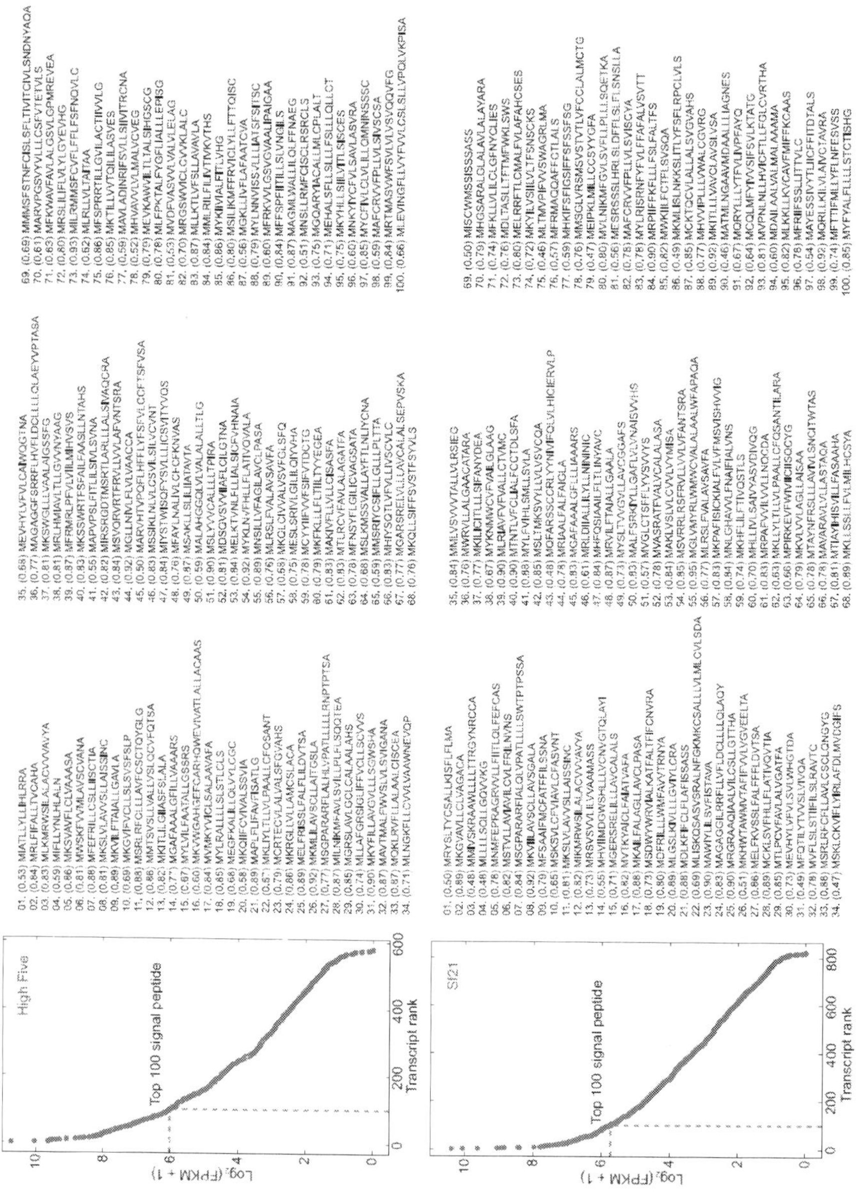

Figure 6: Highly expressed predicted signal peptides. Plot on the left are the expression values of those transcripts which have a predicted signal peptide.

Sequences on the right are the top 100 signal peptide sequence with the signalP score in the brackets

MATERIALS AND METHODS

RNA extraction and library construction

High Five cell line in this study was purchased from Life technologies, USA. Cells were cultured in serum free medium containing 0.5% Penicillin-Streptomycin (Gibco 15140-122) for 24 h at 27°C in suspension at a shaking speed of 110 rpm. When cell density reached 2.4×10^6/mL, 2×10^7 cells were collected by centrifugation. Total RNA was extracted with QIAGEN RNeasy Mini Kit (QIAGEN, 74104) immediately after cell collection. RNA quality was examined using Agilent 2100 RNA chip (standard setting). Poly-A tailed RNA was enriched and used to construct sequencing library with Illumina TruSeq RNA Sample Prep Kit (Illumina, RS-122-2001), following standard instruction. RNA-seq library was sequenced on Illumina HiSeq 2000 platform.

Assembly and statistics

After we got the raw reads, quality control was performed to remove poor-quality reads with in-house QC script. Then we used Trimmomatic (v0.32) (Bolger et al. 2014) software to trim low quality ends of all reads. Reference transcriptome was assembled with trimmed reads using Trinity (v2.0.6) (Grabherr et al. 2011; Haas et al. 2013). To reduce the assembly redundancy, cd-hit-est (v4.6.1) (Fu et al. 2012) was used to cluster reads with 90% identity. At this time, the remaining contigs are the so called 'unigenes'. Coding peptide sequences were predicted with Trans Decoder.

We used the SwissProt database to check the integrity of transcripts to evaluate the quality of our assembly result. All transcript sequences were aligned to SwissProt using blastx (Camacho et al. 2009), and only the most similar target was kept with the e-value cutoff of 1×10^{-20}. Length coverage of aligned transcript was examined (Fig. 1A).

Function annotation

Function annotation was performed at both transcript and protein level.

All transcripts were aligned to SwissProt, TrEMBL90 and NCBI nr database with blastx (Camacho et al. 2009; UniProt Consortium 2015). Predicted protein sequences were aligned to SwissProt, TrEMBL90 and EggNOG 4.1 with blastp (Powell et al. 2014). We also aligned protein sequences to Pfam28 database with hmmsearch (Finn et al. 2014). Subsequently, we annotated Gene Ontology (GO) and related pathways with blast2go and KOBAS (Conesa and Götz 2008; Xie et al. 2011). The GO annotation is presented in figure with WEGO (Ye et al. 2006). We also used SignalP (v4.1) (Petersen et al. 2011) for signal peptide prediction. Since some protein sequences are 5′ truncated or predicted starting from an upstream point, we manually checked the prediction of top expression of transcripts containing signal peptides.

To compare the constituent of glycogenes, we collected glycogene sequences from GGDB as reference (Narimatsu 2004). Published Sf21 transcriptome was downloaded from TSA database with accession No. GCTM00000000 (Kakumani et al. 2015). Protein sequences were aligned to this reference with blastp. Blast results of glycogenes were manually checked by aligning sequences to the TrEMBL90 and nr databases, to eliminate false positive hits. Classification by function, mechanism and structure were based on information from GGDB and CAZy database (Narimatsu 2004; Lombard et al. 2014).

Codon usage evaluation

We used our predicted coding sequences in codon usage evaluation. CDS sequences from other species were downloaded from Ensembl and NCBI genome databases. Only those sequences longer than 100 amino acids (300 bp CDS sequence) were used for calculation. All codons were counted with our script by shifting a simulated reading frame from 5′ end to 3′ end. With the count data of all codons, we calculated RSCU value according to the formula as described (Cannarozzi and Schneider 2012).

Expression analysis

To evaluate the transcript abundance, Bowtie (v1.0.0) and RSEM (v1.2.15) were used for sequence alignment and calculation of TPM (Transcripts Per Million) and FPKM (Fragments Per Kilobase of transcript per Million mapped reads) (Langmead et al. 2009; Li and Dewey 2011). This value could measure the transcripts abundance with transcript length, thus we can compare it among different genes or samples.

ACKNOWLEDGMENTS

We thank Dr. Yun Zhang and the High-Throughput Sequencing Center of Peking University for assistance with Illumina sequencing. This work was supported by the National Basic Research Program (973 Program) (No. 2014CB910102 to X.S.).

REFERENCES

1. Bolger AM, Lohse M, Usadel B (2014) Trimmomatic: a flexible trimmer for Illumina sequence data. Bioinformatics 30:2114–2120. doi:10.1093/bioinformatics/btu170
2. Breitbach K, Jarvis DL (2001) Improved glycosylation of a foreign protein by Tn-5B1-4 cells engineered to express mammalian glycosyltransferases. Biotechnol Bioeng 74:230–239. doi:10.1002/bit.1112
3. Brondyk WH (2009) Selecting an appropriate method for expressing a recombinant protein. Methods Enzymol 463:131–147. doi:10.1016/S0076-6879(09)63011-1
4. Camacho C, Coulouris G, Avagyan V et al (2009) BLAST+: architecture and applications. BMC Bioinformatics 10:421. doi:10.1186/1471-2105-10-421
5. Cannarozzi GM, Schneider A (eds) (2012) Codon Evolution. Oxford University Press
6. Castilho A (ed) (2015) Glyco-Engineering. Springer New York, New York, NYG
7. Chaney JL, Clark PL (2015) Roles for Synonymous Codon Usage in Protein Biogenesis. Annu Rev Biophys 44:143–166. doi:10.1146/annurev-biophys-060414-034333
8. Conesa A, Götz S (2008) Blast2GO: A comprehensive suite for functional analysis in plant genomics. Int J Plant Genomics 2008:619832. doi:10.1155/2008/619832
9. Davis TR, Trotter KM, Granados RR, Wood HA (1992) Baculovirus Expression of Alkaline Phosphatase as a Reporter Gene for

Evaluation of Production, Glycosylation and Secretion. Bio/Technology 10:1148–1150. doi:10.1038/nbt1092-1148

10. Finn RD, Bateman A, Clements J et al (2014) Pfam: the protein families database. Nucleic Acids Res 42:D222–D230. doi:10.1093/nar/gkt1223
11. Fu L, Niu B, Zhu Z et al (2012) CD-HIT: accelerated for clustering the next-generation sequencing data. Bioinformatics 28:3150–3152. doi:10.1093/bioinformatics/bts565
12. Grabherr MG, Haas BJ, Yassour M et al (2011) Full-length transcriptome assembly from RNA-Seq data without a reference genome. Nat Biotechnol 29:644–652. doi:10.1038/nbt.1883
13. Haas BJ, Papanicolaou A, Yassour M et al (2013) De novo transcript sequence reconstruction from RNA-seq using the Trinity platform for reference generation and analysis. Nat Protoc 8:1494–1512. doi:10.1038/nprot.2013.084
14. Hollister JR, Jarvis DL (2001) Engineering lepidopteran insect cells for sialoglycoprotein production by genetic transformation with mammalian 1,4-galactosyltransferase and 2,6-sialyltransferase genes. Glycobiology 11:1–9. doi:10.1093/glycob/11.1.1
15. Hollister J, Grabenhorst E, Nimtz M et al (2002) Engineering the Protein N-Glycosylation Pathway in Insect Cells for Production of Biantennary, Complex N-Glycans †. Biochemistry 41:15093–15104. doi:10.1021/bi026455d
16. Hollister JR, Shaper JH, Jarvis DL (1998) Stable expression of mammalian beta 1,4-galactosyltransferase extends the N-glycosylation pathway in insect cells. Glycobiology 8:473–480
17. Holm L (1986) Codon usage and gene expression. Nucleic Acids Res 14:3075–3087
18. Jarvis DL (2003) Developing baculovirus-insect cell expression systems for humanized recombinant glycoprotein production. Virology 310:1–7. doi:10.1016/S0042-6822(03)00120-X
19. Kakumani PK, Malhotra P, Mukherjee SK, Bhatnagar RK (2014) A draft genome assembly of the army worm, Spodoptera frugiperda. Genomics 104:134–143. doi:10.1016/j.ygeno.2014.06.005
20. Kakumani PK, Shukla R, Todur VN et al (2015) De novo

transcriptome assembly and analysis of Sf21 cells using illumina paired end sequencing. Biol Direct 10:44. doi:10.1186/s13062-015-0072-7

21. Kost TA, Condreay JP, Jarvis DL (2005) Baculovirus as versatile vectors for protein expression in insect and mammalian cells. Nat Biotechnol 23:567–575. doi:10.1038/nbt1095
22. Lairson LL, Henrissat B, Davies GJ, Withers SG (2008) Glycosyltransferases: structures, functions, and mechanisms. Annu Rev Biochem 77:521–555. doi:10.1146/annurev.biochem.76.061005.092322
23. Langmead B, Trapnell C, Pop M, Salzberg SL (2009) Ultrafast and memory-efficient alignment of short DNA sequences to the human genome. Genome Biol 10:R25. doi:10.1186/gb-2009-10-3-r25
24. Li B, Dewey CN (2011) RSEM: accurate transcript quantification from RNA-Seq data with or without a reference genome. BMC Bioinformatics 12:323. doi:10.1186/1471-2105-12-323
25. Li H, Durbin R (2010) Fast and accurate long-read alignment with Burrows-Wheeler transform. Bioinformatics 26:589–595. doi:10.1093/bioinformatics/btp698
26. Lombard V, Golaconda Ramulu H, Drula E et al (2014) The carbohydrate-active enzymes database (CAZy) in 2013. Nucleic Acids Res 42:D490–D495. doi:10.1093/nar/gkt1178
27. Narimatsu H (2004) Construction of a human glycogene library and comprehensive functional analysis. Glycoconj J 21:17–24. doi:10.1023/B:GLYC.0000043742.99482.01
28. Olczak M, Olczak T (2006) Comparison of different signal peptides for protein secretion in nonlytic insect cell system. Anal Biochem 359:45–53. doi:10.1016/j.ab.2006.09.003
29. Petersen TN, Brunak S, von Heijne G, Nielsen H (2011) SignalP 4.0: discriminating signal peptides from transmembrane regions. Nat Methods 8:785–786. doi:10.1038/nmeth.1701
30. Powell S, Forslund K, Szklarczyk D et al (2014) eggNOG v4.0: nested orthology inference across 3686 organisms. Nucleic Acids Res 42:D231–D239. doi:10.1093/nar/gkt1253

31. Soejima Y, Lee J, Nagata Y et al (2013) Comparison of signal peptides for efficient protein secretion in the baculovirus-silkworm system. Open Life Sci 8:1–7. doi:10.2478/s11535-012-0112-6
32. UniProt Consortium (2015) UniProt: a hub for protein information. Nucleic Acids Res 43:D204–D212. doi:10.1093/nar/gku989
33. Vandenborre G, Smagghe G, Ghesquière B et al (2011) Diversity in Protein Glycosylation among Insect Species. PLoS One 6:e16682. doi:10.1371/journal.pone.0016682
34. Vaughn JL, Goodwin RH, Tompkins GJ, McCawley P (1977) The Establishment of Two Cell Lines from the Insect Spodoptera frugiperda (Lepidoptera; Noctuidae). In Vitro 13:213–217
35. von Heijne G, Abrahmsén L (1989) Species-specific variation in signal peptide design. Implications for protein secretion in foreign hosts. FEBS Lett 244:439–446
36. Wickham TJ, Davis T, Granados RR, et al Screening of insect cell lines for the production of recombinant proteins and infectious virus in the baculovirus expression system. Biotechnol Prog 8:391–6. doi: 10.1021/bp00017a003
37. Xie C, Mao X, Huang J et al (2011) KOBAS 2.0: a web server for annotation and identification of enriched pathways and diseases. Nucleic Acids Res 39:W316–W322. doi:10.1093/nar/gkr483
38. Xu C, Ng DTW (2015) Glycosylation-directed quality control of protein folding. Nat Rev Mol Cell Biol 16:742–752. doi:10.1038/nrm4073
39. Ye J, Fang L, Zheng H et al (2006) WEGO: a web tool for plotting GO annotations. Nucleic Acids Res 34:W293–W297. doi:10.1093/nar/gkl031

CHAPTER

13

A SIMPLE AND EFFECTIVE METHOD FOR ULTRASTRUCTURAL ANALYSIS OF MITOSIS IN DROSOPHILA S2 CELLS

Anton Strunov[a,b], Lidiya V. Boldyreva[a], Gera A. Pavlova[a,c], Alexey V. Pindyurin[a,b,d], Maurizio Gatti[a,e], Elena Kiseleva[a,b]

[a]Institute of Molecular and Cellular Biology, Siberian Branch of RAS, Novosibirsk, 630090, Russia

[b]Institute of Cytology and Genetics, Siberian Branch of RAS, Novosibirsk, 630090, Russia

[c]Kazan Federal University, Kazan, 420008, Russia

[d]Novosibirsk State University, Novosibirsk, 630090, Russia

[e]IBPM CNR and Department of Biology and Biotechnology, Sapienza University of Rome, Rome, 00185, Italy

ABSTRACT

The *Drosophila* S2 tissue culture cells are a widely used system for studies on mitosis. S2 cells are particularly sensitive to gene silencing by RNA interference (RNAi), allowing targeted inactivation of mitotic genes. S2 cells are also well suited for high-resolution light microscopy analysis of mitosis in fixed cells, and can be easily immunostained to detect mitotic components. In addition, S2 cells are amenable to transformation with plasmid encoding fluorescently tagged mitotic proteins, allowing *in vivo* analysis of their behavior throughout cell division. However, S2 cells have not been widely used for transmission electron microscopy (TEM) analysis, which provides ultrastructural details on the morphology of the mitotic apparatus that cannot be obtained with high-resolution confocal microscopy. Here, we describe a simple method for the ultrastructural analysis of mitosis in *Drosophila* S2 cells.

- Our method, which involves fixation and sectioning of a cell pellet, provides excellent preservation of mitotic structures and allows analysis of a higher number of mitotic divisions per sample, compared to correlative light-electron microscopy.
- Dividing cells are randomly oriented within the pellet and are sectioned along different planes, providing all-around information on the structure of the mitotic apparatus.

METHOD DETAILS

Preliminary notes

All procedures should be performed at room temperature (23 ± 2 °C) unless otherwise specified. Refrigerated solutions should be allowed to reach room temperature before use. If more than one specimen is processed, the number of cells and all the volumes of solutions should be adjusted accordingly. Each phase of the protocol is carried out without pauses, permitting a precise estimation of the experimental timing beforehand. Cell harvesting and fixation procedures take 5–6 h and are followed by an overnight post-fixation step. Drying the specimen and its embedding in a resin takes the next 2 days, followed by 2–3 additional days required for resin polymerization. The resin-embedded specimen can be then stored indefinitely at room temperature before sectioning.

Cell culture handling

The S2 cells used here have been grown in the laboratory of one of the authors (MG) since 1997 and have been employed in several RNAi-based studies (e.g., [1,2]). Since 1997, the line has been frozen 4 times. After each thawing, the cells have been propagated for 2–3 months and frozen again. The cells examined in this study are from aliquots frozen in 2004 (fourth freezing) in the laboratory of MG and cultured for 2 months at the

IMCB in Novosibirsk. The karyotype of our S2 cells is slightly different from those of the S2-DRSC and S2R+ cells, although the three lines share several marker chromosomes [2–4]. It is therefore unlikely that our S2 line is a derivative of the S2R+ line, which has been first described in 1998 [5]. In addition, our S2 cells do not grow attached to the surface of the flasks as do the S2R+ cells. Thus, we believe that our S2 cell line is one of the many sub-lines derived from the original Schneider's 2 line [6].

Cells are maintained at 25 °C in Schneider's Insect Medium (Sigma #S0146) supplemented with 10% Fetal Bovine Serum (Gibco #10270; heat-inactivated for 1 h at 65 °C). The cell density should be kept below 6–8 × 10^6 cells/ml to avoid formation of cell aggregates (Fig. 1) [7]. To increase the proportion of dividing cells, it is advisable to split the culture 1:2 with fresh medium one day before the fixation procedure. The mitotic index (percentage of mitotic cells) in the specimens used to illustrate the protocol was 6.05 ± 0.77 (mean ± SEM; calculated by examining samples of 1000 cells from 4 independent experiments).

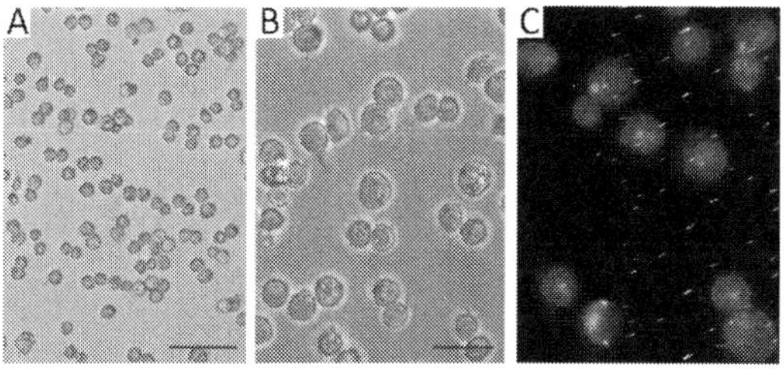

Figure 1: (A, B) Live *Drosophila* S2 cells in culture as seen under a phase-contrast light microscope (A and B, 20× and 40× magnification, respectively). (C) Fixed metaphase and interphase cells obtained according to [2]. Cells are stained with DAPI (DNA, blue), anti-α-tubulin antibodies (green) and antibodies against the centrosomal marker DSpd2 (red) [8] (40× magnification). Scale bars: (A), 50 μm; (B, C), 20 μm (For interpretation of the references to colour in this figure legend, the reader is referred to the web version of this article.).

Fixation of cell suspension

Cells should be harvested immediately before fixation and treated gently (both the suspension and the pellet) during all steps. The optimal cell number per specimen is 4–5 × 10^6. The prefixation solution (2.5%

glutaraldehyde in the culture medium) should be prepared immediately before addition to the cell suspension: 1 volume of 25% glutaraldehyde stock solution (Sigma #G5882) is added to 9 volumes of the culture medium (see "Cell culture handling"). It is crucial to add the prefixation solution to suspended cells dropwise. The fixation solution (2.5% glutaraldehyde in 0.1 M sodium cacodylate buffer, pH 7.4) should also be made immediately before addition to the cell suspension: 1 volume of 25% glutaraldehyde stock solution (Sigma #G5882) is added to 9 volumes of freshly prepared 0.1 M sodium cacodylate buffer (2.14 g of cacodylic acid sodium salt trihydrate (AppliChem #A2140) is dissolved in 100 ml of deionized water, and pH is adjusted to 7.4 with approximately 2 ml of 0.2 M HCl). After every centrifugation step during the fixation procedure, the supernatant should be removed very carefully avoiding disturbing the pellet.

Use the following step-by-step protocol:

- Harvest cells using a cell scraper and suspend them by repeatedly pipetting up and down. Measure the cell density using a cell counter or a hemacytometer, transfer 4–5 × 10^6 cells into a plastic tube, and spin down at 200 *g* for 5 min in a bucket rotor.
- Pour out the supernatant and completely resuspend the cell pellet in the small amount of medium left (~100–300 μl) by tapping the tube with a finger.
- Add 1 ml of the prefixation solution dropwise and mix by careful tilting the tube several times.
- Incubate cells in the prefixation solution for 15 min with continuous tilting the tube back and forth (~100 ° from its vertical position), with a frequency of 20 oscillations per min in an appropriate mixer (e.g., Intelli-Mixer RM-1, ELMI, Latvia).
- Centrifuge cells at 200 *g* for 5 min in a bucket rotor. Remove the supernatant.
- Add 1 ml of the fixation solution and completely resuspend cells by slow and gentle pipetting.

- Incubate cells in the fixation solution for 2 h with continuous tilting the tube back and forth (~100 ° from its vertical position), with a frequency of 20 oscillations per min.
- Centrifuge cells at 200 *g* for 5 min in a bucket rotor. Remove the supernatant.
- Wash the pellet three times (5 min per wash) without disturbing it with 1 ml of 0.1 M sodium cacodylate buffer (pH 7.4) (see above), and then centrifuge at 200 *g* for 5 min in a bucket rotor.
- Post-fix the pellet for 1 h, continuously shaking (in vertically placed tube, on a regular laboratory horizontal shaker) at 20 rpm in 1% solution of osmium tetroxide in 0.1 M sodium cacodylate buffer (pH 7.4; see above), containing few crystals of potassium ferricyanide ($K_3[Fe(CN)_6]$) that makes the solution yellowish. After this step, no further centrifugation is needed.
- Carefully wash the pellet three times without disturbing it with 1 ml of distilled water and incubate overnight at 4 °C in 1% water solution of uranyl acetate (Serva #77870).

Drying and embedding in resin

In the second day of the protocol, the cell pellet is washed once with distilled water and then gradually dehydrated it in ethanol series (in 1 ml of 30% and 50% ethanol for 7 min, in 70% and 96% ethanol for 10 min, and in 100% ethanol for 20 min) and then in acetone (twice, in 1 ml for 20 min) continuously shaking at 20 rpm during each incubation, and finally embedded in Agar 100 Resin (Agar Scientific #AGR1031; hereafter resin). The embedding step is crucial for the preparation of the specimen, since the size of the cell pellet is bigger (~7 mm^3) than that conventionally used for TEM preparations (~1 mm^3). Thus, proper resin infiltration is required.

Use the following step-by-step embedding protocol:

- Incubate the cell pellet in 1 ml of 1:3 (v/v) mixture of resin: acetone for 1 h, with continuous horizontal shaking (in a vertically placed tube on a regular laboratory horizontal shaker) at 20 rpm.

- Remove the solution, add 1 ml of 1:1 (v/v) mixture of resin: acetone, and incubate for 1.5 h with continuous horizontal shaking at 20 rpm.
- Remove the solution, add 1 ml of 3:1 (v/v) mixture of resin: acetone and incubate for 2 h with continuous horizontal shaking at 30 rpm.
- Remove the solution, add 1 ml of pure resin (starting from this step, keep open the lid of the tube) and incubate it for 1–2 h with continuous horizontal shaking at 30 rpm.
- Replace the resin with 1 ml of fresh resin and incubate overnight with continuous horizontal shaking at 30 rpm.
- In the third day of the protocol, replace the resin again with 0.5 ml of fresh resin, and incubate in a desiccator overnight to get rid of bubbles.
- Transfer the specimen to 60 °C for 2–3 days for polymerization of the resin.

Cutting sections and electron microscopy analysis

Use the following step-by-step protocol:

- Release the polymerized block with the cell pellet from the 1.5-ml tube by cutting its plastic wall with a lancet.
- Fasten the block in a specimen holder and manually cut a few sections with a razor blade from its very tip, so as to reach the cell layer.
- Obtain semi-thin sections (500 nm) using an ultramicrotome (ultracut E, Reihert-Jung, Austria or a similar ultramicrotome) and collect sections on a drop of distilled water placed on a slide; dry the water with a spirit lamp to attach the sections to the slide.
- Stain the sections by covering them with a drop of 1% toluidine blue (Sigma #89640) dissolved in 1% borax (Sigma #71997); heat the drop with a spirit lamp for 1–2 min (without letting them boil or dry completely) and wash them with distilled water.

- Find the region of interest under a light microscope and shape it with a razor blade into a pyramid. Use an ultramicrotome (ultracut UCT, Leica, Austria) to obtain 60 nm sections with a diamond knife (Diatome #DU4515).
- Transfer sections to a 150 mesh hexagonal copper grid (SPI Supplies #2850C-XA) pretreated with acetone for 20–30 min and let them dry for a few hours.
- Sections could be stained with a drop of lead citrate for 4 min to achieve a good contrast for visualization of cellular membranes; however, a good contrast is also obtained using crystals of potassium ferricyanide during post-fixation.
- Cover sections with a thin layer of carbon using a carbon evaporator unit (JEE 4B, JEOL, Japan) to make them conductive and stable under the electron beam.
- Analyze sections under an electron microscope. The images presented here were obtained using a JEM-100SX electron microscope (JEOL, Japan) and a film (Agfa #EB19H).

Troubleshooting and tips

- Cells clumps in the culture may lead to an irregular cell density in the pellet. Avoid an excessively high density of cells in the culture (not more than 6–8 × 10^6 cells/ml) to prevent cell clumping (see [7]).
- The fixation procedure should be gentle but fast; keeping the correct timing of the steps is crucial. Thus, it is advisable to proceed with no more than 4–6 specimens at once.
- After the first centrifugation, the cell pellet should be carefully resuspended by gentle shaking or pipetting. The prefixation solution should be added dropwise only when the cells are completely resuspended in the small volume of medium left after pouring out the supernatant.
- Prefixation and fixation solutions must be at room temperature (23 ± 2 °C), as lower temperatures may cause ineffective fixation and depolymerize the microtubules [9, 10].
- Extensive washing of the cell pellet can remove some cells from the pellet surface. The washing solution should therefore

be removed very gently leaving just a little bit of it above the pellet. The new solution should be added very slowly, drop by drop, along the wall of the tube. The pellet should never dry out.

- The osmium tetroxide penetration into the pellet should be carefully controlled. Sometimes, the bottom of the big (~7 mm^3) cell pellet remains unfixed and fails to blacken like the rest of the pellet. In this case, the pellet should be gently flipped over with a toothpick to make the unfixed area accessible to osmium tetroxide. In addition, in this case, the incubation time should be extended up to 15 min.
- Semi-thin sections can be used for detecting cells at the desired mitotic stage (Fig. 2). Cells in different mitotic phases can be identified by light microscopy from the presence of specific features. We usually stain semi-thin section with toluidine blue. However, mitotic cells might also be detected by phase contrast microscopy in unstained sections. It should be recalled that *Drosophila* mitosis is semi-closed, with the nuclear envelope persisting throughout prometaphase and at least in part also during metaphase [11]. The vast majority of cells are in interphase; they are clearly recognized because of their round and light nucleus with dense nucleoli (Fig. 2A). Cells in late prophase/prometaphase exhibit morphologically irregular (often slightly elongated) nuclear envelopes, an opaque nucleoplasm (similar to the cytoplasm) and condensed chromosomes (Fig. 2B, C). Metaphases are characterized by chromosome alignment at the cell's equator and a strong reduction, or even the absence, of the nuclear envelope (Fig. 2D). Anaphases are elongated cells containing two groups of segregating chromosomes not surrounded by a nuclear envelope (Fig. 2E). The main feature of telophases is the cytokinetic constriction between the two daughter cells that remain connected by the midbody (Fig. 2F). Thus, light microscopy observation allows the choice of a region of interest in a block, which can then be trimmed to obtain serial sections for TEM analysis. This procedure significantly reduces the time required for collecting the data that are necessary for the ultrastructural analysis of the mitotic phase under study.

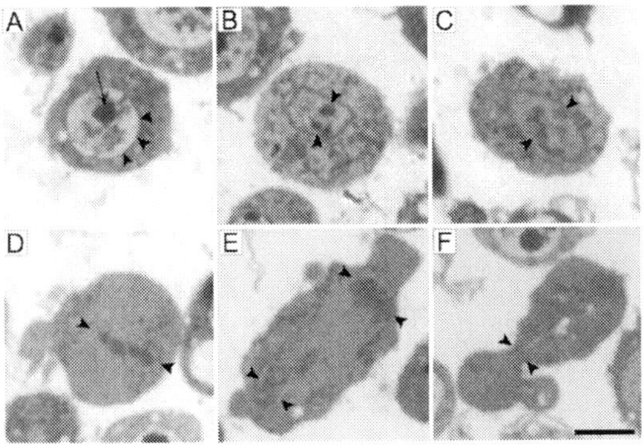

Figure 2: Phases of *Drosophila* S2 cells mitosis observed in semi-thin sections stained with toluidine blue under a light microscope (100× magnification). (A) Interphase, with a light round nucleus (arrowheads) and a dense nucleolus (arrow). (B, C) Late prophase/prometaphase figures showing chromatin condensation (arrowheads) and variations in the nuclear envelope morphology. Note that the nucleus is no longer lighter than the cytoplasm. (D) Metaphase showing well-aligned chromosomes (arrowheads). (E) Anaphase showing two sets of segregating chromosomes (arrowheads). (F) Telophase displaying the cytokinetic constriction and the midbody (arrowheads). Scale bar: 5 μm.

METHOD VALIDATION

As mentioned previously, a successful preparation of mitotic cells for TEM analysis depends on the accurate execution of several major steps: fixation, dehydration and embedding in a resin. The quality of embedding is very important and can be checked by the analysis of semi-thin sections of the pellet under a light microscope. In case of proper fixation and infiltration of the resin into the pellet, the cells are well separated and lie at a distance of at least 3–4 μm (Fig. 3A, B). Inaccurate embedding is revealed by a high cell density in the pellet (with distances of less than 1 μm between neighboring cells) and a clearcut border between the cell pellet and the resin (Fig. 3C, D).

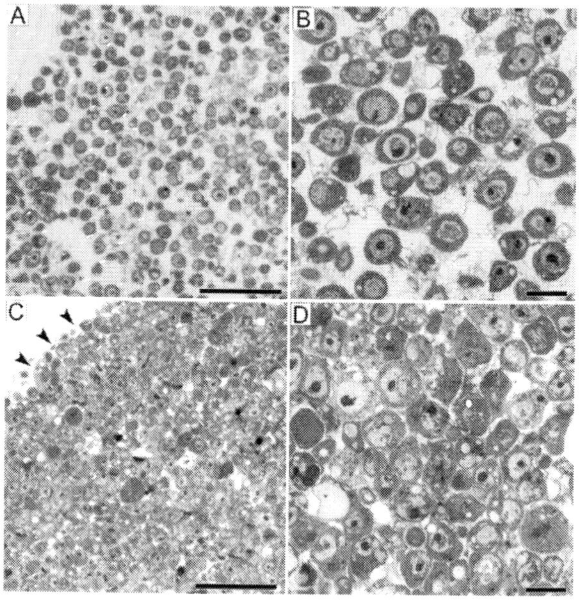

Figure 3: Quality of the cell pellet under a light microscope. Correct (A, B) and incorrect (C, D) embedding of the pellet into the resin. Note the differences in cell densities and the sharp border between the pellet and the resin (arrowheads in C). Scale bars: 50 μm (A, C), 10 μm (B, D).

Proper ultrastructure preservation can only be checked under an electron microscope. Dividing cells are easily detected, as they show compact electron-dense chromosome masses and partially disassembled nuclear envelope. As shown in Fig. 4, our protocol provides excellent preservation of fine cellular structures such as microtubules (MTs) and kinetochores. Importantly, sectioning a pellet from a cell suspension provides both longitudinal and transverse sections of the mitotic apparatus, whereas sectioning of an adherent cell monolayer mostly provides longitudinal sections. Transverse sections permit observation and quantification of structural details that are difficult to define in longitudinal sections. For example, transverse sections are particularly useful for visualization of the microtubule bundles that interact with the kinetochores and the proteinaceous structures that connect MTs (Fig. 4E, F).

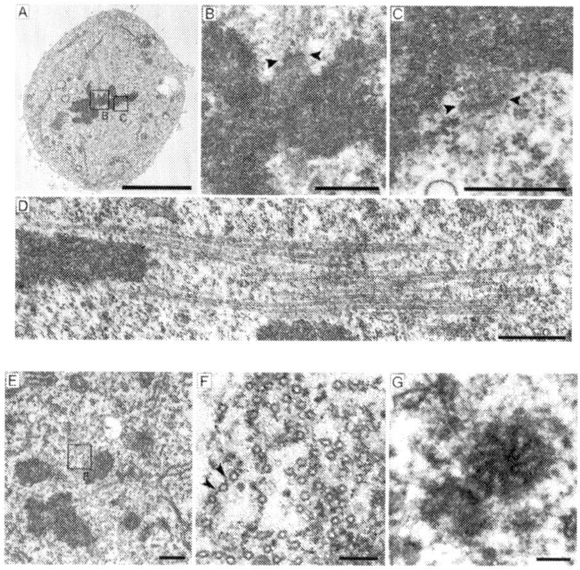

Figure 4: Ultrastructure of *Drosophila* S2 cells undergoing mitosis. (A–C) metaphase; B and C are the enlarged insets indicated in A, showing the ultrastructure of kinetochores (arrowheads) and attached MTs. (D) Fine structure of a kinetochore fiber from a prometaphase. (E) Transverse section of the spindle from a cell in prometaphase. (F) High magnification of MTs from the inset in E; note the proteinaceous bridges (arrowheads) between MTs. (G) Ultrastructure of mother and daughter centrioles. Scale bar: 5 μm (A), 0.5 μm (B–E), 0.1 μm (F, G).

Background

Drosophila S2 cells serve as an excellent model system for a variety of cells biological analyses and high-throughput screens at genome-wide level [12]. S2 cells are a spontaneously immortalized cell line from a primary culture of *Drosophila* embryonic cells; their characteristics suggest that they are derived from macrophages [6]. S2 cells are very easy to grow and highly susceptible to RNAi. Importantly, in these cells there is no interferon response like in mammalian cells and RNAi can be induced with long double-stranded RNAs (dsRNAs) that are directly incorporated into the cells from the medium. The easy of RNAi-mediated gene silencing has made S2 cells an ideal system for phenotypic analysis of the consequences of inactivation of specific mitotic genes. In addition, S2 cells have been often used to perform genome-wide RNAi-based screens aimed at the identification of new mitotic functions as well as screens aimed at investigating the mitotic functions of selected gene groups, such as those encoding kinesins, actin-binding proteins, kinases or phosphatases (e.g., [2,13–17]).

Despite the widespread usage of *Drosophila* S2 cells, only a few studies analyzed their mitotic structures at the ultrastructural level (e.g., [18–20]). Most protocols for electron microscopy investigation of *Drosophila* cell cultures use flat embedding of cell monolayers [18–21]. This is an excellent procedure for correlative light-electron microscopy, but is not convenient for analyses that require collection of a large amount of data. Collecting TEM data on multiple mitotic cells is not an easy task, because *Drosophila* tissue culture cells are very difficult to synchronize (e.g., [21,22]). Here, we describe a straightforward approach for ultrastructural investigation of mitosis in *Drosophila* S2 cells. Our approach is based on pelleting a cell suspension by centrifugation; the cell pellet is then fixed and embedded in resin for TEM analysis. Protocols for correlative light-electron microscopy allow examination of a few dividing cells per specimen [23], whereas sectioning of a cell pellet permits detection of a relatively high number of mitotic cells (~1 dividing cell per every 100 cells). In addition, since fluorescent imaging can induce cytotoxic and genotoxic effects [24,25], TEM preparations obtained by correlative light-electron microscopy protocols might reflect some of these effects. Our protocol does not include any treatment of the cells before fixation and usually results in excellent preservation of membranes, microtubules, kinetochores, and cell organelles. In a previous study, it was suggested that centrifugation might cause mechanical stress upon the cells, but this issue was not investigated [21]. We compared the morphologies of intact cells from a monolayer and pelleted cells exposed to mild centrifugation (200*g* for 5 min), and did not find any differences in their ultrastructure (data not shown).

An advantage of our method is that the analysis of semi-thin sections by light microscopy permits selection of cells at the desired mitotic stage, allowing the operator to focus on the process of interest to obtain quantitative information. Another advantage of the method described here is the possibility to observe mitotic spindles from different angles. Cells growing in a monolayer usually form spindles with the long axis parallel to the substrate and therefore most sections of these spindles are longitudinal. In contrast, sectioning the pellets provides many transverse sections of the spindle. An analysis of transverse section is particularly useful to define the arrangement of MTs and their interaction with kinetochores.

ACKNOWLEDGEMENTS

We thank Leonid V. Omelyanchuk for discussion and useful suggestions. This work was supported by the grant from the Ministry of Education and Science of Russian Federation no. 14.Z50.31.0005 to MG. Microscopic analysis was carried out at the Multiple-access Center for Microscopy of Biological Subjects (Institute of Cytology and Genetics, Novosibirsk, Russia).

REFERENCES

1. M.P. Somma, B. Fasulo, G. Cenci, E. Cundari, M. Gatti, Molecular dissection of cytokinesis by RNA interference in Drosophila cultured cells, Mol. Biol. Cell. 13 (7) (2002) 2448–2460.
2. M.P. Somma, F. Ceprani, E. Bucciarelli, V. Naim, V. De Arcangelis, R. Piergentili, et al., Identification of Drosophila mitotic genes by combining co-expression analysis and RNA interference, PLoS Genet. 4 (7) (2008) e1000126.
3. M.P. Somma, B. Fasulo, G. Siriaco, G. Cenci, Chromosome condensation defects in barren RNA-interfered Drosophila cells, Genetics 165 (3) (2003) 1607–1611.
4. H. Lee, C.J. McManus, D.Y. Cho, M. Eaton, F. Renda, M.P. Somma, et al., DNA copy number evolution in Drosophila cell lines, Genome Biol. 15 (8) (2014) R70.
5. S. Yanagawa, J.S. Lee, A. Ishimoto, Identification and characterization of a novel line of Drosophila Schneider S2 cells that respond to wingless signaling, J. Biol. Chem. 273 (48) (1998) 32353–32359.
6. I. Schneider, Cell lines derived from late embryonic stages of Drosophila melanogaster, J. Embryol. Exp. Morph. 27 (2) (1972) 353–365.
7. S.L. Rogers, G.C. Rogers, Culture of Drosophila S2 cells and their use for RNAi-mediated loss-of-function studies and

immunofluorescence microscopy, Nat. Protoc. 3 (4) (2008) 606–611.

8. M.G. Giansanti, E. Bucciarelli, S. Bonaccorsi, M. Gatti, Drosophila SPD-2 is an essential centriole component required for PCM recruitment and astral-microtubule nucleation, Curr. Biol. 18 (4) (2008) 303–309.
9. L.U. Cassimeris, P. Wadsworth, E.D. Salmon, Dynamics of microtubule depolymerization in monocytes, J. Cell Biol. 102 (6) (1986) 2023–2032.
10. G.A. Pavlova, J.A. Galimova, J.V. Popova, A.F. Munzarova, A.V. Razuvaeva, A.L. Alekseeva, et al., Factors governing the pattern of spindle microtube regrowth after tubulin depolymerization, Tsitologia 58 (4) (2016) 300–303.
11. S. Güttinger, E. Laurell, U. Kutay, Orchestrating nuclear envelope disassembly and reassembly during mitosis, Nat. Rev. Mol. Cell Biol. 10 (3) (2009) 178–191.
12. S.E. Mohr, RNAi screening in Drosophila cells and in vivo, Methods 68 (1) (2014) 82–88.
13. G. Goshima, R.D. Vale, The roles of microtubule-based motor proteins in mitosis: comprehensive RNAi analysis in the Drosophila S2 cell line, J. Cell Biol. 162 (6) (2003) 1003–1016.
14. U.S. Eggert, A.A. Kiger, C. Richter, Z.E. Perlman, N. Perrimon, T.J. Mitchison, et al., Parallel chemical genetic and genomewide RNAi screens identify cytokinesis inhibitors and targets, PLoS Biol. 2 (12) (2004) e379.
15. M. Bettencourt-Dias, R. Giet, R. Sinka, A. Mazumdar, W.G. Lock, F. Balloux, et al., Genome-wide survey of protein kinases required for cell cycle progression, Nature 432 (7020) (2004) 980–987.
16. F. Chen, V. Archambault, A. Kar, P. Lio, P.P. D'Avino, R. Sinka, et al., Multiple protein phosphatases are required for mitosis in Drosophila, Curr. Biol. 17 (4) (2007) 293–303.
17. G. Goshima, R. Wollman, S.S. Goodwin, N. Zhang, J.M. Scholey, R.D. Vale, et al., Genes required for mitotic spindle assembly in Drosophila S2 cells, Science 316 (5823) (2007) 417–421.
18. H. Maiato, P. Sampaio, C.L. Lemos, J. Findlay, M. Carmena, W.C.

Earnshaw, et al., MAST/Orbit has a role in microtubulekinetochore attachment and is essential for chromosome alignment and maintenance of spindle bipolarity, J. Cell Biol. 157 (5) (2002) 749–760.

19. H. Maiato, P.J. Hergert, S. Moutinho-Pereira, Y. Dong, K.J. Vandenbeldt, C.L. Rieder, et al., The ultrastructure of the kinetochore and kinetochore fiber in Drosophila somatic cells, Chromosoma 115 (6) (2006) 469–480.
20. K.J. VandenBeldt, R.M. Barnard, P.J. Hergert, X. Meng, H. Maiato, B.F. McEwen, Kinetochores use a novel mechanism for coordinating the dynamics of individual microtubules, Curr. Biol. 16 (12) (2006) 1217–1223.
21. A. Debec, C. Marcaillou, Structural alterations of the mitotic apparatus induced by the heat shock response in Drosophila cells, Biol. Cell 89 (1) (1997) 67–78.
22. Z. Lipinszki, P. Wang, R. Grant, C. Lindon, N.S. Dzhindzhev, P.P. D'Avino, et al., Affinity purification of protein complexes from Drosophila embryos in cell cycle studies, Methods Mol. Biol. 1170 (2014) 571–588.
23. C.L. Rieder, G. Cassels, Correlative light and electron microscopy of mitotic cells in monolayer cultures, Methods Cell. Biol. 61 (1999) 297–315.
24. J. Ge, D.K. Wood, D.M. Weingeist, S. Prasongtanakij, P. Navasumrit, M. Ruchirawat, et al., Standard fluorescent imaging of live cells is highly genotoxic, Cytometry A 83 (6) (2013) 552–560.
25. M.K. Daddysman, M.A. Tycon, C.J. Fecko, Photoinduced damage resulting from fluorescence imaging of live cells, Methods Mol. Biol. 1148 (2014) 1–17.

CHAPTER

14

EFFECTS OF HORMONE AGONISTS ON SF9 CELLS, PROLIFERATION AND CELL CYCLE ARREST

Maeva Giraudo[1,2,] Je´ ro^ me Califano[1,3], Fre´de´ rique Hilliou[1] , Trang Tran[1,4,] Nathalie Taquet[1,5], Rene´ Feyereisen[1] , Gae lle Le Goff[1]

[1]Institut National de la Recherche Agronomique, UMR 1301 Interactions Biotiques et Sante´ Ve´ge´tale, Centre National de la Recherche Scientifique, UMR 6243, Universite´ de Nice Sophia Antipolis, Sophia-Antipolis, France,

[2]UMR 6023 CNRS-Universite´ Blaise Pascal, Ba^t. Biologie A – Campus des Ce´zeaux, Aubie`re, France,

[3]De´partement des affaires re´glementaires, Grasse, France,

[4]Lanaud Gestion-Po^le de Lanaud, Boisseuil, France,

[5]Bioimagerie, Villeneuve Loubet, France

INTRODUCTION

Growth and development are controlled by two major hormones in insects, the steroid 20-hydroxyecdysone (20E) and the sesquiterpenoid juvenile hormone (JH) [1]. The cross-talk between these two hormones regulates all stages from egg-larva-pupa to adult. A high level of 20E is required to initiate all developmental transitions and JH determines the nature of the moult [2]. JH is necessary for larval moulting and growth [3]. The signalling action of these hormones involves nuclear receptors. If the mode of action of 20E is well-known, that of JH remains more enigmatic. 20E exerts its action through binding to a nuclear receptor heterodimer consisting of an ecdysone receptor (EcR) and ultraspiracle (USP) which is the insect ortholog of retinoid-X-receptor from vertebrates [4]. The complex regulates expression of target genes by binding to gene promoter regions. In Drosophila, it was shown that 20E linked to its receptor activates early genes among which are the transcription factor regulators, the Broad complex (BR-C), E74 and E75 [5], [6]. It is those transcription factors that in turn regulate late genes that have direct effector roles (including affecting cell death, cellular proliferation, differentiation and cuticle production). Several receptor candidates for JH exist including MET (Methoprene tolerant) a member of the bHLH-PAS transcription factor family [7] and USP [8].

MET can bind JH at physiological concentrations [9] whereas USP was shown to bind JH with low affinity, at concentrations at least 100 times lower than expected for a nuclear receptor [10]. However the situation is complex and it is difficult to generalize findings on Met and USP from one insect group to another. Indeed, *Met* has a close paralog in Drosophila, germ cell expressed (*gce*) [11]; hence Met-null mutants are fully viable [12]. The *Met/gce* duplication is recent and the two paralogs are found in the Drosophila genus but are not found in mosquitoes [13]. In other insects, *Met* has only one ortholog and in *Tribolium castaneum* its depletion by RNAi causes premature pupal morphogenesis [14]. A phylogenetic study of USP receptors shows that there are two types of receptor in arthropods, one having lost the ability to bind a ligand as in *Bemisia tabaci* (Hemiptera) and *T. castaneum* (Coleoptera) and another still able to bind a ligand in Diptera and Lepidoptera [15]. Moreover, understanding of the molecular signalling mechanism downstream of JH binding to its putative receptor remains limited. Two transcription factors, the Broad complex (BR-C) and Krüppel homolog 1 (Kr-h1) seem to play an important role [16]–[19]. Minakuchi et al. (2009) have proposed a model in the red flour beetle *T. castaneum* whereby Kr-h1 works downstream of Met at the larval stage and downstream of Met but upstream of BR-C in the pupa, allowing the inhibition of metamorphosis in one case or its initiation in the other [20]. It was recently shown that the crosstalk between 20E and JH signalling pathways could be mediated by a nuclear receptor co-activator, the steroid receptor co-activator in *T. castaneum* and its homolog in the mosquito *Aedes aegypti*AaFISC [21], [22]. This receptor interacts with EcR and Met respectively in presence of each hormone; however, its own role in the regulation of hormone responses needs further studies.

These hormonal receptors (EcR, USP and Met) are also the target for insecticides which act by disrupting insect development. Agonist hormone insecticides are of growing interest because some have selective toxicity, they are potent against pest insects and less or non toxic for beneficial insects, mammals, fishes and birds [23]. Among 20E agonists are diacylhydrazines, a non-steroidal agonist family, having insecticide activity by binding to the EcR-USP receptors. This family of compounds provokes a premature moult that leads to the death of the insect and they are only acting on larvae. The activity spectrum of these compounds varies within insect orders and is directly correlated to receptor affinity for the insecticide [23], [24]. For example, methoxyfenozide is more effective against Lepidoptera [25]. The other insecticides that mimic hormone action are juvenile hormone agonists (JHA), initially designed to be metabolically stable JH analogs. Their precise molecular target is less well-known due to the still controversial mode of action of JH. JHA block insects at an intermediate stage during development, making them unable to emerge as normal adults. They also disrupt reproduction in insects where JH is gonadotropic. Methoprene was the first successfully used JHA [26] and it is more effective against dipteran insects compared to Lepidoptera.

Cell lines can be a useful tool to understand insecticide mode of action. Several members of the diacylhydrazines have been tested on insect cell lines, showing an inhibition of cellular proliferation. This is the case for the Drosophila Kc cell treated by RH-5849 and tebufenozide [27], [28]. Similar effects on cell growth arrest have also been observed with these compounds in the lepidopteran cell line IAL-PID2 from the imaginal wing disks of *Plodia interpunctella* [29], as well as in an epithelial cell line from *Chironomus tentans* [30]. Further studies on IAL-PID2 with tebufenozide have shown a G2/M arrest with an induction of mRNA transcripts for EcR and USP associated with a decrease in the expression of cyclin B, one of the proteins involved in cell cycle control [31]. Effects on cell proliferation were also reported for the JH agonists methoprene and fenoxycarb on IAL-PID2 [32], but the molecular mechanism leading to this arrest was not clarified.

In this study we were interested in the effects of insecticides that mimic hormone action on the *Spodoptera frugiperda* Sf9 cell line. The toxicity of methoxyfenozide and methoprene was evaluated. Both insecticides inhibit cellular proliferation. Flow cytometry analysis showed a distinctly different action between these compounds with a G2/M arrest after methoprene treatment, whereas methoxyfenozide induced a slight accumulation in G1. To investigate the differential molecular mode of action of these hormone agonists, we have performed microarray experiments and followed the expression of nuclear receptors by real-time quantitative PCR (RT-qPCR). Our results suggest two different signalling pathways in response to methoxyfenozide and methoprene treatments.

RESULTS

Toxicological effects of methoxyfenozide and methoprene on Sf9 cells

Cell viability was determined by the MTT test after an insecticide exposure of 24, 48 and 72 hours. Methoxyfenozide had almost no effect at 10 nM, the lowest concentration tested, but already a marked effect at 100 nM (Fig. 1A). Increased insecticide concentrations did not significantly modify the cell viability. The IC_{50} is below 100 nM at 72 h and cannot be calculated at 24 and 48 h. Figure 1B shows the results for methoprene. Almost no effect was observed up to 25 μM, whatever the length of treatment (90% of cells were viable). For a concentration range between 50 and 75 μM, toxicity was low (after 72 h, still 70% of cells were viable). The calculated IC_{50} at 48 and 72 h was 184.2±4.8 μM and 86.3±9.8 μM respectively. The two hormone agonists induced cell death and methoxyfenozide was more potent than methoprene by a toxicological factor of about 1,000.

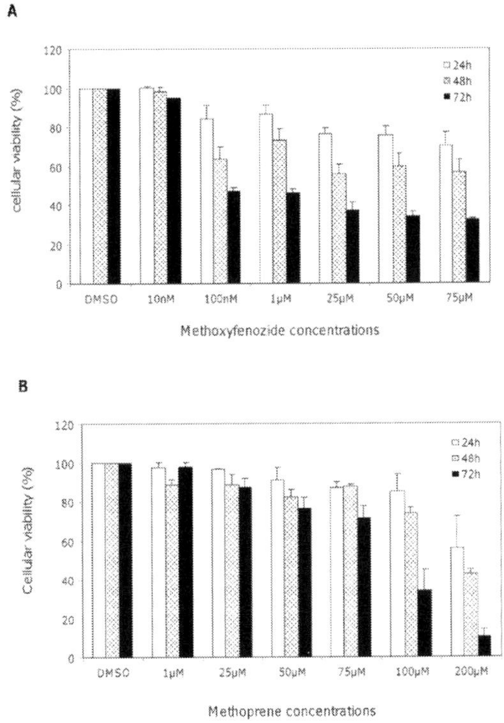

Figure 1: Methoxyfenozide and methoprene toxicity to Sf9 cells.

Cell viability was followed by the MTT test at 24, 48 and 72 h post-treatment of Sf9 cells by methoxyfenozide (A) or methoprene (B). Data are the mean of three independent experiments with SE.

Cellular proliferation inhibition

The effect of these insecticides on cellular proliferation was monitored. Cells in the DMSO control grew to a density of 15 to 17.10^5 cells/ml at 72 h (Fig. 2). Cell density at 24 h and 72 h in DMSO was significantly different, indicating that cells had proliferated. A normal and significant growth of the cells treated by methoxyfenozide was observed at the lower concentration (10 nM), however, all the other tested concentrations had an antiproliferative effect (Fig. 2A). At 50 μM of methoxyfenozide, the number of cells remained stable for 3 days. Methoprene had no effect at 1 μM with a cell density equivalent to the DMSO control. An inhibition of cell proliferation was observed in the concentration range between 50 and 100 μM, with no significant difference between numbers of cells at 24 h or 72 h (Fig. 2B). Proliferation arrest was reversible in both cases after removal of the insecticide (data not shown). Methoprene and methoxyfenozide therefore both caused an arrest of cell proliferation in a dose-dependent and reversible manner.

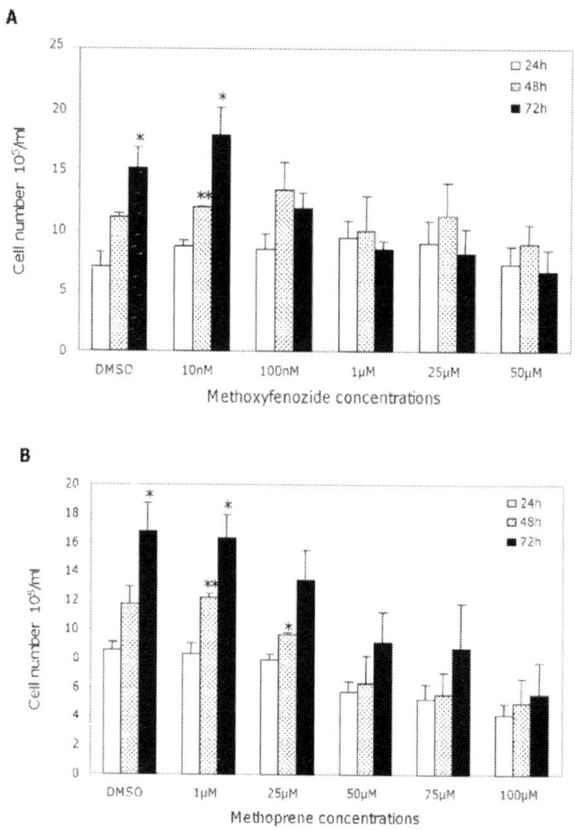

Figure 2. Effects of methoxyfenozide and methoprene on Sf9 cellular proliferation.

Cells were counted 24, 48 and 72 h post-insecticide treatment, (A) methoxyfenozide, (B) methoprene. Data are the mean of three independent experiments with SE. A *t*-test was performed to determine significance of the results (* p-Value<0.05, ** p-Value<0.01).

Distinct phases of arrest in the cell cycle

In order to evaluate in which phase of the cell cycle the cells were arrested after insecticide treatment, we used flow cytometry. In cells treated with methoxyfenozide, a slight but significant (p-Value>0.01) accumulation occurred in the G0/G1 phase compared to the DMSO control (Table 1). In cells treated with methoprene, cells were arrested in the G2/M phase with 66.37% of cells in this stage (Table 1) and values in the three different phases were significantly different from the DMSO control. Cells were also treated with 25 μM methoprene acid as a negative control. Methoprene acid lacks the isopropyl ester group of methoprene and has therefore no JH agonist activity. Cell percentages in the cell cycle phases were similar in methoprene acid and DMSO-treated cells, indicating that methoprene acid did not affect the cell cycle at a concentration where methoprene does.

Table 1. Cell cycle analysis by flow cytometry after 24 h of insecticide treatment

	G0/G1	S	G2/M	debris
0.4% DMSO control	21.77±0.86	14.85±0.44	51.48±2.14	11.78±1.02
Methoxyfenozide 100 nM	26.40±0.43*	13.23±0.33	47.13±0.78	13.17±1.03
Methoprene 25 μM	12.10±0.85*	9.4±0.95*	66.37±1.28*	12.13±0.39
Methoprene acid 25 μM	21.00±0.36	13.07±0.24	55.33±0.43	9.87±0.17

Molecular pathways involved in the cell response to hormone agonist treatments

The phenotypic effects were similar between Sf9 cells treated by methoprene or methoxyfenozide with an arrest of the cell growth and cell morphology identical to that of control cells (data not shown). However, the insecticides had different molecular effects with an accumulation in the G0/G1 or G2/M phase for methoxyfenozide and methoprene, respectively. The molecular genetic regulation leading to this arrest still needs to be elucidated. We therefore compared the transcriptional effect of each hormone agonist by using a specifically designed oligonucleotide microarray. We chose the first concentration of insecticide having a significant effect on cell proliferation, i.e 100 nM for methoxyfenozide and 25 μM for methoprene, in order to focus on physiological effects and avoid toxicological effects.

Genes were considered as differentially regulated by the insecticide if their expression ratio was >1.5 or <0.66 and P value <0.05. Sequences of these genes were analysed in Blast2Go to assign them Gene Ontology

(GO) terms and then classified in biological process level 3 (Table 2). The sequences for which no homology was found by blastx were submitted to a blastn in Butterflybase and classified in "hypothetical protein" category if homology was found with the sequence of another lepidopteran transcript, while the sequences restricted to *S. frugiperda*were put in a "hypothetical transcripts" category.

Table 2. Classification of genes regulated in Sf9 cells after treatment by methoxyfenozide or methoprene according to GO terms, blast2Go annotation level 3

	methoxyfenozide		methoprene	
Biological process	**up-regulated**	**down-regulated**	**up-regulated**	**down-regulated**
biosynthetic process	2	1	4	3
cellular component organisation	3	0	1	3
cell cycle	0	0	1	1
cellular metabolic process	8	4	9	4
establishment of localization	2	1	3	1
multicellular organismal development	1	1	2	2
regulation of biological process	2	1	2	3
reproductive process	2	0	1	2
response to stress	0	2	5	2
transcription/translation	4	1	7	1
transport	2	1	2	1
hypothetical protein	0	2	4	0
hypothetical transcript	2	3	15	7
unknown function	0	1	0	0

After methoxyfenozide treatment, 26 genes were differentially expressed with 14 overexpressed and 12 down-regulated (Table 3). Looking in more detail at the list of differentially regulated genes may help to identify the potential function of genes involved in the molecular and cellular effects of methoxyfenozide. Therefore, genes were classified in more precise categories of Gene Ontology. The most up-regulated gene codes for a vacuolar ATPase subunit B that is involved in the transport of proton across the membrane. The most down-regulated gene is an aldehyde oxidase. Several genes with function in translation and transcription were overexpressed as well as genes encoding calcium-dependent proteins, like cadherin and calreticulin.

Table 3. Microarray data for selected genes after Sf9 treatment by methoxyfenozide

Gene description and putative function	adhoc	ratio	P value
Actin cytoskeleton			
gelsolin	40151	1.59	0.0406
Carbohydrate metabolism			
6-phosphogluconolactonase	27223	2.39	0.0102
Catabolism process			
3-hydroxyisobutyryl Coenzyme A hydrolase	25654	0.64	0.0083
hydroxyphenylpyruvate dioxygenase	39567	1.64	0.0486
Cell adhesion			
cadherin	26462	6.32	0.0098
Chaperone proteins			
calreticulin	40650	4.47	0.0168
Proteolysis			
von Hippel-Lindau tumour suppressor protein	24976	0.18	0.0014
chymotrypsin	25557	2.9	0.0386
Reproductive protein			
vitellogenin	25793	1.69	0.0149
Response to stress/Detoxification			
heat shock protein 90	39008	0.57	0.0440
aldehyde oxidase	34664	0.14	0.0436
Sugar synthesis			
chondroitin sulfate synthase	44313	3.55	0.0254
Translation/Transcription			
polyadenylate binding protein 2	36158	6.19	0.0102
p27BBP/eIF6	33876	3.58	0.0252
dead box RNA helicase	25789	3.11	0.0422
Transport			
vacuolar ATPase subunit B	34648	9.44	0.0059
vacuolar ATPase subunit C	40849	0.64	0.0131

In the case of methoprene, 55 genes were differentially regulated with 39 over-expressed and 16 down-regulated (Table 4). There was no overlap between the genes regulated by methoxyfenozide and methoprene. The main category of genes up-regulated by methoprene is the hypothetical transcripts category (15 genes). The next categories are the genes involved in cellular metabolic process (9 genes) and transcription/translation (7 genes). These categories are also the most populated for cells treated by methoxyfenozide. Five genes upregulated by methoprene belonged to the response to stress category. The most induced (mitochondrial ribosomal protein L49) and most repressed (mitochondrial translational release factor 1 like) genes belong to the category of genes with functions in translation and transcription. This category has the highest number of genes regulated by methoprene (8 genes), followed by the class of genes with functions in response to stress and detoxification (7 genes). Several genes involved in spindle assembly are also over-expressed.

Table 4. Microarray data for selected genes after Sf9 treatment by methoprene

Gene description and putative function	adhoc	ratio	P value
Amino acids biosynthetic process			
phosphoserine phosphatase	25880	0.62	0.0022
Extracellular matrix protein			
hemicetin like protein 1	25674	1.7	0.0355
Immune protein			
scolexin B like	34835	1.77	0.0116
Mitotic cell cycle checkpoint			
14-3-3 epsilon protein	41120	0.62	0.0022
Phospholipid biosynthetic process			
choline/ethanolamine kinase	36601	0.54	0.0273
Polyamine synthesis			
S-adenosylmethionine decarboxylase	44296	1.71	0.0151
Regulation of Rab GTPase activity			
Tbc1 domain family	39160	0.57	0.0026
Response to stress/Detoxification			
carboxylesterase	40600	3.89	3.75E-07
prophenoloxidase activating factor	26333	2.55	0.0086
aldehyde dehydrogenase 7 family member A1	25971	1.89	0.0019
pheromone degrading enzyme 2	34999	1.59	0.0172
apolipoprotein D precursor	35472	1.55	0.0119
DNAJ-1	38331	0.65	0.0456
uridine diphosphate glucosyltransferase	35495	0.55	0.0422
Spindle assembly			
microtubule associated protein RP/EB family 3	44445	2.86	0.0441
kinesin like protein	36395	2.75	0.0110
beta-tubulin cofactor E	36883	2.07	0.0152
Structural constituent of cuticule			
cuticle protein 1 like	38620	0.61	0.0462
Translation/Transcription			
mitochondrial ribosomal protein L49	27227	4.62	3.83E-06
60S ribosomal protein L31	25976	1.77	0.0019
coiled-coil-helix-coiled-coil helix domain containing 8	34522	1.67	7.20E-05
ribosomal protein L10	38355	1.55	0.0139
spt3 associated factor 42	41154	1.52	0.0044
mitochondrial translational release factor 1 like	40890	0.43	0.0035
tRNA splicing endonuclease 2	35120	1.9	1.29E-05
bip2 like	41062	1.5	0.0050
Transport			
phosphate transport protein	38062	1.85	0.0017
translocase of inner mitochondrial membrane	40155	0.61	0.0079
Vesicle trafficking			
exocyst complex component 6	36460	1.5	0.0340

Expression of hormone receptors

Microarrays were not sensitive enough to detect transcripts of the nuclear receptors EcR or USP, and we did not have probes for Met on our array. Indeed we were unable to find the Met sequence in Spodobase (http://www.spodobase.univ-montp2.fr/Spodobase). Therefore, we used RT-qPCR approaches to study the expression of EcR and USP in Sf9 cells treated with methoxyfenozide or methoprene and compared to DMSO treated cells. Methoxyfenozide significantly induced the expression of both receptors, but methoprene only induced USP (Fig. 3).

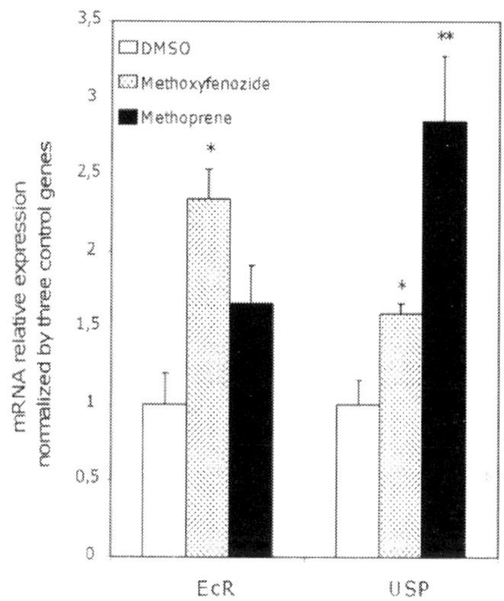

Figure 3: Transcript abundance of hormone nuclear receptors measured by RT-qPCR after insecticide treatment.

A *t*-test was performed to determine significance of the results (* p-Value<0.05, ** p-Value<0.01).

DISCUSSION

Cell lines provide a useful tool to investigate the molecular mode of action of insecticides. The *Spodoptera frugiperda* Sf9 cell line is probably the most widely used for high level expression of recombinant proteins [33]. Moreover, *S. frugiperda* from which the Sf9 cell line is derived is a major crop pest, a polyphagous insect able to feed on many different host plants. We decided to use this cell line to determine the effects of two insecticides which are agonists of major hormones controlling insect development, 20E and JH.

Methoxyfenozide acts as an agonist of 20E by binding to the EcR/USP receptor complex. Its affinity for the receptor in Lepidoptera is 420 times higher than that of the insect moulting hormone [23], [25]. However, the affinity varies within insect orders, it is more potent on the lepidopteran cell line *Plodia interpunctella* than on the *Drosophila* Kc cells [23]. Methoprene has a different spectrum of activity, being very effective against dipteran insects but less so against Lepidoptera [34]. These different potencies are correlated to data obtained in our toxicological tests, where methoxyfenozide acts at a lower concentration than methoprene. However it seems that there are two populations of cells in the Sf9 line. One very sensitive to methoxyfenozide dies at doses 10 to 100 nM, which is in accordance to data obtained on another cell line Se4 from *Spodoptera exigua* [35], and another one more resistant for which stronger doses of 1 to 75 μM had no effect. Nevertheless these two putative cell sub-populations have the same sensitivity for methoprene.

We show that the two insecticides inhibit cellular proliferation. Treatment by tebufenozide, another diacylhydrazine, leads to the same effects in two other lepidoptera cell lines, IAL-PID2 from *Plodia interpunctella* [31] and Se4 from *Spodoptera exigua* [36]. Methoprene and another analog of JH, fenoxycarb, significantly inhibit cell proliferation of the IAL-PID2 cell line [32]. Auzoux-Bordenave *et al.* (2005) reported that tebufenozide arrested the cell cycle in G2/M in the same cell line. The moulting hormone itself, 20E, causes an arrest in G2 in Kc cell [37] and in IAL-PID2 cells [38] whereas arrest occurs in the G1 phase in mosquito C7-10 cells [39]. The accumulation of cells in a given phase of the cell cycle is usually reported 10 to 24 h after treatment by hormones or analogues [31], [39]. We chose to perform our flow cytometry and microarray experiments at 24 h. In Sf9 cells, we show that inhibition of cellular proliferation associated with methoxyfenozide treatment induces a modest accumulation in G1 phase. In contrast, methoprene blocks the cells in G2/M. Cell accumulation in G1 or G2 correspond to an arrest at the two cell cycle checkpoints: in one case cells are stopped before DNA synthesis (G1) and in the other, entrance in mitosis is prevented (G2/M). When tested at the same concentration as methoprene, methoprene acid has no effect on cellular proliferation, which suggests that methoprene arrest of the cell cycle is due to its role as hormone mimic.

We have investigated insecticide mode of action at the molecular level in order to understand pathways leading to this cellular arrest. We used a custom microarray consisting of 9,773 probes of *S. frugiperda*, which represents approximately 67% of the genome compared to the lepidopteran model genome of *Bombyx mori* [40]. Several studies have examined the effects of hormones on cell lines [41], [42]. In addition, Mosallanejad *et al.* [43] reported microarray data on *Drosophila* cell line S2 resistant to methoxyfenozide. However, to date no study has been performed by microarray on cell lines following hormone agonist treatment. In our study, very few genes are differentially expressed after treatment with methoxyfenozide (26 genes) or methoprene (55 genes). It may therefore correspond to a more physiological response linked to the regulation of cell cycle rather than a generalized stress response to the xenobiotic. The cell cycle consists of four distinct phases: in the G1 phase cells grow and cyclin D is expressed. In the S phase DNA replication occurs and cyclins E and A predominate. In the G2 phase, degradation of cyclin E and accumulation of cyclin B occur. This phase is followed by mitosis and cell division [44]. The progression through the cell cycle is controlled by cyclins and cyclin dependent kinases (CDK). The cyclins form complexes with CDK, CDK4 for cyclin D, CDK2 for cyclin E. Our array only has one probe for cyclin A, but none for other cyclins. No change in cyclin A levels was detected, although Mottier *et al.* reported a significant decrease in the expression of cyclin A and B after a 20E treatment of IAL-PID2 cells, the level of both cyclins remaining very low between 12 and 36 h post-treatment [38]. On the other hand, we observed some differentially expressed transcripts that are consistent with an arrest of the cell cycle: as overexpression of different genes involved in the spindle assembly after methoprene treatment. Indeed a biosynthetic step occurs during the G2 phase, mainly involving the production of microtubules which are required during the process of mitosis. Genes such as cadherin are overexpressed in our methoxyfenozide experiments and these are overexpressed in cells that had stopped to proliferate [45]. Among the genes affected by methoxyfenozide, we find genes coding for different subunits of vacuolar ATPase. They are differentially regulated, with subunit B upregulated and C downregulated. Insect vacuolar ATPase (V-ATPase) consists of two functional parts, the peripheral catalytic

V1 complex composed of eight different subunits (from A to H) that hydrolyzes ATP and the integral membrane V0 complex consisting of four different subunits (a,c,d,e) that transports protons across the membrane [46]. These two parts can disassemble and reassemble depending on conditions which regulate V-ATPase activity [47], [48]. Both subunits can bind actin filament [49], [50], but subunit C is the only subunit that can be phosphorylated [51]. Another specificity of subunit C is its release in the cytosol upon dissociation of the two complexes [52], [53]. Clearly, subunit C has its own properties and can be a good candidate to mediate signalling pathway [54]. Opposite regulation of subunits B and C may therefore not be surprising. Moreover, promoter studies have revealed different regulatory elements between the two subunits' genes in *M. sexta* [48]. Down regulation of V-ATPase in apical globelet cell of *Manduca sexta* during moulting and starvation was also suggested [47], [55]. These data may indicate a possible involvement of hormone on the regulation of V-ATPase expression.

Genes and pathways involved in the various stages of the cell cycle progression were identified in a study on human HeLa cells using small interfering RNAs to target >95% of the protein coding genes [56]. Several concordant observations can be made between genes shown to be essential in that study and genes differentially regulated in our study. For example, several ribosomal proteins, kinesin, DNA-J have been shown to be essential for G2/M progression and are differentially regulated after methoprene treatment. Similarly, eIF, ATPase and dead box RNA helicase are essential in G1 phase and are differentially regulated after methoxyfenozide treatment.

More information concerning our microarray results (unknown genes) should be obtained with the forthcoming release of the *S. frugiperda* genome. Arrest of cell proliferation occurs at two checkpoints by distinct gene regulatory mechanism, at our level of sensitivity there are no genes in common. The recent discovery of a receptor co-activator able to bind EcR or Met depending on hormone concentration may explain why genes are up or down regulated: this co-activator may be a possible link for the cross-talk in these two signalling pathways [21], [22]. The presence of such a receptor co-activator remains unproven in the Sf9 cells. Further work is required to understand more precisely the mode of action of hormone agonists.

MATERIALS AND METHODS

Cell culture

The Sf9 cells (from Invitrogen), derived from the pupal ovarian tissue of *S. frugiperda*, were cultured in a flask at 27°C in monolayer with the insect-Xpress protein free medium (Lonza). Cell density was determined by Malassez haemocytometer counts and cell viability was evaluated by methylene blue (1 mg/ml, v/v) staining. Prior to experiments, cells were seeded onto 6 well plates at 5.10^5 cells/ml and left at 27°C for adhesion. Attached cells were then treated for 24 h with different concentrations of methoxyfenozide, methoprene, methoprene acid (all in 0.4% DMSO) or with 0.4% DMSO alone.

MTT assay of cell viability

Sf9 cells were seeded in 96-well culture plates and treated for 24, 48 and 72 hours with increasing concentrations of methoprene and methoxyfenozide. Cells in culture were then loaded with MTT (0.5 mg/ml) and incubated at 27°C for 2 hours. Cell homogenates were used to measure absorbance at 570 nm using a microplate reader (SpectraMax, Molecular Devices).

Cell cycle analysis

Cellular DNA content was determined by staining cells with propidium iodide and measuring fluorescence (FACSCalibur, Becton Dickinson). The Sf9 cells were incubated for 24 hours with methoxyfenozide or methoprene then resuspended and fixed on ice for 30 minutes with 70% ethanol/PBS (10 mM Na2HPO4, 138 mM NaCl, 2.7 mM KCl, pH 7.4). The fixed cells were incubated for 20 min at 37°C in a solution containing 50 μg/ml RNAse and 50 μg/ml propidium iodide. For each cell population, 10,000 cells were analysed by FACS and the percentage of cells in a specific phase of the cell cycle was determined with the propidium iodide DNA staining technique [57]. Cells were classified in G0/G1, S and G2/M phases depending on the intensity of the fluorescence peaks.

RNA extraction

Total RNA was extracted from cells of a well of the 6 wells plate using Trizol Reagent (Invitrogen Life technologies). Extractions were performed on three independent biological replicates.

Microarray experimental design

Our oligonucleotides were designed from 79148 ESTs sequences of eight different tissues of *S. frugiperda* (http://www.spodobase.univ-montp2.fr/Spodobase/). Using the assembly analysis (programme CAP3), we obtained 10,092 contigs and singletons from these ESTs. Our *S. frugiperda* microarray consists of 9,773 60-mers oligonucleotides synthesized by Sigma-Aldrich that were designed to match unique contigs or singletons and to suit our hybridization conditions (GC content average 46% and average Tm of 86.8°C). Each comparison consisted of six slides, three biological replicates hybridized with dye swap (fully balanced dye swap) and duplicate spots. cDNA were synthesized from 7 µg of total RNA and labelled with the dyes Cy3-dCTP and Cy5-dCTP (Amersham) using the ChipShot direct labeling system (Promega) according to the manufacturer›s instructions. The microarray were hybridized with cDNA prepared as described by Le Goff *et al.* [58] and scanned using a GenePixPro scanner (Axon,version 3.01). Experimental data and associated microarray designs have been deposited in the NBCI Gene Expression Omnibus (GEO) (http://www.ncbi.nlm.nih.gov/geo/) under serie GSE30937 and platform record GPL8717 using Mediante database for data transfer [59].

Data analysis

We used the Bioconductor suite of statistical packages [60]: *limma* [61] for our data analysis. The expression intensity was obtained by subtracting the background intensity from the foreground intensity for each non-flagged spot (all flagged spots were eliminated). The expression data were normalized by the use of the within-array normalization with the "loess method" and the between-array normalization using the "quantile method" [62]. The linear model for series of arrays and empirical Bayes method were then applied for assessing differential expression [63]. The false discovery rate of the p-value for multiple tests was controlled by using the Benjamini-Hochberg method. Differentially expressed genes were selected if the absolute value of log2-fold-change was greater than 1, the adjusted p-value below 0.01 and the average intensity greater than twice the average background. In order to provide an overall measure of evidence of differential expression, we used Fisher's method for combining adjusted p-values from independent tests of significance of duplicate spots [64].

Quantitative real-time PCR

Total RNA (1 μg) was reverse transcribed using the iScript cDNA Synthesis kit (Biorad). RTqPCR reactions were carried out on an Opticon monitor 2 (Biorad) using the qPCR Mastermix plus for SYBR Green I no ROX (Eurogentec). The PCR conditions were as follows: 50°C for 2 min, 95°C for 10 min, followed by 40 cycles of 95°C for 30 s, 60°C for 30 s and 72°C for 30 s. Each reaction was performed in technical triplicates and the mean of the three independent biological replicates was calculated. All results were normalized using mRNA level of three control genes (RpL4, L18 and G6PD) and relative expression values were calculated in R using the RqPCRAnalysis package developed in our laboratory (Hilliou and Tran, manuscript in preparation). Primer sequences and PCR efficiencies are listed in the Table 5.

Table 5. Primers used in RT-qPCR.

Name	primer sequence	fragment length	PCR efficiency %
EcR-F	5'-AAGCCTTTACCGGAGGATGT-3'	82	95
EcR-R	5'-TACCAGCGTCGTCTCATGTC-3'		
USP-F	5'-CGTATCAACCGACCTCCACT-3'	135	102
USP-R	5'-CCAGCTCAAGGGAACTGAAG-3'		
G6PD-F	5'-GGCCCTGTGGCTAACAGAAT-3'	142	99
G6PD-R	5'-CATCGTCTCTACCAAAAGGCTTC-3'		
L18-F	5'-CGTATCAACCGACCTCCACT-3'	126	104
L18-R	5'-AGGCACCTTGTAGAGCCTCA-3'		
RpL4-F	5'-CAACAAGAGGGGTTCACGAT-3'	149	99
RpL4-R	5'-GCACGATCAGTTCGGGTATC-3'		

ACKNOWLEDGMENTS

We thank Julie Cazareth, Institut de Pharmacologie Moleculaire et Cellulaire Sophia Antipolis, for her technical help in the flow cytometry experiments.

AUTHOR CONTRIBUTIONS

Conceived and designed the experiments: MG GLG RF. Performed the experiments: MG JC NT GLG. Analyzed the data: MG JC FH GLG. Contributed reagents/materials/analysis tools: FH TT. Wrote the paper: GLG RF.

REFERENCES

1. Truman JW, Riddiford L (1999) The origins of insect metamorphosis. Nature 401: 447–452.JW TrumanL. Riddiford1999The origins of insect metamorphosis.Nature401447452
2. Dubrovsky EB (2005) Hormonal cross talk in insect development. Trends Endocrinol Metab 16: 6–11.EB Dubrovsky2005Hormonal cross talk in insect development.Trends Endocrinol Metab16611
3. Riddiford LM (1996) Juvenile hormone: the status of its "status quo" action. Arch Insect Biochem Physiol 32: 271–286.LM Riddiford1996Juvenile hormone: the status of its "status quo" action.Arch Insect Biochem Physiol32271286
4. Yao TP, Segraves WA, Oro AE, McKeown M, Evans RM (1992) Drosophila ultraspiracle modulates ecdysone receptor function via heterodimer formation. Cell 71: 63–72.TP YaoWA SegravesAE OroM. McKeownRM Evans1992Drosophila ultraspiracle modulates ecdysone receptor function via heterodimer formation.Cell716372
5. Karim FD, Thummel CS (1992) Temporal coordination of regulatory gene expression by the steroid hormone ecdysone. EMBO J 11: 4083–4093.FD KarimCS Thummel1992Temporal coordination of regulatory gene expression by the steroid hormone ecdysone. EMBO J1140834093
6. Thummel CS (2001) Molecular mechanisms of developmental timing in C. elegans and Drosophila. Dev Cell 1: 453–465.CS Thummel2001Molecular mechanisms of developmental timing in C. elegans and Drosophila.Dev Cell1453465
7. Ashok M, Turner C, Wilson TG (1998) Insect juvenile hormone resistance gene homology with the bHLH-PAS family of transcriptional regulators. Proc Natl Acad Sci U S A 95: 2761–2766.M. AshokC. TurnerTG Wilson1998Insect juvenile hormone resistance gene homology with the bHLH-PAS family of

transcriptional regulators.Proc Natl Acad Sci U S A9527612766

8. Jones G, Sharp PA (1997) Ultraspiracle: an invertebrate nuclear receptor for juvenile hormones. Proc Natl Acad Sci U S A 94: 13499–13503.G. JonesPA Sharp1997Ultraspiracle: an invertebrate nuclear receptor for juvenile hormones.Proc Natl Acad Sci U S A941349913503
9. Miura K, Oda M, Makita S, Chinzei Y (2005) Characterization of the Drosophila Methoprene -tolerant gene product. Juvenile hormone binding and ligand-dependent gene regulation. FEBS J 272: 1169–1178.K. MiuraM. OdaS. MakitaY. Chinzei2005Characterization of the Drosophila Methoprene -tolerant gene product. Juvenile hormone binding and ligand-dependent gene regulation.FEBS J27211691178
10. Jones G, Wozniak M, Chu Y, Dhar S, Jones D (2001) Juvenile hormone III-dependent conformational changes of the nuclear receptor ultraspiracle. Insect Biochem Mol Biol 32: 33–49.G. JonesM. WozniakY. ChuS. DharD. Jones2001Juvenile hormone III-dependent conformational changes of the nuclear receptor ultraspiracle.Insect Biochem Mol Biol323349
11. Moore AW, Barbel S, Jan LY, Jan YN (2000) A genomewide survey of basic helix-loop-helix factors in Drosophila. Proc Natl Acad Sci U S A 97: 10436–10441.AW MooreS. BarbelLY JanYN Jan2000A genomewide survey of basic helix-loop-helix factors in Drosophila. Proc Natl Acad Sci U S A971043610441
12. Wilson TG, Ashok M (1998) Insecticide resistance resulting from an absence of target-site gene product. Proc Natl Acad Sci U S A 95: 14040–14044.TG WilsonM. Ashok1998Insecticide resistance resulting from an absence of target-site gene product.Proc Natl Acad Sci U S A951404014044
13. Baumann A, Fujiwara Y, Wilson TG (2010) Evolutionary divergence of the paralogs Methoprene tolerant (Met) and germ cell expressed (gce) within the genus Drosophila. J Insect Physiol 56: 1445–1455.A. BaumannY. FujiwaraTG Wilson2010Evolutionary divergence of the paralogs Methoprene tolerant (Met) and germ cell expressed (gce) within the genus Drosophila.J Insect Physiol5614451455
14. Konopova B, Jindra M (2007) Juvenile hormone resistance gene

Methoprene-tolerant controls entry into metamorphosis in the beetle *Tribolium castaneum*. Proc Natl Acad Sci U S A 104: 10488–10493.B. KonopovaM. Jindra2007Juvenile hormone resistance gene Methoprene-tolerant controls entry into metamorphosis in the beetle *Tribolium castaneum*.Proc Natl Acad Sci U S A1041048810493

15. Iwema T, Billas IM, Beck Y, Bonneton F, Nierengarten H, et al. (2007) Structural and functional characterization of a novel type of ligand-independent RXR-USP receptor. EMBO J 26: 3770–3782.T. IwemaIM BillasY. BeckF. BonnetonH. Nierengarten2007Structural and functional characterization of a novel type of ligand-independent RXR-USP receptor.EMBO J2637703782
16. Konopova B, Jindra M (2008) Broad-Complex acts downstream of Met in juvenile hormone signaling to coordinate primitive holometabolan metamorphosis. Development 135: 559–568.B. KonopovaM. Jindra2008Broad-Complex acts downstream of Met in juvenile hormone signaling to coordinate primitive holometabolan metamorphosis.Development135559568
17. Minakuchi C, Zhou X, Riddiford LM (2008) Kruppel homolog 1 (Kr-h1) mediates juvenile hormone action during metamorphosis of Drosophila melanogaster. Mech Dev 125: 91–105.C. MinakuchiX. ZhouLM Riddiford2008Kruppel homolog 1 (Kr-h1) mediates juvenile hormone action during metamorphosis of Drosophila melanogaster.Mech Dev12591105
18. Parthasarathy R, Tan A, Bai H, Palli SR (2008) Transcription factor broad suppresses precocious development of adult structures during larval-pupal metamorphosis in the red flour beetle, *Tribolium castaneum*. Mech Dev 125: 299–313.R. ParthasarathyA. TanH. BaiSR Palli2008Transcription factor broad suppresses precocious development of adult structures during larval-pupal metamorphosis in the red flour beetle, *Tribolium castaneum*.Mech Dev125299313
19. Suzuki Y, Truman JW, Riddiford LM (2008) The role of Broad in the development of *Tribolium castaneum*: implications for the evolution of the holometabolous insect pupa. Development 135: 569–577.Y. SuzukiJW TrumanLM Riddiford2008The role of Broad in the development of *Tribolium castaneum*: implications for the evolution of the holometabolous insect pupa.Development135569577

20. Minakuchi C, Nimiki T, Shinoda T (2009) Kruppel homolog 1, an early juvenile hormone-response gene downstream of Methoprene-tolerant, mediates its anti-metamorphic action in the red flour beetle *Tribolium castaneum.* Dev Biol 325: 341–350.C. MinakuchiT. NimikiT. Shinoda2009Kruppel homolog 1, an early juvenile hormone-response gene downstream of Methoprene-tolerant, mediates its anti-metamorphic action in the red flour beetle *Tribolium castaneum*.Dev Biol325341350

21. Li M, Mead EA, Zhu J (2011) Heterodimer of two bHLH-PAS proteins mediates juvenile hormone-induced gene expression. Proc Natl Acad Sci U S A 108: 638–643.M. LiEA MeadJ. Zhu2011Heterodimer of two bHLH-PAS proteins mediates juvenile hormone-induced gene expression.Proc Natl Acad Sci U S A108638643

22. Zhang Z, Xu J, Sheng Z, Sui Y, Palli SR (2011) Steroid Receptor Co-activator Is Required for Juvenile Hormone Signal Transduction through a bHLH-PAS Transcription Factor, Methoprene Tolerant. J Biol Chem 286: 8437–8447.Z. ZhangJ. XuZ. ShengY. SuiSR Palli2011Steroid Receptor Co-activator Is Required for Juvenile Hormone Signal Transduction through a bHLH-PAS Transcription Factor, Methoprene Tolerant.J Biol Chem28684378447

23. Dhadialla TS, Carlson GR, Le DP (1998) New insecticides with ecdysteroidal and juvenile hormone activity. Annu Rev Entomol 43: 545–569.TS DhadiallaGR CarlsonDP Le1998New insecticides with ecdysteroidal and juvenile hormone activity.Annu Rev Entomol43545569

24. Smagghe G, Dhadialla TS, Lezzi M (2002) Comparative toxicity and ecdysone receptor affinity of non-steroidal ecdysone agonists and 20-hydroxyecdysone in Chironomus tentans. Insect Biochem Mol Biol 32: 187–192.G. SmaggheTS DhadiallaM. Lezzi2002Comparative toxicity and ecdysone receptor affinity of non-steroidal ecdysone agonists and 20-hydroxyecdysone in Chironomus tentans.Insect Biochem Mol Biol32187192

25. Carlson GR, Dhadialla TS, Hunter R, Jansson RK, Jany CS (2001) The chemical and biological properties of methoxyfenozide, a new insecticidal ecdysteroid agonist. Pest Manag Sci 57: 115–119. GR CarlsonTS DhadiallaR. HunterRK JanssonCS Jany2001The chemical and biological properties of methoxyfenozide, a new

insecticidal ecdysteroid agonist.Pest Manag Sci57115119

26. Henrick CA, Staal GB, Siddall JB (1973) Alkyl 3,7,11-trimethyl-2,4-dodecadienoates, a new class of potent insect growth regulators with juvenile hormone activity. J Agri Food Chem 21: 354–359. CA HenrickGB StaalJB Siddall1973Alkyl 3,7,11-trimethyl-2,4-dodecadienoates, a new class of potent insect growth regulators with juvenile hormone activity.J Agri Food Chem21354359
27. Mikitani K (1996) A new nonsteroidal chemical class of ligand for the ecdysteroid receptor 3, 5-di-tert-butyl-4-hydroxy-N-isobutyl-benzamide shows apparent insect molting hormone activities at molecular and cellular levels. Biochem Biophys Res Commun 227: 427–432.K. Mikitani1996A new nonsteroidal chemical class of ligand for the ecdysteroid receptor 3, 5-di-tert-butyl-4-hydroxy-N-isobutyl-benzamide shows apparent insect molting hormone activities at molecular and cellular levels.Biochem Biophys Res Commun227427432
28. Wing KD (1988) RH 5849, a nonsteroidal ecdysone agonist: effects on a Drosophila cell line. Science 241: 467–469.KD Wing1988RH 5849, a nonsteroidal ecdysone agonist: effects on a Drosophila cell line.Science241467469
29. Silhacek DL, Oberlander H, Porcheron P (1990) Action of RH 5849, a non-steroidal ecdysteroid mimic, on *Plodia interpunctella* (Hubner) *in vivo* and *in vitro*. Arch Insect Biochem Physiol 15: 201–212.DL SilhacekH. OberlanderP. Porcheron1990Action of RH 5849, a non-steroidal ecdysteroid mimic, on *Plodia interpunctella* (Hubner) *in vivo* and *in vitro*.Arch Insect Biochem Physiol15201212
30. Quack S, Fretz A, Spindler-Barth M, Spindler KD (1995) Receptor affinities and biological responses of non-steroidal ecdysteroid agonists on the epithelial cell line from *Chironomus tentans* (Diptera: Chironomidae). Eur J Entomol 92: 341–347.S. QuackA. FretzM. Spindler-BarthKD Spindler1995Receptor affinities and biological responses of non-steroidal ecdysteroid agonists on the epithelial cell line from *Chironomus tentans*(Diptera: Chironomidae).Eur J Entomol92341347
31. Auzoux-Bordenave S, Solvar M, Queguiner I, Bozzolan F, Mottier V, et al. (2005) Comparative effects of a non-steroidal ecdysone

agonist RH-5992 and 20-hydroxyecdysone in a lepidopteran cell line (IAL-PID2). Insect Biochem Mol Biol 35: 1033–1042.S. Auzoux-BordenaveM. SolvarI. QueguinerF. BozzolanV. Mottier2005Comparative effects of a non-steroidal ecdysone agonist RH-5992 and 20-hydroxyecdysone in a lepidopteran cell line (IAL-PID2).Insect Biochem Mol Biol3510331042

32. Oberlander H, Leach CE, Shaaya E (2000) Juvenile hormone and juvenile hormone mimics inhibit proliferation in a lepidopteran imaginal disc cell line. J Insect Physiol 46: 259–265.H. OberlanderCE LeachE. Shaaya2000Juvenile hormone and juvenile hormone mimics inhibit proliferation in a lepidopteran imaginal disc cell line.J Insect Physiol46259265
33. Atmann F, Staudacher E, Wilson IB (1999) Insect cells as hosts for the expression of recombinant glycoproteins. Glycoconj J 2: 109–123.F. AtmannE. StaudacherIB Wilson1999Insect cells as hosts for the expression of recombinant glycoproteins.Glycoconj J2109123
34. Staal GB (1975) Insect growth regulators with juvenile hormone activity. Annu Rev Entomol 20: 417–460.GB Staal1975Insect growth regulators with juvenile hormone activity.Annu Rev Entomol20417460
35. Swevers L, Soin T, Mosallanejad H, Iatrou K, Smagghe G (2008) Ecdysteroid signaling in ecdysteroid-resistant cell lines from the polyphagous noctuid pest *Spodoptera exigua*. Insect Biochem Mol Biol 38: 825–833.L. SweversT. SoinH. MosallanejadK. IatrouG. Smagghe2008Ecdysteroid signaling in ecdysteroid-resistant cell lines from the polyphagous noctuid pest *Spodoptera exigua*.Insect Biochem Mol Biol38825833
36. Decombel L, Tirry L, Smagghe G (2005) Action of 24-epibrassinolide on a cell line of the beet armyworm, *Spodoptera exigua*. Arch Insect Biochem Physiol 58: 145–156.L. DecombelL. TirryG. Smagghe2005Action of 24-epibrassinolide on a cell line of the beet armyworm, *Spodoptera exigua*.Arch Insect Biochem Physiol58145156
37. Stevens B, Alvarez CM, Bohman R, O'Connor JD (1980) An ecdysteroid-induced alteration in the cell cycle of cultured Drosophila cells. Cell 22: 675–682.B. StevensCM AlvarezR.

BohmanJD O'Connor1980An ecdysteroid-induced alteration in the cell cycle of cultured Drosophila cells.Cell22675682

38. Mottier V, Siaussat D, Bozzolan F, Auzoux-Bordenave S, Porcheron P, et al. (2004) The 20-hydroxyecdysone-induced cellular arrest in G2 phase is preceded by an inhibition of cyclin expression. Insect Biochem Mol Biol 3: 51–60.V. MottierD. SiaussatF. BozzolanS. Auzoux-BordenaveP. Porcheron2004The 20-hydroxyecdysone-induced cellular arrest in G2 phase is preceded by an inhibition of cyclin expression.Insect Biochem Mol Biol35160
39. Gerenday A, Fallon AM (2004) Ecdysone-induced accumulation of mosquito cells in the G1 phase of the cell cycle. J Insect Physiol 50: 831–838.A. GerendayAM Fallon2004Ecdysone-induced accumulation of mosquito cells in the G1 phase of the cell cycle.J Insect Physiol50831838
40. The International Silkworm Genome Consortium (2008) The genome of a lepidopteran model insect, the silkworm *Bombyx mori*. Insect Biochem Mol Biol 38: 1036–1045.The International Silkworm Genome Consortium2008The genome of a lepidopteran model insect, the silkworm *Bombyx mori*.Insect Biochem Mol Biol3810361045
41. Gauhar Z, Sun LV, Hua S, Mason CE, Fuchs F, et al. (2009) Genomic mapping of binding regions for the Ecdysone receptor protein complex. Genome Res 19: 1006–1013.Z. GauharLV SunS. HuaCE MasonF. Fuchs2009Genomic mapping of binding regions for the Ecdysone receptor protein complex.Genome Res1910061013
42. Willis DK, Wang J, Lindholm JR, Orth A, Goodman WG (2010) Microarray analysis of juvenile hormone response in *Drosophila melanogaster* S2 cells. J Insect Sci 10: 66.DK WillisJ. WangJR LindholmA. OrthWG Goodman2010Microarray analysis of juvenile hormone response in *Drosophila melanogaster* S2 cells.J Insect Sci1066
43. Mosallanejad H, Badisco L, Swevers L, Soin T, Knapen D, et al. (2010) Ecdysone signaling and transcript signature in Drosophila cells resistant against methoxyfenozide. J Insect Physiol 56: 1973–1985.H. MosallanejadL. BadiscoL. SweversT. SoinD. Knapen2010Ecdysone signaling and transcript signature

in Drosophila cells resistant against methoxyfenozide.J Insect Physiol5619731985

44. Fallon AM, Gerenday A (2010) Ecdysone and the cell cycle: Investigations in a mosquito cell line. J Insect Physiol 56: 1396–1401.AM FallonA. Gerenday2010Ecdysone and the cell cycle: Investigations in a mosquito cell line.J Insect Physiol5613961401
45. Crossin KL, Tai MH, Krushel LA, Mauro VP, Edelman GM (1997) Glucocorticoid receptor pathways are involved in the inhibition of astrocyte proliferation. Proc Natl Acad Sci U S A 94: 2687–2692.KL CrossinMH TaiLA KrushelVP MauroGM Edelman1997Glucocorticoid receptor pathways are involved in the inhibition of astrocyte proliferation.Proc Natl Acad Sci U S A9426872692
46. Beyenbach KW, Wieczorek H (2006) The V-type H+ ATPase: molecular structure and function, physiological roles and regulation. J Exp Biol 209: 577–589.KW BeyenbachH. Wieczorek2006The V-type H+ ATPase: molecular structure and function, physiological roles and regulation.J Exp Biol209577589
47. Sumner JP, Dow JA, Earley FG, Klein U, Jager D, et al. (1995) Regulation of plasma membrane V-ATPase activity by dissociation of peripheral subunits. J Biol Chem 270: 5649–5653.JP SumnerJA DowFG EarleyU. KleinD. Jager1995Regulation of plasma membrane V-ATPase activity by dissociation of peripheral subunits.J Biol Chem27056495653
48. Wieczorek H, Grber G, Harvey WR, Huss M, Merzendorfer H, et al. (2000) Structure and regulation of insect plasma membrane H(+) V-ATPase. J Exp Biol 203: 127–135.H. WieczorekG. GrberWR HarveyM. HussH. Merzendorfer2000Structure and regulation of insect plasma membrane H(+)V-ATPase.J Exp Biol203127135
49. Holliday LS, Lu M, Lee BS, Nelson RD, Solivan S, et al. (2000) The amino-terminal domain of the B subunit of vacuolar H+-ATPase contains a filamentous actin binding site. J Biol Chem 275: 32331–32337.LS HollidayM. LuBS LeeRD NelsonS. Solivan2000The amino-terminal domain of the B subunit of vacuolar H+-ATPase contains a filamentous actin binding site.J Biol Chem2753233132337
50. Vitavska O, Wieczorek H, Merzendorfer H (2003) A novel role

for subunit C in mediating binding of the H+-V-ATPase to the actin cytoskeleton. J Biol Chem 278: 18499–18505.O. VitavskaH. WieczorekH. Merzendorfer2003A novel role for subunit C in mediating binding of the H+-V-ATPase to the actin cytoskeleton.J Biol Chem2781849918505

51. Voss M, Vitavska O, Walz B, Wieczorek H, Baumann O (2007) Stimulus-induced phosphorylation of vacuolar H(+)-ATPase by protein kinase A. J Biol Chem 282: 33735–33742.M. VossO. VitavskaB. WalzH. WieczorekO. Baumann2007Stimulus-induced phosphorylation of vacuolar H(+)-ATPase by protein kinase A.J Biol Chem2823373533742

52. Kane PM (2000) Regulation of V-ATPases by reversible disassembly. FEBS Lett 469: 137–141.PM Kane2000Regulation of V-ATPases by reversible disassembly.FEBS Lett469137141

53. Merzendorfer H, Reineke S, Zhao XF, Jacobmeier B, Harvey WR, et al. (2000) The multigene family of the tobacco hornworm V-ATPase: novel subunits a, C, D, H, and putative isoforms. Biochim Biophys Acta 1467: 369–379.H. MerzendorferS. ReinekeXF ZhaoB. JacobmeierWR Harvey2000The multigene family of the tobacco hornworm V-ATPase: novel subunits a, C, D, H, and putative isoforms.Biochim Biophys Acta1467369379

54. Wieczorek H, Beyenbach KW, Huss M, Vitavska O (2009) Vacuolar-type proton pumps in insect epithelia. J Exp Biol 212: 1611–1619.H. WieczorekKW BeyenbachM. HussO. Vitavska2009Vacuolar-type proton pumps in insect epithelia.J Exp Biol21216111619

55. Graf R, Harvey WR, Wieczorek H (1996) Purification and properties of a cytosolic V1-ATPase. J Biol Chem 271: 20908–20913.R. GrafWR HarveyH. Wieczorek1996Purification and properties of a cytosolic V1-ATPase.J Biol Chem2712090820913

56. Mukherji M, Bell R, Supekove L, Wang Y, Orth AP, et al. (2006) Genome-wide functional analysis of human cell-cycle regulators. Proc Natl Acad Sci U S A 103: 14819–14824.M. MukherjiR. BellL. SupekoveY. WangAP Orth2006Genome-wide functional analysis of human cell-cycle regulators.Proc Natl Acad Sci U S A1031481914824

57. Crissman HA, Mullaney PF, Steinkamp JA (1975) Methods and

applications of flow systems for analysis and sorting of mammalian cells. Methods Cell Biol 9: 179–246.HA CrissmanPF MullaneyJA Steinkamp1975Methods and applications of flow systems for analysis and sorting of mammalian cells.Methods Cell Biol9179246

58. Le Goff G, Hilliou F, Siegfried B, Boundy S, Wajnberg E, et al. (2006) Xenobiotic response in *Drosophila melanogaster*: sex dependence of P450 and GST gene induction. Insect Biochem Mol Biol 36: 674–682.G. Le GoffF. HilliouB. SiegfriedS. BoundyE. Wajnberg2006Xenobiotic response in *Drosophila melanogaster*: sex dependence of P450 and GST gene induction.Insect Biochem Mol Biol36674682
59. Le Brigand K, Barbry P (2007) Mediante: a web-based microarray data manager. Bioinformatics 23: 1304–1306.K. Le BrigandP. Barbry2007Mediante: a web-based microarray data manager. Bioinformatics2313041306
60. Gentleman RC, Carey VJ, Bates DM, Bolstad B, Dettling M, et al. (2004) Bioconductor: open software development for computational biology and bioinformatics. Genome Biol 5: R80.RC GentlemanVJ CareyDM BatesB. BolstadM. Dettling2004Bioconductor: open software development for computational biology and bioinformatics. Genome Biol5R80
61. Wettenhall JM, Smyth GK (2004) limmaGUI: a graphical user interface for linear modeling of microarray data. Bioinformatics 20: 3705–3706.JM WettenhallGK Smyth2004limmaGUI: a graphical user interface for linear modeling of microarray data. Bioinformatics2037053706
62. Yang YH, Thorne NP (2003) Normalization for two-color cDNA microarray data. In: Goldstein DR, editor. Science and Statistics: A Festschrift for Terry Speed, IMS Lecture Notes - Monograph Series. Beachwood, OH: Institute of Mathematical Statistics. pp. 403–418.YH YangNP Thorne2003Normalization for two-color cDNA microarray data.DR GoldsteinScience and Statistics: A Festschrift for Terry Speed, IMS Lecture Notes - Monograph SeriesBeachwood, OHInstitute of Mathematical Statistics403418
63. Smyth GK (2004) Linear models and empirical bayes methods for assessing differential expression in microarray experiments. Stat

Appl Genet Mol Biol 3: Article3.GK Smyth2004Linear models and empirical bayes methods for assessing differential expression in microarray experiments.Stat Appl Genet Mol Biol3Article3

64. Hess A, Iyer H (2007) Fisher's combined p-value for detecting differentially expressed genes using Affymetrix expression arrays. BMC Genomics 8: 96.A. HessH. Iyer2007Fisher's combined p-value for detecting differentially expressed genes using Affymetrix expression arrays.BMC Genomics896

CHAPTER

15

FUNCTIONAL PROCESSING AND SECRETION OF CHIKUNGUNYA VIRUS E1 AND E2 GLYCOPROTEINS IN INSECT CELLS

Stefan W Metz[1], Corinne Geertsema[1], Byron E Martina[2], Paulina Andrade[1], Jacco G Heldens[3], Monique M van Oers[1], Rob W Goldbach[1], Just M Vlak[1] and Gorben P Pijlman[1]

[1]Laboratory of Virology, Wageningen University, Droevendaalsesteeg 1, 6708 PB Wageningen, the Netherlands.

[2]Erasmus MC, Department of Virology, 3000 CA Rotterdam, the Netherlands.

[3]Nobilon International BV, Boxmeer, the Netherlands.

ABSTRACT

Background

Chikungunya virus (CHIKV) is a mosquito-borne, arthrogenic *Alphavirus* that causes large epidemics in Africa, South-East Asia and India. Recently, CHIKV has been transmitted to humans in Southern Europe by invading and now established Asian tiger mosquitoes. To study the processing of envelope proteins E1 and E2 and to develop a CHIKV subunit vaccine, C-terminally his-tagged E1 and E2 envelope glycoproteins were produced at high levels in insect cells with baculovirus vectors using their native signal peptides located in CHIKV 6K and E3, respectively.

Results

Expression in the presence of either tunicamycin or furin inhibitor showed that a substantial portion of recombinant intracellular E1 and precursor E3E2 was glycosylated, but that a smaller fraction of E3E2 was processed by furin into mature E3 and E2. Deletion of the C-terminal transmembrane domains of E1 and E2 enabled secretion of furin-cleaved, fully processed E1 and E2 subunits, which could then be efficiently purified from cell culture fluid via metal affinity chromatography. Confocal laser scanning microscopy on living baculovirus-infected *Sf*21 cells revealed that full-length E1 and E2 translocated to the plasma membrane, suggesting similar posttranslational processing of E1 and E2, as in a natural CHIKV infection. Baculovirus-directed expression of E1 displayed fusogenic activity as concluded from syncytia formation. CHIKV-E2 was able to induce neutralizing antibodies in rabbits.

Conclusions

Chikungunya virus glycoproteins could be functionally expressed at high levels in insect cells and are properly glycosylated and cleaved by furin. The ability of purified, secreted CHIKV-E2 to induce neutralizing antibodies in rabbits underscores the potential use of E2 in a subunit vaccine to prevent CHIKV infections.

BACKGROUND

Chikungunya virus (CHIKV) is an arthropod-borne (arbo)virus that causes epidemics in Africa, India and South-East Asia [1]. Recent outbreaks in Italy in 2007 [2] and autochthonous transmission events in France in 2010 [3] exemplify the threat of continued spread of CHIKV in the Western world, which correlates with the concurrent expanding distribution of its insect vector. CHIKV is maintained in a sylvatic transmission cycle of mosquitoes, rodents and primates, with *Aedes aegyti* as the primary vector. However, the responsible vector causing the severe CHIKV epidemic on the Reunion Islands in 2005/2006 was *Ae. albopictus*[4]. This vector switch made the virus endemic in more temperate regions and resulted in the first European cases (Italy, 2007) of transmission by local populations of *Ae. albopictus*[1, 5].

CHIKV (family *Togaviridae*: genus *Alphavirus*) has a single-stranded positive-sense RNA genome, which varies slightly in length between different isolates, but is approximately 11,800 nts long [6] and encodes two open reading frames (ORF). The RNA is encapsidated in a nucleocapsid of approximately 40 nm in diameter [7]. The nucleocapsid is tightly enveloped by a host-derived lipid bilayer (envelope) supporting the virus-encoded envelope proteins. Eighty glycoprotein spikes are C-terminally anchored within the viral envelope and are exposed on the surface of virions and of infected cells.

The nonstructural proteins required for viral RNA replication are directly translated from the 5' two-thirds region of the viral genome. The structural polyprotein is translated from a viral subgenomic mRNA (sgRNA), located at the 3'one-third part of the genome [8, 9]. The five structural proteins (capsid, E3, E2, 6K, E1) are translated as a single polyprotein, from which capsid (C) is autocatalytically cleaved off to encapsidate new plus-strand RNA molecules. The envelope polyprotein precursor E3-E2-6K-E1 is then translocated to the endoplasmatic reticulum (ER). Host signalases process the polyprotein at the N- and C-terminal end of the 6K peptide, resulting in E3E2 (also known as precursor E2: PE2), 6K and E1 [10], all anchored in the ER membrane. After this proteolytic cleavage, E3E2 and E1 will eventually form heterotrimers in the early Golgi compartment. Subsequently, E3E2 undergoes a furin-dependent maturation cleavage in the trans-Golgi system at the consensus cleavage signal R-X-(K/R)-R. This furin cleavage is not a prerequisite for virion assembly [11]. The hetero-trimeric spikes consisting of E2 and E1 facilitate cell receptor recognition, cell entry via pH-dependent endocytosis and support viral budding [9].

The major clinical symptoms of a CHIKV infection are febrile illness and severe joint pains [12]. Recently, macrophages were identified as being key players in CHIKV infection, persistence and pathogenesis. Macrophage-derived pro-inflammatory products are strongly involved in the muscle and joint immunopathological findings after alphavirus infection [13, 14]. Currently, there are no specific treatments for CHIKV infections and no licensed vaccine for any alphavirus is available for human use. During an infection with the alphavirus type species Sindbis virus (SINV) neutralizing antibodies are generally directed against E2 and to a lesser extent to E1. This holds true for other alphaviruses as well, suggesting that E1 and E2 are conserved among alphaviruses as epitope donors [15, 16]. Therefore glycoproteins E1 and E2 serve as the major targets in the development of a (subunit) vaccine against CHIKV infections. A recently developed, experimental CHIKV vaccine based on virus-like particles (VLPs) containing both E1 and E2 induced a protective immune response in non-human primates [17]. While this VLP approach may be a way forward in the development of a CHIKV vaccine, the described method of transfecting large DNA plasmids into mammalian cells remains challenging in terms of upscaling. Therefore, a subunit approach may be better compatible with industrial operations [18, 19].

In this study, the baculovirus-insect cell expression system was used to study recombinant CHIKV glycoprotein subunit formation. This production system is a safe and efficient way of expressing heterologous proteins on a large scale in eukaryotic cells [20] and the proven technology has resulted in several commercially available veterinary vaccines [21], including a veterinary subunit vaccine targeting Classical Swine Fever [18] and a human vaccine against cervical cancer (Cervarix, GlaxoSmithKline) [22]. A recombinant human vaccine against influenza virus (FluBlok) Biological License Application is currently in the process of evaluation by the Food and Drug Administration in the USA [19]. Recombinant protein expression by baculoviruses is based on the use of the strong polyhedrin promoter and the exchange of the polyhedrin gene for the heterologous gene of interest. This late phase baculoviral protein is expressed in large amounts in infected cells, but is not essential for baculovirus replication [23]. Furthermore, protein expression in insect cells allows post-translational modifications, accurate folding and efficient secretion [21, 24]. Drosophila cells have recently been used to efficiently express and process CHIKV-glycoprotein complexes [25], yet there are few reports on successful expression of alphavirus structural polyproteins with recombinant baculoviruses [26, 27], and there are no studies that describe the expression of individual full-length alphavirus E1 and/or E2 glycoproteins. Here, we investigated the expression of individual

CHIKV E1 and E2 glycoproteins in insect cells and analyzed the routing and post-translation modifications of these glycoproteins as well as the fusogenic properties of recombinant E1. In addition, purification methods were developed for secreted forms of these glycoproteins, of which the ability to generate a neutralizing antibody response was investigated.

RESULTS

Expression of CHIKV-E2 and -E1 glycoproteins in *Sf* 21 insect cells by recombinant baculoviruses

Four recombinant baculoviruses were generated to express membrane attached and secreted versions of CHIKV glycoproteins E1 and E2 (Ac-6KE1, Ac-6KE1ΔTM, Ac-E3E2 and Ac-E3E2ΔTM) (Figure 1A). The glycoprotein coding sequences were cloned downstream of the polyhedrin promoter in an *Autographa californica*multicapsid nucleopolyhedrovirus (AcMNPV) bacmid from which the promoters and open reading frames of the p10, cathepsin and chitinase genes were deleted (See Materials and Methods). All four glycoprotein constructs were equipped with a C-terminal 6 × his tag. *Sf*21 insect cells were infected with the respective recombinant baculoviruses at a MOI of 10 $TCID_{50}$ units per cell and cells were harvested 72 h post infection (hpi), which appeared to be the optimal time point for harvesting (data not shown). Recombinant protein content in the cell fraction was analyzed by Coomassie brilliant blue (CBB)-staining and western analysis using α-His monoclonal antibodies (mabs) and α-E1 and α-E2 polyclonal antisera (Figure 1B).

CBB staining of *Sf* 21 cell lysates showed that CHIKV-E2 and -E1 were expressed at very high levels. The size of the proteins observed after infection with Ac-6KE1 and Ac-6KE1ΔTM (Figure 1B, lanes 1 and 2) closely matched the predicted molecular mass of 6KE1 and 6KE1ΔTM of 54.2 kDa and 52.1 kDa, respectively. CHIKV 6KE1 was not detected by western analysis using α-His mabs, but was recognized well by α-E1 polyclonal antiserum. This is most likely caused by the proteolysis of the C-terminal his-tag of CHIKV 6KE1. Expression analysis of insect cells infected with Ac-E3E2 and Ac-E3E2ΔTM (Figure 1B, lane 3 and 4) resulted in two polypeptides for each construct. The smallest ones correspond to the predicted molecular mass of unprocessed E3E2 (54.6 kDa) and E3E2ΔTM (49.2 kDa), respectively. The larger ones, estimated to be 3-4 kDa larger in size, are possibly glycosylated forms of E3E2 and E3E2ΔTM, respectively. When the cell fractions were analyzed with the α-E2 polyclonal antiserum, a third, smaller protein was seen, which most likely represents furin-processed, glycosylated E2 and E2ΔTM, respectively. To allow secretion of E1 and E2 into the cell culture fluid, the C-terminal transmembrane domains of E1 and E2 were deleted (constructs Ac-6KE1ΔTM and Ac-E3E2ΔTM). Medium fraction analysis shows that the removal of the C-terminal TM of E1 and E2 indeed resulted in secretion of E1 and E2 (Figure 1B, bottom, lanes 2 and 4).

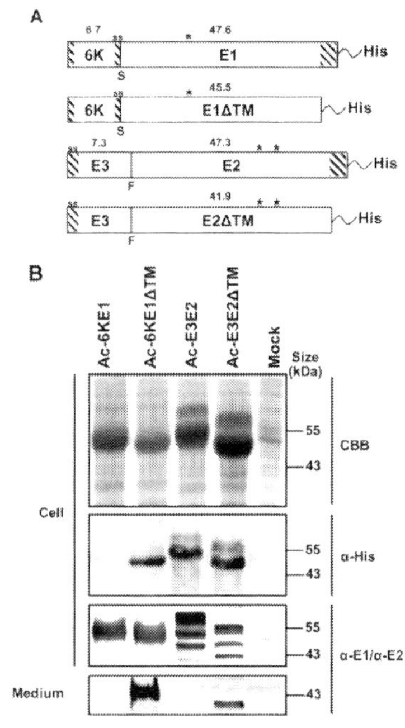

Figure 1: CHIKV E1 and E2 expression using recombinant baculoviruses. A) Schematic representation of four CHIKV glycoprotein constructs expressed from recombinant baculoviruses. The shaded areas represent transmembrane domains or signal peptides (ss), asterisks indicate predicted N-glycosylation sites, S and F indicate signalase and furin cleavage sites, respectively. The predicted molecular mass of the glycoproteins is indicated. All constructs were equipped with a C-terminal 6 × His-tail. B) Gene expression in the cell and in medium fractions of infected Sf21 cells was analyzed by Coomassie brilliant blue-staining and western blotting using α-His mabs and rabbit α-E1 and rabbit α-E2 polyclonal antisera.

Glycosylation status of baculovirus expressed CHIKV-E1 and -E2

To analyze whether the size difference observed between the upper and lower bands of E3E2 or E3E2ΔTM in CBB and western blots (Figure 1B) can be attributed to glycosylation of E2 a tunicamycin assay was performed to analyze the N-glycosylation status of both glycoproteins. It is common knowledge that alphavirus E1 and E2 are N-linked glycosylated, but the number of glycosylation sites can vary among species [28, 29, 30, 31]. CHIKV-E1 is predicted to be glycosylated at N141 and CHIKV-E2 at N263 and N273 [32]. *Sf*21 cells were infected with the 4 recombinant baculoviruses, respectively, (Figure 1A) and incubated in the presence or absence of tunicamycin, harvested at 72 hpi and separated from the medium fraction. Whole cell lysates were analyzed by western blotting (Figure 2A) and by Periodic Acid Schiff staining (Figure 2B), which stains glycosyl groups on proteins [33].

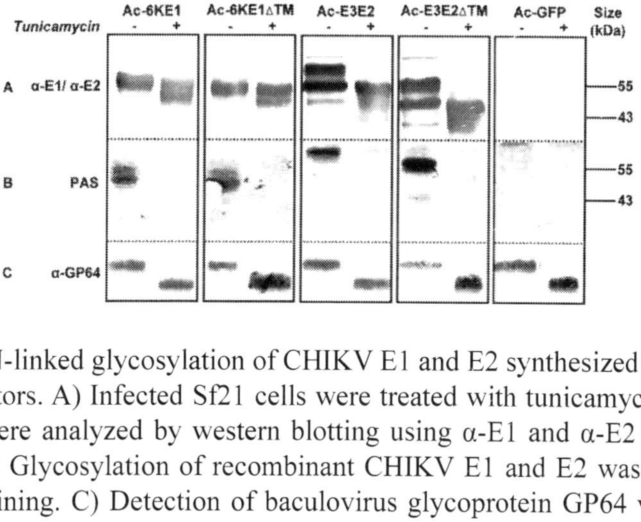

Figure 2: N-linked glycosylation of CHIKV E1 and E2 synthesized with baculovirus vectors. A) Infected Sf21 cells were treated with tunicamycin and cell fractions were analyzed by western blotting using α-E1 and α-E2 polyclonal antisera. B) Glycosylation of recombinant CHIKV E1 and E2 was confirmed by PAS-staining. C) Detection of baculovirus glycoprotein GP64 verified tunicamycin activity.

Expression of 6KE1 and 6KE1ΔTM in tunicamycin treated *Sf* 21 cells resulted in a band with lower molecular mass (Figure 2A, left). This suggests that E1 is at least partially glycosylated. PAS-staining of 6KE1 and 6KE1ΔTM resulted in protein bands with a size similar to the proteins detected by western blotting and confirmed that CHIKV-E1 was glycosylated. Western blot detection using α-E2 polyclonal antiserum showed that infection with Ac-E3E2 and Ac-E3E2ΔTM resulted in the triple-band pattern previously observed (Figure 1B). The minor fourth upper-band that was detected is only observed occasionally and may represent E2 aggregates. Upon tunicamycin treatment of Ac-E3E2 and Ac-E3E2ΔTM-infected cells, the E2-specific polypeptides with the highest molecular mass (~58 and ~55 kDa, respectively) disappeared, suggesting that these are the N-glycosylated forms of E3E2 and E3E2ΔTM (Figure 2A). PAS-staining, which is absent in the tunicamyin treated lanes, confirmed this result. Thus it can be concluded that the bands of ~55 kDa and ~49 kDa in infections with Ac-E3E2 and Ac-E3E2ΔTM, respectively, represent non-glycosylated E3E2 and E3E2ΔTM proteins. The smaller polypeptides of ~50 kDa and ~44 kDa, correspond in size to glycosylated, furin-cleaved E2 and E2ΔTM, respectively. PAS-staining on the Ac-E3E2ΔTM infected cell fraction also stains this additional protein of ~44 kDa (Figure 2B), suggesting that glycosylated, furin-cleaved E2ΔTM is indeed produced, albeit in relatively low abundance. Immunodetection of the baculovirus GP64 envelope protein was used as a positive control (Figure 2C). GP64 is essential for baculovirus budding and cell entry, and is known to be heavily glycosylated [34, 35] at 4 of the 5 predicted positions [36]. As expected, a significant decrease (~6 kDa) in molecular mass was observed as a result of tunicamycin treatment. GP64 is best detected on highly concentrated budded virus preparations and more difficult in insect cells. PAS staining is not sensitive enough to detect glycans on minor amounts of GP64, compared to overexpressed CHIKV glycoproteins, which explains the lack of PAS staining in cells infected the Ac-GFP control virus (Figure 2B, right).

These results show that a substantial fraction of CHIKV E1 and E2 proteins expressed by recombinant baculoviruses in *Sf* 21 insect cells is N-glycosylated and that only a minor fraction of the glycosylated E3E2 precursor is processed by furin into mature, N-glycosylated E2.

Furin processing of recombinant CHIKV-E3E2 and -E3E2ΔTM

During natural alphavirus glycoprotein maturation, E3 is released from E3E2 via cleavage by a furin-like protease [11]. However, the triple-band patterns found during infection with Ac-E3E2 and Ac-E3E2ΔTM (Figure 2A, right) suggest that the bulk of intracellular E3E2 is incompletely processed by furin into the individual E3 and E2 proteins. To investigate the level of furin-processing and to confirm the nature of the smallest polypeptide in the triple-band pattern, *Sf* 21 cells were infected with Ac-E3E2 and Ac-E3E2ΔTM in the presence or absence of furin inhibitor.

Western blot analysis of Ac-E3E2 infected cells shows that the smallest polypeptide (~44 kDa) detected by anti-E2 antiserum disappeared in the presence of furin inhibitor (Figure 3A, left) indicating that this represents the mature E2, which according to the previous experiment was also N-glycosylated (Figure 2). As expected, all E2-specific polypeptides were retained in the cell fraction (Figure 3B, left) unless the TM was deleted (Figure 3B, right) in accordance with the experiments showing that E2 is only secreted when the C-terminal TM-domain was deleted (Figure 1B). It should be noted that furin-processed E2ΔTM could hardly be observed in the cell fraction, which might suggest that the protein is secreted shortly after it is cleaved by furin in the trans-Golgi complex. Interestingly, when furin cleavage was inhibited in Ac-E3E2ΔTM-infected cells, the N-glycosylated fraction of E3E2ΔTM was still secreted into the culture fluid (Figure 3B, right). It can therefore be concluded that secretion of E2ΔTM is not dependent on furin processing.

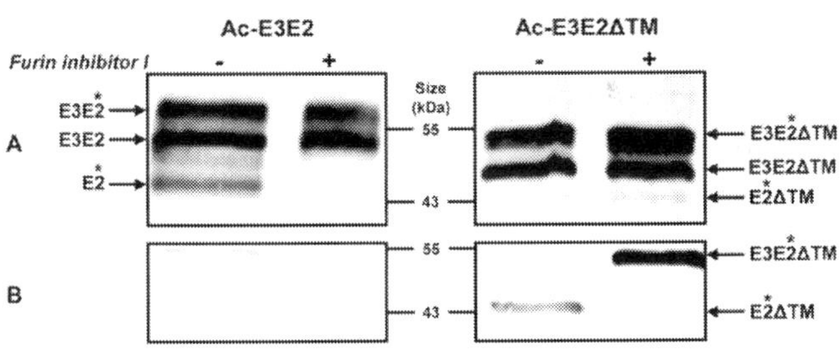

Figure 3: Furin inhibiton assay on Ac-E3E2 and Ac-E3E2ΔTM infected *Sf* 21 cells. Sf21 cells were infected with Ac-E3E2 and Ac-E3E2ΔTM in the absence (-) or presence (+) of furin inhibitor. The cell fraction (A) and the concentrated medium fraction (B) were analyzed by western blotting using α-E2 polyclonal antiserum. Glycosylated proteins are indicated with an asterisk.

Secretion of recombinant CHIKV-E1ΔTM and -E3E2ΔTM from insect cells

The recombinant baculovirus expression analysis and the furin inhibition assay showed that CHIKV E1ΔTM and E2ΔTM are secreted from *Sf* 21 cells when they are expressed with their native signal peptide 6K and E3, respectively (Figures 1B and 3B). To investigate whether other heterologous signal peptides might have a stronger effect on protein secretion, 6K and E3 were both exchanged for the honey bee melittin signal peptide (HBM), which has been shown to improve both expression and secretion of heterologous proteins in baculovirus infected cells [37]. In other constructs, the first 41 amino acids from 6K were deleted as they are not part of the signal peptide necessary for ER translocation of E1. The remaining 6K signal sequence was cloned upstream of E1ΔTM (Ac-6K^{-40}E1ΔTM) and of E2ΔTM (Ac-6K^{-40}E2ΔTM). The level of intracellular or secreted E1 and E2 was compared by western blotting. A control lane of CHIKV-infected Ap61 mosquito cells was loaded to indicate the size of native E1 and E2 proteins.

The replacement of 6K for HBM upstream of E1ΔTM resulted in lower intracellular E1 levels (Figure 4A), while the secretion level was also dramatically decreased (Figure 4B). Deletion of the first 41 amino acids from 6K resulted in equal secretion levels compared to the native complete signal peptide 6K. It could be concluded that the signal sequence in 6K is sufficient for efficient secretion of E1ΔTM. When the same signal peptides were exchanged with E3 to test whether E2ΔTM secretion could be enhanced, all constructs were expressed successfully (Figure 4A), but medium fraction analysis showed that neither the HBM nor the 6K^{-40} signal sequence could improve E2ΔTM protein secretion. This indicates that the CHIKV-6K and E3 signal peptides direct E1ΔTM and E2ΔTM secretion better than other well studied signal peptides like the HBM signal sequence. These results suggest that the CHIKV signal peptides could potentially serve as strong signal peptides for secretion of other heterologous proteins.

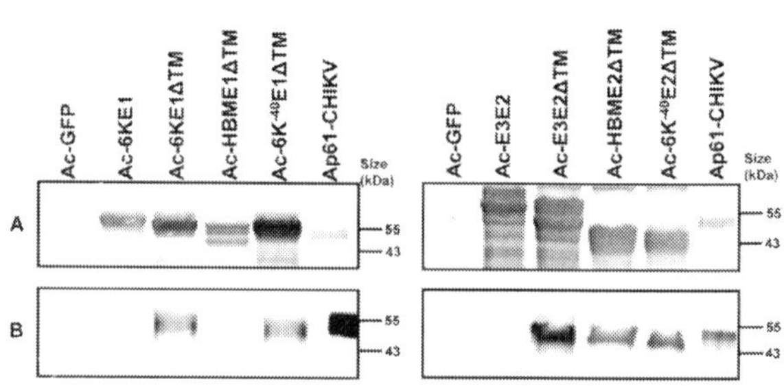

Figure 4: CHIKV E1ΔTM and E2 protein secretion. Both CHIKV E1ΔTM and E2ΔTM were expressed with E3, HBM or 6K^{-40} as signal sequence. The cell

fraction A) and the medium fraction B) were analyzed by western blot using α-E1 and α-E2 polyclonal antibodies. CHIKV infected *Ap* 61 mosquito cell-lysate was used as a control.

Purification of secreted recombinant CHIKV-E1ΔTM and -E2ΔTM

Both Ac-6KE1ΔTM and Ac-E3E2ΔTM constructs were tagged with a C-terminal polyhistidine-tail for detection and purification purposes. Medium fraction analyses showed that E1ΔTM and E2ΔTM were effectively secreted from the cell, which enables protein purification by Co^{2+}-histidine interaction. Total medium fractions were loaded onto Talon® spin columns and bound proteins were eluted with 150 mM imidazol. CBB analyses of the three elution fractions (Figure 5A) showed a dramatic increase in concentration of both E1ΔTM and E2ΔTM compared to the non-purified fractions, indicating that both recombinant proteins were efficiently purified from the medium fraction. Western analyses using the polyclonal antisera (Figure 5B) confirmed that the purified fractions indeed contained E1ΔTM and E2ΔTM and that multiple elution steps are required to completely elute the recombinant CHIKV glycoproteins from the column. Two bands were detected in the elution fractions of Ac-E3E2ΔTM (Figure 5B, right). The protein band with the highest molecular mass most likely represents glycosylated E3E2ΔTM, which complies with the finding that secretion of E2ΔTM is not dependent on furin cleavage (Figure 3B). Protein concentrations in the elution fractions were determined by Bradford assay and resulted in a total yield of 38 and 30 mg/l secreted protein for CHIKV-E1ΔTM and CHIKV (E3)E2ΔTM, respectively, at a cell concentration of 6.7 × 10^5 Sf21 cells/ml. Metal affinity purification of secreted CHIKV-E1ΔTM and -E2ΔTM resulted in a protein recovery of ~60% and >95%, respectively.

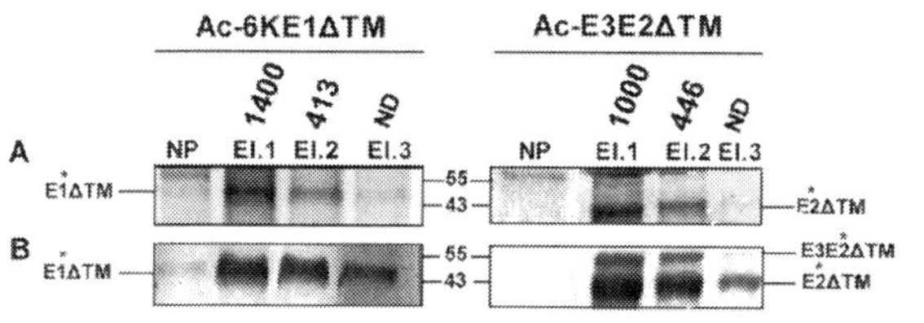

Figure 5: Purification of secreted, recombinant CHIKV E1ΔTM and E2ΔTM proteins. *Sf* 21 cells were infected with Ac-6KE1ΔTM and Ac-E3E2ΔTM and

the respective serum-free medium fractions were loaded onto Talon® spin columns. Bound proteins were eluted 3 times with 150 mM imidazol. The elution fractions were analyzed with CBB (A) and western blotting, using α-E1 and α-E2 polyclonal antisera (B). NP represents non-purified fractions and El.1-3 indicate the subsequent elution fractions. Total protein concentration of the elution fractions (1 ml each) are noted above in μg/ml.

Surface expression of CHIKV-E1 and -E2 in *Sf* 21 cells

Alphavirus glycoproteins are, during a natural infection, expressed at the surface of the host cell to allow budding of progeny virus [7]. From the foregoing experiments it became clear that baculovirus expressed CHIKV E1 and E2 are (at least partially) glycosylated and secreted from *Sf* 21 cells when their C-terminal TM domain is deleted. It was also observed that E3E2 is (at least partially) processed by furin, suggesting that the posttranslational processing of CHIKV glycoproteins in baculovirus-infected cells may resemble the processing during a natural CHIKV infection [10]. If this is the case, E1 and E2 are expected to be transported to the cell membrane during baculovirus infection and be exposed at the cell surface.

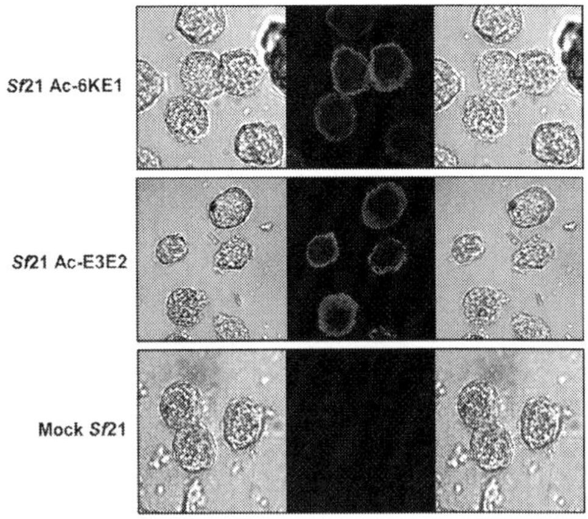

Figure 6: Surface expression of CHIKV E1 and E2 on the plasma membrane. *Sf* 21 cells were infected with recombinant baculoviruses and living cells were immunostained against E1 and E2. Confocal microscopy reveals the presence of CHIKV-E1 and-E2 at the surface of infected *Sf* 21 cells.

Therefore, non-permeable, living *Sf* 21 cells infected with Ac-6KE1 or Ac-E3E2 were subjected to immunofluorescence using polyclonal E1 and E2 antisera [38]. Positive staining indicated that the glycoproteins are exposed at the surface of the cell (Figure 6). From the ring-like structure observed by confocal microscopy (Figure 6, top and middle) it could be concluded that both E1 and E2 are translocated to the plasma membrane when expressed in insect cells. This ring-like structure was not observed in mock infected cells (Figure 6, bottom).

Fusogenic activity of CHIKV-E1 expressed in *Sf* 21 cells

This study has shown that production of CHIKV E1 and E2 insect cells using baculovirus vectors leads to glycosylation, (partial) furin processing and plasma membrane translocation and/or secretion of E1 and E2. The individually expressed glycoproteins appear to follow similar processing steps as compared to wildtype CHIKV infections [10], which might imply that the recombinant glycoproteins retained their functionality. The following experiment was designed to test the fusogenic activity of baculovirus expressed E1, which in CHIKV infection regulates fusion during endocytosis in a pH-dependent manner, with a pH-optimum of 5.5 [39]. Since alphavrus fusion is cholesterol dependent [40] and insect are cholesterol auxotrophs, *Sf* 9-ET cells were cultured in cholesterol-supplemented Sf900 II insect medium (pH = 6.4) and were infected with Ac-6KE1, Ac-E3E2 and Ac-GFP. *Sf* 9-ET is a transgenic cell line expresing eGFP under the control of the polyhedrin promoter, thereby only allowing eGFP expression during baculovirus replication [41]. In this way, baculovirus infected cells could be easily visualized. Therefore, *Sf* 9-ET cells infected with Ac-GFP express higher amounts of GFP, compared to Ac-6KE1 and Ac-E3E2 infected *Sf* 9-ET cells. Cells were treated with acidified medium of pH = 5.8, pH = 5.5 and pH = 5.0. Cells were screened for syncytia formation 4 h post treatment. Syncytia formation after induction with medium of pH6.4, pH = 5.8 or pH = 5.5 was only observed in cells producing CHIKV-E1 at 72 h.p.i (Figure 7, right). In contrast, cells infected with Ac-E3E2 (Figure 7, middle) or Ac-GFP (Figure 7, left) did not form syncytia, unless they were induced with acidified medium of pH = 5.0 (Figure 7, bottom). At this low pH, baculovirus GP64 induces syncytia formation [42]. Thus, expression of E1 is correlated with syncytium formation and E1 is functionally active as viral fusion protein in a defined pH range. The fusogenicity of CHIKV E1 is in the same order of magnitude to what has been found for the major envelope fusion protein F of baculoviruses with this particular assay suggesting that E1 is highly fusogenic [43]. These data concerning pH-dependent syncytia formation exclude the possibility that the syncytia formation in this assay at pH values of 5.8 and 5.5 was caused by the baculovirus infection itself. These findings are in line with the surface expression of CHIKV-E1 (Figure 6) and indicate that E1 retains its fusogenic activity when expressed in *Sf*-cells.

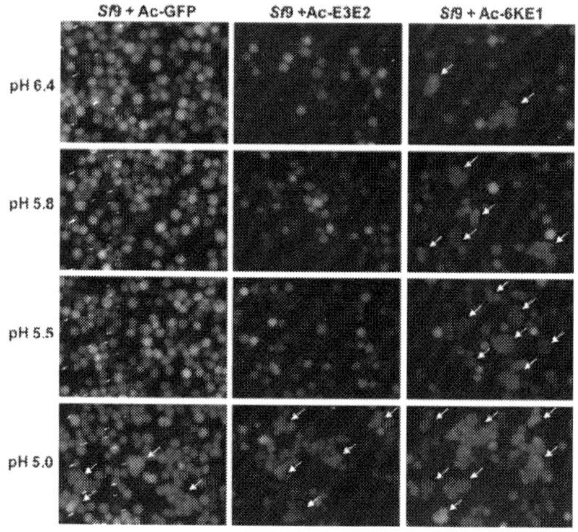

Figure 7: Syncytia formation after surface expression of CHIKV-E1 in insect cells. *Sf* 9-ET cells in cholesterol supplemented Sf900 II medium (pH = 6.4) were infected with Ac-6KE1, Ac-E3E2 and Ac-GFP. Cells were incubated 72 h post infection with medium of pH = 5.8, pH = 5.5 and pH = 5.0 for 2 min. Pictures were taken 4 h post induction. Syncytia are indicated with arrows.

Virus neutralization test

Purified CHIKV-E1ΔTM and -E2ΔTM (Figure 5) were used to generate polyclonal antisera in rabbits. To analyse the neutralizing activity of the polyclonal α-E1 and α-E2, the antisera were incubated with 100 $TCID_{50}$/ml CHIKV and subsequently incubated for 4 days with BHK-21 cells. Total suppression of cytopathic effect was used as a microscopic marker for virus neutralization. Rabbits vaccinated with CHIKV-E2ΔTM developed neutralizing antibody titers against CHIKV, up to a dilution of 1:40. No neutralizing antibodies were measured in sera of the animals vaccinated with CHIKV-E1ΔTM. The sera of CHIKV infected mice was used as a control and neutralized CHIKV up to 1:320 dilution. These results show that purified CHIKV-E2ΔTM, expressed by recombinant baculoviruses, is able to induce neutralizing antibodies in rabbits.

DISCUSSION

The continued spread of CHIKV worldwide, the associated threat for invasion of Europe and other Western countries, and the economic impact of a CHIKV epidemic highlights the demand for an effective antiviral therapy and/or vaccine. Several attempts have been undertaken to design such a vaccine and include the development of a life attenuated CHIKV strain and, more recently, a VLP-based approach [17]. VLPs have safety profiles very similar to those of subunits and immunogenic properties comparable to killed vaccines. It has been shown that there is great potential for the baculovirus expression system to generate effective (subunit) vaccines [21] and several vaccines produced in this system are on the market or in the process of registration. Furthermore, CHIKV is transmitted by mosquitoes and replicates to high viral titers in mosquitoes cells, thus it might be beneficial to express CHIKV proteins in insect cells [44]. In this research, vaccine candidates were generated consisting of individual CHIKV glycoproteins expressed in insect cells using recombinant baculoviruses.

CHIKV glycoprotein genes were cloned downstream of the polyhedrin promoter of baculovirus *Ac* MNPV to analyze individual expression of E1 and E2 with their native signal peptides and in the presence or absence of their respective C-terminal transmembrane domains. Results of SDS-PAGE and western blot analyses indicated that all proteins were expressed at high levels in *Sf* 21 cells and that a substantial fraction of these proteins was processed in a similar fashion as during a natural CHIKV infection. The glycosylation patterns and resulting size changes found for CHIKV E1 and E2 appear to correspond with the postulated number of glycosylation sites in E1 and E2 (1 and 2, respectively). Our findings are in agreement with results obtained in glycoprotein expression studies for other alphaviruses and for the related *Rubellavirus*, that used recombinant baculoviruses [26, 27, 45, 46].

We have shown that the glycoproteins travel through the ER, which is very sensitive to any homeostatic alterations and disturbances. Such ER stress is induced by protein misfolding, considerable protein overproduction and inhibition of N-linked glycosylation [47]. Generally, two baculovirus proteins cathepsin and chitinase are abundantly expressed and the latter tends to accumulate in the ER [48], thereby clogging up the ER and competing with recombinant proteins. Even though chitinase is deleted from the recombinant baculovirus backbone, it appears that the CHIKV glycoproteins are expressed in such massive amounts, that the unfolded protein response (UPR), which is normally induced in response to ER-stress [49], is not sufficient to mitigate ER stress, as concluded from the fact that a fraction of the 6KE1, 6KE1ΔTM, E3E2 and E3E2ΔTM appear in their unprocessed forms. Incomplete glycosylation and retention in the ER of intracellular viral glycoproteins expressed at very high levels in baculovirus-infected cells is common [50], but this did not compromise membrane localization and subunit secretion.

While glycosylation of recombinant CHIKV-E1 and -E2 was relatively efficient, expression analysis with the baculoviruses Ac-E3E2 and -E3E2ΔTM indicated that only a small fraction was cleaved by cellular furin. This, however, did not prevent secretion of the uncleaved, glycosylated E3E2 precursor as observed in the furin inhibition experiment and during metal-affinity chromatographay purification of secreted E2 subunits. We can therefore conclude that not furin activity, but rather the accumulation and retention of non-glycosylated E3E2 precursors in the ER, is limiting subunit secretion. This phenomenon has been well described for high level expression in insect cells of CSFV-E2, a similar glycoprotein from a different virus [50].

Although E3 plays a major role during CHIKV replication in the formation of E1 and E2 heterodimers [10], and the presence of uncleaved E3E2 in progeny alphavirus particles induces defects in virus production [51], it is not known if the presence of E3 has negative effects on the stability, functionality or antigenicity of E2 subunits. Considering that alphavirus virions with incorporated uncleaved E3E2 are able to bind efficiently to the cell surface, which is modulated by E2 [52], we expect that E3E2 is still sufficiently immunogenic. In fact, a recent report has shown that E3 also harbours protective epitopes [53]. Future studies might further explicate the effect of glycosylation on subunit antigenicity and epitope presentation.

Confocal microscopy showed that E1 and E2 are present at the surface of *Sf* 21 cells. Since cells are able to secrete uncleaved, but glycosylated E3E2ΔTM, it is likely that cells expressing E3E2 display the protein in its glycosylated configuration on the surface of the cell. The reason that displayed or secreted proteins are all glycosylated, might be explained simply by the fact that glycosylation occurs prior to entering of the secretory pathway. Cells infected with Ac-6KE1 were able to form syncytia, indicating that E1 protein displayed at the surface was able to induce membrane fusion. Subjecting Ac-6KE1, Ac-E3E2 and Ac-GFP infected *Sf* 9-ET cells to a range of acidified Sf900 II medium (pH = 5.0, 5.5, 5.8 and 6.4) excluded the possibility that syncytia formation was induced at pH 5.8 by the baculovirus fusion protein GP64, which is also expressed at the surface of infected cells [42]. The pH of Sf900 II medium was in the same range as it would be during endocytosis (pH5.5-pH5.8). Surprisingly, fusogenic activity was also shown at a pH = 6.4, which is expected to be slightly out of the fusogenic range of E1. This phenomenon was most likely caused by the addition of cholesterol to the culture medium. Cholesterol is known to be an important factor in alphavirus fusion [40], and a supplement activates CHIKV-E1 to be fusogenic. Whether its fusogenicity means that E1 is present in the form of homotrimers on the surface of the cell needs experimental confirmation.

The virus neutralization test shows that the rabbit polyclonal antiserum elicited against purified CHIKV-E2ΔTM is able to neutralize CHIKV. This is the first time that a secreted form of CHIKV-E2, expressed by the recombinant baculovirus-insect cell expression system elicits neutralizing antibodies in rabbits. No neutralizing antibodies were detected in the serum of E1ΔTM vaccinated rabbits, which is not entirely unexpected since E1 is partially covered by E2 in mature virions [54]. This is clear proof for antigenicity of the E2 subunit vaccine candidate, which will now be further studied in vaccination trials in an animal model. The fact that expression of CHIKV-glycoproteins in insect cells results in correct processing similar to the processing found during wildtype infections, proposes it to be a very useful and promising expression system in the generation and development of alphavirus subunit vaccines. In addition, proteins appear to retain their original function, paving the way for recombinant baculoviruses to be used in functionality studies in insect cells.

CONCLUSIONS

In conclusion, this study has shown that expression of full-length CHIKV 6KE1 and E3E2 glycoproteins using baculovirus vectors in insect cells leads to glycosylation, furin processing, plasma membrane translocation of E1 and E2, and CHIKV-E1 retains its functional activity as a membrane fusion protein. In addition, the deletion of the C-terminal transmembrane domain enables secretion of glycoprotein E1 and E2 independent of furin cleavage. The purified, glycosylated and secreted CHIKV E2 subunit induced neutralizing antibodies in rabbits and can now be tested for its ability to provide protection against CHIKV challenge in an animal model.

METHODS

Cells and viruses

Spodoptera frugiperda (*Sf* 21) insect cells (Invitrogen) were maintained in tissue culture flasks (Greiner) as a monolayer culture in Grace's insect medium (Invitrogen), supplemented with 10% fetal bovine serum (FBS, Hyclone). *Sf* 9-easy titer (ET) cells [41] were maintained as a monolayer culture in Sf900 II (Invitrogen) serum free medium. The Bac-to-Bac baculovirus expression system (Invitrogen) based upon *Autographa californica*multicapsid nucleopolyhedrovirus (AcMNPV) was used to generate recombinant baculoviruses. The bacmid backbone AcΔcc contained a deletion of the promoters and large parts of the ORFs of cathepsin and chitinase, thereby leaving flanking essential genes intact [55]. Furthermore, the p10 promoter and ORF were deleted from the AcΔcc backbone by replacing these elements with a zeocin resistance marker by Lambda Red recombination [56, 57]. This zeocin resistance marker was flanked by modified loxP sequences as described by [58] for bacterial genome modifications. This allowed the subsequent removal of the marker gene with Cre recombinase.

Baculovirus titers were determined by end point dilution assay and expressed as tissue culture infectious dose 50 ($TCID_{50}$) per ml. Viral RNA of Chikungunya virus isolate S27 (Tanzania, 1953) provided by Erasmus Medical Center (EMC), Rotterdam, was used as a source for cDNA synthesis.

CHIKV-S27 virus stocks were produced in BHK-21 cells in a BSL3 laboratory. Input virus was removed by extensive washing, and supernatant containing infectious virus was harvested four days post infection (pi) and filtered. Virus titration assays were performed on BHK-21 cell monolayers in 96 wells plate and titer was calculated using the Karber method. Virus stock titers were expressed as $TCID_{50}$ per ml. CHIKV stocks containing 10^7 $TCID_{50}$ /ml were aliquoted and stored at -80°C.

Construction of recombinant baculoviruses encoding CHIKV glycoproteins

Amplicons of CHIKV 6KE1 and E3E2 were generated using SuperScript One-Step RT-PCR with Platinum Taq (Invitrogen), cloned in pGEM-Teasy vectors (Promega) and sequenced (Eurofins Operon, Germany). CHIKV cDNA sequences encoding 6KE1, 6KE1ΔTM, E3E2 and E3E2ΔTM (Figure 1) were PCR amplified from these plasmids using extended primers (6K-F; ggggacaagtttgtacaaaaaagcaggctta ggatccaccatggccacataccaagag gctgc, E3-F; ggggacaagtttgtacaaaaaagcaggctta ggatccaccatgagtcttgccat cccagttatg, E1-R; ggggaccactttgtacaagaaagctgggta aagcttctaatgatgatg atgatgatgcatgtgcctgctgaacgacacgc, E1ΔTM-R; ggggaccactttgtacaagaa agctggg taaagcttctaatgatgatgatgatgatgcatccatgacatcgccgtagcgg, E2-R;ggggaccacttt gtacaagaaagctgggta aagcttctaatgatgatgatgatgatgctgcag cgctttagctgttctgatgc and E2ΔTM-R; ggggaccactttgtacaagaaagctgggta aagcttctaatgatgatgatgatgatgctgcagcagctcataataatacagaa) that introduce AttB recombination sites (underlined) to enable Gateway® cloning (Invitrogen). Resulting PCR products were cloned into pDONR207 donor plasmid (Invitrogen) and subsequently transferred to the pFastBac1 derivative pDEST8 (Invitrogen). Recombinant baculoviruses encoding the CHIKV protein constructs were generated using the Bac-to-Bac baculovirus expression system (Invitrogen), resulting in Ac-6KE1, Ac-6KE1ΔTM, Ac-E3E2 and Ac-E3E2ΔTM.

Protein analysis

For analysis of protein expression, 6 × 10^6 *Sf* 21 cells were seeded into 75 cm^2 culture flasks. Cells were infected with recombinant virus at a multiplicity of infection (MOI) of 10 $TCID_{50}$ units per cell. Cells were harvested 72 hpi and washed twice in 1 ml phosphate buffered saline

(PBS). Finally, cells were resuspended in 500μl PBS and stored at -20°C. Whole cell lysates were analyzed by sodium dodecyl sulphate polyacrylamide gel electrophoresis (SDS-PAGE) and Coomassie brilliant blue staining. A lysate of CHIKV-infected Ap61 mosquito cells was used as positive control. Proteins were transferred to an immobilon transfer membrane (Millipore) for Western blot (WB) analyses. Membranes were blocked in PBS + 0.1% Tween-60 (PBST) containing 3% skim milk for 1 h at 37°C. Next, the membranes were washed with PBST and incubated with rabbit polyclonal antiserum (supplied by Nobilon International BV) at a dilution of 1:15,000 for 1 h at 37°C. After incubation, membranes were washed and treated with 1:3,000 diluted Alkaline Phosphatase (AP) conjugated goat anti-rabbit IgG antibodies (Sigma) for 45 min at 37°C. Proteins were detected by NBT/BCIP staining (Roche).

Tunicamycin assay

A tunicamycin assay was performed to analyze the glycosylation status of recombinant CHIKV E1 and E2. *Sf* 21 cells were infected with recombinant baculoviruses at an MOI of 10 $TCID_{50}$ units per cell. Cells were incubated at 27°C in the presence of 10 μg/ml tunicamycin (Sigma) and harvested 72 hpi. Protein glycosylation was visualized using Western blotting and Periodic acid Schiff staining. Proteins were blotted on an immobilon transfer membrane (Millipore) and soaked in PAS solution (1% perjodium acid in 3% acetic acid) for 15 min at room temperature (RT). The membrane was washed several times in water and was incubated in the dark for 15 minutes at RT with Schiff›s reagent. Finally, the membrane was washed for 5 min in 50% Na_2CO_3.

Furin inhibition assay

Sf 21 cells were infected with Ac-E3E2 and E3E2ΔTM at an MOI of 10 in Grace's insect medium (Invitrogen) without FBS. Cells were washed twice with medium and incubated with medium containing 50 μM of Furin Inhibitor I (Calbiochem). Cells were harvested at 72 hpi and the medium fraction was separated from the cell fraction by centrifugation. Protein processing was analyzed by Western blotting.

Immunostaining of baculovirus infected *Sf* 21 cells

Sf 21 cells were infected with Ac-E3E2 and Ac-6KE1 at a MOI of 10 to determine surface expression of CHIKV-E1 and -E2. Cells were harvested 48 hpi and washed with PBS. Next, the cells were incubated for 1 h at RT with PBS+5%FBS containing 1:5,000 diluted rabbit α-CHIKV-E1 and rabbit α-CHIKV-E2 polyclonal antibodies. Cells were washed 3 times and incubated for 1 h at RT with PBS+5%FBS containing 1:1,000 diluted goat-anti-rabbit-Alexa Fluor 488 (Invitrogen). Cells were analyzed using laser scanning confocal microscopy on a Zeiss LSM 510 Meta, Axiovert 100 m (Zeiss) with Argon laser (488 nm) and images were analyzed with Zeiss LSM image browser.

Syncytium formation assay

The syncytium formation assay was performed by infecting *Sf* 9-easy titer (ET) cells [41] with Ac-E3E2,Ac-6KE1 at an MOI of 10 in Sf900 II medium (pH = 6.4), supplemented with cholesterol (0.2 mg/ml, Sigma). A recombinant AcMNPV expressing GFP (Ac-GFP) was used as a negative control. Syncytium formation was induced 72 hpi, by subjecting infected cells for 2 minutes to acidified medium with pH = 5.8, pH = 5.5 and pH = 5.0. Syncytium formation was scored 4 h post induction using fluorescence microscopy.

Purification of secreted CHIKV-E1 and -E2 subunits from insect cells

Sf 21 cells were infected with Ac-6KE1ΔTM and Ac-E3E2ΔTM at an MOI of 10 in Grace's insect medium without FBS and were incubated for 72 h at 27°C. The recombinant proteins were purified from the medium fraction using Talon® spin columns (Clontech) according to the manufacturer›s protocol. Bound protein fractions were eluted with 150 mM imidazol and total protein content was determined via a Bradford protein assay (Biorad) according to the manufacturer›s procedure.

Virus neutralization test

Virus neutralizing antibody titers in sera of rabbits vaccinated with purified CHIKV-E1ΔTM and E2ΔTM, were determined as follows. Serial two-fold dilutions of heat-inactivated rabbit sera were prepared in triplicate in 96-wells plate and 100 $TCID_{50}$ of CHIKV suspension was added to each well. After one hour of incubation at 37°C, 1×10^4 BHK-21 cells were added to each well and plates were incubated for another four days. Neutralizing titers were determined microscopically and expressed as the reciprocal of the highest serum dilution still giving 100% suppression of cytopathic effect.

ACKNOWLEDGEMENTS

Han van den Bosch (Nobilon International BV) and Albert Osterhaus (EMC) are acknowledged for initiating the project, their continued interest and advice. Dowty Movita constructed the AcMNPV deletion bacmid and Leon Oosterik helped with cloning of the HBM constructs. We would like to thank Lizzy Buijs for generating polyclonal antibodies against CHIKV-E1 and -E2. This work was performed within the framework of Dutch Top Institute Pharma, project "Development of recombinant live and subunit vaccines against Chikungunya virus infections".

REFERENCES

1. Powers AM, Logue CH: Changing patterns of chikungunya virus: re-emergence of a zoonotic arbovirus. *Journal of General Virology* 2007, 88: 2363-2377. 10.1099/vir.0.82858-0
2. Enserink M: Infectious diseases. Chikungunya: no longer a third world disease. *Science* 2007, 318: 1860-1861. 10.1126/science.318.5858.1860
3. Gould E, Gallian P, De Lamballerie X, Charrel R: First cases of autochthonous dengue fever and chikungunya fever in France: from bad dream to reality! *Clinical Microbiology and Infection* 2010, 12: 1702-1704.
4. Tsetsarkin KA, Vanlandingham DL, McGee CE, Higgs S: A single mutation in chikungunya virus affects vector specificity and epidemic potential. *PLoS Pathog* 2007, 3: 1895-1906.
5. Rezza G, Nicoletti L, Angelini R, Romi R, Finarelli AC, Panning M, Cordioli P, Fortuna C, Boros S, Magurano F: Infection with chikungunya virus in Italy: an outbreak in a temperate region. *The Lancet* 2007, 370: 1840-1846. 10.1016/S0140-6736(07)61779-6
6. Khan AH, Morita K, Parquet MC, Hasebe F, Mathenge EGM,

Igarashi A: Complete nucleotide sequence of chikungunya virus and evidence for an internal polyadenylation site. *Journal of General Virology* 2002, 83: 3075.

7. Strauss JH, Strauss EG: The alphaviruses: gene expression, replication, and evolution. *Microbiology and Molecular Biology Reviews* 1994, 58: 491-562.
8. Simmons DT, Strauss JH: Replication of Sindbis virus. I. Relative size and genetic content of 26 s and 49 s RNA. *J Mol Biol* 1972, 71: 599-613. 10.1016/S0022-2836(72)80026-3
9. Schlesinger S, Schlesinger MJ: Togaviridae: the viruses and their replication. *Fields virology* 2001, 1: 895-916.
10. Kuhn RJ: Togaviridae: the viruses and their replication. *Philadelphia, PA: Lippincott, Williams and Wilkins* 2007, 1001-1022.
11. Zhang X, Fugere M, Day R, Kielian M: Furin processing and proteolytic activation of Semliki Forest virus. *J Virol* 2003, 77: 2981-2989. 10.1128/JVI.77.5.2981-2989.2003
12. Solignat M, Gay B, Higgs S, Briant L, Devaux C: Replication cycle of chikungunya: a re-emerging arbovirus. *Virology*2009, 393: 183-197. 10.1016/j.virol.2009.07.024
13. Lidbury BA, Rulli NE, Suhrbier A, Smith PN, McColl SR, Cunningham AL, Tarkowski A, van Rooijen N, Fraser RJ, Mahalingam S: Macrophage-derived proinflammatory factors contribute to the development of arthritis and myositis after infection with an arthrogenic alphavirus. *J Infect Dis* 2008, 197: 1585-1593. 10.1086/587841
14. Labadie K, Larcher T, Joubert C, Mannioui A, Delache B, Brochard P, Guigand L, Dubreil L, Lebon P, Verrier B, *et al.*: Chikungunya disease in nonhuman primates involves long-term viral persistence in macrophages. *J Clin Invest* 2010, 120: 894-906. 10.1172/JCI40104
15. Strauss EG, Stec DS, Schmaljohn AL, Strauss JH: Identification of antigenically important domains in the glycoproteins of Sindbis virus by analysis of antibody escape variants. *J Virol* 1991, 65: 4654-4664.
16. Vrati S, Fernon CA, Dalgarno L, Weir RC: Location of a major antigenic site involved in Ross River virus neutralization. *Virology* 1988, 162: 346-353. 10.1016/0042-6822(88)90474-6

17. Akahata W, Yang ZY, Andersen H, Sun S, Holdaway HA, Kong WP, Lewis MG, Higgs S, Rossmann MG, Rao S, Nabel GJ: A virus-like particle vaccine for epidemic Chikungunya virus protects nonhuman primates against infection. *Nat Med*2010, 16: 334-338. 10.1038/nm.2105

18. Bouma A, de Smit AJ, de Kluijver EP, Terpstra C, Moormann RJ: Efficacy and stability of a subunit vaccine based on glycoprotein E2 of classical swine fever virus. *Vet Microbiol* 1999, 66: 101-114. 10.1016/S0378-1135(99)00003-6

19. Cox MM, Hollister JR: FluBlok, a next generation influenza vaccine manufactured in insect cells. *Biologicals* 2009, 37:182-189. 10.1016/j.biologicals.2009.02.014

20. Kost TA, Condreay JP, Jarvis DL: Baculovirus as versatile vectors for protein expression in insect and mammalian cells. *Nat Biotechnol* 2005, 23: 567-575. 10.1038/nbt1095

21. van Oers MM: Vaccines for viral and parasitic diseases produced with baculovirus vectors. *Adv Virus Res* 2006, 68:193-253.

22. Paavonen J, Lehtinen M: Introducing human papillomavirus vaccines - questions remain. *Ann Med* 2008, 40: 162-166. 10.1080/07853890701802404

23. Smith GE, Vlak JM, Summers MD: Physical Analysis of Autographa californica Nuclear Polyhedrosis Virus Transcripts for Polyhedrin and 10,000-Molecular-Weight Protein. *J Virol* 1983, 45: 215-225.

24. Vialard JE, Arif BM, Richardson CD: Introduction to the molecular biology of baculoviruses. *Methods Mol Biol* 1995, 39:1-24.

25. Voss JE, Vaney MC, Duquerroy S, Vonrhein C, Girard-Blanc C, Crublet E, Thompson A, Bricogne G, Rey FA: Glycoprotein organization of Chikungunya virus particles revealed by X-ray crystallography. *Nature* 2010, 468: 709-712. 10.1038/nature09555

26. Oker-Blom C, Summers M: Expression of Sindbis virus 26S cDNA in Spodoptera frugiperda (Sf9) cells, using a baculovirus expression vector. *Journal of virology* 1989, 63: 1256-1264.

27. Hodgson LA, Ludwig GV, Smith JF: Expression, processing, and immunogenicity of the structural proteins of Venezuelan equine encephalitis virus from recombinant baculovirus vectors. *Vaccine* 1999, 17: 1151-1160. 10.1016/S0264-410X(98)00335-1

28. Burke D, Keegstra K: Carbohydrate structure of Sindbis virus glycoprotein E2 from virus grown in hamster and chicken cells. *J Virol* 1979, 29: 546-554.

29. Rice CM, Strauss JH: Nucleotide sequence of the 26S mRNA of Sindbis virus and deduced sequence of the encoded virus structural proteins. *Proc Natl Acad Sci USA* 1981, 78: 2062-2066. 10.1073/pnas.78.4.2062

30. Knight RL, Schultz KL, Kent RJ, Venkatesan M, Griffin DE: Role of N-linked glycosylation for sindbis virus infection and replication in vertebrate and invertebrate systems. *J Virol* 2009, 83: 5640-5647. 10.1128/JVI.02427-08

31. Simizu B, Yamamoto K, Hashimoto K, Ogata T: Structural proteins of Chikungunya virus. *J Virol* 1984, 51: 254-258.

32. Blom N, Sicheritz-Pontén T, Gupta R, Gammeltoft S, Brunak S: Prediction of post-translational glycosylation and phosphorylation of proteins from the amino acid sequence. *Proteomics* 2004, 4: 1633-1649. 10.1002/pmic.200300771

33. Kapitany RA, Zebrowski EJ: A high resolution PAS stain for polyacrylamide gel electrophoresis. *Analytical biochemistry*1973, 56: 361-369. 10.1016/0003-2697(73)90202-9

34. Monsma SA, Oomens AG, Blissard GW: The GP64 envelope fusion protein is an essential baculovirus protein required for cell-to-cell transmission of infection. *J Virol* 1996, 70: 4607-4616.

35. Pijlman GP, Roode EC, Fan X, Roberts LO, Belsham GJ, Vlak JM, van Oers MM: Stabilized baculovirus vector expressing a heterologous gene and GP64 from a single bicistronic transcript. *J Biotechnol* 2006, 123: 13-21. 10.1016/j.jbiotec.2005.10.022

36. Jarvis D, Wills L, Burow G, Bohlmeyer D: Mutational analysis of the N-linked glycans on Autographa californica nucleopolyhedrovirus gp64. *Journal of virology* 1998, 72: 9459-9469.

37. Tessier DC, Thomas DY, Khouri HE, Laliberte F, Vernet T: Enhanced secretion from insect cells of a foreign protein fused to the honeybee melittin signal peptide. *Gene* 1991, 98: 177-183. 10.1016/0378-1119(91)90171-7

38. Kaba SA, Hemmes JC, van Lent JW, Vlak JM, Nene V, Musoke AJ, van Oers MM: Baculovirus surface display of Theileria parva

p67 antigen preserves the conformation of sporozoite-neutralizing epitopes. *Protein Eng* 2003, 16: 73-78. 10.1093/proeng/gzg004

39. Wu SR, Haag L, Sjoberg M, Garoff H, Hammar L: The dynamic envelope of a fusion class II virus. E3 domain of glycoprotein E2 precursor in Semliki Forest virus provides a unique contact with the fusion protein E1. *J Biol Chem*2008, 283: 26452-26460. 10.1074/jbc.M801470200
40. Kielian MC, Helenius A: Role of cholesterol in fusion of Semliki Forest virus with membranes. *J Virol* 1984, 52: 281-283.
41. Hopkins R, Esposito D: A rapid method for titrating baculovirus stocks using the Sf-9 Easy Titer cell line.*Biotechniques* 2009, 47: 785-788. 10.2144/000113238
42. Volkman LE: The 64K envelope protein of budded Autographa californica nuclear polyhedrosis virus. *Curr Top Microbiol Immunol* 1986, 131: 103-118.
43. Long G, Pan X, Vlak JM: Absence of N-linked glycans from the F2 subunit of the major baculovirus envelope fusion protein F enhances fusogenicity. *J Gen Virol* 2007, 88: 441-449. 10.1099/vir.0.82418-0
44. Metz S, Pijlman G: Arbovirus vaccines; opportunities for the baculovirus-insect cell expression system. *Journal of Invertebrate Pathology* 2011, 107: S16-S30.
45. Oker-Blom C, Blomster M, Osterblad M, Schmidt M, Akerman K, Lindqvist C: Synthesis and processing of the rubella virus p110 polyprotein precursor in baculovirus-infected Spodoptera frugiperda cells. *Virus Res* 1995, 35: 71-79. 10.1016/0168-1702(94)00079-R
46. Cho B, Jeon BY, Kim J, Noh J, Kim J, Park M, Park S: Expression and Evaluation of Chikungunya Virus E1 and E2 Envelope Proteins for Serodiagnosis of Chikungunya Virus Infection. *Yonsei Med J* 2008, 49: 828-835. 10.3349/ymj.2008.49.5.828
47. Kaufman R: Stress signaling from the lumen of the endoplasmic reticulum: coordination of gene transcriptional and translational controls. *Genes & development* 1999, 13: 1211-1233. 10.1101/gad.13.10.1211
48. Saville GP, Patmanidi AL, Possee RD, King LA: Deletion of the Autographa californica nucleopolyhedrovirus chitinase KDEL

motif and in vitro and in vivo analysis of the modified virus. *J Gen Virol* 2004, 85: 821-831. 10.1099/vir.0.19732-0

49. Rutkowski D, Kaufman R: A trip to the ER: coping with stress. *Trends in cell biology* 2004, 14: 20-28. 10.1016/j.tcb.2003.11.001
50. van Oers MM, Thomas AA, Moormann RJ, Vlak JM: Secretory pathway limits the enhanced expression of classical swine fever virus E2 glycoprotein in insect cells. *J Biotechnol* 2001, 86: 31-38. 10.1016/S0168-1656(00)00403-X
51. Ozden S, Lucas-Hourani M, Ceccaldi P, Basak A, Valentine M, Benjannet S, Hamelin J, Jacob Y, Mamchaoui K, Mouly V: Inhibition of Chikungunya Virus Infection in Cultured Human Muscle Cells by Furin Inhibitors. *Journal of Biological Chemistry* 2008, 283: 21899-21908. 10.1074/jbc.M802444200
52. Lobigs M, Zhao HX, Garoff H: Function of Semliki Forest virus E3 peptide in virus assembly: replacement of E3 with an artificial signal peptide abolishes spike heterodimerization and surface expression of E1. *J Virol* 1990, 64: 4346-4355.
53. Parker MD, Buckley MJ, Melanson VR, Glass PJ, Norwood D, Hart MK: Antibody to the E3 Glycoprotein Protects Mice Against Lethal Venezuelan Equine Encephalitis. *J Virol* 2010, 84: 12683-12690. 10.1128/JVI.01345-10
54. Li L, Jose J, Xiang Y, Kuhn RJ, Rossmann MG: Structural changes of envelope proteins during alphavirus fusion. *Nature*2010, 468: 705-708. 10.1038/nature09546
55. Kaba SA, Salcedo AM, Wafula PO, Vlak JM, van Oers MM: Development of a chitinase and v-cathepsin negative bacmid for improved integrity of secreted recombinant proteins. *J Virol Methods* 2004, 122: 113-118. 10.1016/j.jviromet.2004.07.006
56. Datsenko KA, Wanner BL: One-step inactivation of chromosomal genes in Escherichia coli K-12 using PCR products.*Proc Natl Acad Sci USA* 2000, 97: 6640-6645. 10.1073/pnas.120163297
57. Pijlman GP, Dortmans JC, Vermeesch AM, Yang K, Martens DE, Goldbach RW, Vlak JM: Pivotal role of the non-hr origin of DNA replication in the genesis of defective interfering baculoviruses. *J Virol* 2002, 76: 5605-5611. 10.1128/JVI.76.11.5605-5611.2002

58. Suzuki N, Nonaka H, Tsuge Y, Inui M, Yukawa H: New multiple-deletion method for the Corynebacterium glutamicum genome, using a mutant lox sequence. *Appl Environ Microbiol* 2005, 71: 8472-8480. 10.1128/AEM.71.12.8472-8480.2005

CHAPTER

16

EPIGENOMIC CONSEQUENCES OF IMMORTALIZED PLANT CELL SUSPENSION CULTURE

Milos Tanurdzic[1], Matthew W. Vaughn[1], Hongmei Jiang[2], Tae-Jin Lee[3], R. Keith Slotkin[1], Bryon Sosinski[3], William F. Thompson[4], R. W. Doerge[5], Robert A. Martienssen[1]

[1]Cold Spring Harbor Laboratory, Cold Spring Harbor, New York, United States of America,

[2]Department of Statistics, Northwestern University, Evanston, Illinois, United States of America,

[3]Department of Horticultural Science, North Carolina State University, Raleigh, North Carolina, United States of America,

[4]Departments of Plant Biology, Genetics, and Crop Science, North Carolina State University, Raleigh, North Carolina, United States of America,

[5]Departments of Statistics and Agronomy, Purdue University, West Lafayette, Indiana, United States of America

ABSTRACT

Plant cells grown in culture exhibit genetic and epigenetic instability. Using a combination of chromatin immunoprecipitation and DNA methylation profiling on tiling microarrays, we have mapped the location and abundance of histone and DNA modifications in a continuously proliferating, dedifferentiated cell suspension culture of *Arabidopsis*. We have found that euchromatin becomes hypermethylated in culture and that a small percentage of the hypermethylated genes become associated with heterochromatic marks. In contrast, the heterochromatin undergoes dramatic and very precise DNA hypomethylation with transcriptional activation of specific transposable elements (TEs) in culture. High throughput sequencing of small interfering RNA (siRNA) revealed that TEs activated in culture have increased levels of 21-nucleotide (nt) siRNA, sometimes at the expense of the 24-nt siRNA class. In contrast, TEs that remain silent, which match the predominant 24-nt siRNA class, do not change significantly in their siRNA profiles. These results implicate RNA interference and chromatin modification in epigenetic restructuring of the genome following the activation of TEs in immortalized cell culture.

AUTHOR SUMMARY

Cultured plant cells are genetically and epigenetically unstable. We investigated the epigenomic consequences of long-term plant cell culture and found that some genes show an increase in DNA methylation, reminiscent of immortalized animal cell lines and cancer cells. By contrast, in the heterochromatic portion of the genome, some transposable elements (TE), undergo dramatic and very precise loss of DNA methylation and transcriptional activation. The reactivated TEs in culture are also accompanied by a production of 21-nucleotide (nt) small RNAs. In contrast, TEs that remain methylated and silent retain the predominant class of 24-nt small RNAs, and do not change significantly in their small RNA profiles. Our results implicate RNA interference in epigenetic restructuring of the genome following the activation of TEs in immortalized cell culture.

INTRODUCTION

More than half a century has passed since the concept and practice of plant cell culture was first introduced [1]. Unlike most animal cells, plant cells can change from one differentiated state, representing a committed developmental program, to a completely different one via a transition through a dedifferentiated state typical of callus tissue [2]. This process is achieved by varying concentrations and relative proportions of two major plant growth regulators (auxin and cytokinin) in the growth medium [3]. Under appropriate conditions, callus cells can continue to grow in "immortalized" suspension cultures, which can be maintained indefinitely without differentiation. Plant cells grown in culture exhibit unprecedented levels of genetic and epigenetic instability [1]. Through changes in gene activity, plant cells are able to respond to the challenges presented by tissue culture conditions and continue to divide according to internal and external cues. Epigenetic regulation plays an important role. For example, hormone habituation is a process during which plant cells in culture shift their requirement for exogenous growth regulators between the auxo- and autotrophic states [4]. The epigenetic nature of this shift is demonstrated by the fact that upon plant regeneration from habituated cell culture and reculturing, the original requirement for growth factors is present again in the reestablished cell culture. In addition to the reversibility, this switch occurs at rates that are orders of magnitude higher than mutations rates in plants [5].

The culturing of plant cells induces heritable variation known as somaclonal variation. It has been observed in numerous instances for a variety of plant species, and given the early promise of cell culture for advances in agriculture, the stochastic nature of somaclonal variation poses a significant challenge [6,7]. Genetic analyses of somaclonal variants in maize, rice, tobacco, and other plant species identified novel transposon insertions into genes as an important cause of somaclonal variation [8–14]. Additionally, heritable epialleles are known to occur in plants [15–18], and there is recent evidence that meiotically heritable epialleles with high reversion rates can be induced in plant cell culture [19]. Plant cells in culture undergo changes in both DNA methylation of the repetitive portions of plant genomes [20–24], as well as at individual TEs [25,26]. Reactivation of retrotransposon *Athila*, a highly repetitive component of the pericentromeric heterochromatin in Arabidopsis thaliana [27], was also reported in a cell suspension culture of continuously dividing dedifferentiated *Arabidopsis* cells [28,29]. TEs can serve as insertional mutagens of plant genomes, whereas widespread activation can lead to a wide array of chromosomal rearrangements. These rearrangements can, in turn, lead to misregulation of genes and transposons [1].

Plant genomes contain numerous and diverse TEs [30,31], most of which are transcriptionally silent [27]. In Arabidopsis thaliana, and in other plants, silencing is achieved by both DNA methylation as well as by histone deacetylation and H3 lysine 9 dimethylation, which is targeted by small RNAs (sRNAs) in the 21–24-nucleotide (nt) range [32–34]. In addition, this mutually enforcing set of chromatin modifications allows for the silenced states of TEs to be inherited throughout normal cell divisions and plant development. When silencing is perturbed by mutations in DNA methyltransferases, RNA interference (RNAi) machinery, or histone modifiers, TEs can become reactivated [32,35–38].

With the advent of chromatin immunoprecipitation (ChIP), microarray and deep sequencing technologies, it is now possible to map DNA and histone modifications genome-wide [39–41]. In animal systems, these approaches have uncovered histone modifications specific to promoters, such as histone H3 trimethylation at lysine4 (H3K4me3), to gene coding regions H3K4me2 and H3K36me3, to polycomb-regulated genes (H3K27me3), and to repeats and TEs (H3K9me3 and H4K20me3) [42–44]. This epigenomic information has been used to describe chromatin states specific for different stages of differentiation from pluripotent embryonic stem cells, as well as in cancer cell lines [45–48].

Comparable chromosome-wide and genome-wide epigenomic profiles in plants provided the first glimpses into the plant epigenome. DNA methylation in *Arabidopsis* has a bimorphic distribution: dense cytosine methylation in all three sequence contexts (CpG, CpNpG, and CpNpN) characterizes TEs and other repetitive DNA, whereas short clusters of CpG methylation are found in over 25% of gene coding regions [32,49–51]. The role of euchromatic, genic DNA methylation is not known, but it is extremely variable between accessions of Arabidopsis thaliana and, as a rule, is not accompanied by heterochromatic histone H3K9 dimethylation or the presence of sRNAs [51]. Additionally, the coding regions of *Arabidopsis*genes are enriched in H3K4me2 and H3K27me3 [32,52,53]. In contrast, heterochromatic transposon DNA methylation is strongly correlated with histone H3K9me2 and associated small interfering RNA (siRNA) [32,54].

Beyond a role in transposon silencing, epigenetic modifications play a role in plant gene regulation and development. Gene imprinting in the endosperm [55] and the regulation of vernalization [56,57] are examples of epigenetic control in plant development that are regulated by DNA and histone methylation, respectively. However, epigenetic changes that occur in rapidly dividing dedifferentiated cells during plant cell culture have not yet been described.

Using a combination of ChIP and DNA methylation profiling on tiling microarrays, and high-throughput sRNA sequencing, we have mapped the location and abundance of histone and DNA modifications in rapidly dividing, dedifferentiated *Arabidopsis* suspension cells, and compared them with gene expression in suspension cells and in callus tissue. We have found dramatic DNA hypomethylation and activation of specific TEs, as well as hypermethylation in genic regions that may be responsible for the emergence of epialleles. DNA methyltransferase and most other epigenetic modifiers are up-regulated in cell culture, indicating that loss of targeting rather than modification is responsible for TE activation. Gain of 21-nt siRNA and associated expression of the *Argonaute2* gene family member may accompany this loss of targeting.

RESULTS

Chromatin Maps of Plant Cell Suspension Culture

Chromosome-wide maps of DNA and histone modifications were created by microarray hybridization to a custom *Arabidopsis* chromosome 4 tiling array [51]. The 22,761 probes interrogate most of *Arabidopsis* chromosome 4, including the repetitive pericentromeric heterochromatin, as well as the heterochromatic knob on the short arm of chromosome 4 [39]. We profiled covalent modifications of DNA and histones as well as sRNA populations in cultured suspension cells of the Columbia ecotype that have been in continuous propagation for several years (I. Meier, unpublished data). The cells were cultured in 1× B5 medium with minimal organics supplemented with 3% sucrose and 1.1 mg/l 2,4-D with a 1:1 and 1:9 final dilution for the 16-h (cells7) and the 96-h (cells4) culture regimens, respectively, every 7 d. Flow sorting of DAPI-stained nuclei at 96 h postculture split showed that a vast majority of these cells were diploid and in the G1 stage of the cell cycle. In addition, BrdU incorporation results showed that the cells7 sample had close to three times higher BrdU incorporation relative to the cells4 sample (Figure S1). We also tested the ability of this long-term cell suspension culture to redifferentiate into shoots by growing microcalli from the suspension culture on solid shoot-inducing medium. In contrast to freshly induced calli, which regenerated shoots after 6 wk under the same conditions, cell suspension culture-derived calli continued to proliferate without any differentiation or shoot regeneration.

Cytosine methylation was determined by comparing McrBC-digested genomic DNA to an undigested sample of the same DNA by two-color hybridization to the chromosome 4 tiling array [32]. The presence of 5mC was detected as a statistically significant difference in hybridization signal in the untreated sample relative to the McrBC-treated sample, using a linear model that partitions the variation into both biological and technical sources. Samples from two different time points during the cell suspension culture cycle of 7 d (cells4 and cells7, as noted above) were analyzed in quadruplicate. Upon testing for differences between the two DNA methylation profiles, almost no change in DNA methylation (false discovery rate [FDR] significance level $\alpha = 0.01$) occurs during each culture passage. Subsequent statistical analyscs bcnefited from combining these two experiments, thus treating the two DNA samples as biological replicates (Figure 1).

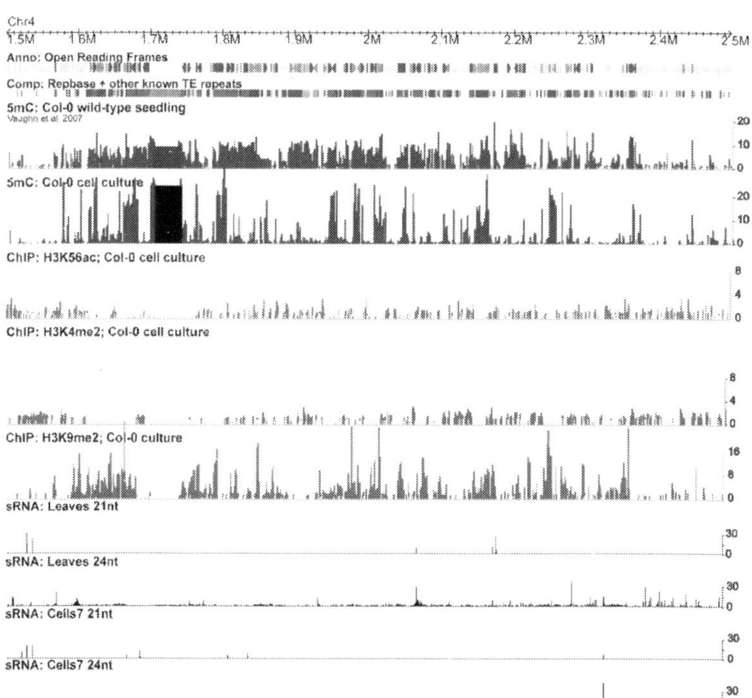

Figure 1: Heterochromatic Modification of Chromosome 4 in *Arabidopsis* Cell Suspension Cultures

One megabase encompassing the heterochromatic knob on the short arm of chromosome 4 and the flanking euchromatic region are displayed. Genes are depicted in yellow, DNA transposons are in red and orange, and retrotransposons are in green in the annotation (Anno) and Repbase tracks (Comp). DNA methylation profiles of cell suspension culture and the reference Col seedling DNA methylation profile [51] are presented as red bars. Tiles with significant enrichment of H3K4me2 and H3K9me2 are shown in purple, while those tiles associated with H3K56ac are shown in green. A 100-bp–resolution sliding window analysis of sRNA frequencies, corrected for sRNA library size and genomic copy number, for 21- and 24-nt size fractions is shown.

ChIP-chip was performed by hybridization of DNA immunoprecipitated with antibodies against H3K4me2, H3K9me2, and histone H3 acetylated at lysine56 (H3K56ac) [58]. Each immunoprecipitated and amplified sample was hybridized in triplicate against the input DNA sample, using two-color array hybridization. Statistical analysis identified tiles with significant enrichment of a particular histone modification (Figure 1).

We found significant genome-wide hypomethylation of heterochromatic DNA in cell suspension culture (Figure 1) as compared to Columbia seedling DNA [51]. Significant differences in DNA methylation were also seen in the euchromatic portions of chromosome 4, in which a gain of methylation in cell suspension culture was the most prevalent trend, typically within the transcribed regions of genes ("gene-body methylation") where CpG methylation occurs.

Specific Families of TEs Lose Heterochromatic Modifications and Are Transcriptionally Activated in Cultured Cells

The chromatin maps of the heterochromatic portion of chromosome 4 showed that significant loss of DNA methylation occurred in cell suspension culture relative to differentiated leaf cells (Figure 1). Although very few regions lost DNA methylation completely, a majority of TEs that comprise heterochromatin became hypomethylated (Figure 1). Contrary to gene-body DNA methylation, which is correlated with the presence of H3K4me2, heterochromatic DNA methylation in *Arabidopsis* is associated with H3K9me2 and with the presence of sRNAs, which direct this histone modification. DNA hypomethylation, loss of H3K9me2, and in most cases, gain of H3K4me2, were specific to families of TEs. Most retrotransposons, including *Athila*, and *Copia* elements, as well as DNA transposons such as helitrons, *AtMu* elements, and a few *Vandal* and *AtENSPM* elements, lose heterochromatic marks irrespective of their location on the chromosome. In contrast, gypsy class *Atlantys* and *AtGP* long terminal repeat (LTR) retrotransposons and other DNA transposons such as most of the *Vandal* and *AtENSPM* elements retain heterochromatic marks (Figure S2). This observation is particularly striking when two repetitive elements are found juxtaposed with a sharp boundary between the loss and retention of heterochromatic marks corresponding to the boundaries of coding regions of the TEs (Figure 1).

Transcriptional reactivation of TEs was analyzed by microarray analysis using Affymetrix *Arabidopsis* ATH1 microarrays. Cells7 samples were compared with 2-wk-old seedlings grown on plates under standard conditions, and 6-wk-old *Arabidopsis* calli generated from cotyledons grown on MS plates supplemented with a combination of cytokinin and auxin at callus-inducing concentrations. ATH1 arrays, in addition to bona fide genic probes, have close to 1,200 probe sets specific to various *Arabidopsis* TEs, of which approximately 200 map to chromosome 4 (R. K. Slotkin and R. A. Martienssen, unpublished data). A majority of oligonucleotide probes on this array are specific to a single TE, or a small number of closely related TEs, which allowed for precise mapping of TE transcripts to their chromosomal location(s).

Weak expression of a few TEs was seen in callus, but a strong up-regulation of a much larger number of TEs was found in suspension cells, relative to both seedling and callus samples (Figure 2). A total of 142 (12%) out of 1,155 TE-specific probes on the array were up-regulated 1.5-fold or higher in cell suspension culture. Only 22 and 56 TEs (<2% and <5% of the total TE probes) were expressed in seedling or callus samples, respectively (Figure 2), none of which were in common with cell suspension samples. The transcriptional reactivation of selected TEs located on *Arabidopsis* chromosome 4 was confirmed by reverse-transcription PCR (RT-PCR) (Figure S2).

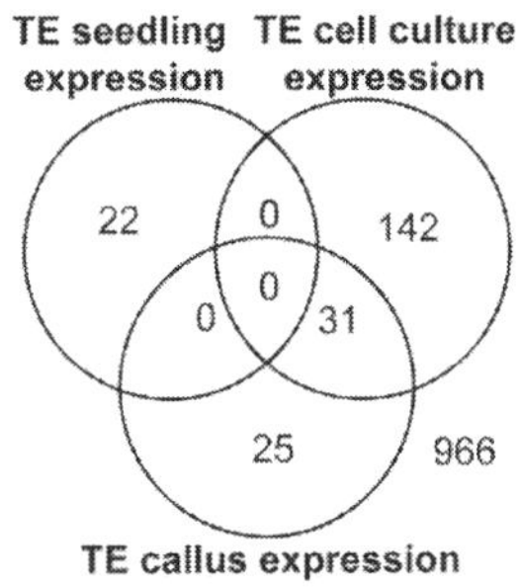

Figure 2: TE Expression Differences between Cell Suspension Culture, Seedling, and Callus Tissues

A Venn diagram of transcripts with significant TE homology that show significant overaccumulation (normalized value ≥ 1.5) in ATH1-based microarray expression profiles of cell suspension culture, seedling, and callus samples. There is no overlap between samples and a much higher degree of TE activity in cell suspension culture.

Small RNA Profiles in Suspension Cell Culture

We further investigated the specificity of epigenomic changes in *Arabidopsis* cell suspension culture heterochromatin by sequencing sRNA from the two cell suspension culture samples (cells7 and cells4), as well as from leaf and callus samples. In total, we sequenced 1,910,836 and 3,075,388 sRNAs matching the *Arabidopsis* Col-0 reference genome from the cells7 and cells4 libraries, respectively, as well as 1,737,719 and 2,070,923 sRNAs from the leaf and callus libraries, respectively. We tested the technical reproducibility by resequencing the cells4 library, which showed an excellent match in relative size frequency distribution (unpublished data).The overall size distribution of siRNAs from the four libraries showed a typical distribution with the relative frequency maxima at 21 nt and 24 nt, when known microRNAs (miRNAs) were removed from the libraries (Figure 3). Whereas the 24-nt class was the most abundant size class, an overall decrease in the relative frequency of 24-nt sRNAs was noted in the cell suspension culture library, cells4. This prompted further analysis of the sRNA data to identify genomic regions esponsible for this reduction.

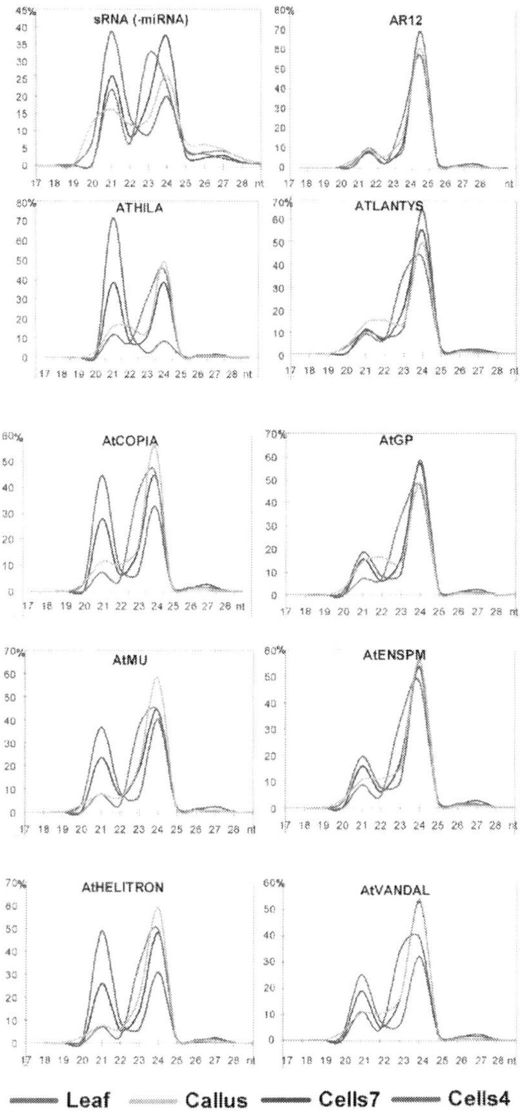

Figure 3. Size Distribution of sRNA Sequences in Cell Suspension Culture (Cells4 and Cells7), Col-0 Leaves, and Col-0 Callus

The relative frequencies of all sRNAs (with all known miRNA sequences removed) according to their size for each of the small RNA libraries, as well as only for those sRNAs matching particular families of *Arabidopsis* repeats (*AR12*, *Athila*, *Atlantys*, *Helitron*, *AtGP1*, *AtMu*, *AtENSPM*, *AtCopia*, and *AtVandal*).

Each sRNA sequence from the dataset was associated with a particular Repbase repeat class representing all the major families of TEs in the *Arabidopsis* genome, and we examined the size distribution of those classes of sRNAs (Figure 3). This analysis showed that when we compared the relative size frequency of sRNAs specific to particular TE families (Repbase repeat classes), two patterns emerged. The sRNAs specific to those TEs that lose heterochromatic marks and become transcriptionally reactivated in cell suspension culture (*Athila*, *AtCopia*, *AtMu*, and *Helitron* elements) showed a major shift in relative frequency from 24 nt to 21 nt in the cells4, but also in cells7 libraries, relative to sRNA profiles of the leaf and callus libraries (Figure 3). In contrast, the sRNAs specific to those TE families that do not lose the heterochromatic marks and remain transcriptionally silent in cell suspension culture (*Atlantys*, *AtGP*, *Vandal*, and *AtENSPM*) retain the relative sRNA size distribution found in the leaves and callus (Figure 3).

We also examined the spatial distribution of sRNAs in cell suspension culture by mapping sRNAs to representative full-length elements representing each of the TE classes listed above. Whereas the distribution and abundance of sRNAs matching a selection of *Atlantys2* (Figure 4), *AtGP1*, and *Vandal14* elements on chromosome 4 did not change significantly in the four sRNA libraries, a major shift was observed in both the size and location of sRNAs mapping to *Athila2* (Figure 4). Importantly, the overrepresented 21-nt *Athila* sRNAs (and the 24-nt sRNAs to a lesser extent) were localized to the coding regions of this element (Figure 4). In addition, the change in relative size frequencies of *Athila* sRNAs observed in cell suspension culture (Figure 3) was found to be due to massive up-regulation in the production of 21-nt sRNAs, rather than a net loss of 24-nt-size sRNAs (Figure 4). The same patterns were observed in four other full-length *Athila* elements located on chromosomes 1, 2, 3, and 5 (unpublished data). Similarly, a *Helitron2* element on chromosome 4 was found to gain 21-nt sRNAs specifically at the 3′ end of the coding region without a loss of 24-nt sRNAs. The only exception to this general trend was the small family of *AtMu* elements in which the increase in the relative frequency of 21-nt sRNAs was accompanied by a significant loss of the 24-nt sRNAs in cell suspension culture.

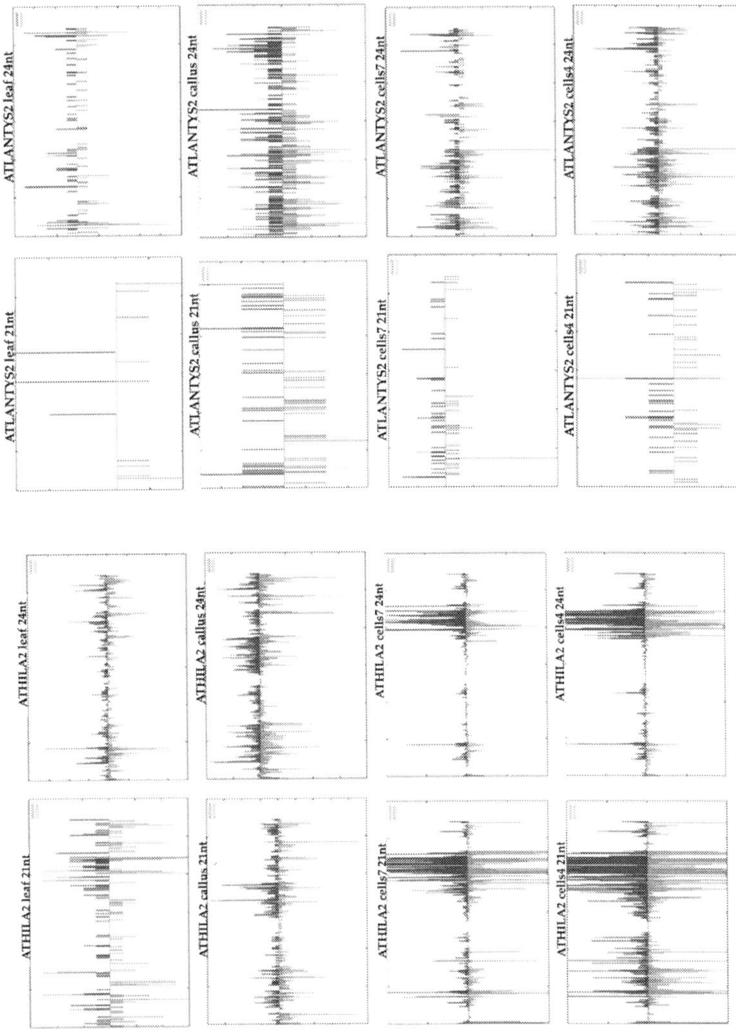

Figure 4. Spatial Distribution and Frequency of sRNA Sequences for *Athila* and *Atlantys* Retrotransposons in the Leaf, Callus, and the Cell Suspension Culture (Cells7 and Cells4) Libraries

A representative full-length *Athila2* and *Atlantys2* TE from chromosome 4 was used as a sequence-matching template to map 21-nt and 24-nt sRNAs with perfect matches, resulting in single-nucleotide coverage of the template sequences. The coverage was plotted on the *y*-axis, for both DNA strands (top strand in red and the bottom strand in green), ranging from 0 to 6, 20, 100, and 100 for *Athila2* 21-nt matches, respectively, and from 0 to 40, 40, 100, and 100 for the *Athila2* 24-nt matches; and from 0 to 3, 2, 4, and 4 for the *Atlantys2* 21-nt matches, and from 0 to 6, 6, 15, and 20 for *Atlantys2* 24-nt matches, respectively. The *x*-axis represents the template sequence in the 5′ to 3′ direction, ranging from 0 to 12,000 bp.

One way that TE families might be differentially regulated by sRNAs is through interaction with different *Argonaute* proteins, which in turn, is determined by the 5′ nucleotide and the sRNA size [59, 60]. We, therefore, compared the first nucleotide frequencies by size, origin, and TE specificity. Size, distribution, and 5′ nucleotide specificity were found to be unchanged in leaves and callus, with the majority of TE sRNAs having A as their 5′ base. However, in the cell suspension culture libraries, the frequency of 5′ A sRNAs is increased in every class of TE that we tested, irrespective of the sRNA size. In addition, sRNAs from TEs that lose heterochromatic silencing also have a decrease in the relative frequencies of sRNAs with 5′ G as their first nucleotide, irrespective of their size. sRNAs specific to *Atlantys*, which remains heterochromatic, have no such shift in 5′ nucleotide preference in cell suspension culture.

Genic Methylation Is Retained in Cell Culture and Is Correlated with Histone H3K4me2 Modification Whereas H3K56ac Marks the Proximal Promoter Regions

In contrast to heterochromatic DNA methylation, which is correlated with transcriptionally silent sequences, gene-body methylation tends to occur within genes that are transcriptionally active [49,51]. To test whether genic DNA methylation coincides with the sequences enriched for H3K4me2, a histone modification characteristic of transcribed regions, the chromosomal distribution of cytosine methylation was compared to H3K4me2 using an "average gene analysis" that utilized all genes represented by multiple tiles and lacking sequence homology to known repetitive elements. For each chromatin modification, the number of tiles detecting enrichment was calculated at 10% intervals relative to the length of each gene, with position 1 being the transcription start site. This was compared with the number expected if distribution of significant features was randomly distributed relative to gene position

(Figure 5A). These comparisons revealed that the distributions of 5mC and H3K4me2 are spatially correlated along gene coding regions, such that both epigenetic marks are depleted from the sequences flanking the gene coding regions (positions 0–1 and 9–10 in Figure 5A), whereas they are specifically enriched, relative to expected average chromosomal distribution of each mark, over the open reading frames of the gene coding regions (positions 1–9 in Figure 5A). In fact, 69% of genic tiles with significant DNA methylation in cell suspension culture are also associated with H3K4me2, compared with only 37% of genic tiles without methylation. This correlation is not unique to cell culture-derived chromatin, since a similar correspondence is observed for the distribution of DNA methylation data in seedling [51] and H3K4me2 methylation in leaf tissue (Figure 5A).

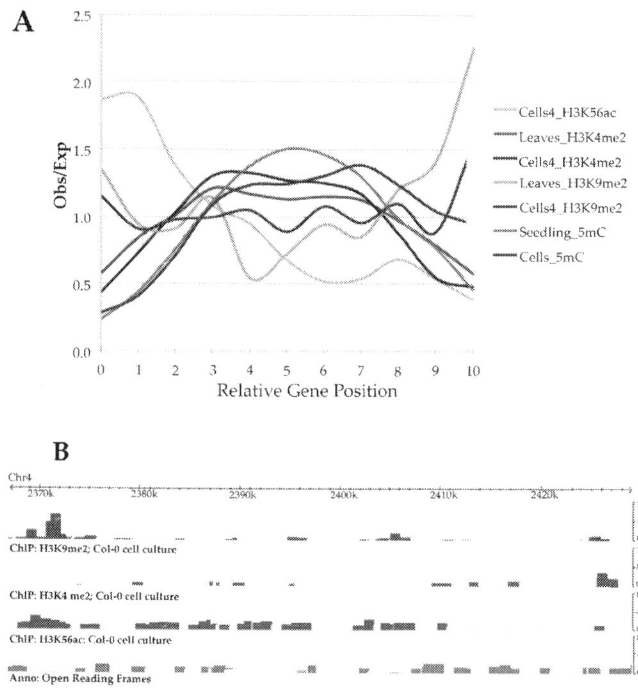

Figure 5: Euchromatic Modification of Chromosome 4 in *Arabidopsis* Cell Suspension Cultures

(A) The number of genic array tiles with no repeat homology detecting significant methylation, H3K56ac, H3K9me2, and H3K4me2 in suspension cell culture was calculated at 10% intervals relative to the length of each gene, and compared with the number expected if these modifications were randomly distributed. Parallel computations were done for leaf H3K4me2 and H3K9me2, and data for seedling DNA methylation were imported from previously published work [51].

(B) Chromatin maps of an approximately 87-kb region of euchromatin from the short arm of chromosome 4. Whereas genic DNA methylation patterns are similar between the cell suspension culture and seedlings [51], H3K9me2 is almost completely lacking from this region. Instead, gene coding regions are enriched in H3K4me2, and this pattern is interrupted by H3K56ac-enriched promoter regions preceding the coding regions in an intercalated pattern.

In contrast to H3K4me2, H3K9me2 is enriched in intergenic regions (Figure 5A) and is negatively correlated with the presence of gene-body DNA methylation. Instead, it is highly enriched in TEs [32], which are found predominantly in intergenic regions. Strikingly, our "average gene analysis" revealed that H3K56 acetylation was uniquely and strongly enriched 0–100 bp upstream of the coding region (Figure 5A). The association of H3K56ac with gene promoter regions is ubiquitous throughout the chromosome (Figure 5B) and does not seem to be restricted to expressed genes. This modification is thus a useful tool for gene-finding algorithms. It is also interesting to note that TEs that gain H3K4me2 are also associated with H3K56ac at their 5′ ends (Figure 1), indicating they may be actively transcribed.

Genes Undergo Hypermethylation in Cell Suspension Culture, Which Can Lead to Epiallele Formation

Both hypomethylation of transposon sequences and hypermethylation of euchromatic gene sequences can lead to the generation of epialleles at specific loci, which have the potential to heritably transmit the transcriptional state imposed by the methylation status of its sequences [61]. Since hypermethylation of the euchromatic portion of chromosome 4 was observed in cell suspension culture relative to seedling DNA methylation [51], we looked at the distribution of this novel DNA methylation with respect to gene coding regions (Figure 6A). Every tile on chromosome 4 was assigned a category depending on whether its sequence was part of a gene, a gene with an associated repeat, or an intergenic region with no known coding potential, and each tile was then counted as methylated if there was any significant DNA methylation detected at that location. The majority of changes in suspension cell culture euchromatin were due to novel DNA methylation relative to seedling DNA methylation [51] in the genic tile classes (46%), although some genes methylated in seedlings lost methylation in cell suspension culture (12%) (Figure 6A). We tested whether this euchromatic hypermethylation is associated with sRNAs, by comparing the gain or loss of DNA methylation to the frequency of sRNAs from leaves and cell suspension culture from the corresponding genomic regions. We found no significant association between the sRNAs and genic methylation on euchromatic DNA hypermethylation ($p = 0.29$). By contrast, the association of sRNAs and heterochromatic DNA methylation was highly significant ($p = 0.001$).

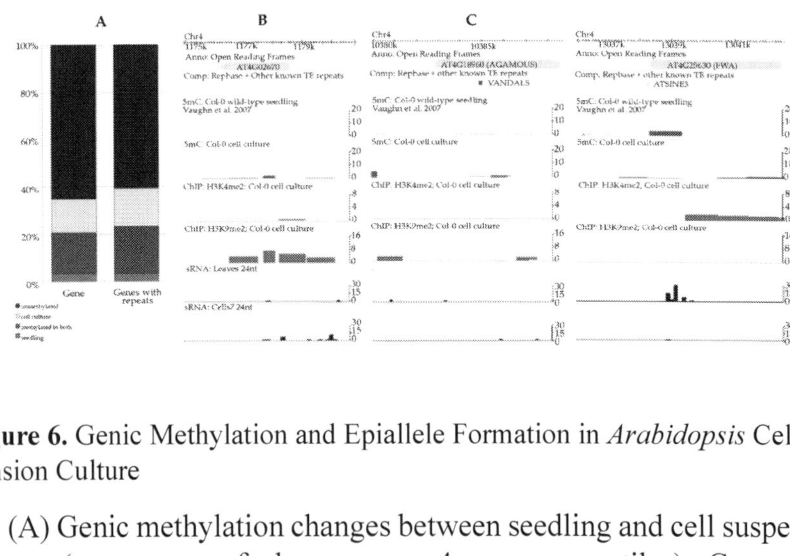

Figure 6. Genic Methylation and Epiallele Formation in *Arabidopsis* Cell Suspension Culture

(A) Genic methylation changes between seedling and cell suspension culture (percentage of chromosome 4 array gene tiles). Green: tiles associated with genic sequences that lose DNA methylation in cell suspension culture; yellow: tiles methylated only in cell suspension culture; red: tiles methylated in both samples; and blue: tiles not methylated in either sample. Chromatin maps of cell suspension culture at the (B) *IDD12(At4g02670)*, (C) *AGAMOUS*, and (D) *FWA* loci. Tiles with DNA methylation are presented as red bars, while those enriched in H3K9me2 and H3K4me2 are shown in purple. Tracks representing the 21-nt and 24-nt sRNA abundance per 100-bp sliding window are shown in black.

We also compared the DNA methylation profiles and the genome-wide gene expression levels using Affymetrix ATH1 array. This comparison revealed major changes in gene expression, with 41% of genes exhibiting statistically significant changes (FDR controlled to 0.05),between cell suspension culture and seedling, and 58% between cell suspension culture and callus (Figure 7). However, these expression changes were not correlated with changes in gene-body methylation between cell suspension culture and seedlings.

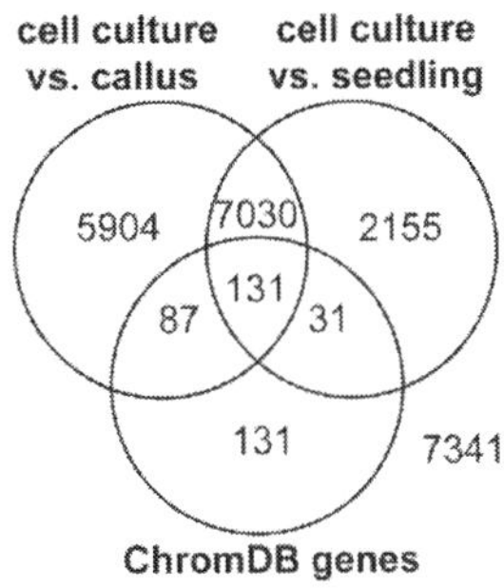

Figure 7. Gene Expression Differences between Cell Suspension Culture and Seedling and Callus Tissues

Significant gene expression differences detected by ATH1 microarray expression profiling between cell suspension culture and seedling or callus samples are represented in this Venn diagram. Also, 65% of genes annotated by ChromDB as having a role in chromatin regulation are differentially expressed in these samples relative to seedling.

In search of hypermethylated genes that might be silenced in cultured cells, we found that 14% of genic tiles that become hypermethylated in cell suspension also gained H3K9me2, doubling the number of doubly marked genes found in seedlings. A total of 89% of these tiles were associated with increased sRNA matching these genes, for a total of 28 genes on chromosome 4 that represent candidates for epiallele formation in the immortalized, proliferating cells of the cell suspension culture. An example of a novel epiallele is a member of the IDD family of zinc finger transcription factors At4g02670, which gains both DNA and H3K9 methylation as well as 24-nt sRNAs in cell suspension culture (Figure 6B). Two *Arabidopsis*genes on chromosome 4 are well known to be epigenetically regulated: the floral regulator *AGAMOUS* (*AG*) and the imprinted homeodomain protein gene *FWA*. A short repeat with limited similarity to the *Vandal* class of transposons in the second intron of *AG* is prone to hypermethylation in *ddm1* and *met1* mutants [17], whereas a short interspersed nuclear element (SINE) element within *FWA* 5′ UTR is hypermethylated and targeted by sRNAs [32,62]. *AG* and *FWA* change in their epigenetic status in cell suspension culture, with the second intron of *AG* gaining a low level of DNA methylation. This is accompanied by methylation of H3K9 in the same region and the appearance of sRNAs 24 nt in size (Figure 6C). The imprinted *FWA*gene is normally expressed only from the maternal chromosome during endosperm development [55]. Along with other TEs, the SINE element within the *FWA* 5′ UTR region becomes hypomethylated in cell suspension culture with a concomitant loss of 24-nt sRNA from the same region (Figure 6D). This is accompanied by strong transcriptional reactivation of *FWA*with a 25-fold increase in cell suspension culture relative to seedling expression and by accumulation of H3K4me2 and low levels of DNA methylation in the *FWA* coding region (Figure 6D).

Genes Involved in Chromatin Regulation Exhibit Cell Culture-Specific Expression Changes

In order to test whether the heterochromatic DNA hypomethylation and changes in histone methylation patterns seen in cell suspension culture may be due to changes in expression of DNA and histone methyltransferases and other genes affecting chromatin modifications, gene expression levels were examined for a list of 380 chromatin regulation-related genes from the ChromDB database (http://www.chromdb.org/). A total of 162 out of 380 genes on this list exhibited significant expression differences between cell suspension culture and Col seedling samples. Of these, 121 were up-regulated and 41 were down-regulated in cell suspension culture. DNA cytosine methyltransferases were overexpressed; from a 2-fold increase for *CMT3*, *DRM1*, and *DRM2* to a 5-fold increase for *MET1*. The chromatin remodeling gene *DDM1* is also required to methylate TEs, and *DDM1* transcripts were six times more abundant in cell suspension culture. Another *SNF2* class chromatin remodeling gene required for gene silencing, *CLSY1*, was also up-regulated in cell suspension culture 3-fold, but *DRD1* did not change in expression levels. The *VIM1* gene, required to guide CpG DNA methylation during replication, was also up-regulated more than 10-fold. Additionally, the DNA demethylase *ROS1* transcript levels were abolished. In summary, genes involved in DNA methylation are up-regulated in cell suspension culture, whereas those involved in DNA demethylation are down-regulated. This likely contributes to genic hypermethylation, but cannot account for TE activation.

Methylation of H3K9me2 in *Arabidopsis* is under control of histone methyltransferases of the Su(var)3–9 type. The *SUVH4* gene (*KYP*) predominates, but *SUVH5* and *SUVH6* have influence over some heterochromatic sequences. The observed loss of H3K9me2 in cell suspension culture could be explained if the transcripts of H3K9 methyltransferases were down-regulated. However, comparison of transcript levels for these three *SUVH* genes showed that only *SUVH6* was down-regulated in cell culture, whereas the major H3K9 methyltransferase *KYP* was up-regulated 3-fold in this sample relative to seedling, with H3K9me2 remaining ubiquitous at heterochromatic regions of the genome. No significant difference in *SUVH5* transcripts was observed. Three putative and closely related H3K9 methyltransferases (*SUVH1*, *SUVH3*, and *SUVH8*) were also up-regulated in cell suspension culture. Histone demethylases and *jumonji* domain family members *JMJ22* and *JMJ30* were up-regulated several fold in the cell suspension culture, whereas *JMJ24* was down-regulated. The *LSD1* family of histone demethylase paralogs in *Arabidopsis*(*FLD*, *LDL*, *LDL2*, and *LDL3*) showed no changes in gene expression. Histone deacetylases followed the same trend, with *HDA6*, *HDA8*, *HDA9*, *HDA17*, *HDA18*, and *HDA19*up-regulated in cell suspension culture, whereas *HDA14* was the only histone deacetylase down-regulated. Overall, these results indicate that a preponderance of genes involved in heterochromatic modifications increase in expression levels in cell suspension culture relative to seedling.

Changes in the pattern, rather than the extent, of DNA and histone methylation in suspension cell culture might be more readily explained by changes in the targeting machinery, which relies on RNAi. We, therefore, looked at the expression levels of genes from the *Dicer-like*, *RNA-dependent RNA Polymerase*, and *Argonaute* gene families as well as the plant-specific RNA polymerase IV (*NRDP1a*). Whereas genes involved in miRNA processing, such as *AGO1* and *DCL1*, were all either down-regulated in cell suspension culture relative to seedling or did not change in expression, those genes that are part of the heterochromatic silencing sRNA pathway, *AGO4*, *NRPD1a*, *RDR2*, and *DCL3*, were all up-regulated in the cell suspension culture at least 2-fold, with the *NRPD1a* transcripts being 18 times more abundant in cell suspension culture. Interestingly, additional members of the *Argonaute* gene family, *AGO2*, *AGO3*, *AGO5*, and *AGO9*, were up to 30-fold up-regulated in cell suspension culture, potentially accounting for changes in sRNA abundance.

DISCUSSION

Differentiated plant cells from various species and tissues can be experimentally manipulated using plant growth factors to dedifferentiate in vitro, resulting in callus tissue, from which somatic embryos and plantlets can later emerge. This process often leads to heritable changes in phenotype, referred to as somaclonal variation [7,19,63]. The patterns and modes of inheritance of somaclonal variation point to misregulation of epigenetic control mechanisms [19,24]. Cultured cells can be propagated indefinitely in suspension as continuously dividing cells reminiscent of immortalized animal cell cultures. The epigenomic chromatin maps of a long-term *Arabidopsis* cell suspension culture that we have generated were radically different from comparable maps of *Arabidopsis* seedling tissue [49–51] from which the cultures were originally derived. Whereas heterochromatic DNA becomes hypomethylated in cell suspension culture, the euchromatic portion of chromosome 4, including many genes, becomes hypermethylated instead. However, genes with remnants of TEs in their promoter or coding sequences can also lose methylation in cell suspension culture relative to *Arabidopsis* seedling samples. For example, the imprinted gene *FWA* is reactivated in cell suspension culture accompanied by loss of methylation in the SINE element that constitutes its promoter [32]. Loss of this methylation generates epialleles [62], and is predicted to contribute to somaclonal variation. Patterns of histone modification and siRNA biogenesis may underlie these changes, as described below.

Euchromatin in Cultured Suspension Cells

We found that most genic DNA methylation is associated with H3K4me2, indicating that gene-body methylation by MET1 might be guided, directly or indirectly, by H3K4me2 (see below).

Gene-body methylation in *Arabidopsis* is dependent on *MET1* [32], is highly variable between *Arabidopsis* accessions, and is positively correlated with elevated gene expression. In support of this idea, the comparison between the distribution of gene-body DNA and histone methylation patterns revealed that H3K4me2 is colocalized with gene-body DNA methylation for over 75% of methylated genes. In yeast, both H3K4me2 and H3K4me3 mark genic regions, but H3K4me3 is characteristic of transcriptionally active genes [64], whereas in plant and mammalian genomes, H3K4me3 is found at promoters [43,65]. In mammalian cells, de novo methylation via DNMT3 complexes cannot be recruited to nucleosomes containing H3K4me3, providing a mechanism. perhaps, for exclusion of DNA methylation from promoters [66].

Whereas H3K4me2 is deposited exclusively within coding regions, we showed that promoters are specifically enriched in H3K56ac while gene coding regions were depleted. In addition to its role in genome stability, the acetylation of Lys56 at histone H3 (H3K56ac) has been shown recently to associate with gene promoter regions in budding yeast [67,68] and in particular with transcriptionally active promoters and the elongating form of RNA Polymerase II in yeast and *Drosophila* [69]. We show for the first time, to our knowledge, that H3K56ac is also present in gene promoter regions in plants, as well as in TEs that have lost heterochromatic marks in cell suspension culture. Acetylation of this particular amino acid located at the first point of contact between the major groove of DNA and the histone octamer may allow access to the core transcriptional machinery [67] as well as replication-coupled nucleosome assembly [70].

Even though most methylated genes are associated with H3K4me2, and are actively transcribed, a small percentage of methylated genes are associated with H3K9me2 in leaves, indicating they are epigenetically silenced. This percentage doubles in cell culture, so that 28 genes from chromosome 4 become hypermethylated and associated with H3K9me2. Close to 90% of these genes are associated with novel 24-nt sRNAs in cell suspension culture, indicating sRNA is the major targeting mechanism for epiallele formation. These genes may also include potential targets of the DNA demethylase ROS1, whose expression is lost in cell culture and is accompanied by hypermethylation of the promoter region of *ROS1* [71], which contains a helitron TE fragment. A third of these genes are significantly down-regulated or silenced and represent good candidates for epigenetically regulated genes affecting various aspects of somaclonal variation and habituation [4,72]. It is important to note that independent plant cell culture lines can vary considerably in their molecular phenotypes. Therefore, analyses of epigenetic changes in independent plant cell culture lines are needed before general conclusions about individual genes can be drawn.

Heterochromatin is activated in cultured suspension cells. In contrast to euchromatin, the pericentromeric heterochromatin of chromosome 4 was massively hypomethylated in cell suspension culture. The majority of these heterochromatic regions consists of various classes of TEs [27], but only some of these classes were hypomethylated, whereas others, such as the LTR retrotransposons *ATLANTYS* and *AtGP*, remained methylated. Loss of DNA methylation was often accompanied by loss of H3K9me2 and in some, but not all cases, gain of H3K4me2. Similar to reactivation of TEs in calli or cell suspension culture of maize, rice, and tobacco [11,14,26,73], as well as *Arabidopsis* [28], we were able to detect transcripts from this group of TEs, though only a subset of the elements with a permissive chromatin landscape was actually transcribed. This very likely reflects the fact that most TEs have undergone genetic changes (mostly deletions and insertions) over evolutionary time, diminishing their coding and transcriptional potential.

These changes are reminiscent of transposon deregulation caused by mutations in *ddm1*, a chromatin remodeling factor [74]. These mutants lose transposon-specific DNA methylation, as well as H3K9me2, accompanied by gain of H3K4me2 [54]. However, *ddm1* mutants affect all families of TEs, whereas cell suspension culture only reactivates specific TEs. Furthermore, 75% of a sample of 380 genes involved in heterochromatin regulation, including *DDM1*, are up-regulated in cell culture, indicating that the silencing machinery is largely intact. Instead, the failure to deliver epigenetic marks to some TEs and not others could be the result of perturbations in targeting these modifications by RNAi and sRNA molecules.

Small Interfering RNA

While sRNAs accumulated from both silenced and reactivated TEs in cell suspension culture, the relative abundance of different size siRNA changed with respect to differentiated tissues. TEs with the strongest reactivation in cell suspension culture, such as *Athila* retrotransposons, as well as some members of the *Helitron*, *Mu*, *Vandal*, and *En/Spm* families of transposons, gained 21-nt siRNA, but lost 24-nt siRNA only rarely. In contrast, TEs that remained silent in cell suspension culture, such as most *Atlantys* and *AtGP* retrotransposons and some *Vandal*and *EnSpm* TEs, produced the 24-nt siRNAs at the same level as they did in vegetative tissue. It would therefore seem that failure to maintain silencing at heterochromatic sequences in cell suspension culture is associated with a shift in relative abundance of 21-nt siRNAs.

The novel 21-nt TE sRNA we have detected in cultured cells show a preference for adenosine (A) at the 5′ end, and a reduction in guanine (G). These changes are diagnostic of the preferential loading of sRNA onto one or more members of the family of *Argonaute* proteins found in *Arabidopsis*. Although the expression levels of AGO4, the *Argonaute* protein known to bind 24-nt heterochromatic sRNA, did not change in cell suspension culture, four *Argonaute*genes whose roles have not yet been defined, *AGO2*, *AGO3*, *AGO5*, and *AGO9*, are massively up-regulated in cell suspension culture, along with PolIVa, which has been implicated in heterochromatic sRNA biogenesis [75,76]. *Arabidopsis* AGO2 has recently been shown to bind 21-nt siRNAs, with a strong preference for A as the first nucleotide of the siRNA [59, 60]. This is precisely the most abundant class of sRNAs in cell suspension culture, and *AGO2* is massively up-regulated in cultured cells. If AGO2 is coupled with the biogenesis of the 21-nt siRNA, then it may preclude the proper targeting by and inclusion of the 24-siRNAs into other AGO complexes, resulting in diminished potential of heterochromatic targeting by sRNAs. Additionally, our data support a minor role for a so-far unidentified AGO protein with a preference for G in targeting H3K9me2 and DNA methylation synergistically with AGO4.

Reprogramming of Epigenetic Modifications in Immortalized Cultured Cells

Reactivation and hypomethylation of TEs is not unique to plant cell culture. *Drosophila* Kc and S2 cultures harbor numerous actively transcribed *copia* and *copia*-like retrotransposons [77, 78] capable of transposition [79, 80]. The reactivation seems independent of the origin of the culture and does not depend on DNA methylation, which is absent from *Drosophila* [81]. However, components of the piRNA pathway required to silence several classes of transposons are not expressed in *Drosophila* cell cultures, perhaps accounting in part for this activation [82]. In immortalized cancer cell cultures, non-LTR Alu SINE elements [83, 84], as well as LINE-1 elements, lose DNA methylation and are reactivated [85–87]. In mouse, hypomethylation of TEs occurs in mutants in the maintenance DNA methyltransferase DNMT1, and as in plants, leads to their transcriptional reactivation [36, 88, 89]. In tumors, genome-wide hypomethylation also results in TE activation ([90, 91], but tumors express DNMT1 at normal or elevated levels [92–94]. These results indicate that, as in plant suspension cell cultures, targeting rather than maintenance of DNA methylation is defective in mammalian cell cultures and tumors [88]. Whereas the majority of CpG islands in mammalian genomes are normally devoid of methylation [95, 96], some CpG islands become hypermethylated in immortalized animal cell cultures [97, 98]. This methylation often leads to transcriptional gene silencing [95, 96], although loss of genic DNA methylation in cell culture can also occur [99–101]. Both the maintenance and de novo DNA methyltransferases are required for CpG hypermethylation in animal cell cultures [102]. Moreover, there is a decline in expression levels of these genes prior to immortalization, which is accompanied by genome-wide DNA hypomethylation, followed by an overexpression in the immortalized cell culture. The resulting hypermethylation [103] is reflected in the gradual increase in CpG hypermethylation of genes in immortalized animal cells [98,104] during the transition from senescent to immortalized cell culture growth [105,106].

Our results show that in *Arabidopsis* cell suspension culture, the expression levels of *MET1*, the maintenance DNA methyltransferase, as well as those of *CMT3* and *DRM2*, follow the same pattern, with a 4-fold drop in expression levels of *MET1* during callus induction, and a

4-fold increase in expression in cell suspension culture, relative to *MET1* expression in seedlings. A corresponding drop in methylation preceding the establishment of the cell suspension culture, would create a need for de novo targeting of methylation, by CMT3 and DRM2, particularly to repeat-containing genic regions [107]. Mistargeted de novo methylation in subsequent cell suspension culture coupled with high levels of MET1 and down-regulation of DNA demethylase ROS1 would then allow for establishment and perpetuation of genic DNA hypermethylation and for the creation of epialleles.

Support for this reprogramming model comes from analysis of the backcross progeny of *met1* mutant plants [36], and subsequent generations [107], which mimic the loss and reestablishment of DNA methylation in cell culture. In backcross plants, methylation patterns are reestablished only sporadically and depend on RNA-dependent DNA and histone H3K9 methylation, directed by 24-nt siRNA. The 21-nt sRNA were found in met1 mutants from the "TSI" repeat [107], which is a truncated form of *Athila* elements, as well as in a broader genome resource sequence of *met1* siRNA [71], although the origin of these siRNA and their persistence in backcrosses was not examined in these studies. Redirection of TE transcripts into a posttranscriptional silencing 21-nt pathway, rather than the transcriptional silencing 24-nt pathway, would account for the failure to reprogram TE DNA methylation in cultured cells. Thus reprogramming of DNA methylation plays a significant role in both animal and plant cell culture, although the role of sRNA in mammalian cell cultures has not yet been examined.

MATERIALS AND METHODS

Plant material

Arabidopsis cell line (ecotype Columbia) was maintained in Gamborg's B5 basal medium with minor organics (Sigma G5893) supplemented with 1.1 mg/l 2,4-D, 3 mM MES, and 3% sucrose. The cells were grown on a shaker at 160 rpm under consistent light condition at 23°C and were subcultured every 7 d with a 1:10 (inoculum:fresh medium) dilution ratio.

The wild-type Col seedlings were grown on MS agar plates under standard conditions (day/light) for 2 wk. Callus samples were obtained through standard plant tissue culture techniques. Cotyledons of

Arabidopsis thaliana, ecotype Columbia, were cut out 7 d after plating on MS plates and transferred onto 1× MS plates supplemented with auxin and cytokinin (0.15 μg/ml IAA and 0.625 μg/ml isopentanyladenine) and 3% (w/v) sucrose. Five weeks later, callus samples 5–8 mm in diameter were collected for gene expression analysis.

Custom chromosome tiling microarray

We used a custom printed genome tiling microarray that comprises approximately 18.6 Mb of Arabidopsis thaliana chromosome 4 sequence (~14% of the entire *Arabidopsis* genome). Approximately 22,000 tiles, on average 1 kb in size, cover the majority of chromosome 4, including the entire heterochromatic knob on the short arm of chromosome 4 and several megabases of pericentromeric heterochromatin. Details of array design and production are described in [51].

DNA methylation analysis

Genomic DNA was isolated from cell suspension culture using Qiagen Plant DNeasy Mini Kit following the manufacturer's protocol (except that genomic DNA was eluted with 10 mM TrisCl [pH 8.0]). Ten micrograms of DNA was sheared to 1–3 kb in size, and half of the sheared DNA was subjected to digestion using McrBC restriction endonuclease, a restriction endonuclease that recognizes 5mC in a variety of sequence contexts (5′ $(G/A)^{m}C(N)_{40-3000}(G/A)^{m}C$ 3′) as described in [51]. Following digestion, both the treated and the untreated DNA sample were briefly run on a 1% agarose gel, and the DNA fragments 0.8–3 kb in size were isolated from the gel slices of both the digested and the undigested DNA sample using QiaexII gel extraction kit. Resulting DNA was amplified using a PCR-based approach described in [58]. Three micrograms of each amplified DNA sample was used in a labeling reaction using the Bioprime Array CGH kit and Cy3 and Cy5 fluorescent dyes. Each labeled DNA sample was eluted in 28 μl of 10 mM TrisCl (pH 8.0), and paired DNA samples were combined in a final volume of 70 μl of 4× SSC and 0.2% SDS. Each microarray hybridization was carried out at 63 °C overnight (14–16 h). Following hybridization, microarrays were washed for 8 min in 1× SSC, 0.2% SDS (warmed up to 50°C), then for 5 min in 1× SSC (at room temperature), and finally in two washes of 0.2× SSC (2 min each).

Microarrays were quickly dried by compressed air and immediately scanned using an Axon 4000B microarray scanner.

DNA methylation levels were measured by cohybridizing differentially labeled McrBC-treated and untreated DNA on a single array. In this way, copy-number differences are not contributing to the difference in hybridization signals on each array. A dye-swap experimental design provided the basis for these chromosome 4 tiling array experiments. DNA methylation was profiled by comparing McrBC-treated samples with untreated samples for both cell culture samples. The experimental design consisted of four replicated dye swaps. Sixteen total arrays were hybridized. Each tile was tested for methylation changes using a simple linear model.

Chromatin immunoprecipitation and array hybridizations

An approximate volume of 5 ml of sedimented cells from cell suspension culture were cross-linked by treating with 1% paraformaldehyde for 15 min. After adding 2 M glycine (100 mM final concentration) and incubating for 5 min to stop the cross-linking, the cells were washed three times with 1× PBS by centrifuging at 1,500 rpm for 5 min. The fixed cells were snap-frozen in liquid nitrogen and stored at −80 °C. Chromatin isolation and immunoprecipitation with antibodies against specific histone modifications (anti-H3K2me2, anti-H3K9me2, and anti-H3K56ac) were done according to [108] with minor modifications (an extra wash at each washing step). Antibodies used were from Upstate/Millipore (anti-H3K4me2: cat# 07–030 [lot# 26335], anti-H3Kme2: cat# 07–212 [lot# 27563], and anti-H3K56ac: cat# 07–677). Immunoprecipitated DNA was amplified, labeled, and hybridized to the custom genome tiling array using the procedures described above for genomic DNA amplification, fluorescent labeling, and microarray hybridization in DNA methylation analysis.

For each immunoprecipitation, we hybridized, amplified, and labeled immunoprecipitated DNA against input DNA using two biological and two technical replicates. As with the DNA methylation experiments, a dye swap provided the experimental design. A total of eight microarrays were hybridized for each immunoprecipitation.

Gene expression microarray analysis

Total RNA from 100 mg of 2-wk-old plate-grown seedlings (Col) and 100 mg of 4-d cell suspension culture was isolated using Qiagen Plant RNeasy Mini Kit. RNA quality was assessed on an Agilent 2100 Bioanalyzer, RNA 6000 Pico Series II Chips (Agilent). Samples were assessed by an RNA integrity number (RIN) score, and those with an RIN score of 7.5 or greater were included. The RNA quantity was assessed by Nanodrop ND-1000 (Nanodrop Technologies). Total RNA was amplified by a modified Eberwine Technique, using a Message Amp II kit (Ambion) for one round of amplification (1). The antisense RNA (aRNA) smear analysis for 3' bias was performed on select samples using Agilent 2100 Bioanalyzer RNA 6000 Nano Series II Chips (Agilent). Samples were then prepared for hybridization, hybridized, washed, and then scanned according to the manufacturer's instructions on ATH1 GeneChips (Affymetrix). Affymetrix QC metrics were used for quality control of the image data.

Small RNA sequencing

The sRNA fractions were isolated from the total RNA sample by size selection on denaturing 15% PAGE gels. RNA molecules between 19 nt and 28 nt in size were eluted from crushed gel slices in four gel volumes of 20 mM Tris (pH 8), 1 mM EDTA, 0.4 M NH_4OAc, 0.5% SDS overnight, with shaking. sRNA was precipitated with glycoblue (Invitrogen) and three volumes of ethanol (−20 °C overnight) followed by ligation of the adenylated 3′ adapter (AMP-5′p = 5′-pCTGTAGGCACCATCAAT-dideoxyC-3′) and then the 5′ adapter (rArCrArCrUrCrUrUrUrCrCr-CrUrArCrArCrGrArCrGrCrUrCrUrUrCrCrGrArUrC) as described in [109]. After size selection using 15% PAGE gel electrophoresis, sRNA libraries were reverser transcribed and PCR amplified using the forward PCR primer (5′-AATGATACGGCGACCACCGAACACTCTTTCCCT ACACGACG-3′) and the reverse PCR primer (5′-CAAGCAGAAGAC-GGCATACGATTGATGGTGCCTACAG-3′) for 22 cycles using proof-reading Taq polymerase. The sRNA libraries were sequenced on an Illumina 1G sequencer using 48 cycles to allow for adequate 3′ adapted masking for precise size determination of the sRNA clones.

RT-PCR

Total RNA was isolated from young rosette leaves of 2-wk-old Columbia plants grown on plates under standard conditions, 4-d cell suspension culture, as well as from cotyledon-derived callus samples described above using Qiagen Plant RNeasy Mini Kit. Five micrograms of total RNA was treated with DNaseI and then reverse transcribed with SuperScript III with oligo dT primer in a 20-μl reaction. Two microliters of 5-fold diluted cDNA reaction was used as a template in the PCR reaction (for a final amount of 100 ng of starting total RNA per each PCR reaction). Thirty cycles of PCR were performed..

Computational analyses

General bioinformatics methods. Histone modification, DNA methylation profiles, and sRNA sequencing data were loaded into a MySQL relational database implementing the Bio::DG::GFF schema, thus facilitating intersecting positional, quantitative, and class-based queries and computations. In addition, array data and genome annotations were displayed visually using a Generic Genome Browser, available for public examination at http://chromatin.cshl.edu/epiculture/.

Genes and noncoding RNAs were annotated based on the TAIR version 6 genome release. TEs and known non-TE classes such as the 180-bp centromeric repeat were annotated based on CENSOR analysis [110] of the *Arabidopsis* genome sequence using version 11.10 of the Repbase database [111]. Anonymous tandem repeats were identified using TandemRepeatsFinder [112], whereas miRNA precursors were localized in the genome by BLAST comparison of precursors sequences provided by Rfam [113] to the *Arabidopsis*genome. Other regions are assumed to be "Intergenic."

Analysis of small RNA sequences

Adapter sequences were identified and removed using the cross_match algorithm (http://www.phrap.org/phredphrapconsed.html) using a minimum match size of 5 and a minimum score of 14. The nonredundant complement of sequences for the cell culture and inflorescence sRNA libraries was determined by a Perl script that indexed unique masked nucleotide sequences and the frequency with which they were identified.

These sequences were aligned to the *Arabidopsis* genome sequence using BLATN [114] with a word size of 8, and further computational analysis was restricted to sRNA that perfectly matched the genome.

In order to examine sRNA distribution across the genome in an unbiased fashion, copy number-corrected sRNA frequencies were calculated based on BLATN-determined sequence identities for nonoverlapping 100-bp windows. These data were converted to UCSC Wiggle format and displayed in graphical form on our genome browser. Copy number correction was used to partition sequencing frequencies for each sRNA across all instances of that sRNA in the genome. It was performed by calculating the number of discrete, perfect matches for each sRNA from a BLAT query against the *Arabidopsis* genome as an estimate of copy number, then dividing the sequencing frequencies of individual sRNA matches by that value.

sRNA size class analysis was performed by parsing results of a genome-wide BLAT analysis to index sRNA matches by sRNA length and the annotation units within which they fell. To facilitate comparison between libraries, results were plotted in Microsoft Excel as percentages of sRNA matches in each size class relative to all matches from all size classes. Results were summarized for the genome as a whole and then partitioned to examine matches to particular transposable elements and classes of repeats.

The physical distribution for each sequence library of sRNAs within representative full-length TEs was determined by performing a BLATN of all sequences in the library against reference sequences. The number of sRNAs on each strand at each base position for the template sequences was computed based on sequencing frequency and the BLATN results and plotted as a histogram using Gnuplot 4.0.

Base preference analysis was performed for all sRNA matching the genome in each sequenced library by counting the number of sRNA sequences with a given nucleotide at each position, then dividing values for each nucleotide by the total number of sRNA sequences. Values from the first base were extracted and used for analyses in this paper. These analyses were performed on all sRNA at large, as well as upon a subset of sRNA in which known miRNAs were excluded, since many miRNAs are known to have specific first-base properties [59].

sRNA matches to miRNAs and their precursors were identified by matching the nonredundant complement of sRNA sequences to FASTA files provided by Rfam [113] using the Patscan algorithm [115] allowing up to two internal mismatches.

sRNA sequences have been deposited in GenBank and the accession numbers can be found at http://chromatin.cshl.edu/epiculture/.

Analysis of ATH1 expression arrays

Affymetrix .CEL files were imported into Agilent GeneSpring GX using the gcRMA algorithm for preprocessing. Significant differences among tissue sources were established using one-way analysis of variance (ANOVA) followed by control of the FDR to 5% using the method of Benjamini and Hochberg [116]. Inter-sample comparisons were extracted from the list of statistically significantly different genes using GeneSpring's built-in filtering tools. The GEO accession numbers for the ATH1 microarray data compiled in Table S5 are available at http://chromatin.cshl.edu/epiculture/.

ACKNOWLEDGMENTS

We thank Christopher Johns and Sohail Khan for help with gene expression profiling, Vincent Colot for chromosome 4 tiling array design and production, and R. W. McCombie, Melissa delaBastide, and Danea Rebolini for their help with small RNA sequencing. We thank our colleagues Linda Hanley-Bowdoin and George Allen for discussions

REFERENCES

1. Lee M, Phillips RL (1988) The chromosomal basis of somaclonal variation. Annu Rev Plant Physiol Plant Mol Biol 39: 413–437.M. LeeRL Phillips1988The chromosomal basis of somaclonal variation. Annu Rev Plant Physiol Plant Mol Biol39413437
2. .Osborne DJ, Mcmanus MT (1986) Flexibility and commitment in plant cells during development. Curr Top Dev Biol 20: 383–396. DJ OsborneMT Mcmanus1986Flexibility and commitment in plant cells during development.Curr Top Dev Biol20383396
3. .Skoog F, Miller CO (1957) Chemical regulation of growth and organ formation in plant tissues cultured in vitro. Symp Soc Exp

Biol 54: 118–130.F. SkoogCO Miller1957Chemical regulation of growth and organ formation in plant tissues cultured in vitro.Symp Soc Exp Biol54118130

4. Meins F (1989) Habituation: heritable variation in the requirement of cultured plant-cells for hormones. Annu Rev Genet 23: 395–408.F. Meins1989Habituation: heritable variation in the requirement of cultured plant-cells for hormones.Annu Rev Genet23395408
5. Meins F, Foster R (1985) Reversible, cell-heritable changes during the development of tobacco pith tissues. Dev Biol 108: 1–5.F. MeinsR. Foster1985Reversible, cell-heritable changes during the development of tobacco pith tissues.Dev Biol10815
6. Evans DA, Sharp WR (1986) Applications of somaclonal variation. Bio-Technology 4: 528–532.DA EvansWR Sharp1986Applications of somaclonal variation.Bio-Technology4528532
7. Jain SM (2001) Tissue culture-derived variation in crop improvement. Euphytica 118: 153–166.SM Jain2001Tissue culture-derived variation in crop improvement.Euphytica118153166
8. Brettell RIS, Dennis ES, Scowcroft WR, Peacock WJ (1986) Molecular analysis of a somaclonal mutant of maize alcohol-dehydrogenase. Mol Gen Genet 202: 235–239.RIS BrettellES DennisWR ScowcroftWJ Peacock1986Molecular analysis of a somaclonal mutant of maize alcohol-dehydrogenase.Mol Gen Genet202235239
9. Dennis ES, Brettell RIS, Peacock WJ (1987) A tissue-culture induced Adh1 null mutant of maize results from a single base change. Mol Gen Genet 210: 181–183.ES DennisRIS BrettellWJ Peacock1987A tissue-culture induced Adh1 null mutant of maize results from a single base change.Mol Gen Genet210181183
10. Brown PTH, Gobel E, Lorz H (1991) Rflp analysis of Zea mays callus cultures and their regenerated plants. Theor Appl Genet 81: 227–232.PTH BrownE. GobelH. Lorz1991Rflp analysis of Zea mays callus cultures and their regenerated plants.Theor Appl Genet81227232
11. Hirochika H, Sugimoto K, Otsuki Y, Tsugawa H, Kanda M (1996) Retrotransposons of rice involved in mutations induced by tissue culture. Proc Natl Acad Sci U S A 93: 7783–7788.H. HirochikaK.

SugimotoY. OtsukiH. TsugawaM. Kanda1996Retrotransposons of rice involved in mutations induced by tissue culture.Proc Natl Acad Sci U S A9377837788

12. Bregitzer P, Halbert SE, Lemaux PG (1998) Somaclonal variation in the progeny of transgenic barley. Theor Appl Genet 96: 421–425.P. BregitzerSE HalbertPG Lemaux1998Somaclonal variation in the progeny of transgenic barley.Theor Appl Genet96421425
13. Grandbastien MA (1998) Activation of plant retrotransposons under stress conditions. Trends Plant Sci 3: 181–187.MA Grandbastien1998Activation of plant retrotransposons under stress conditions.Trends Plant Sci3181187
14. Grandbastien MA, Spielmann A, Caboche M (1989) Tnt1, a mobile retroviral-like transposable element of tobacco isolated by plant-cell genetics. Nature 337: 376–380.MA GrandbastienA. SpielmannM. Caboche1989Tnt1, a mobile retroviral-like transposable element of tobacco isolated by plant-cell genetics.Nature337376380
15. Jacobsen SE, Meyerowitz EM (1997) Hypermethylated SUPERMAN epigenetic alleles in arabidopsis. Science 277: 1100–1103.SE JacobsenEM Meyerowitz1997Hypermethylated SUPERMAN epigenetic alleles in arabidopsis.Science27711001103
16. Cubas P, Vincent C, Coen E (1999) An epigenetic mutation responsible for natural variation in floral symmetry. Nature 401: 157–161.P. CubasC. VincentE. Coen1999An epigenetic mutation responsible for natural variation in floral symmetry.Nature401157161
17. Jacobsen SE, Sakai H, Finnegan EJ, Cao X, Meyerowitz EM (2000) Ectopic hypermethylation of flower-specific genes in Arabidopsis. Curr Biol 10: 179–186.SE JacobsenH. SakaiEJ FinneganX. CaoEM Meyerowitz2000Ectopic hypermethylation of flower-specific genes in Arabidopsis.Curr Biol10179186
18. Kakutani T (2002) Epi-alleles in plants: inheritance of epigenetic information over generations. Plant Cell Physiol 43: 1106–1111.T. Kakutani2002Epi-alleles in plants: inheritance of epigenetic information over generations.Plant Cell Physiol4311061111
19. Meins F, Thomas M (2003) Meiotic transmission of epigenetic changes in the cell-division factor requirement of plant cells. Development 130: 6201–6208.F. MeinsM. Thomas2003Meiotic

transmission of epigenetic changes in the cell-division factor requirement of plant cells.Development13062016208

20. Kaeppler SM, Phillips RL (1993) DNA methylation and tissue culture-induced variation in plants. In Vitro Cell Dev Biol Plant 29P: 125–130.SM KaepplerRL Phillips1993DNA methylation and tissue culture-induced variation in plants.In Vitro Cell Dev Biol Plant29P125130
21. Durante M, Cecchini E, Natali L, Citti L, Geri C, et al. (1989) 5-Azacytidine-induced tumorous transformation and DNA hypomethylation in Nicotiana tissue cultures. Dev Genet 10: 298–303.M. DuranteE. CecchiniL. NataliL. CittiC. Geri19895-Azacytidine-induced tumorous transformation and DNA hypomethylation in Nicotiana tissue cultures.Dev Genet10298303
22. .Kaeppler SM, Phillips RL (1993) Tissue culture-induced DNA methylation variation in maize. Proc Natl Acad Sci U S A 90: 8773–8776.SM KaepplerRL Phillips1993Tissue culture-induced DNA methylation variation in maize.Proc Natl Acad Sci U S A9087738776
23. Arnholdtschmitt B, Herterich S, Neumann KH (1995) Physiological aspects of genome variability in tissue culture. 1. Growth phase-dependent differential DNA methylation of the carrot genome (Daucus-Carota L) during primary culture. Theor Appl Genet 91: 809–815.B. ArnholdtschmittS. HerterichKH Neumann1995Physiological aspects of genome variability in tissue culture. 1. Growth phase-dependent differential DNA methylation of the carrot genome (Daucus-Carota L) during primary culture. Theor Appl Genet91809815
24. Kaeppler SM, Kaeppler HF, Rhee Y (2000) Epigenetic aspects of somaclonal variation in plants. Plant Mol Biol 43: 179–188.SM KaepplerHF KaepplerY. Rhee2000Epigenetic aspects of somaclonal variation in plants.Plant Mol Biol43179188
25. Peschke VM, Phillips RL, Gengenbach BG (1987) Discovery of transposable element activity among progeny of tissue-culture derived maize plants. Science 238: 804–807.VM PeschkeRL PhillipsBG Gengenbach1987Discovery of transposable element activity among progeny of tissue-culture derived maize plants.

Science238804807

26. Brettell RIS, Dennis ES (1991) Reactivation of a silent Ac following tissue-culture is associated with heritable alterations in its methylation pattern. Mol Gen Genet 229: 365–372.RIS BrettellES Dennis1991Reactivation of a silent Ac following tissue-culture is associated with heritable alterations in its methylation pattern.Mol Gen Genet229365372
27. Fransz P, ten Hoopen R, Tessadori F (2006) Composition and formation of heterochromatin in Arabidopsis thaliana. Chromosome Res 14: 71–82.P. FranszR. ten HoopenF. Tessadori2006Composition and formation of heterochromatin in Arabidopsis thaliana. Chromosome Res147182
28. Steimer A, Amedeo P, Afsar K, Fransz P, Scheid OM, et al. (2000) Endogenous targets of transcriptional gene silencing in arabidopsis. Plant Cell 12: 1165–1178.A. SteimerP. AmedeoK. AfsarP. FranszOM Scheid2000Endogenous targets of transcriptional gene silencing in arabidopsis.Plant Cell1211651178
29. May MJ, Leaver CJ (1993) Oxidative stimulation of glutathione synthesis in Arabidopsis thaliana suspension cultures. Plant Physiol 103: 621–627.MJ MayCJ Leaver1993Oxidative stimulation of glutathione synthesis in Arabidopsis thaliana suspension cultures. Plant Physiol103621627
30. Arabidopsis Genome Initiative (2000) Analysis of the genome sequence of the flowering plant Arabidopsis thaliana. Nature 408: 796–815.Arabidopsis Genome Initiative2000Analysis of the genome sequence of the flowering plant Arabidopsis thaliana. Nature408796815
31. Kapitonov VV, Jurka J (1999) Molecular paleontology of transposable elements from Arabidopsis thaliana. Genetica 107: 27–37.VV KapitonovJ. Jurka1999Molecular paleontology of transposable elements from Arabidopsis thaliana.Genetica1072737
32. Lippman Z, Gendrel AV, Black M, Vaughn MW, Dedhia N, et al. (2004) Role of transposable elements in heterochromatin and epigenetic control. Nature 430: 471–476.Z. LippmanAV GendrelM. BlackMW VaughnN. Dedhia2004Role of transposable elements in heterochromatin and epigenetic control.Nature430471476

33. Chan SW, Henderson IR, Jacobsen SE (2005) Gardening the genome: DNA methylation in Arabidopsis thaliana. Nat Rev Genet 6: 351–360.SW ChanIR HendersonSE Jacobsen2005Gardening the genome: DNA methylation in Arabidopsis thaliana.Nat Rev Genet6351360

34. Zilberman D, Henikoff S (2005) Epigenetic inheritance in Arabidopsis: selective silence. Curr Opin Genet Dev 15: 557–562.D. ZilbermanS. Henikoff2005Epigenetic inheritance in Arabidopsis: selective silence.Curr Opin Genet Dev15557562

35. Hirochika H, Okamoto H, Kakutani T (2000) Silencing of retrotransposons in arabidopsis and reactivation by the ddm1 mutation. Plant Cell 12: 357–368.H. HirochikaH. OkamotoT. Kakutani2000Silencing of retrotransposons in arabidopsis and reactivation by the ddm1 mutation.Plant Cell12357368

36. Lippman Z, May B, Yordan C, Singer T, Martienssen R (2003) Distinct mechanisms determine transposon inheritance and methylation via small interfering RNA and histone modification. PLoS Biol 1: e67. Z. LippmanB. MayC. YordanT. SingerR. Martienssen2003Distinct mechanisms determine transposon inheritance and methylation via small interfering RNA and histone modification.PLoS Biol1e67

37. Singer T, Yordan C, Martienssen RA (2001) Robertson's Mutator transposons in A. thaliana are regulated by the chromatin-remodeling gene Decrease in DNA Methylation (DDM1). Genes Dev 15: 591–602.T. SingerC. YordanRA Martienssen2001Robertson's Mutator transposons in A. thaliana are regulated by the chromatin-remodeling gene Decrease in DNA Methylation (DDM1).Genes Dev15591602

38. Miura A, Yonebayashi S, Watanabe K, Toyama T, Shimada H, et al. (2001) Mobilization of transposons by a mutation abolishing full DNA methylation in Arabidopsis. Nature 411: 212–214.A. MiuraS. YonebayashiK. WatanabeT. ToyamaH. Shimada2001Mobilization of transposons by a mutation abolishing full DNA methylation in Arabidopsis.Nature411212214

39. Martienssen RA, Doerge RW, Colot V (2005) Epigenomic mapping in Arabidopsis using tiling microarrays. Chromosome Res 13: 299–308.RA MartienssenRW DoergeV. Colot2005Epigenomic mapping in Arabidopsis using tiling microarrays.Chromosome Res13299308

40. Mockler TC, Chan S, Sundaresan A, Chen HM, Jacobsen SE, et al. (2005) Applications of DNA tiling arrays for whole-genome analysis. Genomics 85: 1–15.TC MocklerS. ChanA. SundaresanHM ChenSE Jacobsen2005Applications of DNA tiling arrays for whole-genome analysis.Genomics85115
41. Bernstein BE, Meissner A, Lander ES (2007) The mammalian epigenome. Cell 128: 669–681.BE BernsteinA. MeissnerES Lander2007The mammalian epigenome.Cell128669681
42. Barski A, Cuddapah S, Cui KR, Roh TY, Schones DE, et al. (2007) High-resolution profiling of histone methylations in the human genome. Cell 129: 823–837.A. BarskiS. CuddapahKR CuiTY RohDE Schones2007High-resolution profiling of histone methylations in the human genome.Cell129823837
43. Bernstein BE, Kamal M, Lindblad-Toh K, Bekiranov S, Bailey DK, et al. (2005) Genomic maps and comparative analysis of histone modifications in human and mouse. Cell 120: 169–181.BE BernsteinM. KamalK. Lindblad-TohS. BekiranovDK Bailey2005Genomic maps and comparative analysis of histone modifications in human and mouse.Cell120169181
44. Schubeler D, MacAlpine DM, Scalzo D, Wirbelauer C, Kooperberg C, et al. (2004) The histone modification pattern of active genes revealed through genome-wide chromatin analysis of a higher eukaryote. Genes Dev 18: 1263–1271.D. SchubelerDM MacAlpineD. ScalzoC. WirbelauerC. Kooperberg2004The histone modification pattern of active genes revealed through genome-wide chromatin analysis of a higher eukaryote.Genes Dev1812631271
45. Bernstein BE, Mikkelsen TS, Xie XH, Kamal M, Huebert DJ, et al. (2006) A bivalent chromatin structure marks key developmental genes in embryonic stem cells. Cell 125: 315–326.BE BernsteinTS MikkelsenXH XieM. KamalDJ Huebert2006A bivalent chromatin structure marks key developmental genes in embryonic stem cells. Cell125315326
46. Mikkelsen TS, Ku M, Jaffe DB, Issac B, Lieberman E, et al. (2007) Genome-wide maps of chromatin state in pluripotent and lineage-committed cells. Nature 448: 553–560.TS MikkelsenM. KuDB JaffeB. IssacE. Lieberman2007Genome-wide maps of chromatin

state in pluripotent and lineage-committed cells.Nature448553560

47. Roh TY, Wei G, Farrell CM, Zhao K (2007) Genome-wide prediction of conserved and nonconserved enhancers by histone acetylation patterns. Genome Res 17: 74–81.TY RohG. WeiCM FarrellK. Zhao2007Genome-wide prediction of conserved and nonconserved enhancers by histone acetylation patterns.Genome Res177481
48. Roh TY, Cuddapah S, Cui KR, Zhao KJ (2006) The genomic landscape of histone modifications in human T cells. Proc Natl Acad Sci U S A 103: 15782–15787.TY RohS. CuddapahKR CuiKJ Zhao2006The genomic landscape of histone modifications in human T cells.Proc Natl Acad Sci U S A1031578215787
49. Zilberman D, Gehring M, Tran RK, Ballinger T, Henikoff S (2007) Genome-wide analysis of Arabidopsis thaliana DNA methylation uncovers an interdependence between methylation and transcription. Nat Genet 39: 61–69.D. ZilbermanM. GehringRK TranT. BallingerS. Henikoff2007Genome-wide analysis of Arabidopsis thaliana DNA methylation uncovers an interdependence between methylation and transcription.Nat Genet396169
50. Zhang XY, Yazaki J, Sundaresan A, Cokus S, Chan SWL, et al. (2006) Genome-wide high-resolution mapping and functional analysis of DNA methylation in Arabidopsis. Cell 126: 1189–1201.XY ZhangJ. YazakiA. SundaresanS. CokusSWL Chan2006Genome-wide high-resolution mapping and functional analysis of DNA methylation in Arabidopsis.Cell12611891201
51. Vaughn MW, Tanurdžić M, Lippman Z, Jiang H, Carrasquillo R, et al. (2007) Epigenetic natural variation in Arabidopsis thaliana. PLoS Biol 5: e174. MW VaughnM. TanurdžićZ. LippmanH. JiangR. Carrasquillo2007Epigenetic natural variation in Arabidopsis thaliana.PLoS Biol5e174
52. Zhang X, Clarenz O, Cokus S, Bernatavichute YV, Pellegrini M, et al. (2007) Whole-genome analysis of histone H3 lysine 27 trimethylation in Arabidopsis. PLoS Biol 5: e129. X. ZhangO. ClarenzS. CokusYV BernatavichuteM. Pellegrini2007Whole-genome analysis of histone H3 lysine 27 trimethylation in Arabidopsis.PLoS Biol5e129
53. Turck F, Roudier F, Farrona S, Martin-Magniette M-L, Guillaume

E, et al. (2007) Arabidopsis TFL2/LHP1 specifically associates with genes marked by trimethylation of histone H3 lysine 27. PLoS Genet 3: e86. F. TurckF. RoudierS. FarronaM-L Martin-MagnietteE. Guillaume2007Arabidopsis TFL2/LHP1 specifically associates with genes marked by trimethylation of histone H3 lysine 27.PLoS Genet3e86

54. Gendrel AV, Lippman Z, Yordan C, Colot V, Martienssen RA (2002) Dependence of heterochromatic histone H3 methylation patterns on the Arabidopsis gene DDM1. Science 297: 1871–1873.AV GendrelZ. LippmanC. YordanV. ColotRA Martienssen2002Dependence of heterochromatic histone H3 methylation patterns on the Arabidopsis gene DDM1.Science29718711873
55. Kinoshita T, Miura A, Choi Y, Kinoshita Y, Cao X, et al. (2004) One-way control of FWA imprinting in Arabidopsis endosperm by DNA methylation. Science 303: 521–523.T. KinoshitaA. MiuraY. ChoiY. KinoshitaX. Cao2004One-way control of FWA imprinting in Arabidopsis endosperm by DNA methylation.Science303521523
56. Bastow R, Mylne JS, Lister C, Lippman Z, Martienssen RA, et al. (2004) Vernalization requires epigenetic silencing of FLC by histone methylation. Nature 427: 164–167.R. BastowJS MylneC. ListerZ. LippmanRA Martienssen2004Vernalization requires epigenetic silencing of FLC by histone methylation.Nature427164167
57. Schmitz RJ, Amasino RM (2007) Vernalization: a model for investigating epigenetics and eukaryotic gene regulation in plants. Biochim Biophys Acta 1769: 269–275.RJ SchmitzRM Amasino2007Vernalization: a model for investigating epigenetics and eukaryotic gene regulation in plants.Biochim Biophys Acta1769269275
58. Lippman Z, Gendrel AV, Colot V, Martienssen R (2005) Profiling DNA methylation patterns using genomic tiling microarrays. Nat Methods 2: 219–224.Z. LippmanAV GendrelV. ColotR. Martienssen2005Profiling DNA methylation patterns using genomic tiling microarrays.Nat Methods2219224
59. Mi SJ, Cai T, Hu YG, Chen Y, Hodges E, et al. (2008) Sorting of small RNAs into Arabidopsis argonaute complexes is directed by the 5′ terminal nucleotide. Cell 133: 116–127.SJ MiT. CaiYG

HuY. ChenE. Hodges2008Sorting of small RNAs into Arabidopsis argonaute complexes is directed by the 5′ terminal nucleotide. Cell133116127

60. Montgomery TA, Howell MD, Cuperus JT, Li DW, Hansen JE, et al. (2008) Specificity of ARGONAUTE7-miR390 interaction and dual functionality in TAS3 trans-acting siRNA formation. Cell 133: 128–141.TA MontgomeryMD HowellJT CuperusDW LiJE Hansen2008Specificity of ARGONAUTE7-miR390 interaction and dual functionality in TAS3 trans-acting siRNA formation. Cell133128141
61. Richards EJ (2006) Inherited epigenetic variation–revisiting soft inheritance. Nat Rev Genet 7: 395–401.EJ Richards2006Inherited epigenetic variation–revisiting soft inheritance.Nat Rev Genet7395401
62. Soppe WJ, Jacobsen SE, Alonso-Blanco C, Jackson JP, Kakutani T, et al. (2000) The late flowering phenotype of fwa mutants is caused by gain-of-function epigenetic alleles of a homeodomain gene. Mol Cell 6: 791–802.WJ SoppeSE JacobsenC. Alonso-BlancoJP JacksonT. Kakutani2000The late flowering phenotype of fwa mutants is caused by gain-of-function epigenetic alleles of a homeodomain gene.Mol Cell6791802
63. Phillips RL, Kaeppler SM, Olhoft P (1994) Genetic instability of plant-tissue cultures: breakdown of normal controls. Proc Natl Acad Sci U S A 91: 5222–5226.RL PhillipsSM KaepplerP. Olhoft1994Genetic instability of plant-tissue cultures: breakdown of normal controls.Proc Natl Acad Sci U S A9152225226
64. Santos-Rosa H, Schneider R, Bannister AJ, Sherriff J, Bernstein BE, et al. (2002) Active genes are tri-methylated at K4 of histone H3. Nature 419: 407–411.H. Santos-RosaR. SchneiderAJ BannisterJ. SherriffBE Bernstein2002Active genes are tri-methylated at K4 of histone H3.Nature419407411
65. Li XY, Wang XF, He K, Ma YQ, Su N, et al. (2008) High-resolution mapping of epigenetic modifications of the rice genome uncovers interplay between DNA methylation, histone methylation, and gene expression. Plant Cell 20: 259–276.XY LiXF WangK. HeYQ MaN. Su2008High-resolution mapping of epigenetic modifications of the

rice genome uncovers interplay between DNA methylation, histone methylation, and gene expression.Plant Cell20259276

66. Ooi SKT, Qiu C, Bernstein E, Li KQ, Jia D, et al. (2007) DNMT3L connects unmethylated lysine 4 of histone H3 to de novo methylation of DNA. Nature 448: 714–U713.SKT OoiC. QiuE. BernsteinKQ LiD. Jia2007DNMT3L connects unmethylated lysine 4 of histone H3 to de novo methylation of DNA.Nature448714U713
67. Xu F, Zhang K, Grunstein M (2005) Acetylation in histone H3 globular domain regulates gene expression in yeast. Cell 121: 375–385.F. XuK. ZhangM. Grunstein2005Acetylation in histone H3 globular domain regulates gene expression in yeast.Cell121375385
68. Rufiange A, Jacques P-E, Bhat W, Robert F, Nourani A (2007) Genome-wide replication-independent histone H3 exchange occurs predominantly at promoters and implicates H3 K56 acetylation and Asf1. Mol Cell 27: 393–405.A. RufiangeP-E JacquesW. BhatF. RobertA. Nourani2007Genome-wide replication-independent histone H3 exchange occurs predominantly at promoters and implicates H3 K56 acetylation and Asf1.Mol Cell27393405
69. Schneider J, Bajwa P, Johnson FC, Bhaumik SR, Shilatifard A (2006) Rtt109 is required for proper H3K56 acetylation: a chromatin mark associated with the elongating RNA polymerase II. J Biol Chem 281: 37270–37274.J. SchneiderP. BajwaFC JohnsonSR BhaumikA. Shilatifard2006Rtt109 is required for proper H3K56 acetylation: a chromatin mark associated with the elongating RNA polymerase II.J Biol Chem2813727037274
70. Li Q, Zhou H, Wurtele H, Davies B, Horazdovsky B, et al. (2008) Acetylation of histone H3 lysine 56 regulates replication-coupled nucleosome assembly. Cell 134: 244–255.Q. LiH. ZhouH. WurteleB. DaviesB. Horazdovsky2008Acetylation of histone H3 lysine 56 regulates replication-coupled nucleosome assembly.Cell134244255
71. Lister R, O'Malley RC, Tonti-Filippini J, Gregory BD, Berry CC, et al. (2008) Highly integrated single-base resolution maps of the epigenome in Arabidopsis. Cell 133: 523–536.R. ListerRC O'MalleyJ. Tonti-FilippiniBD GregoryCC Berry2008Highly integrated single-base resolution maps of the epigenome in Arabidopsis.Cell133523536

72. Pischke MS, Huttlin EL, Hegeman AD, Sussman MR (2006) A transcriptome-based characterization of habituation in plant tissue culture. Plant Physiol 140: 1255–1278.MS PischkeEL HuttlinAD HegemanMR Sussman2006A transcriptome-based characterization of habituation in plant tissue culture.Plant Physiol14012551278

73. Hirochika H (1993) Activation of tobacco retrotransposons during tissue culture. EMBO J 12: 2521–2528.H. Hirochika1993Activation of tobacco retrotransposons during tissue culture.EMBO J1225212528

74. Jeddeloh JA, Stokes TL, Richards EJ (1999) Maintenance of genomic methylation requires a SWI2/SNF2-like protein. Nat Genet 22: 94–97.JA JeddelohTL StokesEJ Richards1999Maintenance of genomic methylation requires a SWI2/SNF2-like protein.Nat Genet229497

75. Herr AJ, Jensen MB, Dalmay T, Baulcombe DC (2005) RNA polymerase IV directs silencing of endogenous DNA. Science 308: 118–120.AJ HerrMB JensenT. DalmayDC Baulcombe2005RNA polymerase IV directs silencing of endogenous DNA. Science308118120

76. Pontes O, Li CF, Nunes PC, Haag J, Ream T, et al. (2006) The Arabidopsis chromatin-modifying nuclear siRNA pathway involves a nucleolar RNA processing center. Cell 126: 79–92.O. PontesCF LiPC NunesJ. HaagT. Ream2006The Arabidopsis chromatin-modifying nuclear siRNA pathway involves a nucleolar RNA processing center.Cell1267992

77. Finnegan DJ, Rubin GM, Young MW, Hogness DS (1977) Repeated gene families in Drosophila melanogaster. Cold Spring Harb Symp Quant Biol 42: 1053–1063.DJ FinneganGM RubinMW YoungDS Hogness1977Repeated gene families in Drosophila melanogaster. Cold Spring Harb Symp Quant Biol4210531063

78. Flavell AJ, Ishhorowicz D (1981) Extrachromosomal circular copies of the eukaryotic transposable element copia in cultured Drosophila cells. Nature 292: 591–595.AJ FlavellD. Ishhorowicz1981Extrachromosomal circular copies of the eukaryotic transposable element copia in cultured Drosophila cells. Nature292591595

79. Potter SS, Brorein WJ, Dunsmuir P, Rubin GM (1979) Transposition

of elements of the 412, copia and 297 dispersed repeated gene families in Drosophila. Cell 17: 415–427.SS PotterWJ BroreinP. DunsmuirGM Rubin1979Transposition of elements of the 412, copia and 297 dispersed repeated gene families in Drosophila. Cell17415427

80. Di Franco C, Pisano C, Fourcade-Peronnet F, Echalier G, Junakovic N (1992) Evidence for de novo rearrangements of Drosophila transposable elements induced by the passage to the cell culture. Genetica 87: 65–73.C. Di FrancoC. PisanoF. Fourcade-PeronnetG. EchalierN. Junakovic1992Evidence for de novo rearrangements of Drosophila transposable elements induced by the passage to the cell culture.Genetica876573

81. Maisonhaute C, Ogereau D, Hua-Van A, Capy P (2007) Amplification of the 1731 LTR retrotransposon in Drosophila melanogaster cultured cells: origin of neocopies and impact on the genome. Gene 393: 116–126.C. MaisonhauteD. OgereauA. Hua-VanP. Capy2007Amplification of the 1731 LTR retrotransposon in Drosophila melanogaster cultured cells: origin of neocopies and impact on the genome.Gene393116126

82. Slotkin RK, Martienssen R (2007) Transposable elements and the epigenetic regulation of the genome. Nat Rev Genet 8: 272–285. RK SlotkinR. Martienssen2007Transposable elements and the epigenetic regulation of the genome.Nat Rev Genet8272285

83. Magewu AN, Jones PA (1994) Ubiquitous and tenacious methylation of the CpG site in codon 248 of the p53 gene may explain its frequent appearance as a mutational hot spot in human cancer. Mol Cell Biol 14: 4225–4232.AN MagewuPA Jones1994Ubiquitous and tenacious methylation of the CpG site in codon 248 of the p53 gene may explain its frequent appearance as a mutational hot spot in human cancer.Mol Cell Biol1442254232

84. Kochanek S, Renz D, Doerfler W (1993) DNA Methylation in the Alu sequences of diploid and haploid primary human-cells. EMBO J 12: 1141–1151.S. KochanekD. RenzW. Doerfler1993DNA Methylation in the Alu sequences of diploid and haploid primary human-cells.EMBO J1211411151

85. Alves G, Tatro A, Fanning T (1996) Differential methylation of

human LINE-1 retrotransposons in malignant cells. Gene 176: 39–44.G. AlvesA. TatroT. Fanning1996Differential methylation of human LINE-1 retrotransposons in malignant cells.Gene1763944

86. Dante R, Dantepaire J, Rigal D, Roizes G (1992) Methylation patterns of long interspersed repeated DNA and alphoid repetitive DNA from human cell-lines and tumors. Anticancer Res 12: 559–563.R. DanteJ. DantepaireD. RigalG. Roizes1992Methylation patterns of long interspersed repeated DNA and alphoid repetitive DNA from human cell-lines and tumors.Anticancer Res12559563
87. Florl AR, Lower R, Schmitz-Drager BJ, Schulz WA (1999) DNA methylation and expression of LINE-1 and HERV-K provirus sequences in urothelial and renal cell carcinomas. Brit J Cancer 80: 1312–1321.AR FlorlR. LowerBJ Schmitz-DragerWA Schulz1999DNA methylation and expression of LINE-1 and HERV-K provirus sequences in urothelial and renal cell carcinomas. Brit J Cancer8013121321
88. Howard G, Eiges R, Gaudet F, Jaenisch R, Eden A (2007) Activation and transposition of endogenous retroviral elements in hypomethylation induced tumors in mice. Oncogene 27: 404–408.G. HowardR. EigesF. GaudetR. JaenischA. Eden2007Activation and transposition of endogenous retroviral elements in hypomethylation induced tumors in mice.Oncogene27404408
89. Walsh CP, Chaillet JR, Bestor TH (1998) Transcription of IAP endogenous retroviruses is constrained by cytosine methylation. Nat Genet20:116–117.CPWalshJRChailletTHBestor1998Transcription of IAP endogenous retroviruses is constrained by cytosine methylation.Nat Genet20116117
90. Schulz WA, Steinhoff C, Florl AR (2006) Methylation of endogenous human retroelements in health and disease. Curr Top Microbiol Immunol 310: 211–250.WA SchulzC. SteinhoffAR Florl2006Methylation of endogenous human retroelements in health and disease.Curr Top Microbiol Immunol310211250
91. Sinibaldi-Vallebona P, Lavia P, Garaci E, Spadafora C (2006) A role for endogenous reverse transcriptase in tumorigenesis and as a target in differentiating cancer therapy. Genes Chromosomes Cancer 45: 1–10.P. Sinibaldi-VallebonaP. LaviaE. GaraciC. Spadafora2006A

role for endogenous reverse transcriptase in tumorigenesis and as a target in differentiating cancer therapy.Genes Chromosomes Cancer45110

92. Szyf M, Detich N (2001) Regulation of the DNA methylation machinery and its role in cellular transformation. Prog Nucleic Acid Res Mol Biol 69: 47–79.M. SzyfN. Detich2001Regulation of the DNA methylation machinery and its role in cellular transformation. Prog Nucleic Acid Res Mol Biol694779
93. Smiraglia DJ, Rush LJ, Fruhwald MC, Dai ZY, Held WA, et al. (2001) Excessive CpG island hypermethylation in cancer cell lines versus primary human malignancies. Hum Mol Genet 10: 1413–1419.DJ SmiragliaLJ RushMC FruhwaldZY DaiWA Held2001Excessive CpG island hypermethylation in cancer cell lines versus primary human malignancies.Hum Mol Genet1014131419
94. Baylin S, Bestor TH (2002) Altered methylation patterns in cancer cell genomes: cause or consequence. Cancer Cell 1: 299–305.S. BaylinTH Bestor2002Altered methylation patterns in cancer cell genomes: cause or consequence.Cancer Cell1299305
95. Antequera F, Boyes J, Bird A (1990) High levels of de novo methylation and altered chromatin structure at Cpg islands in cell lines. Cell 62: 503–514.F. AntequeraJ. BoyesA. Bird1990High levels of de novo methylation and altered chromatin structure at Cpg islands in cell lines.Cell62503514
96. Ordway JM, Bedell JA, Citek RW, Nunberg A, Garrido A, et al. (2006) Comprehensive DNA methylation profiling in a human cancer genome identifies novel epigenetic targets. Carcinogenesis 27: 2409–2423.JM OrdwayJA BedellRW CitekA. NunbergA. Garrido2006Comprehensive DNA methylation profiling in a human cancer genome identifies novel epigenetic targets. Carcinogenesis2724092423
97. Wilson VL, Jones PA (1983) DNA methylation decreases in aging but not in immortal cells. Science 220: 1054–1057.VL WilsonPA Jones1983DNA methylation decreases in aging but not in immortal cells.Science22010541057
98. Meissner A, Mikkelsen TS, Gu HC, Wernig M, Hanna J, et al. (2008) Genome-scale DNA methylation maps of pluripotent

and differentiated cells. Nature 454: 766–U791.A. MeissnerTS MikkelsenHC GuM. WernigJ. Hanna2008Genome-scale DNA methylation maps of pluripotent and differentiated cells. Nature454766U791

99. Feinberg AP, Vogelstein B (1983) Hypomethylation distinguishes genes of some human cancers from their normal counterparts. Nature 301: 89–92.AP FeinbergB. Vogelstein1983Hypomethylation distinguishes genes of some human cancers from their normal counterparts.Nature3018992

100. Costello JF, Fruhwald MC, Smiraglia DJ, Rush LJ, Robertson GP, et al. (2000) Aberrant CpG-island methylation has non-random and tumour-type-specific patterns. Nat Genet 24: 132–138.JF CostelloMC FruhwaldDJ SmiragliaLJ RushGP Robertson2000Aberrant CpG-island methylation has non-random and tumour-type-specific patterns.Nat Genet24132138

101. Kaneda A, Tsukamoto T, Takamura-Enya T, Watanabe N, Kaminishi M, et al. (2004) Frequent hypomethylation in multiple promoter CpG islands is associated with global hypomethylation, but not with frequent promoter hypermethylation. Cancer Sci 95: 58–64.A. KanedaT. TsukamotoT. Takamura-EnyaN. WatanabeM. Kaminishi2004Frequent hypomethylation in multiple promoter CpG islands is associated with global hypomethylation, but not with frequent promoter hypermethylation.Cancer Sci955864

102. Ting AH, McGarvey KM, Baylin SB (2006) The cancer epigenome–components and functional correlates. Genes Dev 20: 3215–3231. AH TingKM McGarveySB Baylin2006The cancer epigenome–components and functional correlates.Genes Dev2032153231

103. Vertino P, Issa J, Pereira-Smith O, Baylin S (1994) Stabilization of DNA methyltransferase levels and CpG island hypermethylation precede SV40-induced immortalization of human fibroblasts. Cell Growth Differ 5: 1395–1402.P. VertinoJ. IssaO. Pereira-SmithS. Baylin1994Stabilization of DNA methyltransferase levels and CpG island hypermethylation precede SV40-induced immortalization of human fibroblasts.Cell Growth Differ513951402

104. Wong DJ, Foster SA, Galloway DA, Reid BJ (1999) Progressive region-specific de novo methylation of the p16 CpG Island in

primary human mammary epithelial cell strains during escape from M0 growth arrest. Mol Cell Biol 19: 5642–5651.DJ WongSA FosterDA GallowayBJ Reid1999Progressive region-specific de novo methylation of the p16 CpG Island in primary human mammary epithelial cell strains during escape from M0 growth arrest.Mol Cell Biol1956425651

105. Jones PA, Wolkowicz MJ, Rideout WM, Gonzales FA, Marziasz CM, et al. (1990) Denovo methylation of the MyoD1 CpG island during the establishment of immortal cell lines. Proc Natl Acad Sci U S A 87: 6117–6121.PA JonesMJ WolkowiczWM RideoutFA GonzalesCM Marziasz1990Denovo methylation of the MyoD1 CpG island during the establishment of immortal cell lines.Proc Natl Acad Sci U S A8761176121

106. .Kulaeva OI, Draghici S, Tang L, Kraniak JM, Land SJ, et al. (2003) Epigenetic silencing of multiple interferon pathway genes after cellular immortalization. Oncogene 22: 4118–4127.OI KulaevaS. DraghiciL. TangJM KraniakSJ Land2003Epigenetic silencing of multiple interferon pathway genes after cellular immortalization. Oncogene2241184127

107. Mathieu O, Reinders J, Caikovski M, Smathajitt C, Paszkowski J (2007) Transgenerational stability of the Arabidopsis epigenome is coordinated by CG methylation. Cell 130: 851–862.O. MathieuJ. ReindersM. CaikovskiC. SmathajittJ. Paszkowski2007Transgenerational stability of the Arabidopsis epigenome is coordinated by CG methylation.Cell130851862

108. Gendrel AV, Lippman Z, Martienssen R, Colot V (2005) Profiling histone modification patterns in plants using genomic tiling microarrays. Nat Methods 2: 213–218.AV GendrelZ. LippmanR. MartienssenV. Colot2005Profiling histone modification patterns in plants using genomic tiling microarrays.Nat Methods2213218

109. Brennecke J, Aravin AA, Stark A, Dus M, Kellis M, et al. (2007) Discrete small RNA-generating loci as master regulators of transposon activity in Drosophila. Cell 128: 1089–1103.J. BrenneckeAA AravinA. StarkM. DusM. Kellis2007Discrete small RNA-generating loci as master regulators of transposon activity in Drosophila.Cell12810891103

110. Kohany O, Gentles AJ, Hankus L, Jurka J (2006) Annotation, submission and screening of repetitive elements in Repbase: RepbaseSubmitter and Censor. BMC Bioinformatics 7: 474.O. KohanyAJ GentlesL. HankusJ. Jurka2006Annotation, submission and screening of repetitive elements in Repbase: RepbaseSubmitter and Censor.BMC Bioinformatics7474

111. Jurka J, Kapitonov VV, Pavlicek A, Klonowski P, Kohany O, et al. (2005) Repbase update, a database of eukaryotic repetitive elements. Cytogenet Genome Res 110: 462–467.J. JurkaVV KapitonovA. PavlicekP. KlonowskiO. Kohany2005Repbase update, a database of eukaryotic repetitive elements.Cytogenet Genome Res110462467

112. Benson G (1999) Tandem repeats finder: a program to analyze DNA sequences. Nucleic Acids Res 27: 573–580.G. Benson1999Tandem repeats finder: a program to analyze DNA sequences.Nucleic Acids Res27573580

113. Griffiths-Jones S, Moxon S, Marshall M, Khanna A, Eddy SR, et al. (2005) Rfam: annotating non-coding RNAs in complete genomes. Nucleic Acids Res 33: D121–D124.S. Griffiths-JonesS. MoxonM. MarshallA. KhannaSR Eddy2005Rfam: annotating non-coding RNAs in complete genomes.Nucleic Acids Res33D121D124

114. Kent WJ (2002) BLAT: the BLAST-like alignment tool. Genome Res 12: 656–664.WJ Kent2002BLAT: the BLAST-like alignment tool.Genome Res12656664

115. Dsouza M, Larsen N, Overbeek R (1997) Searching for patterns in genomic data. Trends Genet 13: 497–498.M. DsouzaN. LarsenR. Overbeek1997Searching for patterns in genomic data.Trends Genet13497498

116. BenjaminiY,HochbergY(1995)ControllingtheFalseDiscoveryRate: a practical and powerful approach to multiple testing. J Roy Statist Soc Ser B 57: 289–300.Y. BenjaminiY. Hochberg1995Controlling the False Discovery Rate: a practical and powerful approach to multiple testing.J Roy Statist Soc Ser B57289300

CHAPTER

17

CLONING OF TRANSGENIC TOBACCO BY-2 CELLS; AN EFFICIENT METHOD TO ANALYSE AND REDUCE HIGH NATURAL HETEROGENEITY OF TRANSGENE EXPRESSION

Eva Nocarova and Lukas Fischer

Department of Plant Physiology, Charles University in Prague, Faculty of Science

ABSTRACT

Background

Phenotypic characterization of transgenic cell lines, frequently used in plant biology studies, is complicated because transgene expression in individual cells is often heterogeneous and unstable. To identify the sources and to reduce this heterogeneity, we transformed tobacco (*Nicotiana tabacum* L.) BY-2 cells with a gene encoding green fluorescent protein (GFP) using *Agrobacterium tumefaciens*, and then introduced a simple cloning procedure to generate cell lines derived from the individual transformed cells. Expression of the transgene was monitored by analysing GFP fluorescence in the cloned lines and also in lines obtained directly after transformation.

Results

The majority (~90%) of suspension culture lines derived from *calli* that were obtained directly from transformation consisted of cells with various levels of GFP fluorescence. In contrast, nearly 50% of lines generated by cloning cells from the primary heterogeneous suspensions consisted of cells with homogenous GFP fluorescence. The rest of the lines exhibited "permanent heterogeneity" that could not be resolved by cloning. The extent of fluorescence heterogeneity often varied, even among genetically identical clones derived from the primary transformed lines. In contrast, the offspring of subsequent cloning of the cloned lines was uniform, showing GFP fluorescence intensity and heterogeneity that corresponded to the original clone.

Conclusion

The results demonstrate that, besides genetic heterogeneity detected in some lines, the primary lines often contained a mixture of epigenetically different cells that could be separated by cloning. This indicates that a single integration event frequently results in various heritable expression patterns, which are probably accidental and become stabilized in the offspring of the primary transformed cells early after the integration event. Because heterogeneity in transgene expression has proven to be a serious problem, it is highly advisable to use transgenes tagged with a visual marker for BY-2 transformation. The cloning procedure can be used not only for efficient reduction of expression heterogeneity of such transgenes, but also as a useful tool for studies of transgene expression and other purposes.

BACKGROUND

Tobacco BY-2 cell line is the most popular and widely used cell line in plant research. Hundreds of scientific papers have been published using this line as a model to study various aspects of plant cell physiology. BY-2 cells are relatively homogenous, allowing studies of cell phenotypes [1]. Moreover, the cells exhibit high growth rate, enabling synchronization of cell divisions and cell-cycle analyses [2, 3]. Being easily transformable either by particle bombardment [4] or by co-cultivation with *Agrobacterium tumefaciens* [5], transgenic derivatives of BY-2 cell line have had high impact in analyses of protein function by ectopic expression, gene knock-outs or translational gene fusions. GFP-tagging of proteins provides viable staining of different cell structures and organelles and analyses of subcellular protein localization [6, 7]. The expression of fluorescent protein constructs can be easily monitored by fluorescence microscopy, whereas expression of non-tagged transgenes cannot be readily detected at individual cell level. In both cases, homogenous and stable expression of transgenes is highly desirable for both molecular/biochemical analyses of the total cell culture and for monitoring the effects of transgene expression in individual cells.

Variation in transgene expression in independent transgenic lines has been repeatedly reported to be related to the sequence of the introduced gene construct, involving RNA-sensing mechanism, the locus of insertion, the number of insertion copies, and the initial level of transgene expression [8, 9, 10, 11, 12, 13].

The impact of the position of the inserted transgene in the chromosomal environment remains unclear, and the reports are partly controversial. In contrast to classical studies [8], Schubert with colleagues reported that the site of insertion had rather marginal effect; the expression of reporter genes under the control of a strong promoter was comparable among independent transgenic *Arabidopsis* plants harbouring the same transgene copy number [11]. However, silenced transgenes integrated into heterochromatin regions were not included in the study due to selection bias, as revealed by subsequent analyses of transgenic plants or cell lines generated without selection pressure [12, 14]. Recently, Fischer with colleagues showed that the integration site significantly influences the sensitivity of the transgene to RNA silencing rather than affecting its initial expression level [15].

In contrast to numerous analyses of independent transgenic lines, much less attention has been paid to analyses of genetically identical clones [16, 17], which could bring valuable information about the variability of transgene expression independently of the positional effect.

Analyses of GFP-tagged transgenic BY-2 cell lines in our laboratory repeatedly produced only a low frequency of lines with well-balanced and stable fluorescence in all cells. In order to analyse the nature and sources of this variability, we transformed tobacco BY-2 cell line with a gene encoding free GFP, which allows simple *in situ* evaluation of transgene expression levels *via* assessment of green fluorescence. The homogeneity and stability of GFP fluorescence was monitored in both the primary *calli* obtained after *Agrobacterium*-mediated transformation and in suspension cultures derived from these *calli*. In order to eliminate high natural heterogeneity in *GFP* expression found in the primary lines, we introduced a simple cloning procedure. In addition to reducing the heterogeneity of GFP expression, the method also offered the opportunity to study the variability of transgene expression in genetically homogeneous clones, thus contributing to understanding of the impact of positional effect in transgene expression.

RESULTS

GFP fluorescence in primary *calli* and suspensions obtained after transformation

About 70% of round-shaped *calli* that were obtained after the *Agrobacterium*-mediated transformation of BY-2 cells displayed GFP fluorescence intensity sufficient for reliable evaluation of its homogeneity. Out of these *calli*, the GFP fluorescence was homogenous over the whole *callus* in only 35 – 50% cases in three independent transformations (Figure 1a; Table 1). The rest of the *calli* contained regions with evidently different levels of GFP fluorescence. Out of these heterogeneous *calli*, ~25% formed separate sectors (Figure 1a) and 36% were mixed in a mosaic arrangement (Figure 1a; Table 1). The frequency of these categories was comparable in all three transformations.

Frequencies of homogeneous and heterogeneous *calli* and suspension cultures derived from these *calli* in primary lines obtained after transformation, and in secondary lines produced by cloning of primary heterogeneous (!) suspensions. Values represent means ± SD (n = 3); data are from three independent transformations; in every replication the number of evaluated lines was 60–80 for *calli* and ~20 for suspensions).

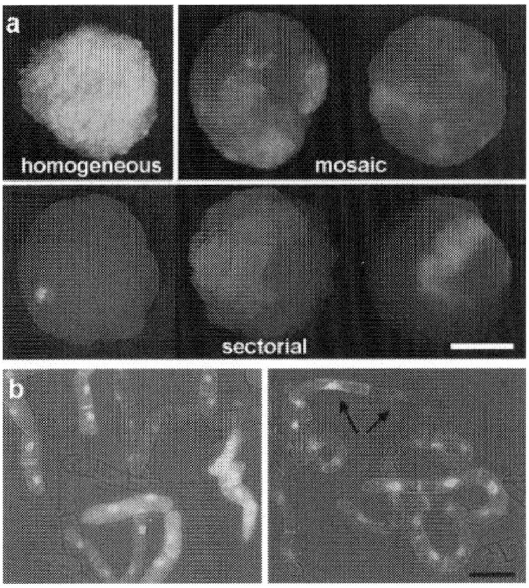

Figure 1: Heterogeneity of GFP fluorescence in BY-2 *calli* and suspension cultures. (a) Primary *calli* obtained after transformation, showing *calli* with homogeneous and heterogeneous *GFP* expression either in mosaic or sectorial arrangements of cell populations with distinct GFP fluorescence. (**b**) Non-homogenous *GFP* expression in suspension cells. The arrows indicate cells with evidently different *GFP* expressions located in a single file. Scale bars: 1 mm for A, 50 μm for B.

Suspension cultures derived from the mixed *calli* contained cells with various GFP fluorescence intensities, as expected. However, also the majority (~70%) of homogenous *calli* gave rise to heterogeneous suspensions (Figure 1b). The cells with varied GFP fluorescence were predominantly in separate cell files, but occasionally were located even within a single cell file (Figure 1b). Classifying cells according to their GFP fluorescence intensities as high, low, or no fluorescence revealed that the proportions among the categories remained stable in the majority of the suspension cultures. Only in few lines (6/3; 1/2) the proportion of cells with high GFP fluorescence gradually declined with time (Figure 2).

Table 1: Frequencies of BY-2 *calli* and suspensions with homogeneous and heterogeneous GFP fluorescence

		Primary suspensions			**Secondary suspensions**	
GFP fluorescence in*callus*	**Primary*calli***	**Homogeneous**	**Heterogeneous**	**Secondary-*calli***	**Homogeneous**	**Heterogeneous**
Homogeneous	39.3% ± 9.7%	29.2% ± 5.3%	70.8% ± 5.3%	93% ± 2.3%	46.3% ± 5.4%	53.7% ± 5.4%
Heterogeneous – mosaic	35.8% ± 13%	0%	100%	7% ± 2.3%	0%	100%
Heterogeneous – sectorial	24.9% ± 6.8%	0%	100%	0%	0%	100%
		11.5% in total	**88.5% in total**		**42.8% in total**	**57.2% in total**

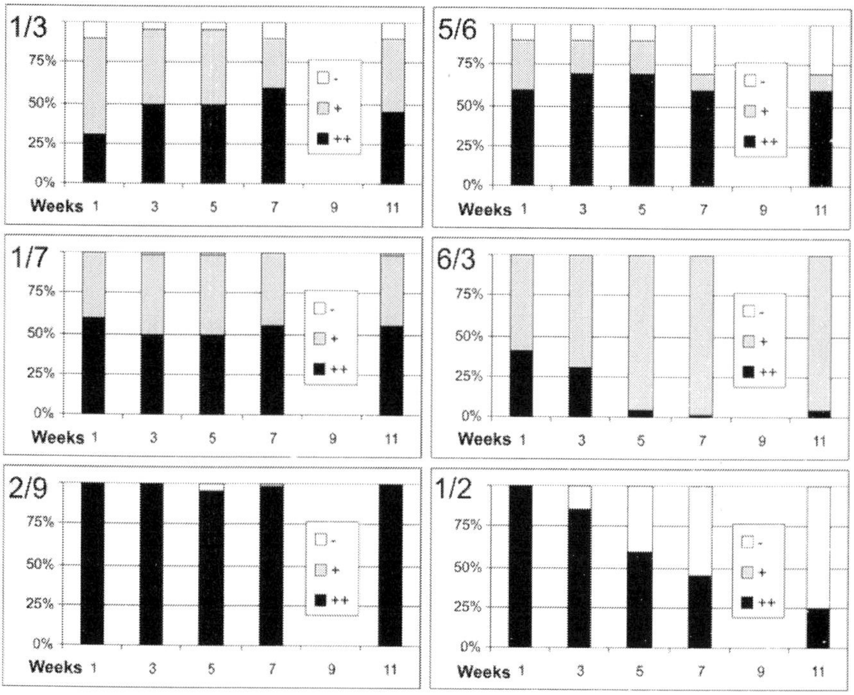

Figure 2: Changes in frequency of GFP-expression categories in selected primary BY-2 suspension cultures with time. GFP-expression categories: ++, strong fluorescence; + weak fluorescence, - no fluorescence.

Cloning of suspension cultures

We introduced a simple and rapid method to generate clones from individual cells or cell files from the suspension cultures as follows: An excess of wild-type BY-2 suspension cells was added to the suspension culture of transformed (kanamycin resistant) cells in stationary phase of growth (Figure 3a). The mixture was then applied onto a Petri dish with solid MS medium containing kanamycin. Within 10 days, macroscopically visible *calli* appeared from individual resistant cells or cell files on the "feeder layer" of the wild-type BY-2 cells (Figure 3b, c). Few days later these "secondary" *calli* reached the size of 1 – 3 mm and could be transferred to a fresh medium for subsequent evaluation of GFP fluorescence homogeneity.

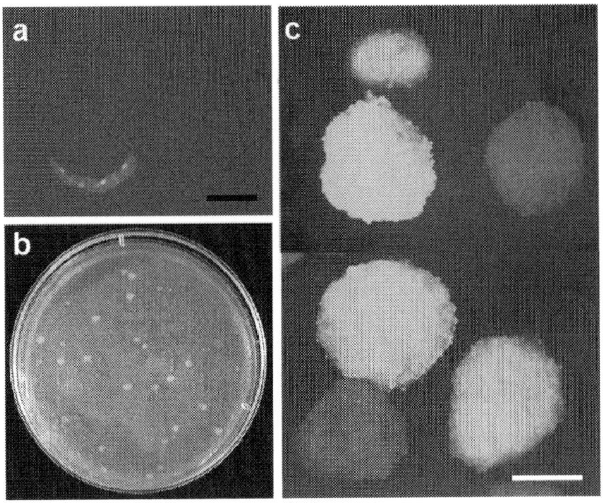

Figure 3: A scheme of the BY-2 cloning procedure. (a) Mixture of transgenic and wild-type lines before plating onto solid media. (b) Cloned *calli* emerging on the feeder layer ~10 days after plating. (**c**) Cloned calli of a heterogeneous line observed with a fluorescence stereomicroscope. Scale bars: 100 μm for A, 1 mm for C.

GFP fluorescence in secondary *calli*and suspensions

Cloning of suspension cultures with heterogeneous GFP fluorescence resulted in secondary *calli*, of which an average 93% gave rise to cell lines with almost exclusively homogenous GFP fluorescence (Table 1). Although the majority of secondary *calli* seemed to be homogenous, more than half of the suspension cultures derived from these *calli* still consisted of cell populations with various GFP levels (Table 1). The frequency of homogenous suspensions varied depending on the original suspension (Figure 4). Some lines (e.g. 1/3, 5/6) gave rise to both homogenous and heterogeneous suspension, whereas in the case of line 1/7 practically all the secondary and tertiary suspensions were homogenous, with either high or low GFP fluorescence intensities. In the case of line 6/3 or a secondary clone 5/6a, the original heterogeneity of GFP fluorescence persisted in all derived clones (Figure 4); even tertiary cloning of these "permanently heterogeneous" lines (e.g. 5/6am) did not diminish their heterogeneity. Generally, subsequent cloning of secondary clones produced almost exclusively homogeneous offspring. Their heterogeneity patterns in terms of proportions of individual GFP fluorescence categories corresponded to those of the original clones (Figure 4). Cloning of suspensions with homogeneous GFP fluorescence consistently gave homogenous subclones (e.g. 1/7d in Figure 4).

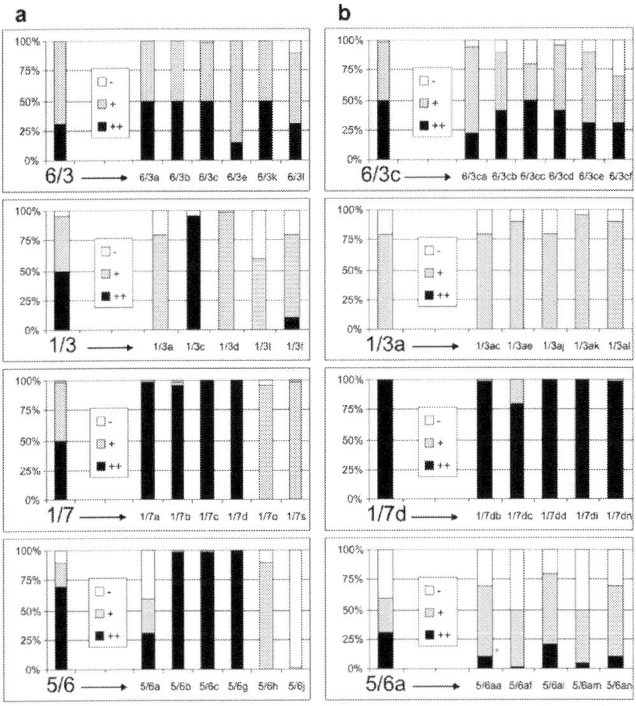

Figure 4: Frequency of GFP-expression categories after the cloning of primary and secondary BY-2 lines. (a) Primary cloning of suspensions obtained from *calli* directly after transformation. (**b**) Secondary cloning of selected subclones. The cloned lines are on the left, with progenies indicated by letters on the right. GFP-expression categories: ++, strong fluorescence; + weak fluorescence, - no fluorescence.

The proportions of GFP fluorescence categories in both hetero- and homogeneous suspensions remained stable for months in almost all cloned secondary suspensions (data not shown).

Molecular analysis of the causes of GFP fluorescence heterogeneity

Analysis of T-DNA insertions in individual clones by Southern hybridisation (Figure 5) revealed that GFP fluorescence heterogeneity could have several causes. Lines 1/7 and 1/3 were composed of two genetically different clones that were separable by cloning, as shown by comparing 1/7d with 1/7o or 1/3f with other 1/3 clones (Figure 5). Other clones that also strongly differed in their proportions of the GFP fluorescence categories (Figure 4) seemed to be genetically identical, as shown by comparing 1/3a, 1/3c and 1/3d or 5/6a, 5/6b, 5/6h and 5/6j samples, where the *GFP* probe hybridised with equally-sized restriction fragments after cleavage with *Hin*dIII or *Bam*HI (Figure 5a). A possible presence of mutations within the *35S* promoter or *GFP*sequence was excluded by sequencing of *35S-GFP* cassettes obtained by PCR amplification; all sequences obtained from individual 5/6 clones 5/6a, 5/6b, 5/6h and 5/6j were identical (data not shown).

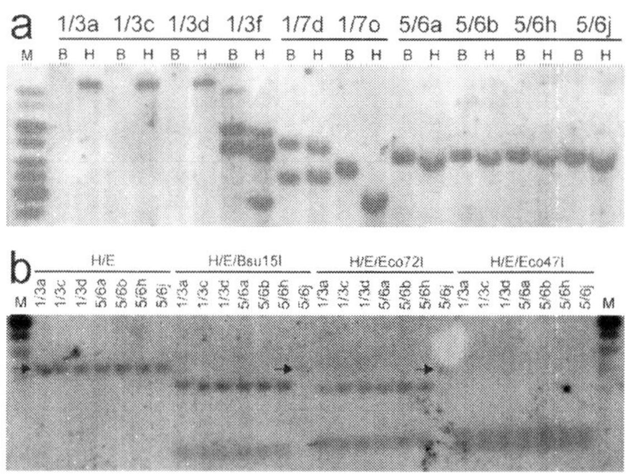

Figure 5: Molecular analysis of selected primary clones. (a) analysis of T-DNA insertions by Southern hybridization of total genomic DNA digested with either *Bam*HI (B) or *Hin*dIII (H). (**b**) Methylation analysis by Southern hybridization of total genomic DNA digested with *Hin*dIII (H) and *Eco*RI (E), cleaving out the *35S-GFP* cassette, and further with methylation-sensitive endocucleases *Bsu*15I, *Eco*72I or *Eco*47I, having restriction sites within the cassette; arrows indicate position of uncleaved *35S-GFP* cassette. The blots were hybridised with DIG-dUTP-labelled *GFP* probe. M, molecular weight ladder.

Genetically identical clones of lines 1/3 and 5/6 were further analysed with respect to DNA methylation. The *35S-GFP* cassette was cleaved out with *Eco*RI and *Hin*dIII, and exposed to the action of several methylation-sensitive enzymes (*Bsu*15I, *Eco*72I and *Eco*47I). Subsequent Southern blot analysis did not reveal any change in cytosine methylation at the analysed restriction sites, with the exception of the completely silenced clone 5/6j, whose DNA seemed to be methylated in the *Bsu*15I and *Eco*72I restriction sites (Figure 5b).

DISCUSSION

Monitoring of GFP fluorescence – a suitable way to estimate *GFP* expression *in situ*

The green fluorescent protein (GFP) is an important reporter molecule for monitoring gene expression and protein localization *in vivo*, *in situ*, and in real-time observation. GFP fluorescence is stable, species-independent, and can be followed non-invasively in living cells where the green fluorescence reports active transcription and translation of the *GFP* gene [18]. Because of its simplicity, monitoring GFP fluorescence has been also routinely used in transgene silencing studies [19, 20]. Although GFP is generally very stable [21], some differences in protein or fluorescence levels may occur due to protein degradation under certain treatments or in certain cell types. Nevertheless, in phenotypically homogeneous cell lines cultured under stable conditions, monitoring GFP fluorescence can be regarded as a suitable method for reliable estimation of *GFP* expression levels.

Sources of *GFP*expression heterogeneity

Analysis of GFP fluorescence in primary suspensions obtained either directly from transformation or from secondary and tertiary clones revealed the coexistence of cell populations with different T-DNA insertions in some primary lines, representing genetic heterogeneity. Although, the majority of *GFP*-expression heterogeneity is most likely determined epigenetically.

Genetic heterogeneity

Southern hybridisation of genomic DNA isolated from selected clones of lines 1/7 and 1/3 clearly documented that even the round-shaped primary *calli* can contain cell populations with different T-DNA insertions. Because the probability of plating independently transformed cell files so closely together was low when considering plating density of ~30 – 50 *calli* per 6 cm-diameter plate, these cells are likely to represent the offspring of independently transformed cells that were located in a single cell file. Considering the large number of plated cells and transformation efficiency of ~0.1 – 0.5%, the results indicate that there could be cell files that are highly susceptible to *Agrobacterium*-mediated transformation. Alternatively, the genetically mixed *calli* could represent the offspring of a single cell transformed in S or G2 phase of the cell cycle with multiple T-DNAs, which then segregated unequally to daughter cells during mitosis.

Epigenetic heterogeneity

Analyses of genetically identical lines have documented developmentally- and environmentally-derived variability in transgene expression [10, 17] and in stress-induced silencing [16]. In our study, *GFP* expression also varied among genetically identical clones even though the cells were phenotypically homogeneous and were cultured under stable conditions. This indicates that factors other than environmental or developmental or stress situations can induce changes in *GFP* expression. Gene expression is generally influenced by specific chromatin marks that may be present on both DNA and associated proteins [22]. Our methylation analysis of the *GF*P gene from selected genetically identical lines revealed that cytosins in analysed restriction sites were methylated only in some restriction sites and only in the line 5/6j with completely silenced *GFP* expression. This indicates that all other observed differences in *GFP* expression levels and heterogeneity are either independent of DNA methylation or methylation occurred in locations other than the selected restriction sites. The epigenetic state of chromatin is heritable through cell division, but can be easily modulated in response to certain triggers. For example, changes at the chromatin level such as cytosine methylation can accompany gene or transgene silencing [22] although the presence of methylated DNA is not necessarily related to the silenced phenotype [23].

Positional effect in the chromosomal environment at the site of transgene integration is known to influence transgene expression [13], although it seems to have much lower impact [11] than previously assumed [8]. Different sites, or arrangements, of T-DNA insertion can influence the accessibility/susceptibility of transgene to epigenetic regulation at either transcriptional or posttranscriptional level [15, 24]; reviewed in

[13]. Specifically in case of integration into heterochromatin region, the epigenetic information is almost regularly reflected in the chromatin structure of the inserted T-DNA, which results in transgene silencing [12, 14, 25]. In our experiments, *GFP* expression often varied strongly among clones with identical T-DNA insertions. For example, a single insertion resulted in completely different transgene expression patterns in genetically identical subclones of line 5/6 (Figure 4). In previous studies, transgene expression was analysed in clonal plant replicates generated long time after the integration event [10, 17]. In contrast, the use of cell cultures and their cloning allowed us to analyse clones/replicates that arose immediately after the transgene integration. Since our results showed that *GFP*-expression patterns were stable and heritable in the cloned lines, the various expression patterns observed among genetically identical clones had to be established and stabilized in the offspring of the primary transformed cells early after the integration event. The process of integration of "naked" T-DNA is known to be accompanied by *de novo* establishment of specific chromatin composition and structure [25]. Our results clearly document that in certain insertion sites the establishment of different epigenetic states/transgene-expression patterns is accidental and independent of chromosomal environment in BY-2 cells.

Heterogeneity in GFP expression derived from the coexistence of either genetically or epigenetically different cells within the primary lines was in some cases resolved by cloning. However, in many clones (e.g. line 6/3; Figure 4) the heterogeneity in *GFP* expression among individual cells persisted and could not be resolved by subsequent cloning, representing a state of „permanent expression heterogeneity". Van Leeuwen with colleagues also observed spatial and temporal variability in the expression of luciferase gene in different leaves and leaf sectors of stably transformed *Petunia* plants [10]. The authors attributed this mosaic character of transgene expression to temporal changes in the accessibility of promoter sequences for transcription factors, or variable levels of these factors in different leaf sectors at a time [10]. The pattern of this transgene expression variability was more or less specific for individual lines [10], similarly to the heterogeneity patterns observed in our lines 6/3c, 1/3a, 5/6a and their subclones (Figure 4). Since the heterogeneity patterns were heritable through subsequent cloning, they could be the result of specific variation in epigenetic states at a certain genomic locus [26].

Finally, the remaining evident cause of heterogeneity or instability of *GFP* expression is silencing [27]. Occurrence of transgene silencing was indicated by the presence of cells with contrasting GFP levels in a single cell file (Figure 1b) and by a gradual decline in the frequency of *GFP*-expressing cells observed in some lines (e.g. line 1/2; Figure 2). Silencing at the transcriptional level in connection with DNA methylation was demonstrated by the detection of methylated cytosin in clone 5/6j (Figure 5b). The role of methylation was confirmed by using the DNA-demethylation drug, 5-azacytidin [28], which reactivated *GFP* expression in several lines after several months of silenced *GFP* expression (e.g. lines 1/2, 5/6j; Nocarova and Fischer, unpublished). Silencing of transgene expression is naturally triggered mainly by high transcript levels [11], but may also be related to changes in the epigenetic status of plant genomic DNA in the process of dedifferentiation [29] that accompanies preparation of transgenic plants and plant cell lines.

Cloning of plant cells – history and future

The first reports of cloning non-transgenic plant cells were published long time ago [30, 31]. The method of cloning transgenic plant cell line introduced in our study has not been, to our knowledge, published and used before. Müller with colleagues described protoplast-based cloning of transgenic wheat lines, although this method was time-consuming and induced high frequency (up to 50% of clonal cells lines) of transgene silencing [16]. It indicates that the process of protoplast formation and regeneration may be accompanied by stress-induced epigenetic changes, causing transgene silencing [16]. The minimal occurrence of silencing in our experiments indicates that the drug selection of resistant cells during the cloning procedure causes significantly little stress. In contrast to protoplast-based cloning, our method does not always produce clones from single cells, because BY-2 cells remain temporarily attached in files. However, as the files originate from single cells, they are genetically homogeneous and *calli* derived from these files represent real clones. Although our cloning procedure did not, against expectations, lead exclusively to lines with homogenous *GFP*expression, clearly the cloning method is an effective way to substantially increase the number of homogenous lines. Whereas only ~10% of the primary cell lines were homogenous just after transformation, the cloning of largely heterogeneous lines produced additional 43% of homogeneous cell lines (Table 1).

In addition to generating homogeneously expressing transgenic lines, the cloning procedure appears to be a suitable tool for detailed analysis of the induction and stabilization of epigenetic changes connected with T-DNA insertion into the plant genomic DNA. Thus, in a modified arrangement, cloning of lines with silenced *GFP* expression using 5-azacytidin-containing media produced clones with reactivated *GFP* expression (Nocarova and Fischer, unpublished). Another possible use of the cloning procedure includes the cloning of epigenetically shifted lines habituated to certain conditions [32]. In transgenic lines carrying a negative selection gene whose expression is lethal under certain treatment, mixing an abundance of such a line with non-transformed line would enable cloning of reversed, non-transformed lines.

CONCLUSION

By analysing GFP fluorescence in tobacco BY-2 cells, we found that expression of *GFP* transgene was highly heterogeneous in the majority of transgenic lines obtained directly from transformation. This heterogeneity had two causes: (1) genetic heterogeneity, namely the presence of cells with different T-DNA insertions; and (2) epigenetic heterogeneity, including transgene silencing, formation of stable epigenetic states early after transformation, and "permanent heterogeneity" with fluctuating changes in *GFP* expression. The genetic heterogeneity and the presence of cells in different but stable epigenetic states was responsible for almost half (43%) of the heterogeneity in the primary lines, and could be resolved by cloning. Because the cloning procedure can significantly increase the frequency/yield of homogenous lines, it is of high general impact for both molecular and biochemical analyses of BY-2 transgenic lines. In order to facilitate a simple way for assessment of transgene expression heterogeneity in both primary and cloned lines, it is highly advisable to use GFP-tagged transgenes. Alternatively, for transgenes that lack a visible, cell-autonomous phenotype the cloning procedure can be used to obtain genetically homogeneous lines with statistically higher chance of homogeneous transgene expression. Analysis of *GFP* expression in primary cell lines and their clones also showed that a single T-DNA insertion often resulted in various heritable transgene expression patterns/epigenetic states. These lines were probably established accidentally and became stabilized in the offspring of the primary transformed cells early after the integration event. Thus, the cloning procedure introduced in this study also appears to be suitable for analysing the sources of variability in transgene expression.

METHODS

Cultivation and transformation of BY-2 cell line

Tobacco cell line BY-2 (*Nicotiana tabacum* L. cv. Bright Yellow 2 [33]) was cultured in modified MS medium [34]. Cells in suspension were subcultured every seventh day (1 ml of cells into 30 ml of liquid media). Stock BY-2 *calli* were maintained on media solidified with 0.7% (w/v) agar and subcultured monthly. The cultures were kept in darkness at 26°C; suspensions were placed on orbital incubator (IKA KS501, IKA Labortechnik, Staufen, Germany; orbital diameter 30 mm). Suspensions were prepared by resuspending of ~1 ml of fresh *calli* in 30 ml of liquid media by repeated pipetting through a cut tip (internal diameter ~5 mm).

Transformation of BY-2 line was performed by a slightly modified protocol introduced by [5]. A 2 ml aliquot of 3-day old BY-2 cells was co-cultivated with 200 μl of an overnight culture of *Agrobacterium tumefaciens* strain C58C1 carrying a helper plasmid *pGV2260* [35] and a modified binary vector *pCP60* [36] (kindly provided by dr. P. Ratet). The T-DNA contained a gene encoding red-shifted green fluorescent protein [37] (kindly provided by ABRC) inserted under the control of *CaMV 35S* promoter with a single enhancer region. The T-DNA further contained neomycin phosphotransferase gene (*NPTII*) driven by nopalin synthase promoter (*pNOS*), which provided kanamycin resistance. After co-cultivation, the cells were washed with 60 ml of 3% sucrose and 20 ml of liquid medium containing 100 μg/ml cefotaxim (CEFTAX, Hikma Farmaceutica, Terrugem, Portugal) in Nalgene filter holder (Nalgene, Rochester NY, USA). Thereafter, the cells were plated onto solid medium containing 50 μg/ml kanamycin and 100 μg/ml cefotaxim. Kanamycin-resistant colonies appeared after 3 to 4 weeks in darkness at 26°C. Transformed *calli* and suspensions were kept on media supplemented with kanamycin (50 µg/ml) for about two months and thereafter they were cultured as described for the BY-2 stock line.

Assessment of *GFP*expression/fluorescence

Round-shaped primary *calli* in size of 1–3 mm (4 weeks after the transformation) were transferred onto fresh media. After additional 2 weeks of cultivation, the homogeneity of *GFP* expression was evaluated as a green fluorescence using a fluorescence stereomicroscope (Leica MZ16F). *Calli* containing a few sectors of different GFP fluorescence intensities separated by sharp borders were classified as mixed *calli* with a "sector arrangement". If the regions with different GFP fluorescence were mixed together without clear borders, or the number of separated regions was higher than approximately five, the arrangement was classified as "mosaic".

In suspension cultures, the homogeneity of *GFP* expression was evaluated using a fluorescence microscope Olympus Provis AX70 equipped with an FITC (U-MWU) filter set. The images were grabbed with a digital TV camera Sony DXC-950P (Sony Corp., Tokyo, Japan) and processed with Lucia image analysis software (Laboratory Imaging, Prague, Czech Republic). The proportions of cells with high, low, or no GFP fluorescence were estimated by evaluating ~100–150 cells. Only lines with clear difference between high and low expression categories are presented in the results section. A suspension was classified as homogenous if the portion of cells with minor classes of *GFP* expression level did not exceed 5% in total.

Cloning of transgenic lines

Four weeks after transformation, the primary *calli* were transferred onto fresh solid medium containing kanamycin. After the next 3 weeks, the *calli* were gently resuspended in liquid medium and cultivated on a rotor shaker for a week. Thereafter, the suspension cells were subcultured (1.5 ml of suspension into 30 ml of fresh medium) and after additional 7 days when the culture reached stationary growth phase, the cells were used for cloning. The transgenic suspension culture was diluted with MS medium in a ratio 1:3 and mixed with 4 ml of similarly prepared wild-type stationary BY-2 culture in a ratio 1:1000. After gentle shaking, 500 μl of this mixture was evenly spread onto the Petri dish (Ø 6 cm) with solidified MS medium containing kanamycin. Clones of individual cells appeared as "secondary" *calli* (approximately 25 per plate) within two weeks.

Molecular Analysis

Total genomic DNA was isolated by Invisorb Spin Plant Mini Kit (Invitek, Berlin, Germany) from 100 mg (fresh weight) of filtered cells. Aliquots of 10 μg DNA were cleaved with *Hin*dIII and *Bam*HI (Fermentas, Burlington, Canada), which cleave the T-DNA in front and behind the *35S* promoter. For methylation analysis the *35S-GFP* cassette was cleaved out with *Hin*dIII and *Eco*RI, and thereafter the DNA was subjected to methylation-sensitive restriction enzymes that cleave within the *35S-GFP* cassette (*Bsu*15I, *Eco*72I, *Eco*47I; Fermentas), and separated on 0.8% agarose gel. Blotting was performed as described in [38]. Hybridisation with PCR-amplified probe of the whole GFP gene, labelled with DIG-dUTP (Roche Molecular Systems, Inc., Mannheim, Germany), was done according to manufacturer's instructions. Autoradiographic detection was done using chemiluminiscent substrate CDP-Star (Tropix, Bedford, USA). Fidelity of the insertions was confirmed by sequencing a PCR amplified *35S-GFP* cassette from total genomic DNA isolated from individual clones of cell line 5/6. PCR was done with *Pfu* polymerase according manufacturers instruction (Fermentas), sequencing was done by Sequencing laboratory, Faculty of Science, Charles University in Prague, Czech Republic).

ACKNOWLEDGEMENTS

We are grateful to Dr. SJ Davis and Dr. RD Vierstra (University of Wisconsin-Madison, USA) and to the Arabidopsis Biological Research Center for providing us with the soluble-modified *RS-GFP* gene, and to Dr. P Ratet (ISV-CNRS, France) for providing us with the binary vector pCP60. Our thanks also belong to Dr. J. Marc for language corrections and Dr. M. Kuthan and Prof. Z. Palkova (Faculty of Science, Charles University in Prague) for letting us analyse GFP fluorescence using a fluorescence stereomicroscope Leica MZ16F in their laboratory. We are also grateful to Prof. Zdenek Opatrny (Faculty of Science, Charles University in Prague) for general support. This work was supported by grants from the Ministry of Education, Youth and Sports of the Czech Rep·ıblic (LC06004, LC06034 and MSM 0021620858).

REFERENCES

1. Nagata T: When I encountered tobacco BY-2 cells!. Tobacco BY-2 Cells. Biotechnology in Agriculture and Forestry. Edited by: Nagata T, Hasezawa S, Inzé D. Berlin Heidelberg New York: Springer, 2004, 53: 1-5.
2. Harashima H, Shinmyo A, Sekine M: Phosphorylation of threonine 161 in plant cyclin-dependent kinase A is required for cell division by activation of its associated kinase. Plant J. 2007, 52: 435-448.
3. Kuthanova A, Fischer L, Nick P, Opatrny Z: Cell cycle phase-specific death response of tobacco BY-2cell line to cadmium treatment. Plant Cell Environ. 2008, 31: 1634-1643.
4. Klein TM, Harper EC, Svab Z, Sanford JC, Fromm ME, Maliga P: Stable genetic-transformation of intact Nicotiana cells by the particle bombardment process. Proc Natl Acad Sci U S A. 1988, 85: 8502-8505.
5. An GH: High-efficiency transformation of cultured tobacco cells. Plant Physiol. 1985, 79: 568-570.
6. Higaki T, Kutsuna N, Sano T, Hasezawa S: Quantitative analysis of changes in actin microfilament contribution to cell plate development in plant cytokinesis. BMC Plant Biol. 2008, 8: 80
7. Joubes J, Inzé D, Geelen D: Improvements of the molecular toolbox for cell cycle studies in tobacco BY-2 cells. Tobacco BY-2 Cells. Biotechnology in Agriculture and Forestry. Edited by: Nagata T, Hasezawa S, Inzé D. Berlin Heidelberg New York: Springer. 2004, 53: 7-20.
8. Pröls F, Meyer P: The methylation patterns of chromosomal integration regions influence gene activity of transferred DNA in Petunia hybrida. Plant J. 1992, 2: 465-475.
9. Iglesias VA, Moscone EA, Papp I, Neuhuber F, Michalowski S, Phelan T, et al: Molecular and cytogenetic analyses of stably and unstably expressed transgene loci in tobacco. Plant Cell. 1997, 9: 1251-1264.

10. van Leeuwen W, Ruttink T, Borst-Vrenssen AWM, Plas van der LHW, Krol van der AR: Characterization of position-induced spatial and temporal regulation of transgene promoter activity in plants. J Exp Bot. 2001, 52: 949-959.
11. Schubert D, Lechtenberg B, Forsbach A, Gils M, Bahadur S, Schmidt R: Silencing in Arabidopsis T-DNA transformants: The predominant role of a gene-specific RNA sensing mechanism versus position effects. Plant Cell. 2004, 16: 2561-2572.
12. Francis KE, Spiker S: Identification of Arabidopsis thaliana transformants without selection reveals a high occurrence of silenced T-DNA integrations. Plant J. 2005, 41: 464-477.
13. Depicker A, Sanders M, Meyer P: Transgene silencing. Plant Epigenetics. Edited by: Meyer P. Blackwell Publishing. 2005, 19: 1-31.
14. Kim SI, Veena, Gelvin SB: Genome-wide analysis of Agrobacterium T-DNA integration sites in the Arabidopsis genome generated under non-selective conditions. Plant J. 2007, 51: 779-791.
15. Fischer U, Kuhlmann M, Pecinka A, Schmidt R, Mette MF: Local DNA features affect RNA-directed transcriptional gene silencing and DNA methylation. Plant J. 2008, 53: 1-10.
16. Müller E, Lorz H, Lutticke S: Variability of transgene expression in clonal cell lines of wheat. Plant Sci. 1996, 114: 71-82.
17. Down RE, Ford L, Bedford SJ, Gatehouse LN, Newell C, Gatehouse JA, et al: Influence of plant development and environment on transgene expression in potato and consequences for insect resistance. Transgenic Res. 2001, 10: 223-236.
18. Kain SR, Adams M, Kondepudi A, Yang TT, Ward WW, Kitts P: Green fluorescent protein as a reporter of gene-expression and protein localization. Biotechniques. 1995, 19: 650-655.
19. Ruiz MT, Voinnet O, Baulcombe DC: Initiation and maintenance of virus-induced gene silencing. Plant Cell. 1998, 10: 937-946.
20. Kalantidis K, Tsagris M, Tabler M: Spontaneous short-range silencing of a GFP transgene in Nicotiana benthamiana is possibly mediated by small quantities of siRNA that do not trigger systemic silencing. Plant J. 2006, 45: 1006-1016.

21. Mirabella R, Franken C, Krogt van der GNM, Bisseling T, Geurts R: Use of the fluorescent timer DsRED-E5 as reporter to monitor dynamics of gene activity in plants. Plant Physiol. 2004, 135: 1879-1887.
22. Henderson IR, Jacobsen SE: Epigenetic inheritance in plants. Nature. 2007, 447: 418-424.
23. Lunerova-Bedrichova J, Bleys A, Fojtova M, Khaitova L, Depicker A, Kovarik A: Trans-generation inheritance of methylation patterns in a tobacco transgene following a post-transcriptional silencing event. Plant J. 2008.
24. Fojtova M, Van Houdt H, Depicker A, Kovarik A: Epigenetic switch from posttranscriptional to transcriptional silencing is correlated with promoter hypermethylation. Plant Physiol. 2003, 133: 1240-1250.
25. Gelvin SB, Kim SI: Effect of chromatin upon Agrobacterium T-DNA integration and transgene expression. Biochim Biophys Acta. 2007, 1769 (5-6): 410-421.
26. Neuhuber F, Park YD, Matzke AJM, Matzke MA: Susceptibility of transgene loci to homology-dependent gene silencing. Mol Gen Genet. 1994, 244: 230-241.
27. Brodersen P, Voinnet O: The diversity of RNA silencing pathways in plants. Trends Genet. 2006, 22: 268-280.
28. Santi DV, Norment A, Garrett CE: Covalent bond formation between a DNA-cytosine methyltransferase and DNA containing 5-azacytosine. Proc Natl Acad Sci U S A. 1984, 81: 6993-6997.
29. Koukalova B, Fojtova M, Lim KY, Fulnecek J, Leitch AR, Kovarik A: Dedifferentiation of tobacco cells is associated with ribosomal RNA gene hypomethylation, increased transcription, and chromatin alterations. Plant Physiol. 2005, 139: 275-286.
30. Muir WH, Hildebrandt AC, Riker AJ: The preparation, isolation, and growth in culture of single cells from higher plants. Am J Bot. 1958, 45: 589-597.
31. Bergmann L: New technique for isolating and cloning cells of higher plants. Nature. 1959, 184: 648-649.

32. Pischke MS, Huttlin EL, Hegeman AD, Sussman MR: A transcriptome-based characterization of habituation in plant tissue culture. Plant Physiol. 2006, 140: 1255-1278.

33. Nagata T, Nemoto Y, Hasezawa S: Tobacco BY-2 cell-line as the Hela-cell in the cell biology of higher-plants. Int Rev Cytol. 1992, 132: 1-30.

34. Murashige T, Skoog F: A revised medium for rapid growth and bio assays with tobacco tissue cultures. Physiol Plantarum. 1962, 15: 473

35. Deblaere R, Bytebier B, Degreve H, Deboeck F, Schell J, Vanmontagu M, et al: Efficient octopine Ti plasmid-derived vectors for Agrobacterium-mediated gene-transfer to plants. Nucleic Acids Res. 1985, 13: 4777-4788.

36. Bolte S, Brown S, Satiat-Jeunemaitre B: The N-myristoylated Rab-GTPase m-Rab(mc) is involved in post-Golgi trafficking events to the lytic vacuole in plant cells. J Cell Sci. 2004, 117: 943-954.

37. Davis SJ, Vierstra RD: Soluble, highly fluorescent variants of green fluorescent protein (GFP) for use in higher plants. Plant Mol Biol. 1998, 36: 521-528.

38. Sambrook J, Fritsch EF, Maniatis T: Molecular cloning: a laboratory manual. Cold Spring Harbor: Cold Spring Harbor Laboratory Press; 1989:2.

CHAPTER

18

ENHANCED CATHARANTHINE AND VINDOLINE PRODUCTION IN SUSPENSION CULTURES OF CATHARANTHUS ROSEUS BY ULTRAVIOLET-B LIGHT

Shilpa Ramani and Chelliah Jayabaskaran

Department of Biochemistry, Indian Institute of Science, Bangalore – 560012, India

ABSTRACT

Suspension cultures of *Catharanthus roseus* were used to evaluate ultraviolet-B (UV-B) treatment as an abiotic elicitor of secondary metabolites. A dispersed cell suspension culture from C. roseus leaves in late exponential phase and stationary phase were irradiated with UV-B for 5 min. The stationary phase cultures were more responsive to UV-B irradiation than late exponential phase cultures. Catharanthine and vindoline increased 3-fold and 12-fold, respectively, on treatment with a 5-min UV-B irradiation.

BACKGROUND

The strong and rapid stimulatory effect of biotic and abiotic elicitors on plant secondary metabolite synthesis attracts now considerable attention and is under intense investigation in many laboratories including ours [**1**-**3**]. Chemical treatments such as osmotic shock by mannitol [**4**], salt stress [**5**], rare earth elements [**5**,**6**] and bioregulators [**7**,**8**] have been reported to increase alkaloid production in *C. roseus* cell cultures; however, these increases are cell-line dependent and this greatly limits their utilization [**5**,**9**]. Although several approaches have been followed in order to increase the accumulation of alkaloids in cell cultures of *C. roseus*, the yields obtained so far are too low to allow commercial production [**7**,**8**]. Efforts are on to look for biotic or abiotic elicitors with more efficient and universal effects on the improvement of indole alkaloids in *C. roseus* cell cultures. Zhao et al. (2000) [**10**] have reported enhanced production of catharanthine in *C. roseus* cell culture by combined elicitor treatment of an *Aspergillum niger* mycelium and tetramethyl ammonium bromide in shake flasks and bioreactors.

In this study, a suspension culture of *C. roseus* was developed and used as a model system to investigate the effects of UV-B on cell growth, cell viability and secondary metabolite accumulation. It was found that UV-B induced accumulation of catharanthine and vindoline without affecting cell growth and viability.

RESULTS

Establishment of a cell suspension culture from leaf explants of *C. roseus*

Cell suspension cultures were established from the leaf explants of *C. roseus*. These cultures were maintained for three years by sub-culturing at weekly intervals to achieve a homogenous population of cells and used in these studies. Cell growth typically exhibited a rapid growth period between day two to day five, and a stationary phase thereafter (Fig. 1). Cell viability during the growth stage was 93–95% between day four and day six as determined by fluorescein diacetate (FDA) staining (data not shown). Growth of cells increased when the cultures were grown under a 16 h light/8 h dark day/night regime while cells grown in the dark showed no growth (Fig. 1).

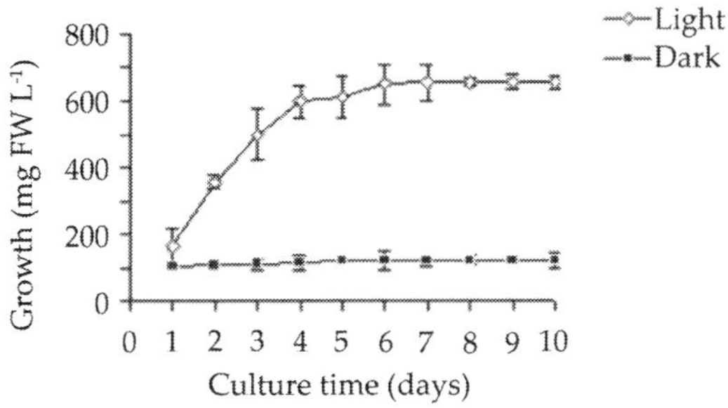

Figure 1: Time-course of growth of *C. roseus* suspension cultures under light v/s dark conditions. Actively growing cells from 4-day-old cultures were inoculated into MS medium supplemented with growth regulators and cultured under 16 h day/8 h dark photoperiod or under complete dark conditions. Cells were harvested on the indicated day and the fresh weight was determined. Values are expressed as the means ± SD (n = 3).

Characteristics of *C. roseus* suspension cultured cells

The morphological investigation of these established cultures revealed compact callus clusters up to three days after inoculation. The compact callus cluster had a diameter of approximately 3.4 cm, yellowish-green colour and a rough surface of soft quality (Fig. 2aand 2b). The aggregates were highly friable and could be easily disrupted by repeated gentle aspiration and release. After three days, the culture became a dispersed turbid cell culture as callus aggregates formed were released continuously into the medium. The aggregates obtained in the shake flasks increased in their size during growth, and by the sixth day dispersed completely giving rise to fine-textured, pale-yellowish colored cultures (Fig. 2b).

The cell length/breadth ratio of suspension-cultured cells was on an average 4.67 in both the control and UV-B-irradiated cells. FDA staining and direct microscopic visualization revealed that the cells were cylindrical in shape with a single plane of division (Fig. 2cand 2d). Cells were seen as a chain due to the division being in a single plane. Though the cells appeared like callus, the callus was friable and separated into free cells.

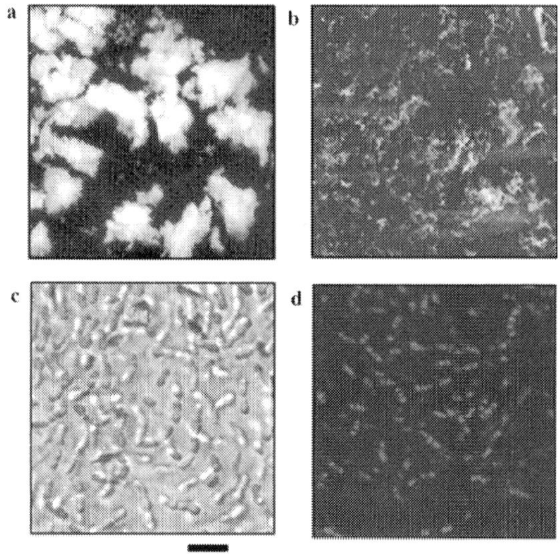

Figure 2: *C. roseus* cell suspension cultures. *A*, Compact callus clusters formed after 3 days of inoculation in the MS liquid medium. ***B***, A dispersed cell suspension culture formed after 6 days of inoculation in the above-mentioned medium. ***C***, Microscopic observation of six-day-old culture: arrows indicate divided cells. ***D***, Fluorescence microscopy (495 nm excitation and 520 nm emission) of FDA-stained cells. Six-day-old cultures were used for microscopic observation and FDA staining. Bar represents 10 μm.

Effect of UV-B on growth of *C. roseus* suspension-cultured cells

By the 6th day, the cells had nearly achieved a stationary phase. We investigated if these cells could respond to UV-B. Figure 3 shows that when four-day-old cultures were irradiated for 5 min with UV-B, the fresh weight of the cells decreased gradually. Visualization of these cells by staining with FDA after 24 h of UV-B irradiation showed that 80–90% of these cells were dead. When six-day-old cultures were irradiated with UV-B for 5 min, there was no decrease in fresh weight till 24 h. Furthermore, no significant effect on the cell viability and growth was observed.

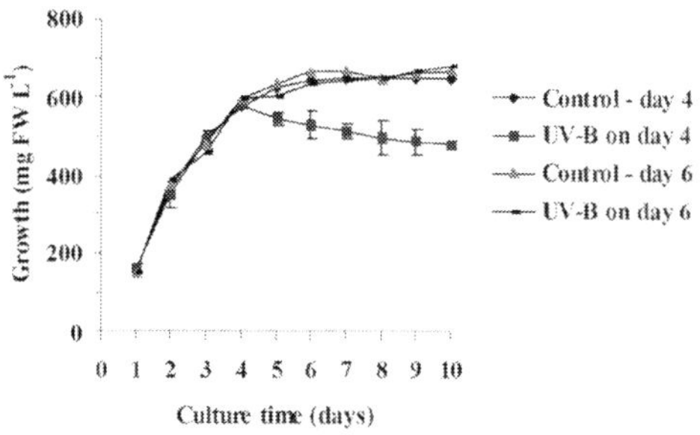

Figure 3: Effect of UV-B irradiation on growth of *C. roseus* suspension cultures. Actively growing cells from 4-day-old cultures were inoculated into fresh MS medium supplemented with growth regulators and cultured under 16 h day/8 h dark photoperiod. Cells were irradiated with UV-B on day 4 or day 6 after sub-culture or left untreated as described in materials and methods. Cell growth was measured every day by determining the fresh weight of the harvested cells. Values are expressed as the means ± SD (n = 3).

UV-B-induced extracellular medium alkalinization

Four- and six-day-old cell suspensions were exposed to UV-B irradiation for 5 min, and extracellular pH changes were measured in the cell suspension medium for 120 min (Fig. 4). The UV-B-induced AR reached a maximum after 10 min of UV-B irradiation in the six-day-old culture and the increase was about 0.8 pH unit. No significant or only slight pH changes were measured in the four-day-old cultures. These results demonstrate that the four-day-old cultures could not effectively mediate a medium alkalinization response and the six-day-old cultures had gained the ability to exhibit medium alkalinization in response to UV-B irradiation.

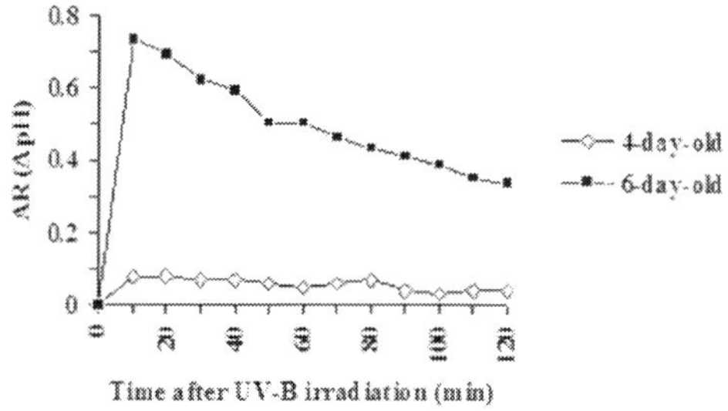

Figure 4: Medium alkalinization of *C. roseus* suspension cultured cells in response to UV-B irradiation. Four- and six-day-old cell suspension cultures were

irradiated with UV-B for 5 min and the pH of the medium was measured at indicated times after the start of irradiation. Alkalinization Response (AR) or ΔpH was measured as described in materials and methods. Experiments were repeated with two different populations of cells and each experiment was carried out in triplicates.

Increased concentration of catharanthine and vindoline on UV-B irradiation

Suspension-cultured cells were irradiated with UV-B and the time-course of concentrations of catharanthine and vindoline was followed up to four days after single dose of irradiation by HPLC analysis of the crude TIAs obtained from the cultured cells (Figure 5). Significant differences in catharanthine and vindoline concentrations between UV-B irradiated and the control cells appeared after 24 h. The highest catharanthine/vindoline concentrations were found at 48 to 72 h after UV-B irradiation. As can be seen in Figure 5, the catharanthine and vindoline concentrations in UV-B irradiated cells were significantly higher than those of the untreated controls. The maximum catharanthine and vindoline concentrations obtained in the UV-B irradiated cultures were 0.12 ± 0.0054 mg/g DW and 0.06 ± 0.0023 mg/g DW, respectively, 48 h after UV-B irradiation, while they were 40 ± 0.2 and 0.51 ± 0.3 μg/g DW, respectively, for the control at that time period.

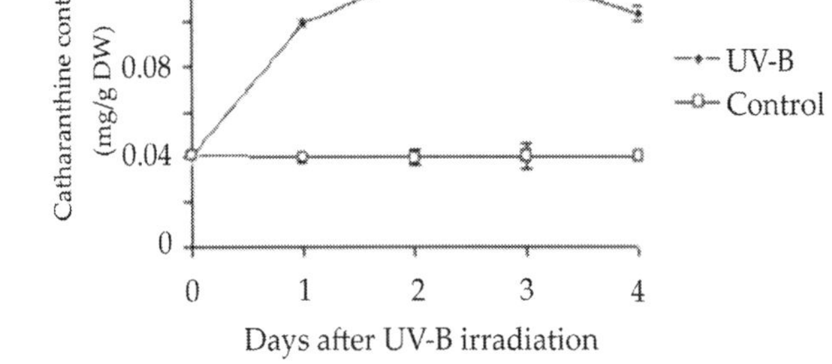

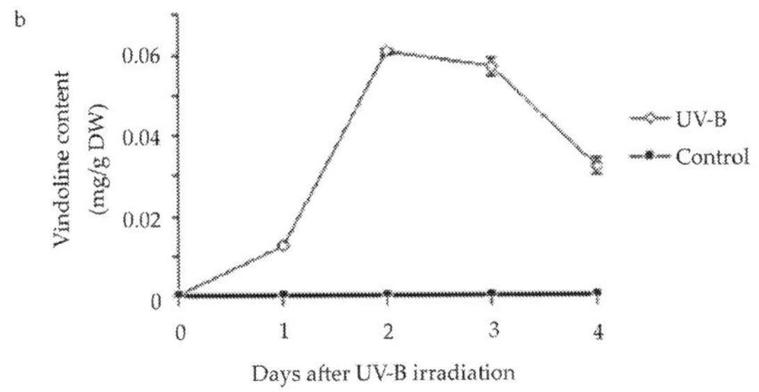

Figure 5: Time-course of catharanthine and vindoline accumulation in cells irradiated with UV-B in *C. roseus* cell suspension cultures. Six-day-old suspension cultured cells of *C. roseus* were irradiated with UV-B light for 5 min or left untreated and harvested at indicated times. ***A***, Catharanthine and ***B***, vindoline accumulation was detected by HPLC as described in materials and methods. Values are represented as the means ± SD (n = 3).

DISCUSSION

The cell suspension cultures in *C. roseus* have been employed for the production of ajmalicine, catharanthine, serpentine and many other pharmaceutically important products [**11**]. However, one major disadvantage with the use of cell suspension cultures in *C. roseus* is the inability of the cultures to produce vindoline and hence vinblastine and vincristine [**12**, **13**]. In the present study, a cell suspension culture was established directly from the leaves, which produced not only increased amounts of catharanthine but also of vindoline under the UV-B irradiated conditions. The experimental results show that cells at the sixth day were more responsive to UV-B irradiation than cells at day four. The results provide evidence that the UV-B induced medium alkalinization in the cell suspension cultures of 6-day-old cells.

Ouwerkerk et al. (1999) [**14**] have reported earlier that UV light induces accumulation of several TIAs as well as expression of the TIA biosynthetic genes in excised *C. roseus* leaves obtained from 3-month old greenhouse-grown *C. roseus* plants. In the present study, catharanthine and vindoline concentrations of UV-B irradiated six-day-old culture cells were checked. On the cell dry weight basis, cells irradiated with UV-B produced 2–3 times the amount of catharanthine and vindoline compared to the non-irradiated cells. Catharanthine accumulation in the UV-B irradiated suspension cultures was observed to be 0.12 ± 0.0054 mg/g DW and was comparable to what was reported in *C. roseus* P2 callus cultures (0.12 ± 0.02 mg/g DW) **[15]**. However, the yield of vindoline in the UV-B irradiated cultures (0.06 ± 0.0023 mg/g DW) was much lower than what has been reported for the P2 callus cultures (0.15 ± 0.02 mg/g DW) **[16]**, but much higher than *A. tumefaciens* transformed suspension cultured cells (~0.017 ± 0.02 mg/g DW) **[17]**.

In conclusion, the cell suspension culture established has many unique features of growth, behaving both like callus and suspension cultures. Our study also shows, for the first time, the role of UV-B light on enhancement of the concentrations of catharanthine and vindoline in *C. roseus* cell suspension cultures. Figure 6 shows the schematic illustration of UV-B signaling in *C. roseus* suspension cells leading to the increase in catharanthine and vindoline levels. Preception of UV-B induces iron fluxes, medium alkalinization, activation of CDPK and NADPH oxidase. This leads to the activation of ROS production and MAPKinases and eventually stimulates to induce TIA biosynthetic gene expression and the accumulation of catharanthine and vindoline. These findings are of importance for further mechanistic study considering the worldwide demand for catharanthine and vindoline.

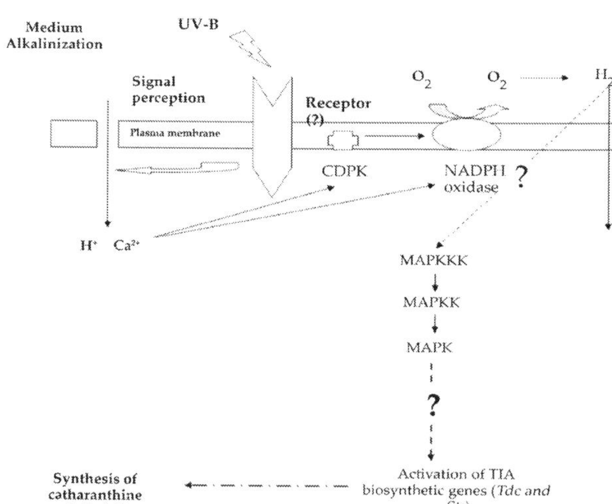

Figure 6: Model for UV-B signaling in *C. roseus*.

MATERIALS AND METHODS

Chemicals

Catharanthine and vindoline were obtained from Shanghai Kangai Biologicals, China. Fluorescein diacetate was obtained from Sigma chemical company, St. Louis, USA. All other chemicals were of the highest purity commercially available.

Establishment of a cell suspension culture, maintenance and culture conditions

Mature leaves were collected from naturally grown plants of *C. roseus* from the campus of Indian Institute of Science, Bangalore. Leaves were cleaned under running tap water, immersed in 0.5% bavistin solution for 20 min and rinsed thoroughly with sterile distilled water. The leaf explants were then treated with 0.01% Tween 20 for 20 min, surface sterilized using an aqueous solution of 0.1% (w/v) mercuric chloride for 5 min, and rinsed extensively with sterile distilled water. The explants were cut into 1 cm^2pieces and inoculated into 200 ml of MS basal medium [**18**] supplemented with 2 mg/l NAA, 0.2 mg/l kinetin and 3% sucrose in a 500 ml Erlenmeyer flask. The pH of the medium was adjusted to 5.8 prior to autoclaving. The culture flasks were capped with cotton plugs and placed on a gyratory shaker at 60 rpm at 25 ± 2°C with a 16 h photoperiod of 40 $\mu molm^{-2}$ s^{-1} using cool white fluorescent tubes. After three weeks of culturing, small callus or cell aggregates were released from swollen explants continuously into the medium. The dispersed cell suspension culture so derived was maintained for three years by sub-culturing at weekly intervals to obtain a homogenous population of cells. The cells were sub-cultured every six days, by inoculating 5 mg/l of fresh weight of cells into 500 ml flasks containing 200 ml medium and cultured under the conditions as mentioned earlier.

Biomass determination

To measure the growth of *C. roseus* cell cultures in shake flasks, fresh weight (FW) of cells was determined by harvesting cells in pre-weighed centrifuge GSI cups from 200 ml culture by centrifugation at 3000 rpm for 15 min. The cups were weighed again, and the fresh weight of the cells was determined as the difference between the total weight and the weight of the empty cup. To determine the dry weight (DW), the cells were harvested as mentioned above and the cells were washed with distilled water, collected, lyophilized and weighed.

UV-B radiation treatment

For all the experiments involving UV-B treatment, four- (late exponential phase of growth) or six-day-old (stationary phase of growth) cultures were harvested and resuspended in six ml of the culture medium. The remaining medium was saved for subsequent use. Three ml of the suspension culture was used for UV-B irradiation and the remaining three ml was used as a control for the same experiment. After a 1 h equilibration period, the cultures were irradiated for 5 min (three ml in 35 mm open petriplates) with a UV-B lamp (Minera lights, UVM 57, San Gabriel, California) by placing at a distance of 2.5 cms from the lamp source. The intensity of UV-B light on the sample surface was 1.26 $\mu W/cm^2$ and the total energy supply was 0.34 J/cm^2. After irradiation, three ml of the suspension cultures (irradiated and control cells) were reconstituted to 100 ml with the saved culture medium in a 250 ml sterile flask and cultured under the same conditions. The cell cultures were harvested at regular intervals and used for further analysis.

Determination of cell viability, cell morphology and cell aspect (length/breadth) ratio

Fluorescence microscopic analysis was used to distinguish viable (fluorescent) from nonviable (nonfluorescent) cells, following staining with fluorescein diacetate (FDA) by the method of Widholm [**19**]. Autofluorescence with unstained cells as control was found to be negligible in all cells. To calculate the cell aspect ratio, 200 μl of six -and four-day-old cell suspension cultures (control and UV-B irradiated cells) were taken, stained with FDA (10 μg/ml) for 1 min and viewed using a confocal microscope. The picture of fluoroscein diacetate stained cells was used to measure the length and breadth of each cell to arrive at the average aspect ratio [**19**]. The cell aspect ratio was measured for 400 cells. To observe the pattern of cell division, late exponential phase cultures were directly mounted on slides and observed at different time points.

Medium alkalinization response (AR) assay

To determine the UV-B-induced medium alkalinization, 1.5 ml of four- and six-day-old suspension cultures were withdrawn and placed directly in 24-well plates. The cells were equilibrated for a period of 1 h to stabilize the pH of the medium and irradiated with UV-B for 5 min. Extracellular pH changes during growth were measured with pH meter in *C. roseus* cell suspensions. Measurements were carried out for a time-period of 120 min and the experiments were repeated twice in triplicates. UV-B induced medium alkalinization response (AR or Δ pH) was calculated as the increase in pH between the untreated controls and the respective UV-B irradiated samples as described previously [**20**].

Terpenoid indole alkaloid extraction and quantification of catharanthine and vindoline by HPLC analysis

The extraction of terpenoid indole alkaloids and quantification of catharanthine and vindoline using HPLC were carried out as described by Schripsema and Verpoorte [**21**]. Briefly, freeze-dried cells (~50 mg) were extracted twice with 5 ml of dichloromethane, and the extraction solutions were combined and concentrated under vacuum. The residues were dissolved in 0.5 ml of HPLC mobile phase [50 mM sodium phosphate pH 3.9: acetonitrile: 2 methoxyethanol (80:15:5 v/v)] and then analyzed by HPLC using Shimadzu LC-6A instrument with shimadzu CR3A integrator and a SPD-6AV UV detector monitoring at 280 nm. The column was a Nucleosil 5 C18 column (250 × 4.6 mm, 5 μm, Novapak). Identification of compounds was based on the retention time and comparisons of their UV spectra with those of the authentic standards. Amounts of catharanthine and vindoline were finally reported as mg g^{-1} DW (dry weight) cells.

ACKNOWLEDGMENTS

The work was supported by grants from Indian Council of Medical Research, and Department of Biotechnology (Genomics Initiative at the Indian Institute of Science, Government of India). We are grateful to Prof. Sunil Podder, Prof. Ramesh Maheshwari, Dr. Nanda Devi, Dr. Latha and Satpal Singh for critical reading of the manuscript. S.R is a recipient of a Research Fellowship from CSIR.

REFERENCES

1. Zhao, J, Davis, LC and Verpoorte, R (2005). Elicitor signal transduction leading to production of plant secondary metabolites. *Biotechnol Adv* 23: 283–333, DOI: https://doi.org/10.1016/j.biotechadv.2005.01.003

2. Verpoorte, R, Alfermann, AWA and Johnson, TS (2007). *Applications of Plant Metabolic Engineering*. Springer, Netherlands.

3. Shilpa, R (2006). Studies on UV-B induced signal transduction events leading to catharanthine/vindoline production in cell suspension cultures of *Catharanthus roseus* and cloning and characterization of the key enzyme-3-deoxyD-arabino-heptulosonate7-phosphate synthase gene involved in tryptophan biosynthesis. *PhD thesis*. Indian Institute of Science, Bangalore, India.

4. Godoy-Hernandez, G and Loyola-Vargas, VM (1991). Effect of fungal homogenate, enzyme inhibitors and osmotic stress on alkaloid content of *Catharanthus roseus* cell suspension cultures. *Plant Cell Rep* 10: 537–540, DOI: https://doi.org/10.1007/BF00232506

5. Smith, JI, Smart, NJ, Misawa, M, Kurz, WGW, Tallevi, SG and DiCosmo, F(1987). Increased accumulation of indole alkaloids by some cell lines of *Catharanthus roseus* in response to addition of vanadyl sulphate. *Plant Cell Rep* 6: 142–145, DOI: https://doi.org/10.1007/BF00269740

6. Zhao, J, Zhu, WH, Hu, Q and He, XW (2000). Enhanced ajmalicine production in *Catharanthus roseus* cell cultures by combined elicitor treatment: from shake-flask to 20-l airlift bioreactor. *Biotechnol Lett* 22: 1221–1226, DOI: https://doi.org/10.1023/A:1005696929724

7. Lee, S-L, Cheng, K-D and Scott, AL (1981). Effect of bioregulators on indole alkaloid biosynthesis in *Catharanthus roseus* cell cultures. *Phytochem* 28: 1841–1843.

8. Simpson, AP and Kelly, SL (1989). Cytochrome P-450 inducer/inhibitor effects on cell cultures of *Catharanthus roseus*. *Plant Sci* 160: 231–236, DOI: https://doi.org/10.1016/0168-9452(89)90171-4

9. Moreno, PRH, van der Heijden, R and Verpoorte, R (1995). Cell and tissue cultures of *Catharanthus roseus*: a literature survey. II.

Updating from 1988 to 1993. *Plant Cell Tiss Organ Cult* 42: 1–25, DOI: https://doi.org/10.1007/BF00037677

10. Zhao, J, Zhu, WH and Hu, Q (2000). Enhanced ajmalicine production in *Catharanthus roseus* cell cultures by combined elicitor treatment: from shake-flask to 20-l airlift bioreactor. *Biotechnol Lett* 22: 509–514, DOI: https://doi.org/10.1023/A:1005616920165
11. Tikhomiroff, C and Jolicoeur, M (2002). Screening of *Catharanthus roseus* secondary metabolites by high-performance liquid chromatography. *J Chromatogr A* 955: 87–93, DOI: https://doi.org/10.1016/S0021-9673(02)00204-2
12. St-Pierre, B and De Luca, V (1995). A cytochrome P-450 monooxygenase catalyzes the first step in the conversion of tabersonine to vindoline in *Catharanthus roseus*. *Plant Physiol* 109: 131–139.
13. Vazquez-Flota, F, De Carolis, ED, Alarco, A-M and De Luca, V (1997). Molecular cloning and characterization of desacetoxyvindoline-4-hydroxylase, a 2-oxoglutarate dependent-dioxygenase involved in the biosynthesis of vindoline in *Catharanthus roseus* (L.) G. Don. *Plant Mol Biol* 34: 935–948, [PubMed] DOI: https://doi.org/10.1023/A:1005894001516
14. Ouwerkerk, PBF and Memelink, J (1999). Elicitor-responsive promoter regions in the tryptophan decarboxylase gene from *Catharanthus roseus*. *Plant Mol Biol* 39: 129–136, DOI: https://doi.org/10.1023/A:1006138601744
15. Zhao, J, Zhu, WH, Hu, Q and Guo, YQ (2001). Compact callus cluster suspension cultures of *Catharanthus roseus* with enhanced indole alkaloid biosynthesis. *Plant Growth Regulation* 37: 68–72.
16. O'Keefe, BR, Mahady, GB, Gills, JJ and Beecher, CWW (1997). Stable vindoline production in transformed cell cultures of *Catharanthus roseus*. *J Nat Prod* 60: 261–264, DOI: https://doi.org/10.1021/np960703n
17. Murashige, T and Skoog, F (1962). A revised medium for rapid growth and bioassay with tobacco tissue cultures. *Plant Physiol* 15: 473–479, DOI: https://doi.org/10.1111/j.1399-3054.1962.tb08052.x
18. Widholm, JM (1972). The use of fluorescein diacetate and

phenosafranine for determining viability of cultured plant cells. *Stain Technol* 47: 189–194.

19. Kim, SW, Jung, KH, Kwak, SS and Liu, JR (1994). Relationship between cell morphology and indole alkaloid production in suspension cultures of *Catharanthus roseus*. *Plant Cell Rep* 14: 23–26, DOI: https://doi.org/10.1007/BF00233292
20. Yalamanchili, RD and Stratmann, J (2000). Ultraviolet-B Activates Components of the Systemin Signaling Pathway in *Lycopersicon peruvianum* Suspension-cultured Cells. *J Biol Chem* 277: 28424–30, DOI: https://doi.org/10.1074/jbc.M203844200
21. Schripsema, J and Verpoorte, R (1992). Search for factors involved in indole alkaloid production in cell suspension cultures of *Tabernaemontana divaricata*. *Tabernaemontana divaricata* 58: 245–249.

CHAPTER

19

ASSESSMENT OF CULTIVATION FACTORS THAT AFFECT BIOMASS AND GERANIOL PRODUCTION IN TRANSGENIC TOBACCO CELL SUSPENSION CULTURES

Nikolay Vasilev[1], Christian Schmitz[1], Ulrike Gro¨ mping[2] , Rainer Fischer[1,3], Stefan Schillberg[1,4]

[1]Department Plant Biotechnology, Fraunhofer Institute for Molecular Biology and Applied Ecology IME, Aachen, Germany,

[2]Department II–Mathematics, Physics and Chemistry, Beuth University of Applied Sciences, Berlin, Germany,

[3]Institute for Molecular Biotechnology, RWTH Aachen University, Aachen, Germany,

[4]Institute for Phytopathology and Applied Zoology, Justus-Liebig University Giessen, Giessen, Germany

ABSTRACT

A large-scale statistical experimental design was used to determine essential cultivation parameters that affect biomass accumulation and geraniol production in transgenic tobacco (*Nicotiana tabacum* cv. Samsun NN) cell suspension cultures. The carbohydrate source played a major role in determining the geraniol yield and factors such as filling volume, inoculum size and light were less important. Sucrose, filling volume and inoculum size had a positive effect on geraniol yield by boosting growth of plant cell cultures whereas illumination of the cultures stimulated the geraniol biosynthesis. We also found that the carbohydrates sucrose and mannitol showed polarizing effects on biomass and geraniol accumulation. Factors such as shaking frequency, the presence of conditioned medium and solubilizers had minor influence on both plant cell growth and geraniol content. When cells were cultivated under the screened conditions for all the investigated factors, the cultures produced ~5.2 mg/l geraniol after 12 days of cultivation in shaking flasks which is comparable to the yield obtained in microbial expression systems. Our data suggest that industrial experimental designs based on orthogonal arrays are suitable for the selection of initial cultivation parameters prior to the essential medium optimization steps. Such designs are particularly beneficial in the early optimization steps when many factors must be screened, increasing the statistical power of the experiments without increasing the demand on time and resources.

INTRODUCTION

Geraniol is an intermediate metabolite in the monoterpenoid-secoiridoid biosynthesis pathway and is used as a flavor and fragrance compound in the agricultural, food and cosmetic industries. It is an ingredient in 43% of fragrances currently on the market [1]. The annual global demand for geraniol was approximately 327 kg in 2008, with a 3–4% annual market growth rate [2]. The three major geraniol manufacturing strategies are biosynthesis (isolation from essential oils), semi-synthesis (chemical conversion of pinenes) and total synthesis from petrochemicals, and the yield varies between 60% (biosynthesis) and 98% (total synthesis) [2].

The enzyme geraniol synthase (GES) is responsible for the biosynthesis of geraniol from geranyl diphosphate [3]–[10]. GES expression in plants may help to reconstitute the terpenoid-indole alkaloid pathway and therefore satisfy the consumer demand for geraniol.

Plant cell and tissue cultures can be used as an alternative to whole plants for the production of valuable natural compounds [11], [12]. Physical factors that have a major impact on the performance of plant-derived cultures include the inoculum state, size and age [11], [13]–[15], the volume of medium in the culture vessel and the shaking speed, both of which influence on the oxygen supply during cultivation [16], [17]. Biological factors also play an important role, including the use of conditioned medium to influence the accumulation of plant cell biomass [18] and the selection of carbohydrates as sources of energy and carbon due to the low levels of CO_2 in the closed-glass cultivation vessels during the photoperiod [19]. Light influences both growth and secondary metabolism in plants, e.g. by regulating the diurnal pattern of monoterpenoid synthase transcript levels and monoterpenoid production [20], [21]. Finally, the equilibrium of production may be influenced by introducing excipients such as cyclodextrins that increase the solubility of sparingly-soluble compounds [22].

All the cultivation factors described above may have a direct influence on the productivity of plant cell suspension cultures in terms of biomass accumulation and the synthesis of secondary metabolites. Typically, a one-factor-at-a-time testing approach is used to find optimal cultivation conditions, but this does not identify interactions between factors that can improve or inhibit productivity. Statistical experimental designs are widely used in the biotechnology industry to avoid the drawbacks of testing single factors while leaving the others unchanged. Instead, statistical experimental designs determine the impact of several nutritional and physical factors simultaneously using full factorial, fractional factorial or response surface designs [18], [23]–[25].

To our knowledge, this is the first report describing a statistical experimental design approach that includes multiple physical factors (inoculum size, shaking frequency, filling volume, light) and biological factors (conditioned medium, carbohydrate source and solubilizers) in one experiment. Our results allowed us to optimize geraniol production in transgenic tobacco suspension cultures expressing plastid-targeted GES from *Valeriana officinalis*.

MATERIALS AND METHODS

Origin of transgenic material and initiation of cell suspension cultures

Cloning of *V. officinalis* GES (*VoGES*) cDNA (GenBank accession: KF951406.1) and stable transformation of tobacco plants was conducted as described previously [3]. *N. tabacum* cv. Samsun NN seeds from a homozygous T_7 line carrying a single *VoGES* expression cassette were selected for initiation of cell suspension cultures. The transgenic seeds were germinated under a 16-h light daily period in sterile conditions on hormone-free MS medium (Duchefa, Netherlands), supplemented with 100 mg/l kanamycin. Next, the intact *in vitro* plants were established on the same medium without antibiotics in plastic transparent containers. Sterile transgenic seedlings were used for the onset of callus culture in Petri dishes containing MS medium with vitamins, supplemented with 0.1 mg/l kinetin and 1 mg/l 1-naphthaleneacetic acid (NAA). Suspended cell cultures were developed subsequently by transferring callus tissue into 10 ml Gamborg's B5 medium plus vitamins, 0.1 mg/l kinetin and 1 mg/l NAA, accommodated in 50 ml TubeSpin bioreactors (Techno Plastic Products AG, Switzerland). Cell suspension cultures were maintained on a gyratory shaker (180 rpm) at 26°C under 16-h illumination (35.6 μmol/s)/8-h darkness and underwent subcultivation every two weeks prior to the start of the screening experimental design. Fresh biomass was collected at the end of the screening cultivation (day 10) and subjected to double filtration under vacuum.

Time curve of *VoGES* suspensions growth was performed in three biological replicates and the samples were collected for analysis every three days after the start of cultivation. Dissolved oxygen was measured online in 250 ml flasks by means of the BPM-60 (BioProcess Monitoring) device (Kühner AG, Switzerland). All the screening or validation experiments were conducted in ISF1-X or in LT-X shaker machines (Kühner AG).

GC-MS analysis

The extraction and quantitation of geraniol in the plant cell cultures was performed as described earlier by Dong et al. [3] with small modifications. Briefly, plant suspension material was filtrated twice under vacuum and subsequently frozen at −20°C. Two-hundred mg plant cell material was taken, ground and resuspended homogenously in 1 ml citrate-phosphate buffer (pH 5.4), prepared by mixing 27.8 ml 0.2 M dibasic sodium phosphate and 22.2 ml 0.1 M citric acid and topping up to 100 ml with water [26]. The samples were ultra-sonificated for 15 min in a water bath at ambient temperature, followed by the addition of 0.5 ml "Viscozyme

L" enzymatic mixture (Sigma-Aldrich, Germany). Next, 1 ml layer of heptane was added to the each sample, containing 10 μg/ml (*Z*)-nerolidol (Sigma-Aldrich) and incubated overnight at 37°C. Next day, each sample was centrifuged at 1750×*g* for 10 min at room temperature. The organic layer was removed and fresh 1 ml heptane was added to sample, vortexed and centrifuged at 1750×*g* for 10 min at room temperature. The last step was repeated once again and all three heptane extracts were pooled, passed through a glass Pasteur pipette provided with a small glass wool plug and ~1.5 cm anhydrous sodium sulfate (Sigma-Aldrich).

The heptane extract was concentrated under nitrogen gas and the geraniol content was measured quantitatively by a GC-MS machine coupled to a QP2010SE quadrupole mass spectrometer (Shimadzu, Japan). The separation was performed on a 30 m length ×0.25 mm internal diameter Zebron ZB-5 ms column (Phenomenex, USA) containing 0.25 μm stationary phase, preceded by a 5-m guard column. The injected 1 μl aliquot of each sample (split mode 1:10, injection port temperature 250°C) was subjected to the following temperature program: 45°C for 1 min followed by a gradient of 10°C per min until the temperature reached 300°C, which was held for a further 7 min. The helium inlet pressure was maintained at a constant column flow rate of 1.0 ml/min.

Geraniol was quantified using the base peak (*m/z*=69) in SIM mode following electric ionization at 1 keV. The recorded mass-spectrum was compared with spectra of a reference standard from the NIST library.

Experimental design and statistical analysis

The experiment was accommodated in four shaker, satisfying the combinatorial states for the factors light intensity and shaking frequency. Temperature was fixed at 26°C based on our preliminary experiments (Table S1) in order to avoid the necessity for additional shaker machines.

We used an experimental design based on an orthogonal array with 72 runs created with the free open source R package DoE.base [27] (for details, see supplementary material). The design was optimized with respect to confounding of low order effects: all main effects are orthogonal to each other (orthogonal array), confounding between main effects and two-factor interactions was minimized, and in a second step, confounding among two-factor interactions was kept as low as possible.

Figure 1 shows a mosaic plot for a triple of design factors that is one of the two triples with the worst-case confounding between main effects and two-factor interactions; even this worst case is reasonably balanced [28].

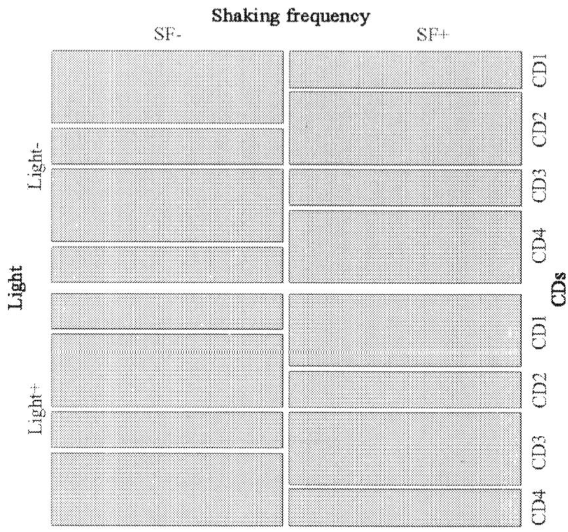

Figure 1: Mosaic plot for three selected factors (light, shaking frequency (SF) and cyclodextrins (CD1-no cyclodextrin; CD2-β-cyclodextrin; CD3-methyl-β-cyclodextrin and CD4-triacetyl-β-cyclodextrin)) showing that any level combination of factors CDs and SF does not determine completely the level of factor "light" which is indicative for limited confounding.

RESULTS

Seven cultivation factors (four 2-level factors, two 3-level factors and one 4-level factor) were screened in a randomized experimental design with limited confounding: light, shaking frequency, inoculum size, filling culture volume, addition of conditioned medium, carbohydrate type (sucrose, glucose and D-mannitol) and cyclodextrin type (β-cyclodextrin, methyl-β-cyclodextrin, triacetyl-β-cyclodextrin and no cyclodextrin) (Table 1). Three responses were measured correspondingly: fresh weight, geraniol content and geraniol yield. The full experimental design with 72 runs in coded values and the measured responses are given in Table S2.

Table 1. Levels of the screened factors

Factor	Levels	Abbreviation
Light	11.50 (μmol/cm²/s)	Lght-
	35.62 (μmol/cm²/s)	Lght+
Shaking frequency	140 rpm	SF-
	180 rpm	SF+
Inoculum size	0.75 g	IS-
	1.40 g	IS+
Filling volume	15 ml	FV-
	20 ml	FV0
	25 ml	FV+
Conditioned medium	100% fresh medium	CM-
	1 ml conditioned medium	CM+
Carbohydrate	Sucrose	Suc
	Glucose	Gluc
	D-Mannitol	Mannit
Cyclodextrin (2 mM)	No cyclodextrin	CD1
	β-cyclodextrin	CD2
	Methyl-β-cyclodextrin	CD3
	Triacetyl-β-cyclodextrin	CD4

Geraniol yield model

The productivity of the plant cell platform depends on both the biomass production and the geraniol content per unit biomass. The multiplication of fresh weight by geraniol content in fresh weight stands for geraniol yield in our production system. The geraniol yield model ($p<0.05$) indicated that the strongest factors with effect on geraniol production are sucrose, filling volume and inoculum size, followed by light (Figure 2).

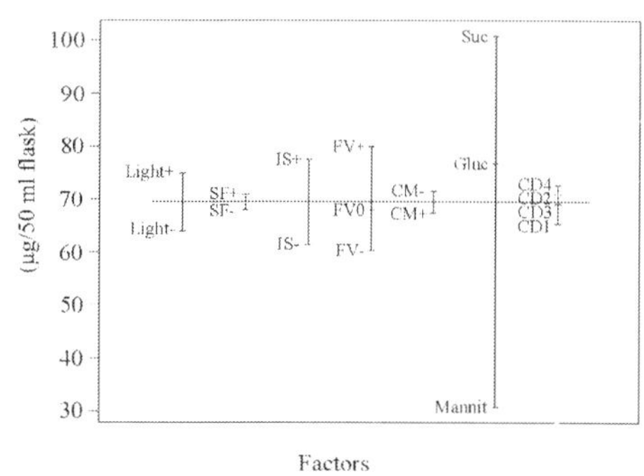

Figure 2: Geraniol yield model plot

The abbreviations are the same as in Table 1.

Biomass model

We created also biomass and geraniol content models ($p<0.05$) in order to demonstrate how plant cell growth and geraniol content influenced the geraniol yield. The biomass plot showed that the carbohydrate source from all screened factors had the strongest influence on biomass production of the transgenic plant cell cultures (Figure 3). The highest growth was achieved when sucrose was used as a carbon source in the cultivation medium. On the contrary, mannitol considerably reduced the plant cell growth when incorporated as a carbohydrate source into the medium. Filling volume and inoculum size exerted also a positive effect on biomass production but in a less profound extent as sucrose. Therefore, sucrose, filling volume and inoculum size affected biomass production in the same manner as geraniol yield. Shaking frequency, light and addition of conditioned medium and cyclodextrins have exhibited insignificant or negligible effect on the growth of plant cell suspensions.

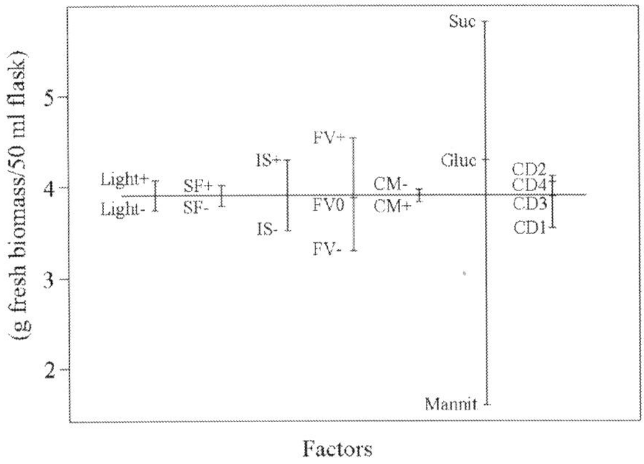

Figure 3. Biomass model plot.

The abbreviations are the same as in Table 1.

Geraniol content model

The screening in terms of the geraniol content identified that the use of mannitol led to the highest geraniol levels accumulated in the tobacco transgenic cell cultures (Figure 4). Light had also a potent effect on geraniol biosynthesis. The rest of cultivation factors did not show a strong impact on geraniol production as seen from the geraniol content model ($p<0.05$).

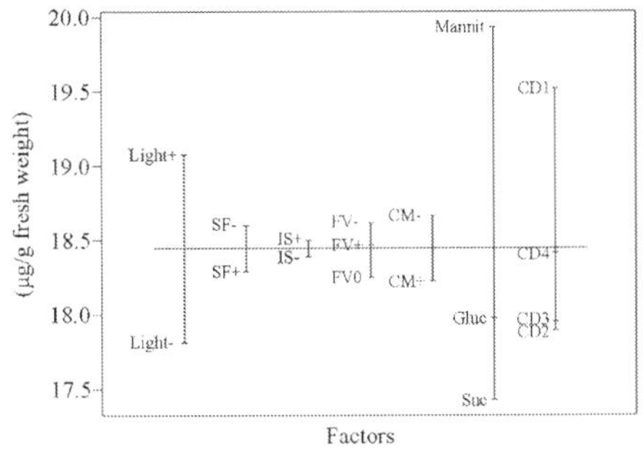

Figure 4. Geraniol content model plot.

The abbreviations are the same as in Table 1.

The pattern of the most significant factors from the biomass model is very similar to the pattern of the strongest factors in the geraniol yield model. This serves as additional evidence that the growth of plant cells contributes to a greater extent to the geraniol yield formation in comparison to the factors acting mainly on the enhancement of geraniol content.

Significant interactions

We also identified several potential two-factor interactions concerning geraniol yield (Table S3). The most significant interactions ($p<0.01$) are filling volume:sugar (Figure 5A), sugar:cyclodextrins (Figure 5B) and light:sugar (Figure 5C). Note that the figures do not show average geraniol yields at each level combination (as a naïve interaction plot would do) but estimated interaction effects from the model with all main effects and 2-factor interactions (obtained with R package effects by Fox 2003 [29]). These are unbiased estimates in the presence of the partial confounding among 2-factor interactions that is present (and in some cases strong) in the experimental design. All three strongest interaction effects involve at least one factor with a strong main effect and are therefore plausible according to the weak effect heredity principle [30]. For two of the strong interactions, both factors have active main effects (strong effect heredity). For instance, Figure 5 shows that an increase of filling volume has a different effect in case mannitol was used as a source of carbohydrate than for the other two sugars. The effect of cyclodextrins on the geraniol yield goes into opposite directions for the sugars mannitol and sucrose, respectively. Light and sucrose potentiate reciprocally their actions on geraniol yield because the corresponding main effects are very strong. Remarkably, shaking frequency also appears to be involved in at least one, possibly several interactions, demonstrating its important role for maximizing geraniol production during plant cell cultivation (Table S3).

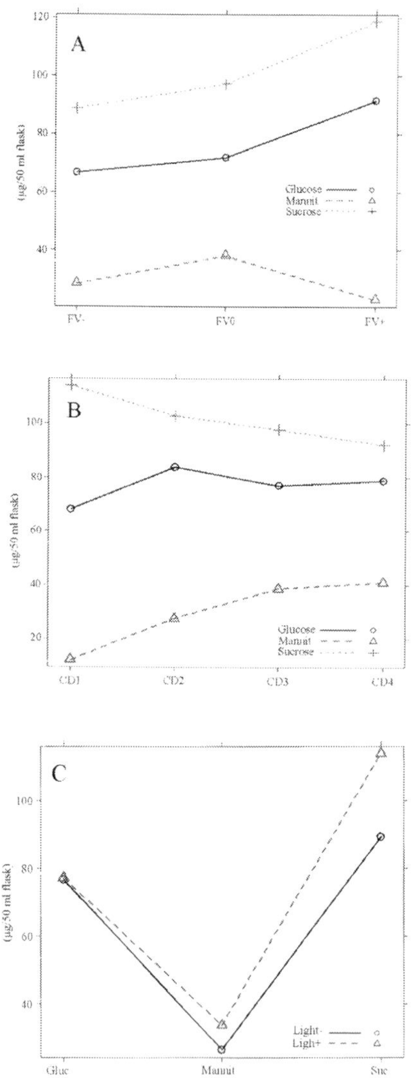

Figure 5: Significant two-factor interactions that affect geraniol yield, detected by the screening design.

The abbreviations are the same as in Table 1. (A) Filling volume:sugar interaction. (B) Sugar:cyclodextrins interaction. (C) Light:sugar interaction.

Time-course dynamics

In the next step, the cultivation of the plant suspension was performed at the favourable levels of factors from the geraniol yield design in order to determine the time dynamics of plant cell productivity in larger shaking flasks (250 ml). We set the screened factors at the beneficial levels in Gamborg's B5-medium supplemented with vitamins and phytohormones in order to keep the corresponding proportions from the geraniol yield design. The following culture conditions were chosen: 20 g/l sucrose; 50% filling volume; 5.6% (w/v) inoculum size; 35.62 μmol/cm^2/s light; 2 mM triacetyl-β-cyclodextrin and 180 rpm shaking frequency. The suspension cultures grew steadily throughout the cultivation period (Figure 6A) and showed signs of cell death after 14 days post inoculation (dpi). The latter was due to the depletion of nutrients and sucrose in the plant cell cultivation medium which is also seen from the reduction of electroconductivity and osmolality (Figure 6B and 6C). Dissolved oxygen levels started to increase after its minimum at 5 dpi and reached its maximum at 12 dpi, which is indicative for the switch from heterotrophic to autotrophic growth (Figure 6D). The maximal geraniol yield in the plant cell cultures at the end of the cultivation period was 5.12 mg/l (Figure 6E) which is commensurate with the highest levels achieved in the screening design (runs 51 and 60 from Table S2) if these results are extrapolated to the same geraniol yield unit (mg/l).

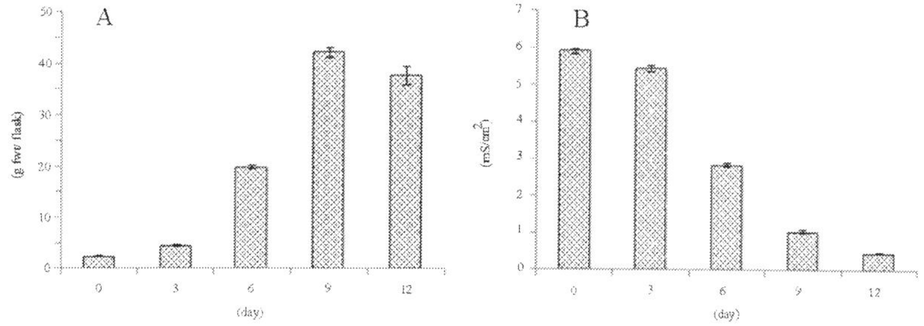

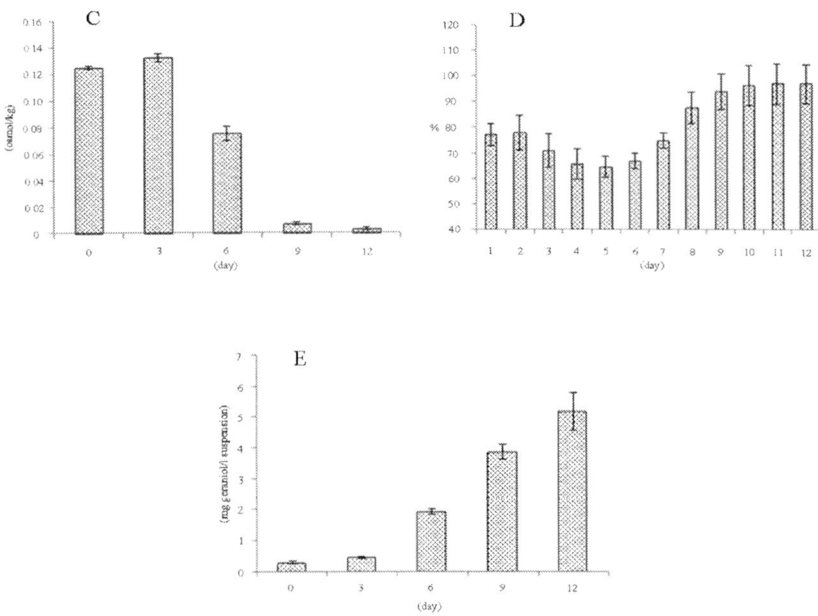

Figure 6: Time-course dynamics.

Cultivation conditions: 250 ml shaking flasks; 20 g/l sucrose; 50% filling volume; 5.6% (w/v) inoculum size; 35.62 (μmol/cm^2/s) light; 2 mM triacetyl-β-cyclodextrin and 180 rpm shaking frequency. (A) Plant cell growth. (B) Electroconductivity. (C) Osmolality. (D) Dissolved oxygen. (E) Geraniol yield.

DISCUSSION

We initiated a multi-factorial experimental design based on orthogonal arrays to investigate the effects of different cultivation factors on biomass accumulation and geraniol production in transgenic tobacco cell cultures grown in shaking flasks. Such a preliminary screening step is ordinarily needed to define the basic conditions of the key cultivation factors prior the main medium optimization, focused on the sequential manipulation of nutrients and elicitors [25]. Furthermore, cultivation factors are known to demonstrate a variation in a species-dependent manner and thus should be adjusted on a case-by-case basis [18]. Finally, being holistic in nature the optimization based on experimental designs demonstrates improved reproducibility because it takes into account the interactive effects of screened factors [31].

Our study demonstrated that the geraniol yield is mainly dependent on the factors that regulate the growth of the plant cell cultures. The factors acting positively on biomass production had a prevailing influence over the factors that are active on geraniol content with regard to the final geraniol yield. Among the tested cultivation parameters, the carbohydrate source had the strongest effect on biomass formation in the transgenic tobacco cultures. Glucose was utilized in a lesser extent than sucrose and the supplementation of the plant cell cultures with mannitol inhibited strongly their growth. Sucrose is known to be the most universal sugar for the plant cell cultivation, followed by glucose, maltose and raffinose [19], whereas the addition of some sugar alcohols like mannitol may cause osmotic stress in the system [19], [32]. Additionally, these sugar alcohols may serve not only as osmotic agents but they also regulate morphogenesis and metabolism [33]. We observed a similar metabolic effect of mannitol as it was the most potent factor stimulating the geraniol biosynthesis. However, the exact mechanism of induction of geraniol metabolism by mannitol remains to be clarified.

Filling volume is one of the operating parameters affecting the oxygen transfer rate in shaking flasks [34]. Filling volume had also an important effect on biomass production and respectively on geraniol yield of *VoGES* tobacco suspension cultures. Increasing the shaking speed and decreasing the medium volume improved biomass production to a certain extent in shake flask culture of *Atropa belladonna* hairy roots [16]. In our tobacco plant cell cultures, shaking speed did not play a significant role on geraniol yield. This indicates that the tobacco suspension cell cultures might not suffer from serious oxygen limitations at the applied shaking frequency and filling volume levels. Our study showed that shaking frequency may not function as a strong single factor but it is involved in several significant interactions, concerning biomass and geraniol yield mainly, and therefore, it should be also taken into consideration in the initial stage of optimization.

Our experiments identified that inoculum size had a moderate effect whilst the addition of conditioned medium had negligible impact on the biomass production of transgenic tobacco suspensions. Multi-factorial analyses using response surface experimental designs revealed that both inoculum size and conditioned medium had a positive effect on the biomass production in oil palm (*Elaeis guineensis*) cultures [18]. The inoculum size and the addition of conditioned medium had also a beneficial effect on alkaloid production in immobilized *Catharanthus roseus* cells [35]. Previous reports showed that the enhancement of inoculum density usually leads to higher biomass productivity but lower growth rate [13], [15]. The high inoculum size can contribute for the elimination or even diminishment of the lag phase during the cultivation [11]. The experimental design in our investigation could also detect a positive influence of high inoculum levels on geraniol yield. This observation is in agreement with earlier reports in literature showing that several enzymes i.e. geraniol-10-hydroxylase, had increased their activity in high-density cultures of *Catharanthus roseus* [36].

Addition of conditioned medium can enhance the cell density and the accumulation of ginseng polysaccharide in *Panax notoginseng* cell cultures [37]. The lack of potent effect of the conditioned medium in our plant cell cultures may be due to various reasons like reduction of osmolarity, decreasing the sugar and nitrogen concentrations, or to a phytohormone alteration because of the conditioned medium addition, although in general conditioned medium may contain some favourable stimulating factors at the same time [18].

The geraniol content model identified light as an important factor for geraniol biosynthesis. Several studies showed previously that light exerts a strong effect on monoterpenoid metabolism by regulating transcript levels of monoterpenoid synthases, precursor availability, constitutive promoter activity, etc. [20], [21], [38]. Furthermore, the interaction we found between light and sucrose might be also indicative for mixed contribution of both autotrophic and heterotrophic carbon metabolism in cultured tobacco cells [39].

We report herein that transgenic tobacco suspensions expressing *V. officinalis* GES could reach geraniol yield of 5.2 mg/l after 12 days cultivation. This accumulation level in non-fully-optimized conditions is close to the geraniol yield produced by yeast cells expressing the *O. basilium* GES [4]. Additional medium optimization may further improve the geraniol yield and turn transgenic plant cell cultures into an attractive production system of valuable natural products, matching the yields achieved by semi-synthesis and total chemical synthesis. Therefore, a combined, empirical study about optimization of plant suspensions productivity is needed, aiming at both physical and chemical factors and experimental statistical designs may serve as a useful tool.

ACKNOWLEDGMENTS

We would like to thank Dr. Harro Bouwmeester (University of Wageningen, the Netherlands) for providing us with *N. tabacum* cv. Samsun NN transgenic seeds, expressing stably *VoGES*. We thank Dr. Richard Twyman for critical reading of the manuscript. None of the authors declares a conflict of interest.

REFERENCES

1. de Roode BM, Oliehoek L, van der Padt A, Franssen MC, Boom RM (2001) Downstream processing of enzymatically produced geranyl glucoside. Biotechnol Prog 17: 881–886.
2. Sathikge ND (2008) Process for the preparation of geraniol and nerol from crude sulphated turpentine stream. M. T. D. Thesis, Nelson Mandela Metropolitan University. Available: http://dspace.nmmu.ac.za:8080/jspui/bitstream/10948/734/1/Ndavheleseni%20David%20Sathikge.pdf. Accessed 2014 June 27.
3. Dong L, Miettinen K, Goedbloed M, Verstappen FW, Voster A, et al. (2013) Characterization of two geraniol synthases from *Valeriana officinalis* and *Lippia dulcis*: Similar activity but difference in subcellular localization. Metab Eng 20: 198–211.
4. Fischer MJ, Meyer S, Claudel P, Bergdoll M, Karst F (2011) Metabolic engineering of monoterpene synthesis in yeast. Biotechnol Bioeng 108: 1883–1892.
5. Fischer MJ, Meyer S, Claudel P, Perrin M, Ginglinger JF, et al. (2013) Specificity of *Ocimum basilicum* geraniol synthase modified by its expression in different heterologous systems. J Biotechnol 163: 24–29.
6. Gutensohn M, Orlova I, Nguyen TT, Davidovich-Rikanati R, Ferruzzi MG, et al. (2013) Cytosolic monoterpene biosynthesis is supported by plastid-generated geranyl diphosphate substrate in transgenic tomato fruits. Plant J 75: 351–363.

7. Iijima Y, Gang DR, Fridman E, Lewinsohn E, Pichersky E (2004) Characterization of geraniol synthase from the peltate glands of sweet basil. Plant Physiol 134: 370–379.
8. Ito M, Honda G (2007) Geraniol synthases from perilla and their taxonomical significance. Phytochemistry 68: 446–453.
9. Simkin AJ, Miettinen K, Claudel P, Burlat V, Guirimand G, et al. (2013) Characterization of the plastidial geraniol synthase from Madagascar periwinkle which initiates the monoterpenoid branch of the alkaloid pathway in internal phloem associated parenchyma. Phytochemistry 85: 36–43.
10. Yang T, Li J, Wang HX, Zeng Y (2005) A geraniol-synthase gene from *Cinnamomum tenuipilum*. Phytochemistry 66: 285–293.
11. Jeong GT, Park DH, Ryu HW, Hwang B, Woo JC (2004) Effects of inoculum conditions on growth of hairy roots of *Panax ginseng* C.A. Meyer. Appl Biochem Biotechnol 113–116: 1193–1203.
12. Smetanska I (2008) Production of secondary metabolites using plant cell cultures. Food Biotechnol 111: 187–228.
13. Cui XH, Chakrabarty D, Lee EJ, Paek KY (2010) Production of adventitious roots and secondary metabolites by *Hypericum perforatum* L. in a bioreactor. Bioresour Technol 101: 4708–4716.
14. Takahashi Y, Hitaka Y, Kino-oka M, Taya M, Tone S (2001) Evaluation of growth property of red beet hairy roots depending on condition of inocula and its application to culture control with fuzzy logic theory. Biochem Eng J 8: 121–127.
15. Zhang CH, Wu JY, He GY (2002) Effects of inoculum size and age on biomass growth and paclitaxel production of elicitor-treated *Taxus yunnanensis* cell cultures. Appl Microbiol Biotechnol 60: 396–402.
16. Kanokwaree K, Doran PM (1997) The extent to which external oxygen transfer limits growth in shake flask culture of hairy roots. Biotechnol Bioeng 55: 520–526.
17. Klockner W, Büchs J (2012) Advances in shaking technologies. Trends Biotechnol 30: 307–314.
18. Gorret N, bin Rosli AK, Oppenheim SF, Willis LB, Lessard PA, et al. (2004) Bioreactor culture of oil palm (*Elaeis guineensis*) and

effects of nitrogen source, inoculum size, and conditioned medium on biomass production. J Biotech 108: 253–263.

19. George EF, Hall MA, De Klerk G-J (2008) Plant propagation by tissue culture. The background. Dordrecht, the Netherlands: Springer. 501 p.
20. Aharoni A, Giri AP, Deuerlein S, Griepink F, de Kogel WJ, et al. (2003) Terpenoid metabolism in wild-type and transgenic *Arabidopsis* plants. Plant Cell 15: 2866–2884.
21. Lu S, Xu R, Jia JW, Pang JH, Matsuda SPT, et al. (2002) Cloning and functional characterization of a beta-pinene synthase from *Artemisia annua* that shows a circadian pattern of expression. Plant Physiol 130: 477–486.
22. Brewster ME, Loftsson T (2007) Cyclodextrins as pharmaccutical solubilizers. Adv Drug Deliver Rev 59: 645–666.
23. Prakash G, Srivastava AK (2005) Statistical media optimization for cell growth and azadirachtin production in *Azadirachta indica* (A. Juss) suspension cultures. Process Biochem 40: 3795–3800.
24. Prakash G, Srivastava AK (2008) Statistical elicitor optimization studies for the enhancement of azadirachtin production in bioreactor *Azadirachta indica* cell cultivation. Biochem Eng J 40: 218–226.
25. Vasilev N, Gromping U, Lipperts A, Raven N, Fischer R, et al. (2013) Optimization of BY-2 cell suspension culture medium for the production of a human antibody using a combination of fractional factorial designs and the response surface method. Plant Biotechnol J 11: 867–874.
26. Pearse AGE (1980) Histochemistry, theoretical and applied: preparative and optical technology. Edinburgh: Churchill Livingstone. 728 p.
27. Grömping U, Amarov B, Xu H (2013) DoE.base: Full factorials, orthogonal arrays and base utilities for DoE packages. Package version 0.25-3. Available: http://CRAN.R-project.org/package=DoE.base. Accessed 26 June 2014.
28. Grömping U (2014) Mosaic plots are useful for visualizing low-order projections of factorial designs. Am Stat 68: 108–116.
29. Fox J (2003) Effect displays in R for generalised linear models. J

Stat Softw 8: 1–27.

30. Yuan M, Joseph VR, Lin Y (2007) An efficient variable selection approach for analyzing designed experiments. Technometrics 49: 430–439.
31. Vaidya BK, Mutalik SR, Joshi RM, Nene SN, Kulkarni BD (2009) Enhanced production of amidase from *Rhodococcus erythropolis* MTCC 1526 by medium optimisation using a statistical experimental design. J Ind Microbiol Biotechnol 36: 671–678.
32. Lipavska H, Vreugdenhil D (1996) Uptake of mannitol from the media by in vitro grown plants. Plant Cell Tiss Org Cult 45: 103–107.
33. Steinitz B (1999) Sugar alcohols display nonosmotic roles in regulating morphogenesis and metabolism in plants that do not produce polyols as primary photosynthetic products. J Plant Physiol 155: 1–8.
34. .Maier U, Büchs J (2001) Characterisation of the gas-liquid mass transfer in shaking bioreactors. Biochem Eng J 7: 99–106.
35. Lee CWT, Shuler ML (2000) The effect of inoculum density and conditioned medium on the production of ajmalicine and catharanthine from immobilized *Catharanthus roseus*cells. Biotechnol Bioeng 67: 61–71.
36. Moreno PR, Schlatmann JE, van der Heijden R, van Gulik WM, ten Hoopen HJ, et al. (1993) Induction of ajmalicine formation and related enzyme activities in *Catharanthus roseus* cells: effect of inoculum density. Appl Microbiol Biotechnol 39: 42–47.
37. Woragidbumrung KO, Sae-Tang P, Yao H, Han J, Chauvatcharin S, et al. (2001) Impact of conditioned medium on cell cultures of *Panax notoginseng* in an airlift bioreactor. Process Biochem 37: 209–213.
38. Schnurr JA, Guerra DJ (2000) The CaMV-35S promoter is sensitive to shortened photoperiod in transgenic tobacco. Plant Cell Rep 19: 279–282.
39. Wolf S, Kalman-Rotem N, Yakir D, Ziv M (1998) Autotrophic and heterotrophic carbon assimilation of in vitro grown potato (*Solanum tuberosum* L.) plants. J Plant Physiol 153: 574–580.

CHAPTER

20

EFFICIENT SECRETION OF RECOMBINANT PROTEINS FROM RICE SUSPENSION-CULTURED CELLS MODULATED BY THE CHOICE OF SIGNAL PEPTIDE

Li-Fen Huang[1], Chia-Chun Tan[1], Ju-Fang Yeh[2], Hsin-Yi Liu[2], Yu-Kuo Liu[3], Shin-Lon Ho[4], Chung-An Lu[2]

[1]Graduate School of Biotechnology and Bioengineering, Yuan Ze University, 135 Yuan-Tung Road, Taoyuan, Taiwan, ROC,

[2]Department of Life Science, National Central University, Taoyuan, Taiwan, ROC,

[3]Graduate Institute of Biochemical and Biomedical Engineering, Chang Gung University, Taoyuan, Taiwan, ROC,

[4]Department of Agronomy, National Chi-Yi University, Chiayi, Taiwan, ROC

ABSTRACT

Plant-based expression systems have emerged as a competitive platform in the large-scale production of recombinant proteins. By adding a signal peptide, αAmy3sp, the desired recombinant proteins can be secreted outside transgenic rice cells, making them easy to harvest. In this work, to improve the secretion efficiency of recombinant proteins in rice expression systems, various signal peptides including αAmy3sp,

CIN1sp, and 33KDsp have been fused to the N-terminus of green fluorescent protein (GFP) and introduced into rice cells to explore the efficiency of secretion of foreign proteins. 33KDsp had better efficiency than αAmy3sp and CIN1sp for the secretion of GFP from calli and suspension-cultured cells. 33KDsp was further applied for the secretion of mouse granulocyte-macrophage colony-stimulating factor (mGM-CSF) from transgenic rice suspension-cultured cells; approximately 76%–92% of total rice-derived mGM-CSF (rmGM-CSF) was detected in the culture medium. The rmGM-CSF was bioactive and could stimulate the proliferation of a murine myeloblastic leukemia cell line, NSF-60. The extracellular yield of rmGM-CSF reached 31.7 mg/L. Our study indicates that 33KDsp is better at promoting the secretion of recombinant proteins in rice suspension-cultured cell systems than the commonly used αAmy3sp.

INTRODUCTION

Plant-based protein expression systems have been successfully used to produce several recombinant proteins [1–4]. To improve the production of recombinant proteins and reduce the labor and time associated with protein purification steps, various plant-based systems, such as hair roots and suspension-cultured cells which are undifferentiated and actively dividing cells growing in liquid medium, have been considered for the direct secretion of recombinant proteins [5–8]. Most secretory proteins contain N-terminal signal peptides, which lead to the proteins being transported into the endoplasmic reticulum (ER) lumen [9,10] in a signal recognition particle (SRP)-dependent manner *via* SRP receptors [11,12]. Then, the signal peptides are cleaved from the precursor proteins in the ER, and the proteins are packed into small vesicles, budding to the Golgi apparatus, and eventually released to extracellular compartments [13]. A signal peptide is a prerequisite for protein secretion. The secretory proteins are correctly folded and some are further modified in the ER and the Golgi apparatus, such as by N-glycosylation, during the secretory process [14–16]. Signal peptides are fused at the N-terminal end of recombinant proteins to ensure the secretion of the recombinant proteins outside of the cell through this default pathway. As examples of this, a signal peptide derived from the resident ER protein calreticulin was applied in a plant hair root system [17, 18]; in addition, an α-amylase signal peptide was used in a rice suspension-cultured cell system [8]. However, it was reported that the recombinant proteins were not secreted efficiently into extracellular space, and a large proportion of recombinant proteins still remained within the cells [19]. This might have been due to the low activity of the applied signal peptide. Influences of signal peptide on protein secretion have been addressed in prokaryote [20], yeast [21] and mammalian cells [22, 23]. Although the detailed mechanism is not clear yet for the correlation between signal peptide and protein secretion, one proposed model is that signal peptide plays an important role to lead recombinant protein into the lumen of the ER [24–26], which is one of limiting steps of protein secretion in a default pathway.

The rice αAmy3/RAmy3D signal peptide has been applied to produce recombinant proteins in the medium of rice suspension-cultured cells. Upon coupling with the *αAmy3/RAmy3D*promoter, the αAmy3/RAmy3D signal peptide (αAmy3sp) is sufficient to ensure the secretion of recombinant proteins into medium by sugar-starved rice suspension-cultured cells [27–34]. The advantage of this system is that the *αAmy3/RAmy3D* promoter is highly inducible and the expressed recombinant protein can be guided extracellularly by αAmy3sp. However, αAmy3sp leads proteins to either extracellular space or to plastids [35]. Signal peptide of another α-amylase, αAmy7/RAmy1A, also drives target proteins into both extracellular space and chloroplasts [36]. Therefore, α-amylase signal peptides including αAmy3sp are not the most appropriate secretion peptides to use in rice suspension-cultured cells for recombinant protein expression.

Granulocyte-macrophage colony-stimulating factor (GM-CSF) is a cytokine that is generated by T cells, macrophages, endothelial cells, and immune-stimulated fibroblasts. GM-CSF exerts biological effects on the immune system, affecting the development and activation of myeloid precursor cells, macrophages, granulocytes, and dendritic cells. Clinical investigation has shown that GM-CSF can be used in a wide range of applications, such as vaccine adjuvant, cancer therapy, and immunotherapy for malignancies [37–40]. Given that GM-CSF has species specificity, mouse GM-CSF (mGM-CSF) is preferentially used in immune- and cancer-related research when mouse is chosen as a model animal.

Most signal peptides consist of 15 to 30 amino acid residues: a few positively charged residues at the N-terminus (n-region), 10–15 hydrophobic amino acid residues in the middle region (h-region), and a signal peptidase cleavage site at the C-terminus (c-region). The c-region usually contains small uncharged amino acid residues, such as glycine or alanine, as the first (-1) and third (-3) amino acid residues upstream of the signal peptidase cleavage site [41]. However, the sequence of amino acids in different signal peptides can vary widely. Therefore, the performance of a signal peptide is not predictable based on its amino acid residue sequence. In this study, to identify an effective signal peptide for application in a recombinant protein production system based on rice suspension-cultured cells, the secretion efficiency of three signal peptides, αAmy3sp, CIN1sp, and 33KDsp, was analyzed in transgenic cell lines by the expression of signal peptide- green fluorescent protcin (GFP) fusion proteins. The percentage of GFP secreted extracellularly was higher when fused with 33KDsp than when fused with the other two signal peptides. Furthermore, 33KDsp was applied to produce mGM-CSF, which resulted in a high level of bioactive mGM-CSF being secreted into the culture medium.

MATERIALS AND METHODS

Plant material

Rice (*Oryza sativa L.*) undifferentiated and actively dividing cells, calli, were derived from seeds of the Tainung No. 67 cultivar as described previously [42]. The calli were used for rice transformation, and transgenic calli were maintained on solid Murashige and Skoog (MS) with 3% (w/v) sucrose and 10 μM 2,4-dichlorophenoxyacetic acid (2,4-D). The transgenic calli were subcultured on same medium every 14 days, until calli were homogenized. For the establishment of rice suspension-cultured cells, calli were incubated in liquid MS medium with 3% (w/v) sucrose and 10 μM 2,4-D. The well-grown calli were transferred into liquid MS medium to generate suspension-cultured cells on a reciprocal shaker at 120 rpm and 26°C in the dark. One milliliter of suspension-cultured cells was subcultured in 25 mL of MS medium every 7 days until suspension cultured cells were homogenized for experiments.

N-terminal protein sequencing

Culture medium of suspension cells was collected and subjected to ammonium sulfate precipitation. Precipitated culture medium proteins (60–80% of ammonium sulfate) were dissolved in a buffer (20 mM Na-acetate, pH 5.2, 10 mM NaCl), and then subjected to dialysis against the same buffer. The culture medium proteins were further purified by fast protein liquid chromatography with Mono S column (GE Healthcare, Waukesha, WI). Collected fractions were examined by 12% SDS-PAGE and expected molecular weight of 33 kDa proteins were pooled and concentrated using a Vivaspin centrifugal concentrator (GE Healthcare). The purified of 33 kDa protein was subjected to N-terminus amino acid microsequencing using protein / peptide sequencer (ABI, Waltham, MA).

Plasmid construction

The 117 bp, 96 bp, and 102 bp DNA fragments corresponding to Amy3sp, CIN1sp, and 33KDsp, respectively, were amplified by PCR using cDNA prepared from rice suspension cultured cells with specific primers (Table 1). These DNA fragments were inserted into the *Bam*HI and *Eco*RI sites of pBluescript II SK(+) (Stratagene, La Jolla, CA) to generate pBS-Amy3sp, pBS-CIN1sp, and pBS-33KDsp. A GFP DNA fragment was excised from pUGFP [43] with *Bam*HI and *Eco*RI restriction enzymes. pAHC18, which contains maize ubiquitin promoter, was digested with *Bam*HI to remove the luciferase DNA fragment. Each DNA fragment encoding the signal peptide was excised from pBS-Amy3sp, pBS-CIN1sp, or pBS-33KDsp, and subcloned together with the GFP DNA fragment into *Bam*HI-digested pAHC18 plasmid to generate pUAmy3-GFP, pUCIN1-GFP, and pU33KD-GFP. For Agrobacterium-mediated gene transformation into rice genome, each resultant plasmid was digested with *Hin*dIII, and then the DNA fragment containing Ubi promoter–signal peptide–GFP was cloned into the binary vector pCAMBIA 1301, which was obtained from Cambia (www.cambia.org), at the *Hin*dIII site to generate pAUAmy3-GFP, pAUCIN1-GFP, and pAU33KD-GFP.

Table 1: Primers used in this study

Primers	Sequence	Purpose
Amy3F	aatggatccatgaagaacaacagcagcttg	Amy3sp-GFP construction
Amy3R	aatgaattcctcccagttgaaaccctggaa	Amy3sp-GFP construction
CIN1F	aatggatccatggggactcggctcttggcg	CIN1sp-GFP construction
CIN1R	aatgaattcccggtggacgacatgggacgc	CIN1sp-GFP construction
33KDF	aatggatccatggcggcattaagccagctg	33KDsp-GFP and 33KD-mGM-CSF construction
33KDR	aatgaattcgatgggatacgtcgtcgccgc	33KDsp-GFP and 33KD-mGM-CSF construction
mGMCSF-F	aatgaattcatggcacccacccgctcaccc	33KD-mGM-CSF construction; RT-PCR
mGMCSF-R	aatggatcctcatttttggactggttttt	33KD-mGM-CSF construction; RT-PCR
actinF	ctgatggacaggttatcacc	RT- PCR
actinR	caggtagcaataggtattacag	RT- PCR

To generate the Ubi promoter-33KD signal peptide-mGM-CSF fusion construct, a 381 bp DNA fragment encoding mGM-CSF was amplified by PCR with specific primers (Table 1) by using pENTR-mGM-CSF [33] as the template. The mGM-CSF DNA fragment was cut with *Bam*HI and *Eco*RI, and incubated together with the *Bam*HI and *Eco*RI double-digested 33KDsp DNA fragment to be subcloned into *Bam*HI-treated pAHC18 to generate pU33KD–mGM-CSF. The resultant plasmid was digested with *Hin*dIII and then cloned into the binary vector pCAMBIA 1301 at the same site to generate pAU33kD-mGM-CSF.

Rice transformation

Four plasmids, pAUAmy3-GFP, pAUCIN1-GFP, pAU33KD-GFP, and pAU33KD-mGM-CSF, were individually introduced into *Agrobacterium tumefaciens* strain EHA105 [44] by electroporation. Rice transformation was performed by *Agrobacterium*-mediated transformation as described previously [42]. Transformed calli were selected on N6 medium with 3% (w/v) sucrose, 10 μM 2,4-D, and 30 mg/L hygromycin.

Genomic PCR

Genomic DNA was isolated from each putative transgenic callus as described previously [45], and 200 ng of rice genomic DNA were used as a template for the PCR reaction with specific primers to amplify the transgene *GFP* and the endogenous gene *Actin* as an internal control DNA (Table 1). The PCR product was analyzed using 1% agarose gels and stained with ethidium bromide.

RT-PCR

Total RNA extraction from rice calli and first-strand cDNA synthesis were performed as described as previously [33]. The first-strand cDNA was diluted 50-fold and then subjected to PCR (22–28 reaction cycles) with *GFP*, *αAmy3*, *CIN1*, *33KD*, and *Act1* gene-specific primers (Table 1).

Protein gel blot analysis

To obtain total intracellular soluble protein, rice calli or suspension-cultured cells were ground to a fine powder in liquid N_2, and then extracted with protein extraction buffer (100 mM K_2HPO_4, pH 7.8, 1 mM EDTA, 7 mM β-mercaptoethanol, 1% (v/v) Triton X-100, and 1% (v/v) glycerol). To obtain total secreted protein released in the liquid medium, the cell culture medium was centrifuged at 18,000 × *g* and 4°C for 15 min in order to remove the cell debris. The protein concentration was determined using Bio-Rad Protein Assay reagent (Bio-Rad, Hercules, CA). Proteins were separated by 12% SDS-PAGE and then blotted onto Hybond-P membrane (GE Healthcare). Commercially available recombinant GFP (Abcam, Cambridge, MA) produced from *E. coli* was used as a positive control. Polyclonal rabbit anti-GFP antibody (Abcam) and Lumi-light Western blotting substrate (Roche, Basel, Switzerland) were used to detect the recombinant GFP. For the detection of recombinant mGM-CSF protein, polyclonal rabbit anti-mGM-CSF antibody (Abcam) was used as described previously [33]. To evaluate the proportion of protein that was secreted to the extracellular environments, 0.5 mL of suspension-cultured cells were incubated in 1 mL of MS with 3% sucrose (+S) or without any carbon source (-S) for various periods. The cell culture medium was collected, and the total soluble intracellular protein was isolated from the rice cells in 1 mL of protein extraction buffer. Both cellular soluble proteins and cell culture medium proteins were loaded into gels at 20 μL per lane for Western blot analysis, and the proportion of protein that was secreted was determined as follows: (target protein in medium / target protein in medium + target protein in the cellular protein fraction) × 100%. The Western blot analysis was repeated 3 times and at least 2 biological replicates were performed for each line.

Fluorometric assays

Total soluble protein was extracted from each transgenic callus or collected from cell culture medium. One hundred microliters of sample were used for GFP fluorescence detection in a Fluoroskan Ascent FL fluorimeter (Thermo, Vantaa, Finland), with excitation at 490 ± 5 nm and emission at 510 ± 5 nm [46]. Fluorescence was normalized using non-transformed cell extract or medium. The fluorometric analyses were performed 3 times and at least 2 biological replicates were applied for each cell line.

MTT analysis

Reduction of 3-(4,5-dimethylthiazol-2-yl)-2,5-diphenyltetrazolium bromide (MTT) was applied to measure NFS-60 cell viability for the murine myeloblastic leukemia cell line, NFS-60, as previously described [33]. The NFS-60 cells were cultured in medium containing rice-produced mGM-CSF for 3 days. In this assay, purple formazan was detected in a spectrophotometer at 545 nm.

RESULTS

The choice of signal peptides for recombinant protein secretion from rice suspension-cultured cells

To identify a better signal peptide than rice αAmy3sp, secretory proteins naturally produced by rice suspension cells are good resources to test. The rice suspension cells were cultured in sugar-containing medium or sugar-free medium, for various periods, and the total protein in the medium was analyzed by SDS-PAGE and silver staining. A strong signal at a molecular weight of approximately 33 kDa was observed in both sugar-containing medium and sugar-free medium (Fig 1). The 33 kDa protein was purified by ammonium sulfate precipitation, and then followed by fast protein liquid chromatography. The protein sequence of the corresponding 33 kDa secretory protein was determined, and was identical to the DUF26-like protein/33KD secretory protein (called Os33KD and encoded by *Os04g0659300*), which has been shown as a secretory protein in the culture medium of rice suspension cells [47]. Thus, the signal peptide of Os33KD, called 33KDsp, was selected to compare the secretion efficiency of recombinant proteins with that of αAmy3sp. In addition to 33KDsp, another rice signal peptide was selected, a rice cell wall invertase (OsCIN1, encoded by *Os02g0534400*) which is present in the cell wall and hydrolyzes sucrose to glucose and fructose [48]; this signal peptide was named CIN1sp. The signal peptide regions of αAmy3, Os33KD, and OsCIN1 were predicted using SignalP 3.0 [49], which putatively showed that 25 amino acids at the N-terminus of αAmy3, 25 amino acids at the N-terminus of Os33KD, and 22 amino acids at the N-terminus of OsCIN1 are the signal peptides (Fig 2).

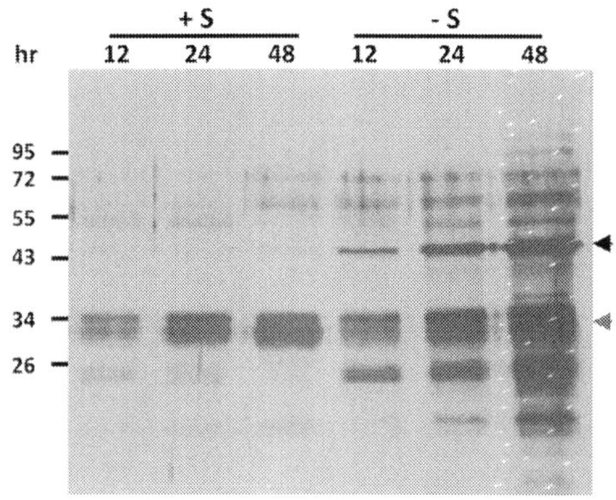

Figure 1: Proteins in medium secreted from suspension-cultured rice cells.

One milliliter of rice suspension cells were cultured in 25 mL of sugar-containing medium (+S) or sugar-free medium (-S), for various time periods. Liquid medium samples of 20 μL were obtained, separated by SDS-PAGE, and subjected to silver staining. The location of α-amylase is indicated by the black arrow, and Os33KD protein is indicated by the grey arrow.

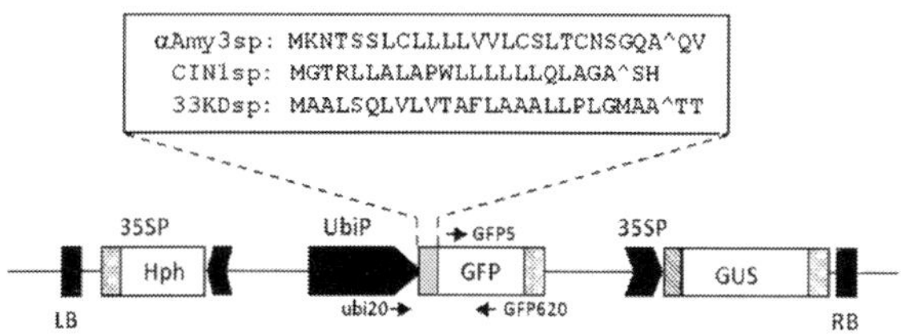

Figure 2: Expression vectors with signal peptides of αAmy3, OsCIN1, or Os-33KD used for rice transformation.

Various signal peptide cDNAs (αAmy3sp, CIN1sp, or 33KDsp) were fused in-frame upstream of green fluorescent protein (GFP) DNA. The fusion genes were engineered to be located downstream of the maize ubiquitin promoter (UbiP). β-glucuronidase (*GUS*) and the hygromycin-resistance gene (*Hph*) were both driven by a CaMV35S promoter (35SP). Constructs contained the left and right borders (LB, RB) flanking the T-DNA that can be transferred into the plant genome via *Agrobacterium*-mediated transformation. Arrows indicate specific primers used in this study. Amino acid sequences for αAmy3sp, CIN1sp, and 33KDsp are shown in the upper box. Carets indicate predicted signal peptide cleavage sites.

Comparison of three different signal peptides in terms of the efficiency of recombinant GFP secretion in transgenic rice calli

To compare the efficacy of the three rice signal peptides for the efficient secretion of recombinant protein in rice cultured cells, and to reduce influence from other unexpected factors on the secretion efficiency of these signal peptides, ubiquitin (*Ubi*) promoter which drives constitutive gene expression both in sugar feeding and starving rice cultured cells was used to control the expression of reporter gene. Three chimeric genes, *Ubi::αAmy3sp-GFP*, *Ubi::CIN1sp-GFP*, and *Ubi::33KDsp-GFP*, containing αAmy3sp, CIN1sp, and 33KDsp, respectively, were generated (Fig 2). These expression cassettes were introduced individually into the rice genome *via Agrobacterium*-mediated transformation. Several transgenic lines were obtained for each signal peptide, and GFP fluorescence was observed as shown in Fig 3A. Nine transgenic lines of each construct were randomly selected to detect GFP levels. We previously demonstrated that the cell viability of suspension-cultured rice cells in sugar containing medium can be maintained 90% at least for 10 days [26]. We further confirmed the result by protein gel analysis, and the band patterns of intracellular soluble proteins were obviously different from cultured medium proteins at 5, 7 and 9 days (Fig 3B). Therefore, the transgenic calli were cultured in MS liquid medium for 5 days, and both cultured calli and medium were collected to prevent artifacts caused by cell lysis. To quantify the level of fluorescence of GFP, a standard curve was used with a concentration range from 0.1 to 3 μg/mL purified recombinant GFP protein (Fig 3C). For each independent cell line, the levels of recombinant GFP proteins in the culture medium and in cellular extracts were measured (Fig 3D; S1 Table), and the proportion of GFP protein secreted was determined by dividing the amount of GFP in the medium by the total GFP. As shown in Fig 3D, wide ranges of intracellular to medium GFP ratios were displayed among the selected lines. For *αAmy3sp-GFP* transgenic cell lines, 37%–95% of total GFP was detected in the medium. For *CIN1spGFP*transgenic cell lines, 0%–75% of total GFP was detected in the medium. However, for *33KDspGFP* transgenic cell lines, 52%–95% of total GFP was detected in the medium, and in the majority of these selected lines, more than 80% of total GFP was present in the culture medium. This result showed that the highest proportion of medium GFP was achieved using 33KDspGFP-containing calli, followed by αAmy3spGFP and then CIN1spGFP, suggesting that the secretion efficiency of 33KDsp is higher than that of αAmy3sp and CIN1sp.

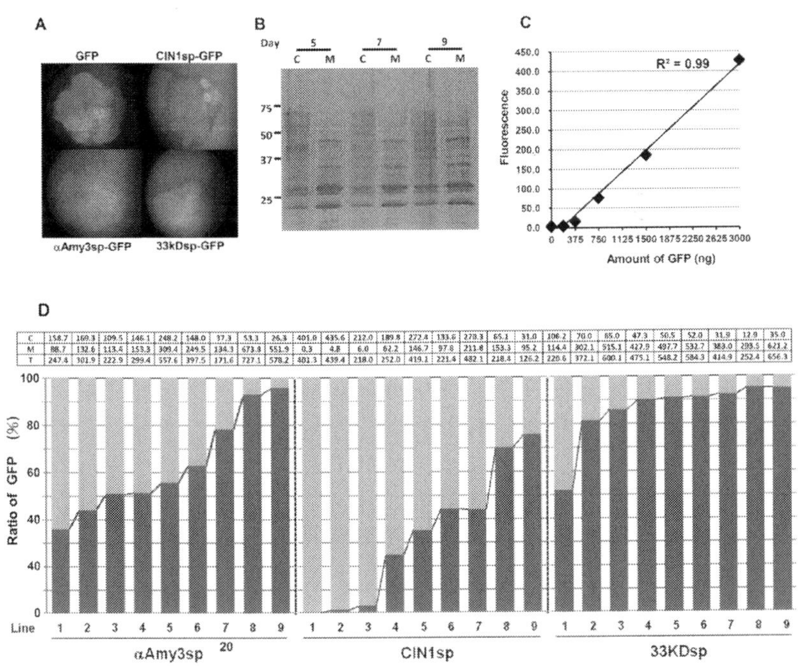

Figure 3: Detection of GFP fluorescence in selected transgenic calli.

A, Putative transgenic calli harboring the GFP gene were grown on hygromycin selection medium and observed by fluorescence microscopy. **B,** Rice calli (cell volume of 0.5 mL) were cultured in 1 mL of MS with sugar medium for 5, 7 and 9 days. Total cellular soluble proteins were isolated in 1 mL of protein extraction buffer from each of the treated cell cultures. Twenty microliters of total cellular protein (C) and 20 μL of total medium protein (M) were subjected to silver staining. **C,** A standard curve for GFP protein amount and fluorescence intensity. **D,** The ratio of GFP protein in cellular extract and cultured medium among various transgenic cell lines. For each analyzed transgenic line, 0.5 mL rice calli were cultured in 1 mL of MS with sugar medium for 5 days. GFP fluorescence signals from the culture medium and total cellular soluble protein were measured. The upper panel indicated the amount of GFP protein, averaged by three replicated fluorescence detections from two biological repeats. C: amount of cellular GFP proteins (μg); M: total amount of GFP proteins in medium (μg); T: amount of total GFP proteins (μg). The standard deviation from each selected transgenic line. The proportion of medium GFP in each independent line is presented as the GFP fluorescence from total GFP in medium compared with that from total GFP. A dark gray bar indicates medium GFP, while a light gray bar indicates intracellular GFP.

Efficiency of three signal peptides for the secretion of recombinant GFP proteins in transgenic suspension-cultured rice cells

To evaluate the secretion efficiency of the three signal peptides further, three independent suspension-cultured cell lines for each signal peptide were established. The mRNA levels of *GFP* in the various suspension cell lines were similar, except for line 2 of *αAmy3sp-GFP*, which had a slightly lower level, and line 3 of *CIN1sp-GFP*, which had a higher level than the other cell lines (Fig 4A). Each selected rice suspension-cultured cell line was cultured in MS medium for 5, 7, and 9 days, after which total cellular soluble protein and culture medium protein were obtained. The relative levels of GFP was detected by Western blot analysis with GFP antibodies, and quantified using Gel-Pro Analyzer (Fig 4B). The GFP proteins were detected in both intracellular and medium total soluble proteins of most suspension cell lines (Fig 4B). Compared with the *αAmy3sp-GFP* suspension cell lines and the *CIN1sp-GFP* suspension cell lines, the *33KDsp-GFP* suspension cell lines accumulated more recombinant GFP proteins in the culture medium, regardless of the duration of the culture period (Fig 4B).

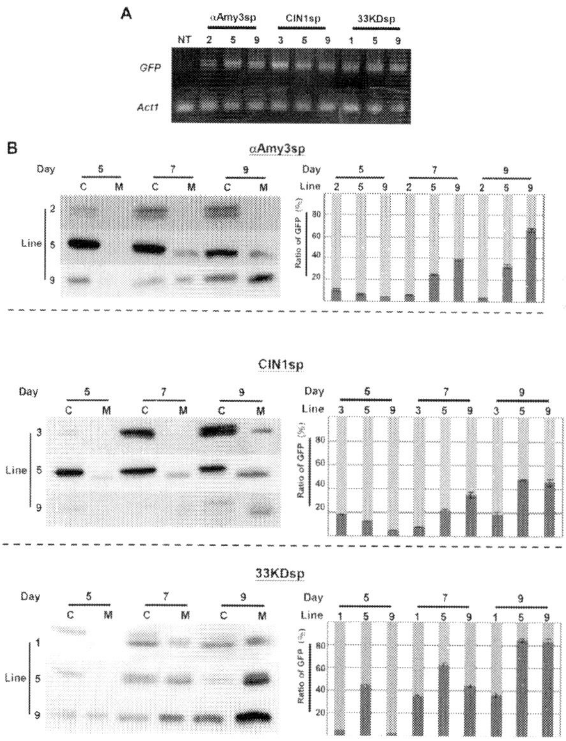

Figure 4: Gene expression of GFP in transgenic rice suspension cell cultures.

A, Detection of GFP mRNA in non-transformed (NT) and selected transgenic rice suspension-cultured cells, including αAmy3sp-2, -5, and -9; CIN1sp-3, -5, and -9; and 33KDsp-1, -5, and -9. Total RNAs were isolated from NT and transgenic rice suspension-cultured cells, and then subjected to RT-PCR with specific *GFP* primers. The rice *Act1* mRNA was used as an internal control. **B,** The proportion of medium GFP protein in various transgenic suspension cell lines. Total soluble proteins were extracted from each cell line, with a cell volume of 0.5 mL, in 1 mL of extraction buffer. A total of 1 mL of culture medium for each line was collected. Twenty microliters of total cellular protein (C) and 20 μL of total extracellular protein (M) were subjected to Western blot analysis using GFP antibodies. GFP proteins within cells and medium were detected, and compared with the total amount of GFP protein. The proportion of medium GFP for each independent cell line is presented as the total GFP in medium compared with the total GFP. A dark gray bar indicates medium GFP, while a light gray bar indicates cellular GFP. Error bars indicate the standard deviation of three Western blot replicates from two biological repeats.

A recombinant protein expression platform in rice suspension cells has been established using a rice α-amylase gene promoter, *αAmy3*p, which drives the production of recombinant protein in sugar-free medium [29]. To evaluate the secretion efficiency of the three signal peptides in sugar-free growth conditions, two suspension-cultured cell lines per signal peptide were incubated in sugar-free medium for 2 and 3 days. As shown in Fig 5, in 2-day-cultured medium, there was substantially more accumulation of recombinant GFP proteins for the *33KDsp-GFP*lines than for the *αAmy3sp-GFP* lines. When suspension-cultured cells were starved of sugar for 3 days, GFP proteins further accumulated in the culture medium, and even more GFP was secreted by the *33KDsp-GFP* cell lines than by *αAmy3sp-GFP*. These results indicate that 33KDsp has higher secretion efficiency than αAmy3sp, and can be applied in a well-developed recombinant protein production platform using sugar-starved rice suspension-cultured cells.

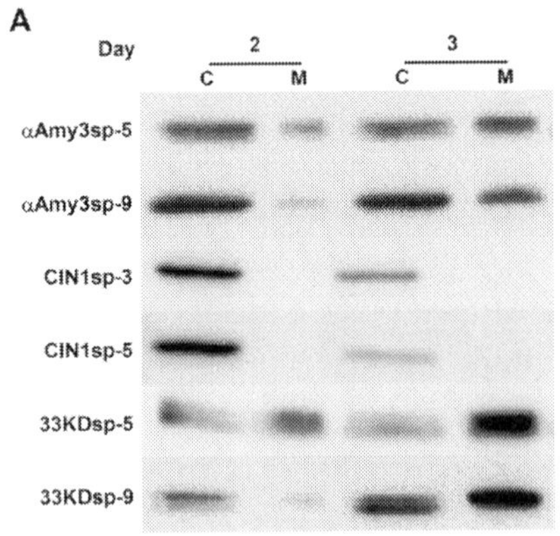

B

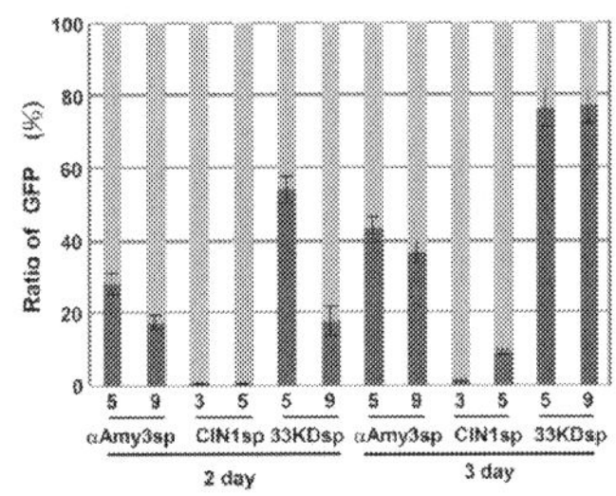

Figure 5: Detection of GFP protein in sugar-starved transgenic rice suspension cell cultures.

A, Rice suspension cells (cell volume of 0.5 mL) were cultured in 1 mL of sugar-free MS medium for 2 and 3 days. Total cellular soluble proteins were isolated in 1 mL of protein extraction buffer from each of the treated cell cultures. Twenty microliters of total cellular protein (C) and 20 μL of total medium protein (M) were subjected to Western blot analysis using GFP antibodies. **B,** The proportion of medium GFP protein in various transgenic suspension cell lines. The amount of GFP within cells and medium was detected, and converted to the total amount of GFP protein. The proportion of medium GFP in each independent line is presented as the total GFP in the medium compared with the total GFP. A dark gray bar indicates medium GFP, while a light gray bar indicates cellular GFP. Error bars indicate the standard deviation of three Western blot replicates from two biological repeats.

Production of bioactive mGM-CSF in the medium of rice suspension-cultured cells using the 33KD signal peptide

To examine whether 33KDsp can be applied for the efficient secretion of other valuable recombinant proteins, mGM-CSF was fused downstream of 33KDsp and expression was driven by the Ubi promoter (Fig 6A). Through the *Agrobacterium*-mediated transformation, several transgenic cell lines were obtained, and line 10 with a high level of *mGM-CSF* mRNA was selected to establish the rice suspension-cultured cell line (Fig 6B). The suspension-cultured cells were incubated in either sugar-containing medium for 5, 7, and 9 days or sugar-free medium for 1, 2, and 3 days, and then subjected to rmGM-CSF monitoring of the total soluble intracellular and medium proteins. As shown in Fig 6C, rice-derived mGM-CSF (rmGM-CSF) was detectable in both cell extract and culture medium, and its level increased with time in sugar-containing medium (Fig 6C). The rmGM-CSF was predominantly present in culture medium, at around 76%–92% of total rmGM-CSF protein (Fig 6C). The yields of rmGM-CSF were 8.5, 24.4, and 31.7 mg/L in sugar-containing medium cultured for 5, 7, and 9 days, respectively (Fig 6E). When the rmGM-CSF production by suspension cell line 10 was monitored upon culture in sugar-free medium for 1, 2, and 3 days, the protein was predominantly detected in the sugar-free culture medium, and the percentage of rmGM-CSF that was secreted reached 90% of the total amount produced (Fig 6D).

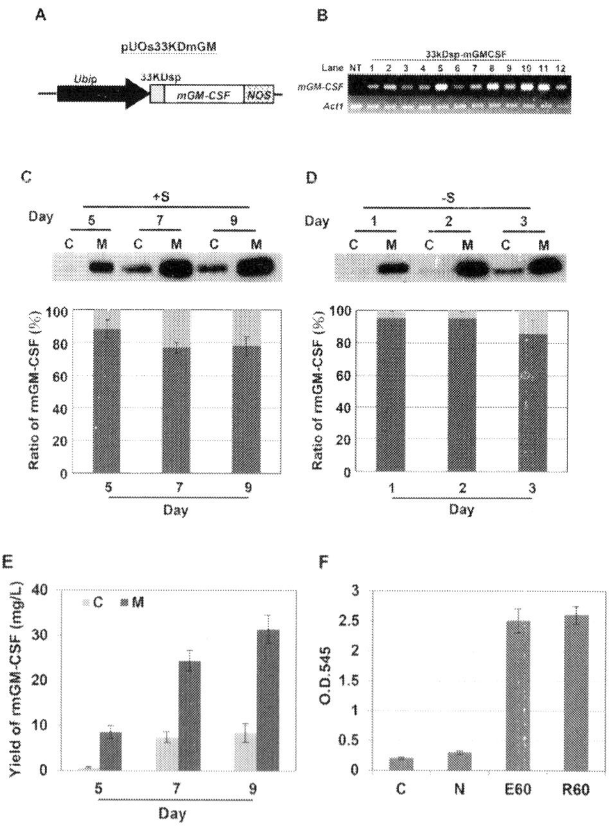

Figure 6: Production of secretory recombinant mGM-CSF in rice cells using the 33KD signal peptide.

A, Expression vector containing a 33KD signal peptide fused with mGM-CSF chimeric gene. The 33kDsp-mGM-CSF fusion gene was engineered to be located downstream of the maize ubiquitin promoter (UbiP). **B**, Detection of mGM-CSF mRNA in non-transformed (NT) and selected 33KDsp-mGMCSF putative transgenic rice calli. Total RNAs were isolated from NT and transgenic rice calli, and then subjected to RT-PCR with specific mGM-CSF primers. The rice *Act1* mRNA was used as an internal control. **C & D,** The proportion of medium rice-derived mGM-CSF (rmGM-CSF) protein in transgenic suspension cell line 10. Suspension cells with cell volume of 0.5 mL were cultured in 1 mL of medium with (C) or without sugar (D). Total soluble proteins were isolated from the suspension cells in 1 mL of protein extraction buffer, and 1 mL of culture medium was collected at various time periods as indicated. Twenty microliters of total cellular proteins (C) and 20 μL of total extracellular proteins (M) were subjected to Western blot analysis using mGM-CSF antibodies. The signals for rmGM-CSF within cells and medium were detected and compared with the total amount of rmGM-CSF protein. The proportion of medium rmGM-CSF is presented as the total rmGM-CSF in the medium compared with the total rmGM-CSF. A dark gray bar indicates medium rmGM-CSF, while a light gray bar indicates cellular rmGM-CSF. Error bars indicate the standard deviation of three Western blot replicates from two biological repeats. **E,**Production of mGM-CSF in transgenic rice suspension cells cultured in sugar-containing medium. The amount of recombinant mGM-CSF produced was measured in culture medium and rice suspension cells at 5, 7, and 9 days. Error bars indicate the standard deviation of three Western blot replicates from two biological repeats. **F,** Analysis of the biological activity of secreted rmGM-CSF from rice suspension cell culture medium at day 9. A murine hematopoietic cell line (NFS-60) was incubated with 60 ng/mL rice-derived mGM-CSF (R60). Commercial mGM-CSF derived from *E. coli* cells (E60; 60 ng/mL) was used as a reference standard. Both sucrose-free MS medium (C) and non-transformed rice culture medium (N) were used as negative controls. The proliferation of NFS-60 cells was measured with the MTT assay system. The error bar represents the standard deviation from triplicate cultures.

The biological activity of rice-derived rmGM-CSF secreted due to the 33KD signal peptide was examined with a colorimetric microassay using murine myeloblastic leukemia NFS-60 cells. The NFS-60 cells were cultured with rice suspension cell- or *E. coli*-derived recombinant mGM-CSF for 3 days. Cell viability was determined using the MTT assay system. Neither sucrose-free MS medium (Fig 6F, column C) nor non-transformed rice culture medium (Fig 6F, column N) could proliferate the NSF-60 cells. In contrast, the viability of NFS-60 cells was increased significantly in the presence of mGM-CSF (Fig 6F). The activities of the two types of mGM-CSF: recombinant mGM-CSF produced from *E. coli* (commercial product as a positive control) or that from rice suspension cells, were similar in terms of the viability of NFS-60 cells that they enabled (Fig 6F).

DISCUSSION

The production of recombinant proteins in cell cultured medium have been conducted in several recombinant protein expression systems. In addition, recombinant proteins that require post-translational modification in order to retain bioactivity or stability can be modified through a series of enzymes located in the ER and the Golgi apparatus before secretion. Rice suspension-cultured cells, used in conjunction with an αAmy3 signal peptide, have been used to successfully produce several recombinant proteins in culture medium. However, the secretion efficiency of the αAmy3 signal peptide in rice suspension cells had not been clarified. In this study, we compared the secretion efficiency of three rice signal peptides, αAmy3sp, CIN1sp, and 33KDsp, in transgenic rice cell lines that expressed signal peptide fused with GFP. The amount of medium GFP as a proportion of the total GFP was used as an indicator of the secretion efficiency. Our results indicated that 33KDsp was more efficient than αAmy3sp and CIN1sp for the secretion of GFP from both rice calli and suspension-cultured cells. 33KDsp was also used to secrete mGM-CSF from transgenic rice suspension-cultured cells, and 76%–92% of the mGM-CSF was detected in the culture medium.

The general features of signal peptides that increase the efficiency of secretion in rice cells are unclear. In this study, although GFP could be secreted out of rice cells, intracellular GFP was still detectable. Thus, the studied signal peptides were not able to cause all GFP to be secreted. However, we cannot rule out possibility that secreted GFP was retained between outside of the cell membrane and the cell wall matrix or in the intracellular space of the cell aggregates.

Plant cell wall porosity has been measured around 3.5–9.2 nm by different measurement methods, thus proteins smaller than 30–50 kDa can easily pass through a plant cell wall [50]. The molecular weight of GFP is 27 kDa, so it can be secreted into cultured medium as expected. In this study, two forms of GFP, 30 kDa and 27 kDa, were identified in Western blot analysis. The 27 kDa GFP was the same as that of the medium GFP. Therefore, we assume that the 30 kDa GFP was due to a failure of the signal peptide to be cleaved. This was supported by the fact that the amount of the 30 kDa GFP was inversely correlated with the secretion efficiency. For example, the 30 kDa GFP was accumulated more in the *CIN1sp-GFP*transgenic cell lines that had a low level of extracellular GFP. In contrast, a low level of the 30 kDa GFP was detected in the *33KDsp-GFP* transgenic cell lines that had high secretion efficiency. These observations suggest that the cleavage of the signal peptide is an important step for protein secretion by rice suspension cells. The three signal peptides studied here all contain classic cleavage sites in the c-region, and the first (-1) and third (-3) amino acids upstream of the cleavage site are small and uncharged amino acids; therefore, primary sequence are not able to offer cleavage efficiency of these signal peptides.

Low affinity between signal peptide and SRP may lead to a failure of secretion in a default pathway. The hydrophobic h-region of the signal peptide forms an α-helix that is bound by the SRP [51] and this step is necessary for secretory proteins to translocate into the lumen of the ER. Then, the signal peptides are removed from secretory proteins in the secretory pathway. On the basis of previous studies on hydrophobic residues of the interleukin-2 signal peptide [26] and the *Metridia* luciferase signal peptide [22] in mammalian cells, secretion efficiency can be improved by increasing the hydrophobicity. The three signal peptides studied here all contain an h-region, and the highest hydrophobicity is exhibited by Amy3sp, followed by CIN1sp and 33KDsp. However, 33KDsp had the highest secretion efficiency, so high hydrophobicity of the h-region was clearly not the primary determinant of this. Instead, it is possible that 33KDsp has specific amino acid residues in the h-region that are important for association with the SRP.

The levels of GFP mRNA did not differ markedly among the selected transgenic cell lines harboring various signal peptides fused to GFP. However, the amounts and secretion ratios of GFP did differ markedly among the different lines, even for the same signal peptide. There is no clear correlation among GFP mRNAs, cellular GFP proteins and secreted GFP proteins, and complicated dynamic cellular activities may interfere the relationship between cellular and secreted GFP. In addition, individual differences among independent transgenic lines also increase the variation. For example, it is common to observe most transgenic lines have various expression levels of transgene, even harboring the same expression cassette. Our results also showed that CIN1sp was ineffective in the rice suspension cell expression system. Both αAmy3sp and 33KDsp were better signal peptides than CIN1sp in our studies, and both *αAmy3* and *Os33KD* gene expression were abundant in the rice cell suspension cultured medium. In contrast, gene expression of *OsCIN1* was strong in leaves [52] and weak in calli [53]. Thus, it would be an interesting topic to study regulatory roles of signal peptides on protein secretion in different types of cells.

In this study, we used the maize *Ubi* promoter, a constitutive expression promoter, to examine the efficiency of secretion promoted by signal peptides in rice suspension-cultured cells. The *Ubi* promoter with 33KDsp was further applied for mGM-CSF production. The bioactive form of rice recombinant mGM-CSF was detected in the culture medium and the maximum yield reached 31.7 mg/L. Our study indicates that application of the constitutive *Ubi* promoter with 33KDsp is recommended in protein expression systems with rice suspension-cultured cells instead of the sugar starvation-inducible *αAmy3* promoter with αAmy3sp.

REFERENCES

1. Ma JK, Drake PM, Christou P. The production of recombinant pharmaceutical proteins in plants. Nat Rev Genet. 2003; 4: 794–805. pmid:14526375
2. Fischer R, Stoger E, Schillberg S, Christou P, Twyman RM. Plant-based production of biopharmaceuticals. Curr Opin Plant Biol. 2004; 7: 152–158. pmid:15003215

3. Paul M, Ma JK. Plant-made pharmaceuticals: leading products and production platforms. Biotechnol Appl Biochem. 2011; 58: 58–67. pmid:21446960
4. Daniell H, Singh ND, Mason H, Streatfield SJ. Plant-made vaccine antigens and biopharmaceuticals. Trends Plant Sci. 2009; 14: 669–679. pmid:19836291
5. Hilton MG, Rhodes MJ. Growth and hyoscyamine production of 'hairy root' cultures of Datura stramonium in a modified stirred tank reactor. Appl Microbiol Biotechnol. 1990; 33: 132–138. pmid:1366453
6. Sharp JM, Doran PM. Strategies for enhancing monoclonal antibody accumulation in plant cell and organ cultures. Biotechnol Prog. 2001; 17: 979–992. pmid:11735430
7. Hellwig S, Drossard J, Twyman RM, Fischer R. Plant cell cultures for the production of recombinant proteins. Nat Biotechnol. 2004; 22: 1415–1422. pmid:15529167
8. Chan MT, Chang HH, Ho SL, Tong WF, Yu SM. Agrobacterium-mediated production of transgenic rice plants expressing a chimeric alpha-amylase promoter/beta-glucuronidase gene. Plant Mol Biol. 1993; 22: 491–506. pmid:8392395
9. von Heijne G. Signal sequences. The limits of variation. J Mol Biol. 1985; 184: 99–105. pmid:4032478
10. von Heijne G, Gavel Y. Topogenic signals in integral membrane proteins. Eur J Biochem. 1988; 174: 671–678. pmid:3134198
11. Gilmore R, Blobel G, Walter P. Protein translocation across the endoplasmic reticulum. I. Detection in the microsomal membrane of a receptor for the signal recognition particle. J Cell Biol. 1982; 95: 463–469. pmid:6292235
12. Walter P, Ibrahimi I, Blobel G. Translocation of proteins across the endoplasmic reticulum. I. Signal recognition protein (SRP) binds to in-vitro-assembled polysomes synthesizing secretory protein. J Cell Biol. 1981; 91: 545–550. pmid:7309795
13. Vitale A, Denecke J. The endoplasmic reticulum-gateway of the secretory pathway. Plant Cell. 1999; 11: 615–628. pmid:10213782
14. Breitling J, Aebi M. N-linked protein glycosylation in the

endoplasmic reticulum. Cold Spring Harb Perspect Biol. 2013; 5: a013359. pmid:23751184

15. Braakman I, Hebert DN. Protein folding in the endoplasmic reticulum. Cold Spring Harb Perspect Biol. 2013; 5: a013201. pmid:23637286
16. .Aebi M. N-linked protein glycosylation in the ER. Biochim Biophys Acta. 2013; 1833: 2430–2437. pmid:23583305
17. Gleba D, Borisjuk NV, Borisjuk LG, Kneer R, Poulev A, Sarzhinskaya M, et al. Use of plant roots for phytoremediation and molecular farming. Proc Natl Acad Sci USA. 1999; 96: 5973–5977. pmid:10339526
18. Borisjuk NV, Borisjuk LG, Logendra S, Petersen F, Gleba Y, Raskin I. Production of recombinant proteins in plant root exudates. Nat Biotechnol. 1999; 17: 466–469. pmid:10331806
19. James EA, Wang C, Wang Z, Reeves R, Shin JH, Magnuson NS, et al. Production and characterization of biologically active human GM-CSF secreted by genetically modified plant cells. Protein Expr Purif. 2000; 19: 131–138. pmid:10833400
20. Ng DT, Sarkar CA. Engineering signal peptides for enhanced protein secretion from Lactococcus lactis. Appl Environ Microbiol. 2013; 79: 347–356. pmid:23124224
21. Eiden-Plach A, Zagorc T, Heintel T, Carius Y, Breinig F, Schmitt MJ. Viral preprotoxin signal sequence allows efficient secretion of green fluorescent protein by Candida glabrata, Pichia pastoris, Saccharomyces cerevisiae, and Schizosaccharomyces pombe. Appl Environ Microbiol. 2004; 70: 961–966. pmid:14766577
22. Stern B, Olsen LC, Tro¨ße C, Ravneberg H, Pryme IF. Improving mammalian cell factories: The selection of signal peptide has a major impact on recombinant protein synthesis and secretion in mammalian cells. Trends Cell Mol Biol. 2007; 2: 1–17.
23. Kober L, Zehe C, Bode J. Optimized signal peptides for the development of high expressing CHO cell lines. Biotechnol Bioeng. 2013; 110: 1164–1173. pmid:23124363
24. Schaaf A, Tintelnot S, Baur A, Reski R, Gorr G, Decker EL. Use of endogenous signal sequences for transient production and efficient secretion by moss (Physcomitrella patens) cells. BMC Biotechnol.

2005; 5: 30. pmid:16271156

25. Knappskog S, Ravneberg H, Gjerdrum C, Trosse C, Stern B, Pryme IF. The level of synthesis and secretion of Gaussia princeps luciferase in transfected CHO cells is heavily dependent on the choice of signal peptide. J Biotechnol. 2007; 128: 705–715. pmid:17316861
26. Zhang L, Leng Q, Mixson AJ. Alteration in the IL-2 signal peptide affects secretion of proteins in vitro and in vivo. J Gene Med. 2005; 7: 354–365. pmid:15619290
27. Hong SY, Kwon TH, Jang YS, Kim SH, Yang MS. Production of bioactive human granulocyte-colony stimulating factor in transgenic rice cell suspension cultures. Protein Expr Purif. 2006; 47: 68–73. pmid:16298143
28. Huang JM, Wu LY, Yalda D, Adkins Y, Kelleher SL, Crane M, et al. Expression of functional recombinant human lysozyme in transgenic rice cell culture. Transgenic Res. 2002; 11: 229–239. pmid:12113455
29. Huang LF, Liu YK, Lu CA, Hsieh SL, Yu SM. Production of human serum albumin by sugar starvation induced promoter and rice cell culture. Transgenic Res. 2005; 14: 569–581. pmid:16245148
30. Kim NS, Yu HY, Chung ND, Shin YJ, Kwon TH, Yang MS. Production of functional recombinant bovine trypsin in transgenic rice cell suspension cultures. Protein Expr Purif. 2011; 76: 121–126. pmid:20951807
31. Kim TG, Baek MY, Lee EK, Kwon TH, Yang MS. Expression of human growth hormone in transgenic rice cell suspension culture. Plant Cell Rep. 2008; 27: 885–891. pmid:18264710
32. Lee SJ, Park CI, Park MY, Jung HS, Ryu WS, Lim SM et al. Production and characterization of human CTLA4Ig expressed in transgenic rice cell suspension cultures. Protein Expr Purif. 2007; 51: 293–302. pmid:17079164
33. Liu YK, Huang LF, Ho SL, Liao CY, Liu HY, Lai YH, et al. Production of mouse granulocyte-macrophage colony-stimulating factor by gateway technology and transgenic rice cell culture. Biotechnol Bioeng. 2012; 109: 1239–1247. pmid:22125231
34. Chen TL, Lin YL, Lee YL, Yang NS, Chan MT. Expression of bioactive human interferon-gamma in transgenic rice cell suspension

cultures. Transgenic Res. 2004; 13: 499–510. pmid:15587273

35. Chen MH, Huang LF, Li HM, Chen YR, Yu SM. Signal peptide-dependent targeting of a rice alpha-amylase and cargo proteins to plastids and extracellular compartments of plant cells. Plant Physiol. 2004; 135: 1367–1377. pmid:15235120
36. Kitajima A, Asatsuma S, Okada H, Hamada Y, Kaneko K, Nanjo Y, et al. The rice alpha-amylase glycoprotein is targeted from the Golgi apparatus through the secretory pathway to the plastids. Plant Cell. 2009; 21: 2844–2858. pmid:19767453
37. Dale DC, Liles WC, Summer WR, Nelson S. Review: granulocyte colony-stimulating factor—role and relationships in infectious diseases. J Infect Dis. 1995; 172: 1061–1075. pmid:7561181
38. Everly JJ, Lonial S. Immunomodulatory effects of human recombinant granulocyte-macrophage colony-stimulating factor (rhuGM-CSF): evidence of antitumour activity. Expert Opin Biol Ther. 2005; 5: 293–311. pmid:15833068
39. Metcalf D. Control of granulocytes and macrophages: molecular, cellular, and clinical aspects. Science. 1991; 254: 529–533. pmid:1948028
40. Hamilton JA, Anderson GP. GM-CSF Biology. Growth Factors. 2004; 22: 225–231. pmid:15621725
41. .Emanuelsson O, Nielsen H, Brunak S, von Heijne G. Predicting subcellular localization of proteins based on their N-terminal amino acid sequence. J Mol Biol. 2000; 300: 1005–1016. pmid:10891285
42. Chen PW, Lu CA, Yu TS, Tseng TH, Wang CS, Yu SM. Rice alpha-amylase transcriptional enhancers direct multiple mode regulation of promoters in transgenic rice. J Biol Chem. 2002; 277: 13641–13649. pmid:11832481
43. Lu CA, Ho TH, Ho SL, Yu SM. Three novel MYB proteins with one DNA binding repeat mediate sugar and hormone regulation of alpha-amylase gene expression. Plant Cell. 2002; 14: 1963–1980. pmid:12172034
44. Hood EE, Gelvin SB, Melchers LS, Hoekema A. NewAgrobacterium helper plasmids for gene transfer to plants. Transgenic Res. 1993; 2: 208–218.

45. Huang CK, Yu SM, Lu CA. A rice DEAD-box protein, OsRH36, can complement an Arabidopsis atrh36 mutant, but cannot functionally replace its yeast homolog Dbp8p. Plant Mol Biol. 2010; 74: 119–128. pmid:20607362

46. Remansi T, Schenk PM, Manners JM, Grof CPL, Elliott AR. A Protocol for the Fluorometric Quantification of mGFP5-ER and sGFP(S65T) in Transgenic Plants. Plant Mol Biol Report. 1999; 17: 385–395.

47. Kim ST, Kang YH, Wang Y, Wu J, Park ZY, Rakwal R. et al. Secretome analysis of differentially induced proteins in rice suspension-cultured cells triggered by rice blast fungus and elicitor. Proteomics. 2009; 9: 1302–1313. pmid:19253284

48. Hirose T, Takano M, Terao T. Cell wall invertase in developing rice caryopsis: molecular cloning of OsCIN1 and analysis of its expression in relation to its role in grain filling. Plant Cell Physiol. 2002; 43: 452–459. pmid:11978873

49. Bendtsen JD, Nielsen H, von Heijne G, Brunak S. Improved prediction of signal peptides: SignalP 3.0. J Mol Biol. 2004; 340: 783–795. pmid:15223320

50. Carpita N, Sabularse D, Montezinos D, Delmer DP. Determination of the pore size of cell walls of living plant cells. Science. 1979; 205: 1144–1147. pmid:17735052

51. Janda CY, Li J, Oubridge C, Hernandez H, Robinson CV, Nagai, K. Recognition of a signal peptide by the signal recognition particle. Nature. 2010; 465: 507–510. pmid:20364120

52. Wang E, Xu X, Zhang L, Zhang H, Lin L, Wang Q, et al. Duplication and independent selection of cell-wall invertase genes GIF1 and OsCIN1 during rice evolution and domestication. BMC Evol Biol. 2010; 10: 108. pmid:20416079

53. Lee ST, Huang WL. Osmotic stress stimulates shoot organogenesis in callus of rice (Oryza sativa L.) via auxin signaling and carbohydrate metabolism regulation. Plant Growth Regul. 2014; 73: 193–204.

CHAPTER

21

IDENTIFICATION AND EXPRESSION ANALYSIS OF METHYL JASMONATE RESPONSIVE ESTS IN PACLITAXEL PRODUCING TAXUS CUSPIDATASUSPENSION CULTURE CELLS

Sangram K Lenka[1], Nadia Boutaoui[1], Bibin Paulose[2], Kham Vongpaseuth[3,4], Jennifer Normanly[2,4], Susan C Roberts[3,4] and Elsbeth L Walker[1,4]

[1]Department of Biology, University of Massachusetts, Amherst, MA 01003, USA.

[2]Department of Biochemistry and Molecular Biology, University of Massachusetts, Amherst, MA 01003, USA.

[3]Department of Chemical Engineering, University of Massachusetts, Amherst, MA 01003, USA.

[4]Plant Biology Graduate Program, University of Massachusetts, Amherst, MA 01003, USA.

ABSTRACT

Background

Taxol® (paclitaxel) promotes microtubule assembly and stabilization and therefore is a potent chemotherapeutic agent against wide range of cancers. Methyl jasmonate (MJ) elicited *Taxus* cell cultures provide a sustainable option to meet the growing market demand for paclitaxel. Despite its increasing pharmaceutical importance, the molecular genetics of paclitaxel biosynthesis is not fully elucidated. This study focuses on identification of MJ responsive transcripts in cultured *Taxus* cells using PCR-based suppression subtractive hybridization (SSH) to identify genes involved in global pathway control.

Results

Six separate SSH cDNA libraries of paclitaxel-accumulating *Taxus cuspidata* P991 cell lines were constructed at three different post-elicitation time points (6h, 18h and 5 day) to identify genes that are either induced or suppressed in response to MJ. Sequencing of 576 differentially screened clones from the SSH libraries resulted in 331 unigenes. Functional annotation and Gene Ontology (GO) analysis of up-regulated EST libraries showed enrichment of several known paclitaxel biosynthetic genes and novel transcripts that may be involved in MJ-signaling, taxane transport, or taxane degradation. Macroarray analysis of these identified genes unravelled global regulatory expression of these transcripts. Semi-quantitative RT-PCR analysis of a set of 12 candidate genes further confirmed the MJ-induced gene expression in a high paclitaxel accumulating *Taxus cuspidata* P93AF cell line.

Conclusions

This study elucidates the global temporal expression kinetics of MJ responsive genes in *Taxus* suspension cell culture. Functional characterization of the novel genes identified in this study will further enhance the understanding of paclitaxel biosynthesis, taxane transport and degradation.

BACKGROUND

Although the supply of most plant-derived secondary metabolic products is through harvest of the whole plant, many of these commercially valuable compounds are produced in slow-growing or non-agriculturally produced species[1]. Harvest from natural sources is often insufficient to provide adequate quantities of the product of interest in a sustainable manner. Plant cell culture provides an alternative production source in which environmental conditions can be more easily controlled, manipulated, and optimized to yield high quantities of these valuable products (*i.e.*, secondary metabolites).

The anticancer drug paclitaxel (registered as a trademark Taxol® by Bristol-Myers Squibb) is a secondary metabolite with a complex chemical structure that has been approved by the Food and Drug Administration (FDA) for the treatment of breast, ovarian and lung cancers, AIDS-related Kaposi›s sarcoma as well as coronary artery disease (http://www.fda.gov). The production of paclitaxel from *Taxus* cell cultures has been achieved by us and others[2, 3, 4, 5, 6, 7, 8, 9, 10, 11]. Plant cell culture technology has received significant attention in meeting the demand for paclitaxel and this method of production of paclitaxel has been licensed by Bristol-Myers Squibb to Phyton Biotech Inc. (Fort Worth, TX).

Paclitaxel is a member of the taxane family of compounds that are produced by all species of yew. Over 400 taxanes have been reported from various *Taxus* species [12]. The structure of paclitaxel was determined by Wani *et al.*, (1971) as a diterpenoid (molecular formula of $C_{47}H_{51}NO_{14}$) with a highly functionalized taxane skeleton [13]. Taxanes vary in the functional groups and side-chain components attached to the tricyclic taxane skeleton. Many steps of the biosynthetic pathway have been elucidated through the isolation of genes and enzymes from *Taxus* tissues and cultured cells[14, 15, 16]. Using molecular approaches, the genes encoding GGPP synthase[17], 10-deacetylbaccatin III-10-O-acetyl transferase[18], taxadiene synthase[19], cytochrome P450 taxane 10β-hydroxylase[20], 3'-N-debenzoyl-2'-deoxyTaxol N-benzoyltransferase[21], baccatin III:3-amino-3-phenylpropanoyl transferase[22], taxane 13α-hydroxylase[23], taxane 2α-O-benzoyltransferase[24], and taxa-4(20),11 (12)-dien-5α-ol-O-acetyl transferase[25] were identified.

The biotic elicitor methyl jasmonate (MJ) is an integral component of the signal transduction process that regulates the inducible defence systems of plants. Exogenous MJ is highly effective at eliciting secondary metabolite accumulation in a variety of plant cell culture systems[26]. Accordingly, paclitaxel accumulation is significantly enhanced upon MJ addition[27, 28, 29, 30, 31].

The non-model gymnosperm *Taxus* is evolutionally diverged and quite distantly related to model angiosperm species. This presents significant challenges to genomics approaches for understanding paclitaxel synthesis and regulation, and necessitates the generation of *Taxus*-specific genetic information in order to establish gene-to-metabolite links. Genetic information for *Taxus* species has begun to emerge from transcriptome sequencing of *Taxus* organs (needles, bark, cambial meristematic cells *etc.*)[32, 33, 34]. However, a differentially-regulated transcripome of paclitaxel accumulating cultured cells in controlled environmental conditions, and in response to secondary metabolite elicitors like MJ, needs to be characterized. Here we describe the results from transcript profiling experiments in *Taxus* culture cells to identify genes involved in global pathway control, including not only potential paclitaxel biosynthetic genes, but also genes that may be involved in MJ-signaling, taxane transport, or taxane degradation. Using a PCR-based suppression subtractive hybridization strategy, MJ responsive ESTs were isolated from the paclitaxel-accumulating P991 *Taxus cuspidata* cell line. Expression analysis of randomly selected differentially-regulated ESTs from the MJ-treated SSH libraries was also performed using macroarrays. Computational functional annotation and Gene Ontology (GO) analysis of up-regulated EST libraries show enrichment of several known paclitaxel biosynthetic genes as well as novel transcripts. RT-PCR analysis of a set of 12 candidate genes further confirmed the MJ-induced gene expression in the paclitaxel accumulating cell line. Hence, this study is useful in expanding our understanding of the molecular genetics of the complex paclitaxel biosynthesis process.

RESULTS

Differential screening of MJ-responsive ESTs

P991 is a well characterized *T. cuspidata* cell line that produces higher amounts of taxanes after MJ elicitation[27, 35]. This cell line was used to detect differentially expressed MJ responsive transcripts with a PCR-based cDNA subtraction method. Six distinct driver/tester combinations were used to construct six different SSH libraries. To generate libraries enriched for up-regulated genes, the combinations of driver/tester were as follows:

- Driver=unelicited control at 6 hour time point; tester=MJ-elicited at 6 hour time point

- Driver=unelicited control at 18 hour time point; tester=MJ-elicited at 18 hour time point
- Driver=unelicited control at 5 day time point; tester=MJ-elicited at 5 day time point

In contrast, to produce SSH libraries enriched for down-regulated genes the opposite driver/tester combination (*i.e.* MJ-elicited driver and unelicited tester) was used with the same cDNA samples. Single pass sequencing was carried out on 288 recombinant clones each from the MJ up- and down-regulated SSH libraries (96 clones each from 6h, 18h and 5 day time point of both the up/down regulated libraries). Among the 576 clones sequenced, 548 were considered for contig assembly and annotation after excluding vector sequence contamination. All the EST sequences from the three up-regulated libraries were used together for contig assembly with CAP3. Similarly, the ESTs from the three down-regulated libraries were used together for asssembly of down-regulated contigs. CAP3 assembly of the ESTs generated 32 MJ up-regulated contigs, 123 singletons and 23 MJ down-regulated contigs, 153 singletons (Table1)[36]. Hence, the up-regulated libraries contain 155 unigenes and 176 unigenes are represented in the down-regulated libraries the mean length of the ESTs obtained from all the libraries was 658 bp, which is adequate for sequence similarity based functional classification.

Table 1: Overview *T. cuspidata* P991 SSH libraries

Features	**Up-regulated libraries**	**Down Regulated libraries**
Number of clones sequenced	288	288
Number of clean ESTs	283	265
Number of non-redundant ESTs	155	176
Total number of contigs	32	23
Total number of singletons	123	153
Number of randomly chosen ESTs used for macroarray analysis	128	128

Table 2: Identity and description of contigs obtained from *T. cuspidata* up-regulated SSH libraries

Contig	Accession No	Match with Accession No	Description	Query coverage (%)	E Value
03	HE799317	ABR16435	Predicted Cytochrome P450	66	5.00E-72
08	HE799322	BAF46017	Putative class I chitinase	35	5.00E-64
21	HE799335	AAR13860	Taxadiene synthase	92	4.00E-61
25	HE799339	ABK26019	Unknown	34	1.00E-42
30	HE799344	BAD90813	Thaumatin-like protein	76	2.00E-37
04	HE799318	AAZ41362	Taxadiene synthase	94	5.00E-30
28	HE799342	XP_002264098	hypothetical protein	49	2.00E-27
26	HE799340	CAI56321	putative leucoanthocyanidin reductase	68	3.00E-25
18	HE799332	XP_002516771	putative Lipoxygenase	59	5.00E-25
15	HE799329	AAS89065	putative taxoid hydroxylase	65	8.00E-22
17	HE799331	ABK21418	Unknown	72	2.00E-23
29	HE799343	ACH59388	putative Chalcone synthase	36	3.00E-21
12	HE799326	AAO85809	putative S-adenosylmethionine synthetase	43	3.00E-20
09	HE799323	BAD10865	predicted 1-aminocyclopropane-1-carboxylic acid oxidase	79	7.00E-18
11	HE799325	ZP_03319418	Hypothetical protein	55	5.00E-16
01	HE799315	AAT73199	predicted taxoid acyltransferase	81	9.00E-16
14	HE799328	XP_002264858	Hypothetical protein	66	9.00E-16
07	HE799321	XP_002973748	Hypothetical protein (XP_002973748)	32	1.00E-15
16	HE799330	XP_001768181	similar to predicted protein (XP_001768181)	47	2.00E-15
27	HE799341	ABR16535	Unknown	16	7.00E-15
02	HE799316	ABR17205	putative DNA binding protein	79	2.00E-13
05	HE799319	ACN40018	Unknown	60	2.00E-09

32	HE799346	ADQ12769	Cytochrome P450	41	3.00E-09
06	HE799320	ABK22210	Unknown	39	8.00E-07
23	HE799337	ADK74829	putative Immuno-modulatory protein	13	3.00E-05
20	HE799334	ABK23181	Unknown	11	3.00E-04
10	HE799324	ABK24846	putative Gibberellin 3-beta-hydroxylase	38	0.001
31	HE799345	ADM79415	similar to Dormancy/ auxin associated-like protein	27	0.001
22	HE799336	YP_260089	Cytotoxin Mcf	41	0.003
19	HE799333	XP_002283086	hypothetical protein	8	0.005
13	HE799327	-	No hits in the data-base	-	-
24	HE799338	-	No hits in the data-base	-	-

Functional classification of the unigenes

Putative functional annotation of the unigenes was assigned by BLASTX analysis against the GenBank non-redundant database for sequence similarity (E-value$\leq 10^{-2}$). Unigenes showing BLASTX homology at E $>10^{-2}$ were designated as unigenes with no match in the database. Approximately 47% of the unigenes fell into this category, and may be specific to *Taxus* species. Approximately 12% of the sequences represented evolutionarily conserved proteins whose function is not known. Many of the known paclitaxel biosynthetic genes were induced, as expected. Several novel transcripts that are similar, but not identical, to known taxane biosynthetic enzymes were identified. These included two genes that are similar to known taxane acyl transferases and five genes that are related to known taxane hydroxylase. Experimental characterization of these novel genes would be needed to definitively determine the exact acylation or hydroxylation reactions catalyzed by the encoded enzymes.

Functional classification of MJ up-regulated unigenes was determined based on GO analysis for biological process, molecular function and cellular component. Among the biological process category, unigenes belonging to oxidation-reduction processes (GO:0055114), responses to stimulus (GO:0050896), small molecule biosynthetic processes (GO:0044283) and paclitaxel biosynthetic processes (GO:0042617) were over-represented. In the molecular function category, oxidoreductase activity (GO:0016491) was most significantly enriched, followed by iron ion binding (GO:0005506) and transferase activity (GO:0016740). Finally, in the cellular component category, proteins associated with intracellular organelles (GO:0043229), membranes (GO:0016020) and cytoplasmic parts (GO:0044444) were most enriched. KEGG pathway analysis of the encoded proteins further indicated MJ-induced paclitaxel, L-methionine and oxylipin biosynthesis pathways in the up-regulated SSH libraries. There were too few truly down-regulated unigenes available (see below) for GO analysis of down regulated transcripts.

Evaluation of MJ-responsive differential expression using macroarrays

Probe sets designed from up- and down-regulated SSH libraries were spotted manually as macroarrays on nylon membranes and hybridized with different RNA targets. Of the 256 randomly chosen ESTs in the macroarray, 81 (71 up- and 10 down- regulated) ESTs showed a significant change in intensity following MJ elicitation. One set of genes showed the highest up-regulation trend at early time points (6h and 18h) following MJ elicitation (Figures1a, b and2), while MJ responsive expression of a separate set of genes exhibited a difference in expression pattern at the late time point (5 day; Figures1c,2). On the other hand, the majority (92.2%) of MJ down-regulated unigenes were found to be false positive SSH clones based on this analysis. Hierarchical clustering analysis of the temporal differential gene expression patterns of the individual ESTs in the macroarray (Figure2) clearly demonstrated the distinctions between the MJ-responsive early, middle and late gene regulation.

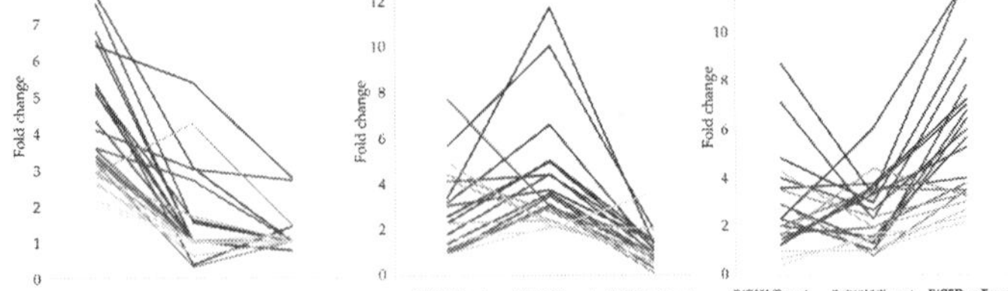

Figure 1: Expression pattern of clones from up-regulated libraries. Macroarrays spotted with randomly selected clones (probes) from all six SSH libraries

were hybridized with labelled RNA targets prepared from mock elicited and elicited cells at three time points: 6h, 18h, and 5 day. Expression profile plots for the set of probes corresponding to each up-regulated library are shown: **a**, 6h library probes **b**, 18h library probes **c**, 5 day library probes. Expression is given as the fold-change in gene expression in the elicited culture over the un-elicited culture. Only results from probes showing a statistically significant change in expression ($P \leq 0.05$) are shown.

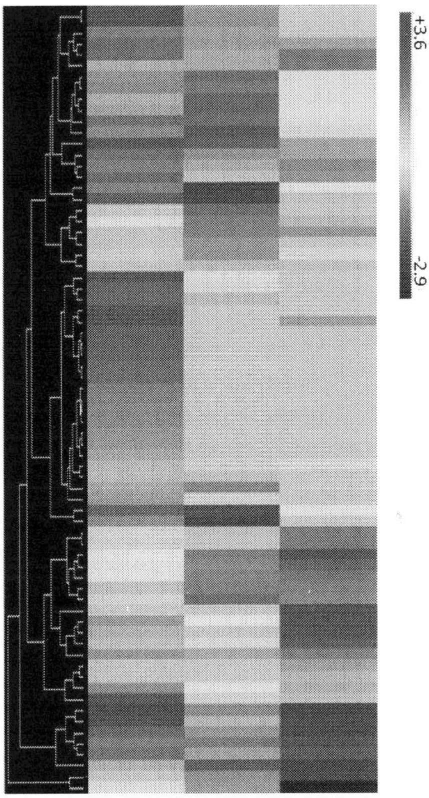

Figure 2: Heat map showing overall expression patterns of differentially regulated genes from up-regulated libraries. Hierarchical clustering of significantly MJ up-regulated genes was performed using average linkage and Euclidean distance as a measurement of similarity. Only results from probes showing a statistically significant change in expression for at least one time point ($P \leq 0.05$) are shown.

Confirmation of MJ-induced expression by RT-PCR

Twelve randomly selected genes from the up-regulated SSH library were evaluated for transcriptional response in another paclitaxel-accumulating line *T. cuspidata* P93AF using semi-quantitative RT-PCR. All 12 genes were up-regulated at 6h after MJ elicitation, further supporting the early overall response observed in the macroarray (Figure3). Transcript levels for flavonol synthase (D5_88H) and the pyridine nucleotide-disulfide oxidoreductase family protein (18R_32H) went down after 6h. Cytochrome P450 (contig 32), unknown (contig 6), ACC oxidase (contig 9), leucoanthocyanidin reductase (contig 26) and lipoxygenase (contig 18) were elevated at both 6h and 18h. However, semi-quantitative RT-PCR could not detect transcripts of these genes on day 5 after elicitation. Transcripts encoding cytochrome P450 (contig3), quinone oxidoreductase-like protein (18R_44D), hypothetical protein (contig 16), dirigent-like protein (18R_17A) and short-chain dehydrogenase/reductase family protein were at elevated levels on day 5 as well as at early time points. Elevated expression of these randomly chosen sequences in the *T. cuspidata* P93AF cell line upon MJ-elicitation further supported that regulated genes were successfully identified using the subtracted cDNA libraries from cell line P991.

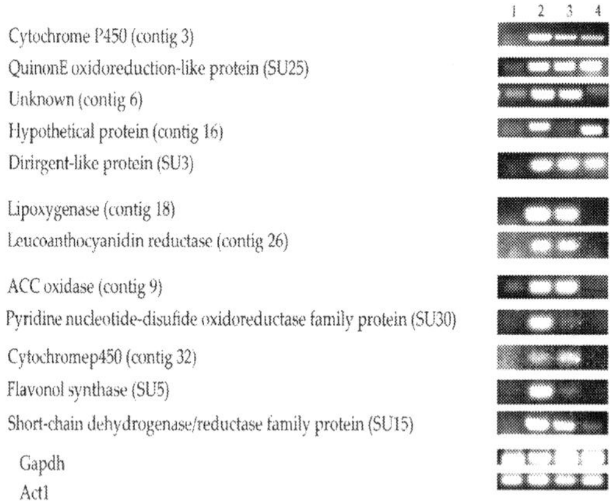

Figure 3: Semi-quantitative RT-PCR analysis of 12 selected up-regulated unigenes. Semi-quantitative RT-PCR was carried out using RNA from the *Taxus cuspidata* P93AF cell line. RNA samples were taken at 6h, 18h and 5 day after MJ elicitation, and from mock elicited cells at the same time points. Lane

1: Mock elicited for 18h, Lane 2-4: elicited with 100 μM MJ for 6h, 18h or 5 day, respectively. *Gapdh* and *Actin1* were used as normalization controls.

DISCUSSION

MJ potentiates secondary metabolite accumulation in a variety of plant suspension cell culture systems, and is now a widely-used augmenting tool for bio-active molecule production[26]. Accordingly, paclitaxel accumulation is significantly enhanced with MJ elicitation in *T. cuspidata* suspension cultured cells[27, 37, 38, 39, 40]. MJ is also implicated in specific induction of a wide range of secondary metabolite genes involved in biotic-stress response and wounding[41]. Conversely, primary metabolic genes such as those involved in photosynthesis, electron transport and cytoskeletal organization are either down-regulated or unaffected by MJ elicitation[42]. MJ elicitation and targeted cloning approaches have been effectively used in *Taxus* to identify specific taxane biosynthetic genes[43, 44]. Therefore, an up-regulated cDNA library of MJ treated cells will presumably contain genes related to secondary metabolism, including paclitaxel biosynthesis.

SSH is a powerful genomics technique that enriches differentially regulated genes and has mostly been used to identify transcripts expressed in contrasting environmental conditions. SSH is a suitable approach in *Taxus*where the genomic information is currently scarce. We used the PCR-based SSH approach to identify MJ responsive (up- and down-regulated) transcripts that precede paclitaxel accumulation in the paclitaxel accumulating *T. cuspidata* P991 cell line. Putative up-regulation of as many as 155 unigenes and down-regulation of 176 unigenes of diverse functional groups suggested involvement of a spectrum of genes regulating paclitaxel biosynthesis, MJ-signaling and response, and potentially, taxane transport and taxane degradation. The up-regulated SSH library contained several previously known positively regulated genes that are likely to be directly involved in paclitaxel synthesis. For instance, putative taxadiene synthase, taxane hydroxylase and taxane acyl transferases having strong sequence similarity with the corresponding known paclitaxel biosynthetic pathway genes were detected in the up-regulated library. Taxadiene synthase catalyses the first committed step of paclitaxel biosynthesis in *Taxus* by cyclization of the linear isoprenoid substrate geranyl geranyl diphosphate (GGPP) to form taxa-4(5),11(12) diene[19]. Four unigenes representing different isoforms of putative taxadiene synthase were up-regulated in the library. It will be interesting to unravel the individual contribution of each of these taxadiene synthase genes towards paclitaxel production. Another paclitaxel biosynthetic gene, taxoid 2-α-hydroxylase, produces 2α-hydroxy taxoids as starting materials for subsequent acylation at the C_2-position of the taxane core that ultimately bears a benzoyl group, an important paclitaxel pharmacophore[45]. Five novel putative taxoid hydroxylases that are similar but not identical to known taxoid hydrolases (taxoid 2-alpha hydroxylase, taxoid 5-alpha hydroxylase, and taxoid 7-beta hydroxylase) were detected in the up-regulated library. Each might function to hydroxylate taxoids at specific positions. Similarly, novel taxane acyl transferases are not only important targets for genetic manipulation to improve paclitaxel production in genetically engineered host systems but also provides a means of attaching modified aroyl groups to taxoid precursors to improve drug efficacy[18, 21, 25, 46]. Characterization of the two newly identified candidate acyl transferases in the library will further suggest roles for these novel enzymes in paclitaxel biosynthesis.

Apart from the paclitaxel biosynthesis genes, up-regulated genes that are potential candidates for regulation of the MJ response were also obtained. One such gene encodes ACC oxidase (1-aminocyclopropane-1-carboxylate oxidase), which is the final step in the synthesis of the plant hormone ethylene[47]. Ethylene is known to be produced following MJ elicitation, and to function in suppression of the MJ signal. An increased production of paclitaxel upon application of ethylene synthesis inhibitors to elicited *Taxus* cultures was reported by Zhang et al., (2003) [48]. Disruption of this ACC oxidase gene may allow the MJ signal to be strengthened, which may lead to increased paclitaxel accumulation.

The library was also enriched with up-regulated genes that presumably have no direct role in paclitaxel accumulation. Good examples are the putative class I chitinase, thaumatin (also known as PR-5); chalcone synthase, and defensin, all of which are defense-related genes that have well-characterized roles in model plant species[49, 50]. These genes are known to be induced by MJ in other plant species and thus are unlikely to be related to the synthesis of taxanes, which are unique to *Taxus* species. For several transcripts, a reliable prediction of general biochemical activity can be made, but predictions neither suggest nor challenge a role in paclitaxel synthesis. Genes in this class include three cytochrome P450 genes that lack significant similarity to the several known taxane biosynthetic pathway P450s, two distinct lipoxygenase genes, and a few other enzyme encoding genes. Any of these genes could participate in the uncharacterized steps of the paclitaxel biosynthetic pathway (e.g., hydroxylation reactions and oxetane ring formation), but could equally well be unrelated to paclitaxel synthesis. Further characterization will be necessary to determine whether these genes have direct relevance to paclitaxel biosynthesis.

A large class of genes (59%) have functions that cannot be predicted based on sequence similarity information. These genes may be involved in important but uncharacterized processes of MJ-signaling, taxane transport or taxane degradation. As many as 47% of these ESTs showing no sequence homology with any of the available Genbank sequences may be specific to *Taxus* species. The other 12% of sequences are similar to genes described as encoding hypothetical proteins or proteins with unknown function. Enrichment of unigenes belonging to various biological process categories (GO), such as oxidation-reduction process (GO:0055114), response to stimulus (GO:0050896), small molecule biosynthetic process (GO:0044283) and paclitaxel biosynthetic process (GO:0042617), provide strong molecular evidence that the SSH-based MJ elicited transcript screening is a good method to identify genes related to paclitaxel metabolism. Nevertheless, this strategy did not enable the isolation of all the known paclitaxel metabolism genes indicating that the collection of MJ-inducible genes identified here is not comprehensive. Sequencing of the total MJ-induced transcriptome will be required to identify transcripts for every gene related to paclitaxel accumulation.

In addition to confirming up-and down-regulated genes, macroarrays identified false positive clones that appeared in the libraries but were not actually differentially regulated in response to MJ. Although most genes identified as being up-regulated were confirmed as such by macroarray analysis, the false positive rate in the down-regulated libraries was 92.2%. SSH libraries of highly complex eukaryotes have often been reported to contain high percentages (30-50%) of false positive clones that escaped subtraction[51, 52]. These high false positive rates are usually attributed to non-specific annealing of primers, partial hybridization of driver that allows tester amplification and non-specific annealing of suppression adaptors[53]. While any of these could have been factors in producing the high false positive rate in our down-regulated libraries, we also note the possibility that there could simply be very few genes that are down-regulated by MJ in the *Taxus* cultures. If few or no down-regulated transcripts occur, then the SSH approach is certain to be unsuccessful, and will identify a random set of highly expressed (but not differentially regulated) genes. Genes encoding cell wall proteins, ribosomal proteins and other highly expressed genes are well represented in the down-regulated libraries. The high false positive rate in the down-regulated libraries here indicates that the genes observed were not actually down-regulated by MJ.

The gene expression kinetics revealed differential temporal regulation of MJ responsive genes. Up-regulation of many of the MJ induced genes at early time points (6h) was highest, followed by middle time point (18h). MJ-responsive up-regulation of most genes, however, did not persist into the latest time point 5 day (Figure2). However, a subset of late-induced genes was observed. In agreement to this study, similar expression trend was also observed previously in MJ-induced temporal expression kinetics of taxol biosynthesis genes in P991 cultured cells[27]. This trend was also observed in the paclitaxel-producing *T. cuspidata* P93AF cell line, as semi-quantitative RT-PCR showed that all 12 genes were up-regulated at 6h but transcripts of only 5 of them were detected on day 5. These 12 genes tested here by RT-PCR do not represent any known genes related to taxol biosynthesis, but rather were selected as novel transcripts identified in this study. Furthermore, use of a second paclitaxel-producing cell line (P93AF) demonstrates the reproducibility of the gene expression changes occurring in *Taxus* cultured cells. Paclitaxel and other taxanes accumulate in the cultured cells at time points after 24h[27]. Therefore, persistent translated gene products of early MJ induced genes appear to play a pivotal role in taxane biosynthesis, while late MJ induced genes may be facilitating other functions such as taxane transport and degradation. Genes involved either in intracellular or extracellular paclitaxel transport and degradation in *Taxus* have not yet been characterized[54], but may be present among the ESTs identified here that could not be assigned functions based on molecular similarity to known proteins.

CONCLUSIONS

Several genes were differentially regulated in response to MJ in *Taxus cuspidata* cell lines. Among these genes, paclitaxel biosynthetic pathway genes were specifically induced in response to MJ. A majority of the up-regulated genes followed specific temporal expression kinetics. SSH was not sensitive enough to isolate tightly regulated down regulated genes. Functional characterization of the novel genes identified in this study will further enhance the understanding of paclitaxel metabolism and its regulation.

METHODS

Cell culture

Taxu s *cuspidata* cell line P991 was used for SSH library preparation and macroarray screening and cell line *Taxu* s *cuspidata* P93AF was used for RT-PCR analysis. All cells were obtained as a gift from Dr. D. Gibson at the U.S. Plant Soil and Nutrition laboratory, USDA (Ithaca, NY). Cell cultures were maintained as described in Nims *et al.*, (2006)[27]. MJ (100 μM final concentration) dissolved in 50% EtOH was added to suspension cultures on day 7 after subculture. Control (unelicited) cells were subjected to mock elicitation by adding an equal volume of 50% EtOH. Cells were harvested from elicited and unelicited cultures at 6 hours, 18 hours and 5 day post-elicitation.

RNA purification

Cells were collected by filtration from the media through Miracloth® (EMD Biosciences, San Diego, CA), flash frozen in liquid nitrogen and stored at -80°C in polypropylene tubes. Total RNA was extracted using guanidinium isothyocyanate and separated through a cesium chloride (CsCl) phase [55]. Poly(A) RNA was extracted using Poly(A) Purist MAG Purification kit (Ambion, Austin, TX). Quality checks of poly (A) RNA preparations were made by PCR using primers specific for 18S rRNA[56]. These primers were: Pt-MGB-18S-f (AGCCTTGCGCTGGCG), Pt-MGB-18S-r (TGCCCTATCAACTTTCGATGGT). Quantification of total RNA and poly(A) RNA was performed with the Ribogreen fluorometric assay (Molecular Probes, Eugene, OR).

Suppression subtractive hybridization (SSH) libraries

Suppression Subtractive Hybridization (SSH) libraries using control and MJ-elicited samples from *Taxus cuspidata* cell line P991 at 6 hours, 18 hours and 5 day post-elicitation were generated using the Clontech PCR-Select cDNA Subtraction Kit (Takara Bio Inc., Clontech, Mountain View, CA) according to the manufacturer's instructions. For each SSH library two different double stranded cDNA pools (a 'tester' and 'driver') were used. Mirror orientation selection (MOS) was then performed to reduce background molecules[53]. Library products were separated on agarose gels. For each library, products below and above 500 base pairs were purified separately using Gel Purification Kit (Qiagen, Valencia, CA) and cloned into TOPO 2-1 vector (Invitrogen, Carlsbad, CA).

Sequence analysis of the cloned ESTs

Plasmids from independent clones were isolated and single pass sequencing was carried out. All EST sequences were screened through VecScreen software (NCBI:http://www.ncbi.nlm.nih.gov) to remove vector sequence contamination. Once cleaned of vector sequence, all the EST sequences from each up-regulated library and from each down regulated library were used separately for contig assembly with CAP3[36]. The ESTs were grouped into contigs and singletons and were termed as unigenes. The unigene sequences were used to search for homology using the default setting of BLASTX program at NCBI. Each unigene was classified as a protein with known function, a protein with unknown function or a protein with no sequence match in the database. Furthermore, sequences with BLASTx hits were annotated according to gene ontology (GO) terms using Blast2GO software (http://www.blast2go.org;[57]). The TrEMBL section of UniProtKB (http://www.uniprot.org) database was used to perform, enzyme class and pathway analysis.

Expression analysis by macroarray

Macroarrays constructed for this study comprised 256 SSH fragments from both the up-regulated and down-regulated libraries (128 up and 128 down). Bacterial cultures were grown overnight in LB media containing 50 μg/mL kanamycin. Plasmid DNA was purified using Qiaprep-miniprep (Qiagen, Valencia, CA). SSH fragments were amplified by PCR using adapter specific primers. Amplification of fragments via PCR was performed using ExTaq Hot Start Polymerase System (Takara Mirus Bio, Madison, WI). PCR conditions were as follows: 95°C for 2 minutes followed by 25X: 30 seconds, 95°C; 30 seconds, 62°C; 2 minutes, 72°C. This was then followed by 1X: 5 minutes, 72°C. PCR products were precipitated, dissolved in water and denatured in 0.01% sodium dodecyl sulfate (SDS), 0.2 N NaOH at 72°C for 15 minutes. Denatured PCR products were printed in duplicate on Zetaprobe membranes (BioRad, Hercules, CA) using hand held 96-pin replicating tool (V&P Scientific, San Diego, CA). Macroarray blots were submerged in neutralization solution (0.5 M Tris pH 7.4, 1.5 M NaCl), then in 6X sodium citrate/ sodium phosphate (SCP) for 1 minute prior to UV cross-linking. Pre-hybridization of the membranes occurred for eight hours in 10% dextran sulfate (average molecular weight of 500,000kD; G.E. Health, Piscataway, NJ), 5.5X SCP, 0.9% *n*-lauroyl sulfate-sodium salt and 0.5 mg/ml heparin. Targets were labelled with ^{32}P dCTP (G.E. Healthcare Bio-Sciences, Piscataway, NJ) using Superscript™ II Reverse Transcriptase (RT) (Invitrogen, Carlsbad, CA) from 300 ng polyA RNA. Hybridization was performed at 65°C for 18H, followed by washing twice with 2X SCP, 0.1% SDS for 15 minutes then twice using 0.2X SCP and 0.2% SDS for 15 minutes at 65°C. Blots were exposed to phosphorimager screen for 24-48H then read by FLA-5000 phosphorimager (Fujifilm Medical Systems Inc., Stamford, CT). Probes were stripped off of blots up to three times using 0.1% SDS at 100°C. Blots were checked for residual hybridization by exposing to film for 24H before re-hybridizing with a different target cDNA.

Signal intensities of each spot were captured using VisualGrid (GPC Biotech Inc, Waltham, MA). Following local background subtraction, the intensity values for each spot were globally normalized using LOWESS (locally weighted polynomial regression [58]) within and across all four technical replicate experiments. The relative fold change expression signal obtained from each clone found significant in one tailed student t-test in response to MJ over mock elicitation were considered for further analysis ($\geq$ 2 fold for up and down). Additionally, relative fold change of three or more independent ESTs corresponding to a particular contig were analysed using Q-test with 95% confidence for finding outlying data points.

Gene expression analysis by RT-PCR

Total RNA was isolated from *Taxus cuspidata* P93AF cell line at 6h, 18h and 5 day after MJ elicitation as well as from unelicited cells (18 h) using RNeasy Plant Mini kit (Qiagen). First strand cDNA was synthesized using Affinityscript™ cDNA synthesis kit (Stratagene) from 3μg total RNA following the manufacture's protocol. Primers. A touchdown PCR was performed for all the transcripts with annealing temperature varying from 64°C to 52°C. The number of cycles used to amplify each fragment was variable in order to detect the corresponding amplicon during linear amplification.

ACKNOWLEDGEMENTS

This study is financially supported by National Institutes of Health (R01 GM070852). The authors would like to thank Dr. Donna Gibson of the US Plant Soil and Nutrition Laboratory of the USDA for the *Taxus* cell cultures.

REFERENCES

1. Farnsworth NR, Akerele O, Bingel AS, Soejarto DD, Guo Z: Medicinal plants in therapy. Bull World Health Organ. 1985, 63 (6): 965-981.
2. Exposito O, Bonfill M, Moyano E, Onrubia M, Mirjalili MH, Cusido RM, Palazon J: Biotechnological production of taxol and related taxoids: current state and prospects. Anticancer Agents Med Chem. 2009, 9 (1): 109-121.
3. Vongpaseuth K, Roberts SC: Advancements in the understanding of Paclitaxel metabolism in tissue culture. Curr Pharm Biotechnol. 2007, 8 (4): 219-236.
4. Frense D: Taxanes: perspectives for biotechnological production. Appl Microbiol Biotechnol. 2007, 73 (6): 1233-1240. 10.1007/s00253-006-0711-0.
5. Tabata H: Production of paclitaxel and the related taxanes by cell suspension cultures of Taxus species. Curr Drug Targets. 2006, 7 (4): 453-461. 10.2174/138945006776359368.

6. Kolewe ME, Henson MA, Roberts SC: Characterization of aggregate size in Taxus suspension cell culture. Plant Cell Rep. 2010, 29 (5): 485-494. 10.1007/s00299-010-0837-5.
7. Vongpaseuth K, Nims E, St Amand M, Walker EL, Roberts SC: Development of a particle bombardment-mediated transient transformation system for Taxus spp. cells in culture. Biotechnol Prog. 2007, 23 (5): 1180-1185.
8. Naill MC, Roberts SC: Culture of isolated single cells from Taxus suspensions for the propagation of superior cell populations. Biotechnol Lett. 2005, 27 (21): 1725-1730. 10.1007/s10529-005-2738-1.
9. Roberts SC, Naill M, Gibson DM, Shuler ML: A simple method for enhancing paclitaxel release from Taxus canadensis cell suspension cultures utilizing cell wall digesting enzymes. Plant Cell Rep. 2003, 21 (12): 1217-1220. 10.1007/s00299-003-0575-z.
10. Pestchanker LJ, Roberts SC, Shuler ML: Kinetics of taxol production and nutrient use in suspension cultures of Taxus cuspidata in shake flasks and a Wilson-type bioreactor. Enzyme Microb Technol. 1996, 19 (4): 256-260. 10.1016/0141-0229(95)00243-X.
11. Fett-Neto AG, Melanson SJ, Nicholson SA, Pennington JJ, Dicosmo F: Improved taxol yield by aromatic carboxylic acid and amino acid feeding to cell cultures of taxus cuspidata. Biotechnol Bioeng. 1994, 44 (8): 967-971. 10.1002/bit.260440813.
12. Zhang MLX, Zhang J, Zhang S, Dong M, Huo C, Shi Q, Gu Y, Cong B: Taxanes from the leaves of Taxus cuspidata. Chem Nat Comp. 2010, 46 (1): 53-58. 10.1007/s10600-010-9523-7.
13. Wani MC, Taylor HL, Wall ME, Coggon P, McPhail AT: Plant antitumor agents. VI. The isolation and structure of taxol, a novel antileukemic and antitumor agent from Taxus brevifolia. J Am Chem Soc. 1971, 93 (9): 2325-2327. 10.1021/ja00738a045.
14. Hezari M, Ketchum RE, Gibson DM, Croteau R: Taxol production and taxadiene synthase activity in Taxus canadensis cell suspension cultures. Arch Biochem Biophys. 1997, 337 (2): 185-190. 10.1006/abbi.1996.9772.
15. Walker K, Croteau R: Taxol biosynthetic genes. Phytochemistry. 2001, 58 (1): 1-7. 10.1016/S0031-9422(01)00160-1.

16. Hefner J, Rubenstein SM, Ketchum RE, Gibson DM, Williams RM, Croteau R: Cytochrome P450-catalyzed hydroxylation of taxa-4(5),11(12)-diene to taxa-4(20),11(12)-dien-5alpha-ol: the first oxygenation step in taxol biosynthesis. Chem Biol. 1996, 3 (6): 479-489. 10.1016/S1074-5521(96)90096-4.
17. Hefner J, Ketchum RE, Croteau R: Cloning and functional expression of a cDNA encoding geranylgeranyl diphosphate synthase from Taxus canadensis and assessment of the role of this prenyltransferase in cells induced for taxol production. Arch Biochem Biophys. 1998, 360 (1): 62-74. 10.1006/abbi.1998.0926.
18. Walker K, Croteau R: Molecular cloning of a 10-deacetylbaccatin III-10-O-acetyl transferase cDNA from Taxus and functional expression in Escherichia coli. Proc Natl Acad Sci U S A. 2000, 97 (2): 583-587. 10.1073/pnas.97.2.583.
19. Wildung MR, Croteau R: A cDNA clone for taxadiene synthase, the diterpene cyclase that catalyzes the committed step of taxol biosynthesis. J Biol Chem. 1996, 271 (16): 9201-9204. 10.1074/jbc.271.16.9201.
20. Schoendorf A, Rithner CD, Williams RM, Croteau RB: Molecular cloning of a cytochrome P450 taxane 10 beta-hydroxylase cDNA from Taxus and functional expression in yeast. Proc Natl Acad Sci U S A. 2001, 98 (4): 1501-1506. 10.1073/pnas.98.4.1501.
21. Walker K, Long R, Croteau R: The final acylation step in taxol biosynthesis: cloning of the taxoid C13-side-chain N-benzoyltransferase from Taxus. Proc Natl Acad Sci U S A. 2002, 99 (14): 9166-9171. 10.1073/pnas.082115799.
22. Walker K, Fujisaki S, Long R, Croteau R: Molecular cloning and heterologous expression of the C-13 phenylpropanoid side chain-CoA acyltransferase that functions in Taxol biosynthesis. Proc Natl Acad Sci U S A. 2002, 99 (20): 12715-12720. 10.1073/pnas.192463699.
23. Jennewein S, Rithner CD, Williams RM, Croteau RB: Taxol biosynthesis: taxane 13 alpha-hydroxylase is a cytochrome P450-dependent monooxygenase. Proc Natl Acad Sci U S A. 2001, 98 (24): 13595-13600. 10.1073/pnas.251539398.

24. Walker K, Croteau R: Taxol biosynthesis: molecular cloning of a benzoyl-CoA:taxane 2alpha-O-benzoyltransferase cDNA from taxus and functional expression in Escherichia coli. Proc Natl Acad Sci U S A. 2000, 97 (25): 13591-13596. 10.1073/pnas.250491997.

25. Walker K, Schoendorf A, Croteau R: Molecular cloning of a taxa-4(20),11(12)-dien-5alpha-ol-O-acetyl transferase cDNA from Taxus and functional expression in Escherichia coli. Arch Biochem Biophys. 2000, 374 (2): 371-380. 10.1006/abbi.1999.1609.

26. Gundlach H, Muller MJ, Kutchan TM, Zenk MH: Jasmonic acid is a signal transducer in elicitor-induced plant cell cultures. Proc Natl Acad Sci U S A. 1992, 89 (6): 2389-2393. 10.1073/pnas.89.6.2389.

27. Nims E, Dubois CP, Roberts SC, Walker EL: Expression profiling of genes involved in paclitaxel biosynthesis for targeted metabolic engineering. Metab Eng. 2006, 8 (5): 385-394. 10.1016/j.ymben.2006.04.001.

28. Kai G, Zhao L, Zhang L, Li Z, Guo B, Zhao D, Sun X, Miao Z, Tang K: Characterization and expression profile analysis of a new cDNA encoding taxadiene synthase from Taxus media. J Biochem Mol Biol. 2005, 38 (6): 668-675. 10.5483/BMBRep.2005.38.6.668.

29. Naill MC, Roberts SC: Flow cytometric analysis of protein content in Taxus protoplasts and single cells as compared to aggregated suspension cultures. Plant Cell Rep. 2005, 23 (8): 528-533. 10.1007/s00299-004-0875-y.

30. Yukimune Y, Tabata H, Higashi Y, Hara Y: Methyl jasmonate-induced overproduction of paclitaxel and baccatin III in Taxus cell suspension cultures. Nat Biotechnol. 1996, 14 (9): 1129-1132. 10.1038/nbt0996-1129.

31. Yukimune Y, Hara Y, Nomura E, Seto H, Yoshida S: The configuration of methyl jasmonate affects paclitaxel and baccatin III production in Taxus cells. Phytochemistry. 2000, 54 (1): 13-17. 10.1016/S0031-9422(00)00006-6.

32. Wu Q, Sun C, Luo H, Li Y, Niu Y, Sun Y, Lu A, Chen S: Transcriptome analysis of Taxus cuspidata needles based on 454 pyrosequencing. Planta Med. 2011, 77 (4): 394-400. 10.1055/s-0030-1250331.

33. da Hao C, Ge G, Xiao P, Zhang Y, Yang L: The First Insight into the Tissue Specific Taxus Transcriptome via Illumina Second

Generation Sequencing. PLoS One. 2011, 6 (6): e21220-10.1371/journal.pone.0021220.

34. Lee EK, Jin YW, Park JH, Yoo YM, Hong SM, Amir R, Yan Z, Kwon E, Elfick A, Tomlinson S: Cultured cambial meristematic cells as a source of plant natural products. Nat Biotechnol. 2010, 28 (11): 1213-1217. 10.1038/nbt.1693.

35. Naill MC, Roberts SC: Flow cytometric identification of Paclitaxel-accumulating subpopulations. Biotechnol Prog. 2005, 21 (3): 978-983.

36. Huang X, Madan A: CAP3: A DNA sequence assembly program. Genome Res. 1999, 9 (9): 868-877. 10.1101/gr.9.9.868.

37. Ketchum REB, Gibson DM, Croteau RB, Shuler ML: The kinetics of taxoid accumulation in cell suspension cultures of *Taxus* following elicitation with methyl jasmonate. Biotechnol Bioeng. 1999, 62 (1): 97-105. 10.1002/(SICI)1097-0290(19990105)62:1<97::AID-BIT11>3.0.CO;2-C.

38. Mirjalili N, Linden JC: Methyl jasmonate induced production of Taxol in suspension cultures of *Taxus cuspidata*: Ethylene interaction and induction models. Biotechnol Prog. 1996, 12: 110-118. 10.1021/bp9500831.

39. Yukimune Y, Tabata H, Higashi Y, Hara Y: Methyl Jasmonate-induced overproduction of paclitaxel and baccatin III in *Taxus* cell suspension cultures. Nat Biotechnol. 1996, 14: 1129-1132. 10.1038/nbt0996-1129.

40. Kim BJ, Gibson DM, Shuler ML: Relationship of viability and apoptosis to taxol production in Taxus sp. suspension cultures elicited with methyl jasmonate. Biotechnol Prog. 2005, 21 (3): 700-707.

41. Weber H: Fatty acid-derived signals in plants. Trends Plant Sci. 2002, 7: 217-224. 10.1016/S1360-1385(02)02250-1.

42. Hermsmeier D, Schittko U, Baldwin IT: Molecular interactions between the specialist herbivore *Manduca sexta*(Lepidoptera, Sphingidae) and its natural host *Nicotiana attenuata*.I. Large-scale changes in accumulation of growth- and defenese-related plant mRNAs. Plant Physiol. 2001, 125: 683-700. 10.1104/pp.125.2.683.

43. Croteau R, Ketchum REB, Long RM, Kaspera R, Wildung MR:

Taxol biosynthesis and molecular genetics. Phytochem Rev. 2006, 5: 75-97. 10.1007/s11101-005-3748-2.

44. Ketchum REB, Croteau R: The Taxus metabolome and the elucidation of the Taxol biosynthetic pathway in cell suspension cultures. Biotechnology in Agriculture and Forestry. Edited by: Saito KD, Dixon RA, Willmitzer L. 2006, Springer-Verlag, Berlin, 291-309.
45. Chau M, Croteau R: Molecular cloning and characterization of a cytochrome P450 taxoid 2alpha-hydroxylase involved in Taxol biosynthesis. Arch Biochem Biophys. 2004, 427 (1): 48-57. 10.1016/j.abb.2004.04.016.
46. Walker K, Ketchum RE, Hezari M, Gatfield D, Goleniowski M, Barthol A, Croteau R: Partial purification and characterization of acetyl coenzyme A: taxa-4(20),11(12)-dien-5alpha-ol O-acetyl transferase that catalyzes the first acylation step of taxol biosynthesis. Arch Biochem Biophys. 1999, 364 (2): 273-279. 10.1006/abbi.1999.1125.
47. Spanu P, Reinhart D, Boller T: Analysis and cloning of the ethylene-forming enzyme from tomato by functional expression of its mRNA in Xenopus laevis oocytes. EMBO J. 1991, 10: 2007-2013.
48. Zhang CH, Wu JY: Ethylene inhibitors enhance elicitor-induced paclitaxel production in suspension cultures of Taxus spp. cells. Enzyme Microb Technol. 2003, 32: 71-77. 10.1016/S0141-0229(02)00266-1
49. van Loon LC, Rep M, Pieterse CM: Significance of inducible defense-related proteins in infected plants. Annu Rev Phytopathol. 2006, 44: 135-162. 10.1146/annurev.phyto.44.070505.143425.
50. Lamb CJ, Dixon RA: Molecular mechanisms underlying induction of plant defence gene transcription. Biochem Soc Symp. 1994, 60: 241-248.
51. Ji SJ, Lu YC, Feng JX, Wei G, Li J, Shi YH, Fu Q, Liu D, Luo JC, Zhu YX: Isolation and analyses of genes preferentially expressed during early cotton fiber development by subtractive PCR and cDNA array. Nucleic Acids Res. 2003, 31 (10): 2534-2543. 10.1093/nar/gkg358.
52. Zheng J, Zhao J, Tao Y, Wang J, Liu Y, Fu J, Jin Y, Gao P, Zhang J, Bai Y: Isolation and analysis of water stress induced genes in

maize seedlings by subtractive PCR and cDNA macroarray. Plant Mol Biol. 2004, 55 (6): 807-823.

53. Rebrikov DV, Britanova OV, Gurskaya NG, Lukyanov KA, Tarabykin VS, Lukyanov SA: Mirror orientation selection (MOS): a method for eliminating false positive clones from libraries generated by suppression subtractive hybridization. Nucleic Acids Res. 2000, 28 (20): E90-10.1093/nar/28.20.e90.
54. Roberts SC: Production and engineering of terpenoids in plant cell culture. Nat Chem Biol. 2007, 3 (7): 387-395. 10.1038/nchembio.2007.8.
55. Chirgwin JM, Przybyla AE, MacDonald RJ, Rutter WJ: Isolation of biologically active ribonucleic acid from sources enriched in ribonuclease. Biochemistry. 1979, 18 (24): 5294-5299. 10.1021/bi00591a005.
56. Dilger M, Felsenstein FG, Schwarz G: Identification and quantitative expression analysis of genes that are differentially expressed during conidial germination in Pyrenophora teres. Mol Genet Genomics. 2003, 270 (2): 147-155. 10.1007/s00438-003-0910-7.
57. Conesa A, Gotz S, Garcia-Gomez JM, Terol J, Talon M, Robles M: Blast2GO: a universal tool for annotation, visualization and analysis in functional genomics research. Bioinformatics. 2005, 21 (18): 3674-3676. 10.1093/bioinformatics/bti610.
58. Cleveland W: Robust locally weighted regression and smoothing scatter plots. J Am Stat Assoc. 1979, 74 (368): 829-836. 10.1080/01621459.1979.10481038.

CITATION

CHAPTER 2

Chugh, R.M., Chaturvedi, M., Yerneni, L.K. (2015) Occurrence and control of sporadic proliferation in growth arrested Swiss 3T3 feeder cells. PLoS One. 10(3), e0122056

CHAPTER 3

Shahla Shahsavandi, Mohammad Majid Ebrahimi, Shahin Masoudi, and Hasan Izadi, "Expression of Factor X in BHK-21 Cells Promotes Low Pathogenic Influenza Viruses Replication," Advances in Virology, vol. 2015, Article ID 675921, 6 pages, 2015. doi:10.1155/2015/675921.

CHAPTER 4

Gulis et al: Optimization of heterologous protein production in Chinese hamster ovary cells under overexpression of spliced form of human X-box binding protein. BMC Biotechnology 2014 14:26. doi:10.1186/1472-6750-14-26

CHAPTER 5

Lin C-Y, Huang Z, Wen W, Wu A, Wang C, Niu L (2015) Enhancing Protein Expression in HEK-293 Cells by Lowering Culture Temperature. PLoS ONE 10(4): e0123562. https://doi.org/10.1371/journal.pone.0123562

CHAPTER 6

Wolun-Cholewa M, Langer K, Szymanowski K, Glodek A, Jankowska A, Warchol W, et al. (2013) An Efficient 3D Cell Culture Method on Biomimetic Nanostructured Grids. PLoS ONE 8(9): e72936. https://doi.org/10.1371/journal.pone.0072936

CHAPTER 7

Fratz EJ, Hunter GA, Ferreira GC (2014) Expression of Murine 5-Aminolevulinate Synthase Variants Causes Protoporphyrin IX Accumulation and Light-Induced Mammalian Cell Death. PLoS ONE 9(4): e93078. https://doi.org/10.1371/journal.pone.0093078

CHAPTER 8

Cassio: Long term culture of MDCK strains alters chromosome content. BMC Research Notes 2013 6:162. doi: 10.1186/1756-0500-6-162

CHAPTER 9

García Münzer DG, Kostoglou M, Georgiadis MC, Pistikopoulos EN, Mantalaris A (2015) Cyclin and DNA Distributed Cell Cycle Model for GS-NS0 Cells. PLoS Comput Biol 11(2): e1004062. https://doi.org/10.1371/journal.pcbi.1004062

CHAPTER 10

Thomassen, Y.E., Rubingh, O., Wijffels, R.H., van der Pol, L.A., Bakker, W.A. (2014) Improved poliovirus D-antigen yields by application of different Vero cell cultivation methods. Vaccine. 32, 2782-2788

CHAPTER 11

Hashimoto et al., Ao38, a new cell line from eggs of the black witch moth, Ascalapha odorata (Lepidoptera: Noctuidae), is permissive for AcMNPV infection and produces high levels of recombinant proteins BMC Biotechnology 2010, 10:50, doi: 10.1186/1472-6750-10-50.

CHAPTER 12

Kai Yu, Yang Yu, Xiaoyan Tang, Huimin Chen, Junyu Xiao &, Xiao-Dong Su, Transcriptome analyses of insect cells to facilitate baculovirus-insect expression, DOI 10.1007/s13238-016-0260-y.

CHAPTER 13

Anton Strunov, Lidiya V. Boldyreva, Gera A. Pavlova, Alexey V. Pindyurin, Maurizio Gatti, Elena Kiseleva, A simple and effective method for ultrastructural analysis of mitosis in Drosophila S2 cells,

CHAPTER 14

Giraudo M, Califano J, Hilliou F, Tran T, Taquet N, Feyereisen R, et al. (2011) Effects of Hormone Agonists on Sf9 Cells, Proliferation and Cell Cycle Arrest. PLoS ONE 6(10): e25708. https://doi.org/10.1371/journal.pone.0025708

CHAPTER 15

Stefan W Metz, Corinne Geertsema, Byron E Martina, Paulina Andrade, Jacco G Heldens, Monique M van Oers, Rob W Goldbach^, Just M Vlak and Gorben P Pijlman, Functional processing and secretion of Chikungunya virus E1 and E2 glycoproteins in insect cells, https://doi.org/10.1186/1743-422X-8-353.

CHAPTER 16

Tanurdzic M, Vaughn MW, Jiang H, Lee T-J, Slotkin RK, Sosinski B, et al. (2008) Epigenomic Consequences of Immortalized Plant Cell Suspension Culture. PLoS Biol 6(12): e302. https://doi.org/10.1371/journal.pbio.0060302

CHAPTER 17

Eva Nocarova and Lukas Fischer, Cloning of transgenic tobacco BY-2 cells; an efficient method to analyses and reduce high natural heterogeneity of transgene expression, doi: 10.1186/1471-2229-9-44.

CHAPTER 18

Ramani S, Jayabaskaran C. Enhanced catharanthine and vindoline production in suspension cultures of Catharanthus roseus by ultraviolet-B light. Journal of Molecular Signaling. 2008;3: Art. 9. DOI: http://doi.org/10.1186/1750-2187-3-9

CHAPTER 19

Vasilev N, Schmitz C, Grömping U, Fischer R, Schillberg S (2014) Assessment of Cultivation Factors that Affect Biomass and Geraniol Production in Transgenic Tobacco Cell Suspension Cultures. PLoS ONE 9(8): e104620. https://doi.org/10.1371/journal.pone.0104620

CHAPTER 20

Huang L-F, Tan C-C, Yeh J-F, Liu H-Y, Liu Y-K, Ho S-L, et al. (2015) Efficient Secretion of Recombinant Proteins from Rice Suspension-Cultured Cells Modulated by the Choice of Signal Peptide. PLoS ONE 10(10): e0140812. https://doi.org/10.1371/journal.pone.0140812

CHAPTER 21

Lenka et al.: Identification and expression analysis of methyl jasmonate responsive ESTs in paclitaxel producing Taxus cuspidata suspension culture cells. BMC Genomics 2012 13:148, doi:10.1186/1471-2164-13-148

INDEX